Essential Concepts of Radio Waves and Fields

Essential Concepts of Radio Waves and Fields

Edited by **Kevin Merriman**

Willford Press

New York

Published by Willford Press,
118-35 Queens Blvd., Suite 400,
Forest Hills, NY 11375, USA
www.willfordpress.com

Essential Concepts of Radio Waves and Fields
Edited by Kevin Merriman

International Standard Book Number: 978-1-68285-293-4 (Hardback)

Printed in China.

Contents

Preface

This book aims to highlight the current researches and provides a platform to further the scope of innovations in this area. This book is a product of the combined efforts of many researchers and scientists from different parts of the world. The objective of this book is to provide the readers with the latest information in the field.

Radio waves are widely used in communication networks and broadcasting. This book is a compilation of chapters that elucidate the most significant concepts related to this field like radio wave propagation, signal transmission and processing, radio wave frequency, emerging applications of radio waves and radio communications, etc. Students and professionals engaged in this field at various levels will find this book full of unexplored and innovative concepts.

I would like to express my sincere thanks to the authors for their dedicated efforts in the completion of this book. I acknowledge the efforts of the publisher for providing constant support. Lastly, I would like to thank my family for their support in all academic endeavors.

Editor

1

Modelling and characterization of a travelling-wave electro-optic modulator on InP

W. Pascher[1], **J. H. den Besten**[2], **D. Caprioli**[2], **R. van Dijk**[3], **X. J. M. Leijtens**[2], **and M. K. Smit**[2]

[1]Allg. und Theoretische Elektrotechnik, FernUniversität, D-58084 Hagen, Germany
[2]Opto-Electronic Devices group, Eindhoven University of Technology, Eindhoven, The Netherlands
[3]TNO Physics and Electronics Laboratory, The Hague, The Netherlands

Abstract. A fast travelling-wave Mach-Zehnder modulator is modelled and designed using a rigorous vectorial analysis. In order to investigate propagation characteristics, velocity and microwave loss, the semiconductor layer stack and the lossy electrodes are modelled using the method of lines. The microwave field distribution is determined, design curves are derived and the cross-sectional dimensions of the modulator are optimized. The loss of the fabricated device agrees very well with small signal measurements up to 40 GHz and HFSS simulations.

1 Introduction

High speed modulators are of great interest for 40 Gb/s transmission systems and beyond. Travelling wave modulators (Walker, 1991) are not limited in frequency by the RC time constant. A Mach-Zehnder modulator has been realized, e.g., on GaAs (Walker, 1995) and on InP (Mörl et al., 1999; Krähenbühl, 1998). However, little is published about the microwave design considerations that are to be taken into account in the design of high-speed travelling wave Mach-Zehnder modulators on InP. Highly accurate design data are necessary for these devices because of the long and costly fabrication process. Hence a rigorous model is needed that includes the main limiting factors: the velocity mismatch between the optical wave and the microwave as well as the microwave loss.

The waveguide structure of the modulator considered here (Fig. 2) is composed of several semiconductor layers with a step index profile and lossy metallic electrodes. Conventional analyses use a model with homogeneous dielectric layers and infinitely thin electrodes, which does not take the conductor losses into account. This yields an error in the microwave phase velocity and an incorrect electric field. In this paper, the model used fully accounts for the microwave and optical properties of the inhomogeneous layer stack. The loss in the modulating electrodes and in the semiconductor layers is considered by means of a complex dielectric constant.

The model structure is analyzed by the method of lines (MoL), which has been efficiently used for the simulation of various planar waveguides for microwave integrated circuits and integrated optics (Pregla and Pascher, 1989; Rogge and Pregla, 1993; Pregla and Pascher, 1995). The MoL takes advantage of the planar layer stack by discretizing the wave equation in the direction parallel to the interfaces of the layers, whereas an analytical solution is retained in the vertical direction. Thus highly accurate results are obtained at a comparatively small numerical effort. In this paper, we present MoL simulations to optimize the electric field across the reverse biased diode of the modulator. The results were confirmed by HFSS (High-Frequency Structure Simulator) and by measurements.

1.1 Modulator layout and cross-section

Since a reverse bias is used and almost no current is flowing in InP-based Mach-Zehnder switches, they can operate at sub-nanosecond speed. If lumped metal contacts are used, i.e. on top of the interferometer arms and on the bottom of the chip, the optical bandwidth of the modulator is limited to a couple of GHz by the capacitance. In a travelling wave design, the capacitance is distributed along the line and is consequently no longer a bandwidth limiting factor. In an ideal design, the modulating microwaves are travelling along the phase-shifter with a low attenuation at the same speed as that of the optical waves.

Figure 2 shows a cross-section of an interferometer arm. The RF-signal is fed to this modulator arm by a coplanar waveguide (CPW). At the beginning of the phase shifting sections, the CPW is transferred into a microstrip by a n^+-buffer layer underneath an optical ridge waveguide. In this ridge the light is confined in a 500 nm thick InGaAsP layer (Q1.25) with a bandgap wavelength of 1.25 μm. The InP-cladding consists of an 200 nm thick intrinsic layer and a

Correspondence to: W. Pascher
(wilfried.pascher@FernUni-Hagen.de)

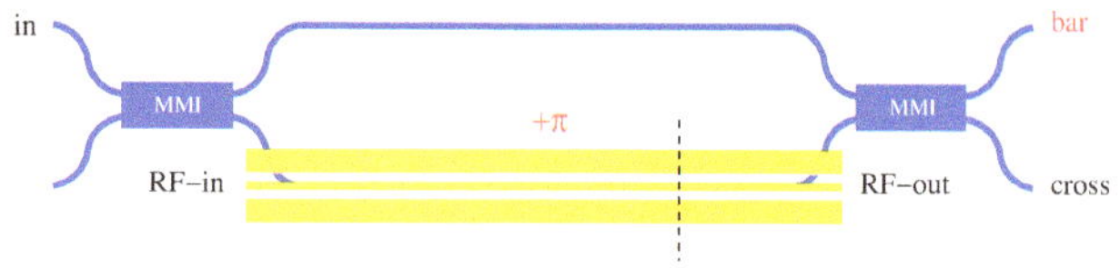

Fig. 1. Top view of a travelling-wave electro-optic modulator. The dashed line locates the cross-section of Fig. 2.

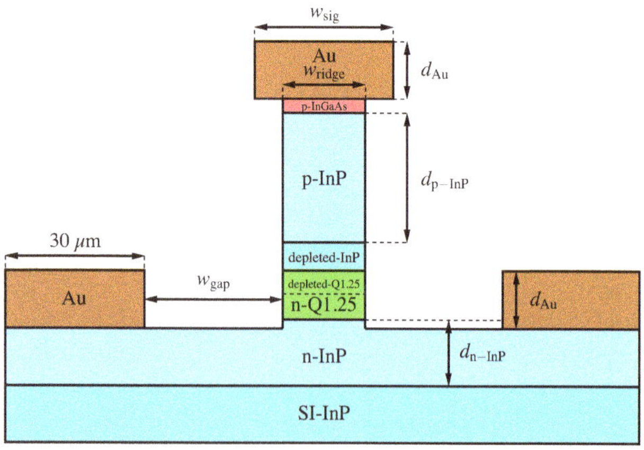

Fig. 2. Cross-section of a travelling-wave electro-optic modulator. The ϵ_r of InP/InGaAsP is 12.55 (the imaginary part is dependent on the doping level).

p-doped layer. Due to the pin-structure of the optical waveguide, the microwave electric field needed for the electro-optic effect will be across the depleted region, which consists of the upper part of the film layer and the intrinsic layer.

2 Vectorial analysis by the method of lines

The analysis is based on the method of lines (MoL) as presented in Pregla and Pascher (1989). The MoL is a semi-analytic approach with analytic solution in *one* coordinate direction, namely perpendicular to the layers and a finite-difference discretization in the other direction. This approach is very well suited for multilayered waveguides as, e.g., for the travelling wave modulator under consideration. The cross-section of a modulator (as in Fig. 2) is partitioned into an "arbitrary" number of layers. These layers are inhomogeneous, also with permittivity steps ($\varepsilon_r = \varepsilon_r(x)$), dielectric loss ($\varepsilon_r$ complex) and metallic loss ($\varepsilon_r = -j\dfrac{\kappa}{\omega\varepsilon_o}$).

The wave equations are established using two scalar potentials ψ_e (electric) and ψ_h (magnetic). We obtain a Helmholtz- and Sturm-Liouville differential equation for each layer

$$\epsilon_r(x)\frac{\partial}{\partial x}\left(\frac{1}{\epsilon_r(x)}\frac{\partial\psi_e}{\partial x}\right)+\frac{\partial^2\psi_e}{\partial y^2} + \left(\epsilon_r(x)k_0^2-k_z^2\right)\psi_e = 0 \tag{1}$$

$$\frac{\partial^2\psi_h}{\partial x^2}+\frac{\partial^2\psi_h}{\partial y^2}+\left(\epsilon_r(x)k_0^2-k_z^2\right)\psi_h=0 \tag{2}$$

where we have assumed wave propagation in z direction according to $\psi_{e,h} \sim \exp(-jk_z z)$.

The analysis steps of the method of lines for the electro-optic modulator are summarized in the flow chart (Fig. 3).

We apply a finite difference discretization in x direction in order to convert to partial differential equations to coupled ordinary ones. In this way we are able to obtain an analytical solution in y direction on discretization lines. Two sets of discretization lines are used, one for the electric potential $\boldsymbol{\psi}_e$ and the other for the magnetic one $\boldsymbol{\psi}_h$. The first derivative with respect to x is expressed using finite differences and the following difference operator $\boldsymbol{D}$ for the magnetic potential $\boldsymbol{\psi}_h$

$$\left.\frac{\partial\psi_h}{\partial x}\right|_{(i+1/2)h} \approx \frac{\psi_{i+1}-\psi_i}{h} \quad \text{or}$$

$$h\frac{\partial\psi_h}{\partial x} \to \underbrace{\begin{bmatrix} 1 & & & \\ -1 & 1 & & \\ & \ddots & \ddots & \\ & & \ddots & \ddots \\ & & -1 & 1 \end{bmatrix}}_{\boldsymbol{D}} \begin{bmatrix}\psi_1\\ \psi_2\\ \vdots\\ \\ \psi_{N-1}\\ \psi_N\end{bmatrix} = \mathbf{D}\boldsymbol{\psi}_h \tag{3}$$

The second derivative is given by

$$h^2\frac{\partial^2\psi_h}{\partial x^2} \to -\boldsymbol{D}^t\boldsymbol{D}\boldsymbol{\psi}_h = -\boldsymbol{P}_h\boldsymbol{\psi}_h$$

with tridiagonal $\boldsymbol{P}_h$ and analogous for the electric potential $\boldsymbol{\psi}_e$. This yields the discretized wave equation (a system of coupled ordinary differential equations)

$$\frac{d^2}{dy^2}\boldsymbol{\psi}_h - \left(k_z^2\boldsymbol{I} - \varepsilon_h + \boldsymbol{P}_h\right)\boldsymbol{\psi}_h = \boldsymbol{0} \tag{4}$$

with a tridiagonal matrix $\mathbf{Q}_h$. In order to solve these equations, we perform a diagonalization, namely by transformation to principal axes:

$$\begin{aligned} \boldsymbol{T}^{-1}\boldsymbol{Q}\boldsymbol{T} &= \boldsymbol{\Gamma}^2 \quad \text{(diagonal)} \\ \boldsymbol{T}^{-1}\boldsymbol{\psi} &= \overline{\boldsymbol{\psi}} \end{aligned} \tag{5}$$

This constitutes two separate eigenvalue problems for $\boldsymbol{\psi}_e$ and $\boldsymbol{\psi}_h$. We end up with the transformed Helmholtz equation

$$\frac{d^2}{dy^2}\overline{\boldsymbol{\psi}} - \boldsymbol{\Gamma}^2\overline{\boldsymbol{\psi}} = \boldsymbol{0} \tag{6}$$

which is decoupled and has an analytical solution in each individual layer

$$\overline{\boldsymbol{\psi}} = \exp(-\boldsymbol{\Gamma}y)\,\mathbf{A} + \exp(\boldsymbol{\Gamma}y)\,\mathbf{B} \tag{7}$$

In the following process the tangential fields are matched on each layer interface and we obtain a formulation of an indirect eigenvalue problem

$$\mathbf{Z}(\varepsilon_{re})\overline{\mathbf{H}}_{tan,M} = \mathbf{0} \tag{8}$$

$$\det(\mathbf{Z}(\varepsilon_{re})) = 0 \tag{9}$$

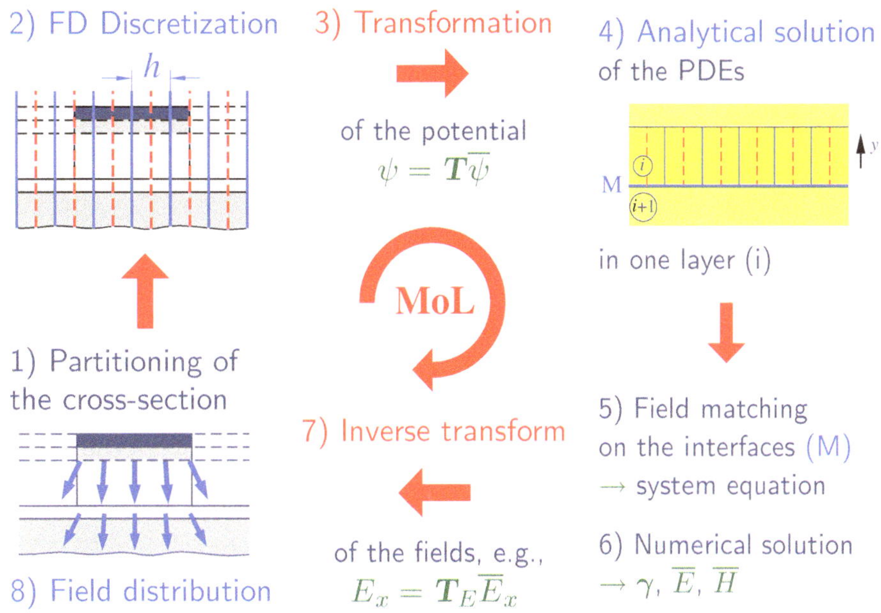

Fig. 3. Flow chart of the analysis steps in the MoL.

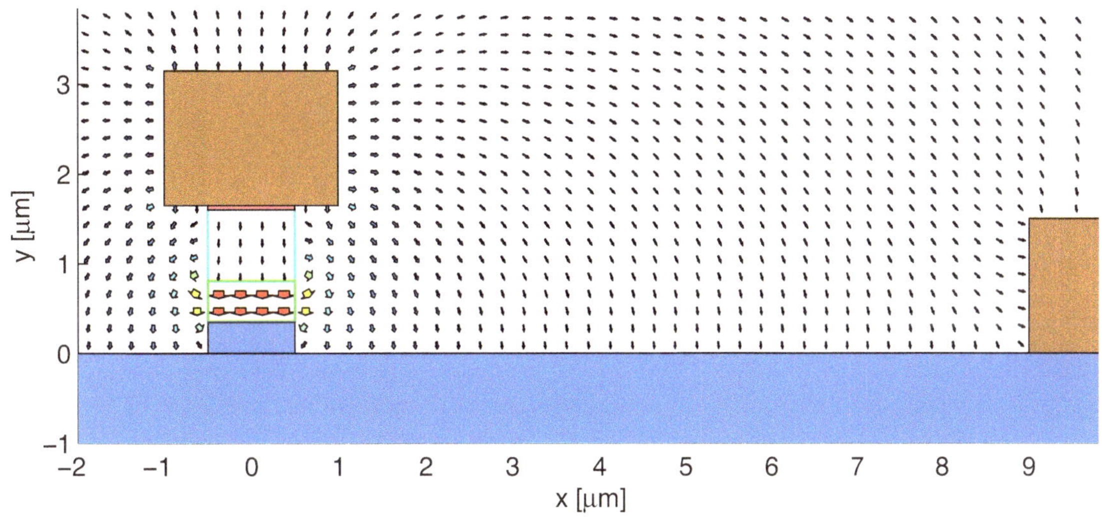

Fig. 4. Microwave electric field (arrow width proportional to field) in the cross-section of the optimized modulator.

which is solved for the the effective permittivity ε_{re} by a numerical search for zeros. This also gives the electromagnetic field $\overline{\mathbf{H}}$, $\overline{\mathbf{E}}$ in transform domain. Finally, an inverse transformation yields the field distribution in the waveguide.

3 Simulation and measurement results

The modulator has been simulated by the MoL according to the cross-section in Fig. 2. First the field distribution is given. Then design curves with respect to various dimensions of the device are presented in order to optimize its performance. Finally, the microwave loss and index are compared with small signal measurements and HFSS (High-Frequency Structure Simulator) simulations.

3.1 Field distribution

The microwave electrical field distribution is presented for the optimized design (see next section), namely in the right half of the cross-section (Fig. 4). It can be seen clearly that the microwave field is concentrated in the depletion region. The wide signal electrode, however, produces a stray field in air, which allows some control of the microwave refractive index. Obviously, the mode encountered is nearly a microstrip mode. The reason lies in the comparitively high conductivity of the n-doped buffer layer, which acts as a kind of ground plane.

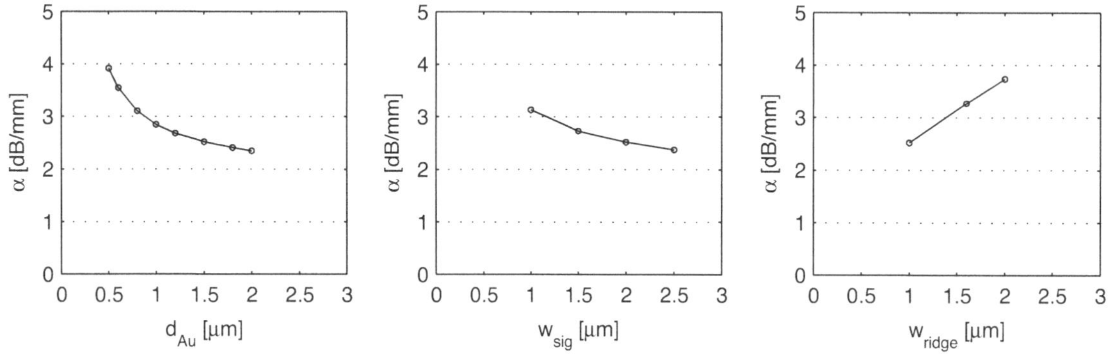

Fig. 5. Attenuation at 40 GHz as a function of gold thickness thickness (left), signal line width (center), and p-cladding thickness (right). Other parameters are as in the optimized design.

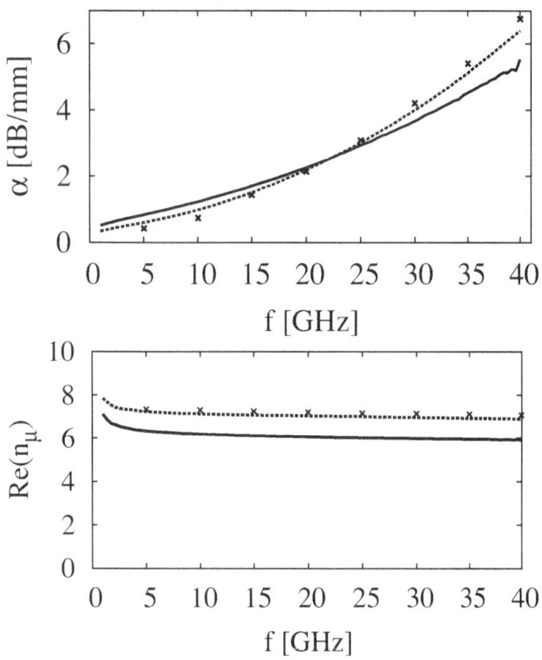

Fig. 6. Measurements (solid line), MoL-simulations (dashed line) and HFSS-simulations (crosses) of the microwave attenuation α (left) and the real part of the microwave index n_μ (right) for a phase shifter with a signal line width w_{sig} of 12 μm.

3.2 Design curves

The biggest electro-optic effect will be obtained if the electric field is highest in the region where the light is confined. To make use of the pin-diode structure of the ridge, doping of the InP-ridge is necessary. Unfortunately, doped semiconductors have a conductivity that is much lower than that of metals. Therefore, the amount of doping in the cladding influences the RF-attenuation due to varying resistivity.

Figure 5 (simulated using the package MOLCAR; Kremer and Pregla, 1997) shows the attenuation is decreasing as the ridge width, and thus the amount of doped material in between the signal line and the depletion-zone edge, is decreased. Figure 5 also shows the effect of the shape of the signal line on the RF-attenuation. A wider signal line width reduces the RF propagation loss, because a bigger part of the electric field is concentrated in the air instead of the semiconductor material. The attenuation is also reduced if the gold thickness is increased. Since the skin depth of Au is about 385 nm at 40 GHz, especially at thin Au thicknesses the attenuation increases dramatically. All plots were simulated at 40 GHz.

3.3 Comparison with measurements

The optimized design as described in the previous sections, has not been fabricated yet. Therefore, we checked measurements on already existing modulators with simulations by the Method of Lines. The left part of Fig. 6 shows the attenuation data of a structure with a ridge width of 4 μm and a signal line width of 12 μm. The measured values for the phase-shifting section have been extracted using the thru-reflect-line (TRL) method (Enge and Hoer, 1979). Depicted are the values measured through small signal (SS-)measurements, calculated by the MoL and calculated by HFSS.

HFSS simulations have been performed using the same structure as for the MOLCAR simulations. HFSS (Ansoft) is a 3-dimensional finite difference EM field solver. The structure is discretised in tetrahedra and the EM field components are calculated on the vertices and edges of these elements using appropriate boundary conditions. The mesh is refined adaptively at the locations where a large gradient exists. The output of a simulation is a.o. the impedance at the ports and the S-parameters. From these the complex propagation constant of the structure is calculated.

Comparison with measured values is very good. It can be noted that the loss curve does not follow a $\sqrt{f}$ law, since the losses are mainly determined by the absorption in the substrate and not by the skin depth of the electrodes.

The right part of Fig. 6 shows the real part of the microwave index of a structure with a signal line width of 12 μm. The high value of the index clearly shows that we are dealing with a slow wave structure which means that the velocity of the microwave is lower than the a wave in the materials considered. This is caused by the doped semiconductor layers.

One of the causes for the offset between simulation and measurement of the effective index could be the lack of reliable data of the used polyimide at RF-frequencies. Also, since the calculated values of the MoL and HFSS agree very well, we believe another source of error is the uncertainty of the doping levels in the different layers, which has not been measured. A different doping profile would not only change the conductivities, but also the thickness of the depletion layer. Attenuation and effective index plots for a structure with a signal line width of $w_{sig} = 6\,\mu$m showed similar behaviour.

4 Conclusion

The method of lines (MoL) has been employed to model and design a fast travelling-wave electro-optic modulator. Field distributions and design curves have been presented and discussed in view of an optimized component with minimum microwave loss. The small-signal measurements of the fabricated device show very good agreement (better than 0.5 dB/mm) with the MoL modelling and with HFSS simulation results.

Acknowledgements. This work has been carried out with the partial funding of the European Union through the project IST 2000-25390 OBANET of the Information Society Technologies Research Programme and the Dutch NRC-Photonics programme.

References

Walker, R. G.: High-speed III-V semiconductor intensity modulators, IEEE J. Quantum Electron., 27, 3, pp. 654–667, Mar. 1991.

Walker, R. G.: GaAs/AlGaAs travelling-wave modulators for mm-wave frequencies, IEE Colloquium Towards Terabit Transmission, pp. 4/1–4/6, 1995.

Mörl, L., Hoffmann, D., Matzen, K., Bornholdt, C., Mekonnen, G. G., and Reier,F.: Traveling wave electrodes for 50 GHz operation of opto-electronic devices based on InP, Proc. IPRM99 conference, pp. 385–388, May 16–20 1999.

Krähenbühl, R.: Electro-Optic Space Switches in InGaAsP/InP for Optical Communication, Hartung-Gorre Verlag, Konstanz, second edition, 202 pp., 1998.

Pregla, R. and Pascher, W.: The Method of Lines, in T. Itoh (ed.): Numerical Techniques for Microwave and Millimeter Wave Passive Structures, New York, USA, J. Wiley Publ., pp. 381–446, 1989.

Rogge, U. and Pregla, R.: Method of Lines for the Analysis of Dielectric Waveguides, J. Lightwave Technol. 11(12), pp. 2015–2020, 1993.

Pregla, R. and Pascher, W.: Vectorial Analysis of Optical Waveguides by the Method of Lines, in Annual Review of Progress in Applied Computational Electromagnetics (ACES), 11, Monterey, USA, pp. 943–950, 1995.

Kremer, D. and Pregla, R.: MOLCAR a Very Efficient and Accurate Design Tool for the Hybrid Analysis of Multilayered Waveguide Structures, in Proc. SPIE Photonics West, Integrated Optics Devices: Potential for Commercialization, San Jose, USA, vol. 2997, pp. 255–262, Feb. 1997.

Enge, G. F. and Hoer, C. A.: Thru-reflect-line: an improved technique for calibrating the dual 6-port automatic network analyzer, IEEE Transactions on Microwave Theory and Techniques, 27, 12, pp. 987–993, Dec. 1979.

2

Stability and conservation properties of transient field simulations using FIT

R. Schuhmann and T. Weiland

Technische Universität Darmstadt, Dept. of Electrical Engineering and Information Technology, Computational Electromagnetics Laboratory (TEMF), Schloßgartenstr. 8, 64289 Darmstadt, Germany

Abstract. Time domain simulations for high-frequency applications are widely dominated by the leapfrog time-integration scheme. Especially in combination with the spatial discretization approach of the Finite Integration Technique (FIT) it leads to a highly efficient explicit simulation method, which in the special case of Cartesian grids can be regarded to be computationally equivalent to the Finite Difference Time Domain (FDTD) algorithm. For stability reasons, however, the leapfrog method is restricted to a maximum stable time step by the well-known Courant-criterion, and can not be applied to most low-frequency applications. Recently, some alternative, unconditionally stable techniques have been proposed to overcome this limitation, including the Alternating Direction Implicit (ADI)-method. We analyze such schemes using a transient modal decomposition of the electric fields. It is shown that stability alone is not sufficient to guarantee correct results, but additionally important conservation properties have to be met.

Das Leapfrog-Verfahren ist ein weit verbreitetes Zeitintegrationsverfahren für transiente hochfrequente elektrodynamischer Felder. Kombiniert mit dem räumlichen Diskretisierungsansatz der Methode der Finiten Integration (FIT) führt es zu einer sehr effizienten, expliziten Simulationsmethode, die im speziellen Fall kartesischer Rechengitter als äquivalent zur Finite Difference Time Domain (FDTD) Methode anzusehen ist. Aus Stabilitätsgründen ist dabei die Zeitschrittweite durch das bekannte Courant-Kriterium begrenzt, so dass das Leapfrog-Verfahren für niederfrequente Probleme nicht sinnvoll angewendet werden kann. In den letzten Jahren wurden alternativ einige andere explizite oder "halb-implizite" Zeitbereichsverfahren vorgeschlagen, u.a. das "Alternating Direction Implicit" (ADI)-Verfahren, die keiner Beschränkung des Zeitschritts aus Stabilitätsgründen unterliegen. Es zeigt sich aber, dass auch diese Methoden im niederfrequenten Fall nicht zu sinnvollen Simulationsergebnissen führen. Wie anhand einer transienten Modalanalyse der elektrischen Felder in einem einfachen 2D-Beispiel deutlich wird, ist die Ursache dafür die Verletzung wichtiger physikalischer Erhaltungseigenschaften durch ADI und verwandte Methoden.

Correspondence to: R. Schuhmann, T. Weiland
(schuhmann/weiland@temf.tu-darmstadt.de)

1 Introduction

Especially in high-frequency field simulations, where one often deals with lossless or at least low-loss structures and a large number of time steps, stability is one of the most important properties of time domain methods, and a required condition for their overall convergence. Here, very often Finite Difference methods (FDTD, Yee (1966)) and the time domain variant of the Finite Integration Technique (FIT, Weiland (1996)) are used, and therein the so-called *leapfrog* (LF) time stepping algorithm. Based on central difference approximations for the time derivatives in Maxwell's equations, it is known to be conditionally stable – ruled by a maximum stable *Courant* time step width Δt_0 – and to conserve the electromagnetic energy in lossless structures, if properly defined (Schuhmann and Weiland (2001)).

Since the time step limitation in the LF scheme is coupled to the resolution of the spatial grid, it sometimes leads to strongly oversampled time signals and thus to a poor efficiency of the overall method. As a remedy, some alternative, 'quasi-implicit' approaches have been proposed, such as the Alternating Direction Implicit (ADI) variant of FDTD (Namiki (2000)), or the family of unconditionally stable algorithms proposed in Kole et al. (2001). However, it has been shown by various authors (Darms (2001); Staker and Piket-May (2001); Garcia et al. (2002)) that such schemes may have severe accuracy problems if the Courant limit of the related LF approach is considerably exceeded. In this paper this dissatisfying result is analyzed using an eigenmode decomposition of the transient fields.

2 Algebraic Formulation

2.1 Basic Equations

We use here the notation of the FIT (Weiland (1977, 1996), where Maxwell's equations are transformed into a set of algebraic equations (linear case, without currents):

$$\mu^{-1}\text{curl}\,\boldsymbol{E} = -\frac{\partial}{\partial t}\boldsymbol{H} \quad \leftrightarrow \quad \mathbf{M}_{\mu}^{-1}\mathbf{C}\widehat{\mathbf{e}} = -\frac{d}{dt}\,\widehat{\mathbf{h}} \tag{1}$$

$$\epsilon^{-1}\text{curl}\,\boldsymbol{H} = \frac{\partial}{\partial t}\boldsymbol{E} \quad \leftrightarrow \quad \mathbf{M}_{\varepsilon}^{-1}\mathbf{C}^{T}\widehat{\mathbf{h}} = \frac{d}{dt}\,\widehat{\mathbf{e}} \tag{2}$$

$$\text{div}\,(\mu\,\boldsymbol{H}) = 0 \quad \leftrightarrow \quad \mathbf{S}\mathbf{M}_{\mu}\widehat{\mathbf{h}} = \mathbf{0} \tag{3}$$

$$\text{div}\,(\epsilon\,\boldsymbol{E}) = \rho \quad \leftrightarrow \quad \widetilde{\mathbf{S}}\mathbf{M}_{\varepsilon}\widehat{\mathbf{e}} = \mathbf{q}. \tag{4}$$

The sparse matrices $\mathbf{C}$ and $\mathbf{S}$ are the topological 'curl'-, and 'source'-operators, respectively, and the vectors $\widehat{\mathbf{e}}$ and $\widehat{\mathbf{h}}$ contain the electric and magnetic voltage-type degrees of freedom on a pair of staggered grids. The material matrices $\mathbf{M}_{\varepsilon}^{-1}$ and $\mathbf{M}_{\mu}^{-1}$ are diagonal and positive definite in the simplest case.

An important property of these equations – which can also be used to derive the FDTD-method – is the *exact* source-free relation of curl-fields,

$$\mathbf{S}\,\mathbf{C} = \mathbf{0} \qquad \text{(primary grid)}, \tag{5}$$

$$\widetilde{\mathbf{S}}\,\mathbf{C}^{T} = \mathbf{0} \qquad \text{(dual grid)}, \tag{6}$$

sometimes referred to as *consistency properties* of the FIT-discretization.

Finally, Eqs. (1) and (2) can be combined to a large system of differential equations for a composite vector $\mathbf{x}$:

$$\frac{d}{dt}\mathbf{x} = \mathbf{A}\mathbf{x} \tag{7}$$

with

$$\mathbf{x} = \begin{pmatrix} \widehat{\mathbf{h}} \\ \widehat{\mathbf{e}} \end{pmatrix} \tag{8}$$

and

$$\mathbf{A} = \begin{pmatrix} \mathbf{0} & -\mathbf{M}_{\mu}^{-1}\mathbf{C} \\ \mathbf{M}_{\varepsilon}^{-1}\widetilde{\mathbf{C}} & \mathbf{0} \end{pmatrix}. \tag{9}$$

The *system matrix* $\mathbf{A}$ can be transformed into a skew-symmetric form using the scaled vectors

$$\mathbf{x}' = \begin{pmatrix} \widehat{\mathbf{h}}' \\ \widehat{\mathbf{e}}' \end{pmatrix} \qquad \text{with} \quad \widehat{\mathbf{h}}' = \mathbf{M}_{\mu}^{1/2}\widehat{\mathbf{h}}, \quad \widehat{\mathbf{e}}' = \mathbf{M}_{\varepsilon}^{1/2}\widehat{\mathbf{e}}. \tag{10}$$

Thus, all eigenvalues of $\mathbf{A}$ are purely imaginary, $\lambda_{A,i} = i\omega_i$, and all eigenvectors of $\mathbf{A}$ are orthogonal to each other (or can be orthogonalized) referring to

$$\langle \mathbf{x}_i, \mathbf{x}_j \rangle = \mathbf{x}_j'^{\,h}\mathbf{x}_i' = \widehat{\mathbf{e}}_j^{h}\mathbf{M}_{\varepsilon}\widehat{\mathbf{e}}_i + \widehat{\mathbf{h}}_j^{h}\mathbf{M}_{\mu}\widehat{\mathbf{h}}_i = \delta_{ij}. \tag{11}$$

2.2 Time Stepping Schemes

2.2.1 Leapfrog (LF) Algorithm

The leapfrog scheme arises from the allocation of the fields on a staggered time axis and the usage of central difference approximations for the time derivatives:

$$\begin{aligned} \frac{d}{dt}\widehat{\mathbf{h}}^{(n+1/2)} &\approx \frac{\widehat{\mathbf{h}}^{(n+1)} - \widehat{\mathbf{h}}^{(n)}}{\Delta t} \\ \frac{d}{dt}\widehat{\mathbf{e}}^{(n+1)} &\approx \frac{\widehat{\mathbf{e}}^{(n+3/2)} - \widehat{\mathbf{e}}^{(n+1/2)}}{\Delta t} \end{aligned} \tag{12}$$

It can be summarized in the *update equations*

$$\mathbf{x}^{(n+1)} = \mathbf{G}_{LF}(\Delta t)\,\mathbf{x}^{(n)}, \qquad \mathbf{x}^{(n)} = \begin{pmatrix} \widehat{\mathbf{h}}^{(n)} \\ \widehat{\mathbf{e}}^{(n+1/2)} \end{pmatrix} \tag{13}$$

with the *iteration matrix*

$$\mathbf{G}_{LF}(\Delta t) = \begin{pmatrix} \mathbf{I} & -\Delta t\mathbf{M}_{\mu}^{-1}\mathbf{C} \\ \Delta t\mathbf{M}_{\varepsilon}^{-1}\widetilde{\mathbf{C}} & \mathbf{I} - \Delta t^{2}\mathbf{M}_{\varepsilon}^{-1}\widetilde{\mathbf{C}}\mathbf{M}_{\mu}^{-1}\mathbf{C} \end{pmatrix}. \tag{14}$$

An important property of the LF-operator is the conservation of electric (and magnetic) charges on the discrete level,

$$\begin{aligned} &\text{div}\,(\mu\boldsymbol{H}^{(n+1)}) = \text{div}\,(\mu\boldsymbol{H}^{(n)}) \\ &\leftrightarrow \mathbf{S}\mathbf{M}_{\mu}\widehat{\mathbf{h}}^{(n+1)} = \mathbf{S}\mathbf{M}_{\mu}\widehat{\mathbf{h}}^{(n)} \end{aligned} \tag{15}$$

$$\begin{aligned} &\text{div}\,(\epsilon\boldsymbol{E}^{(n+3/2)}) = \text{div}\,(\epsilon\boldsymbol{E}^{(n+1/2)}) \\ &\leftrightarrow \widetilde{\mathbf{S}}\mathbf{M}_{\varepsilon}\widehat{\mathbf{e}}^{(n+3/2)} = \widetilde{\mathbf{S}}\mathbf{M}_{\varepsilon}\widehat{\mathbf{e}}^{(n+1/2)} \end{aligned} \tag{16}$$

which can be easily proven using Eq. (13) and the matrix properties Eqs. (5) and (6).

2.2.2 Alternating Direction Implicit (ADI) Algorithm

The ADI-scheme is based on a *splitting of the operator matrix* in two parts,

$$\mathbf{C} = \mathbf{C}_1 + \mathbf{C}_2 \tag{17}$$

both of which are used in alternating order in the update equations. This leads to an update scheme in two half-steps, which can be summarized by an iteration matrix

$$\begin{aligned} \mathbf{G}_{ADI}(\Delta t) = {}& (\mathbf{I} - \frac{\Delta t}{2}\mathbf{Y}_1)^{-1}\,(\mathbf{I} + \frac{\Delta t}{2}\mathbf{Y}_2) \\ & (\mathbf{I} - \frac{\Delta t}{2}\mathbf{Y}_2)^{-1}\,(\mathbf{I} + \frac{\Delta t}{2}\mathbf{Y}_1) \end{aligned} \tag{18}$$

with

$$\mathbf{Y}_{1,2} = \begin{pmatrix} \mathbf{I} & -\frac{\Delta t}{2}\mathbf{M}_{\varepsilon}^{-1}\mathbf{C}_{1,2}^{T} \\ \frac{\Delta t}{2}\mathbf{M}_{\mu}^{-1}\mathbf{C}_{1,2} & \mathbf{I} \end{pmatrix}. \tag{19}$$

For a special choice of the operator splitting Eq. (17), the implicit expressions in Eq. (18) become tridiagonal matrices, which can be efficiently inverted in each time step (see Darms et al. (2003) for more details).

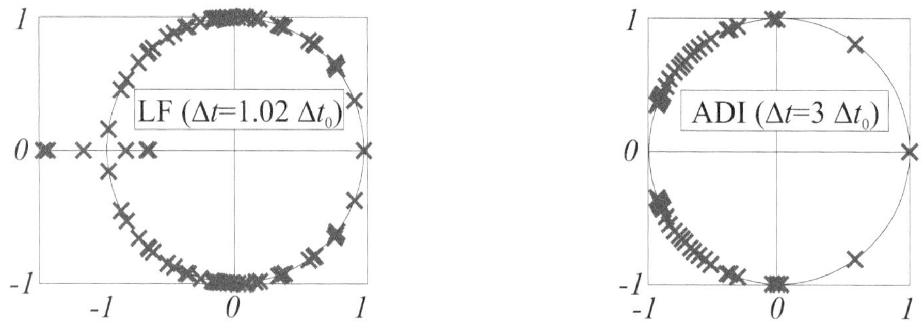

Fig. 1. Complex eigenvalues of the iteration matrices of Leapfrog (LF, left) and ADI (right) for time steps larger than the Courant limit Δt_0. The LF-operator shows some eigenvalues with $|\lambda_{G,lf}| > 1$, causing an instable time integration, whereas all the ADI-eigenvalues lay exactly on the unit cicrle.

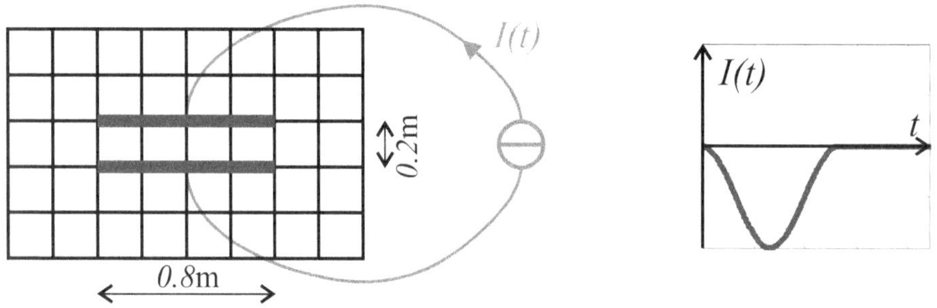

Fig. 2. Model problem: Charging process of a 2D plate capacitor, driven by a 750 kHz current pulse.

2.2.3 Eigensolutions of the Iteration Matrices

For the eigenvalues of these iteration matrices one can find the relations

$$\text{Leapfrog:} \qquad |\lambda_{G,LF}(\Delta t)| = 1 \quad \Leftrightarrow \quad \Delta t \leq \Delta t_0 \tag{20a}$$

$$\text{ADI:} \qquad |\lambda_{G,ADI}(\Delta t)| = 1 \qquad \forall \Delta t > 0, \tag{20b}$$

which are a sufficient condition for the stability of the methods. Whereas the LF method is restricted to time steps below the Courant limit $\Delta t \leq \Delta t_0$, the ADI method is *unconditionally stable* for arbitrary time steps.

In most practical cases the dimension of the iteration matrices is too large to perform a further numerical analysis. For the small test example presented below, however, the matrices are of manageable size, and the results of Eq. (20b) can be visualized as shown in Fig. 1.

For the leapfrog method (left), most of the eigenvalues lay on the unit circle of the complex plane (the stability limit). However, since the time step chosen in this example slightly exceeds the Courant limit ($\Delta t = 1.02\Delta t_0$), some eigenvalues have left the unit circle at $\lambda_G = -1$ and are placed on the negative real axis. The eigenvalues with $\lambda_{G,LF} < -1$ will cause a instable time integration. In the ADI case (Fig. 1, right) all eigenvalues are exactly on the unit circle for arbitrary time steps (here: $\Delta t = 3\Delta t_0$).

Note that for both methods there is a multiple eigenvalue $\lambda_G = 1$, referring to so-called static eigenmodes (electromagnetic fields with eigenfrequency $\omega = 0$), which will not be changed by the time stepping algorithm.

3 Transient Modal Expansion

In the following, the LF and ADI schemes are applied to a 2D (TE) model problem adapted from García et al. (2002). It describes the transient charging process of a simple plate capacitor, driven by a 750 kHz current pulse (cf. Fig. 2).

Figure 3 shows some eigenmodes of the related system matrix $\mathbf{A}$: A static mode (left) with $\nabla \times \boldsymbol{E} = 0$, $\lambda_A = 0$ and $\lambda_G = 1$, a second static mode ('plate mode') describing the desired stationary field solution, and a dynamic mode (right) with $\nabla \times \boldsymbol{E} \neq 0$, $\lambda_A \neq 0$, and $\lambda_G = e^{i\varphi}$.

During the time stepping process the electric fields can now be decomposed into these (and all the other) eigenmodes. The results of this *transient modal expansion*, the expansion coefficients as a function of simulation time, are shown in Fig. 4 (Courant time step $\Delta t = \Delta t_0$) and Fig. 5 (enlarged time step $\Delta t = 3 \cdot \Delta t_0$).

For the Courant time step Δt_0 (Fig. 4) the ADI and LF curves for the plate mode and the dynamic modes are nearly indistinguishable. However, in the ADI simulation arises one more static mode with a magnitude of about 10% at steady state, which cannot be seen in the LF (reference) results, where all the static modes are below numerical noise. For an enlarged $\Delta t = 3\Delta t_0$ (Fig. 5), where no LF results are available any more, the amount of this *parasitic* field reaches

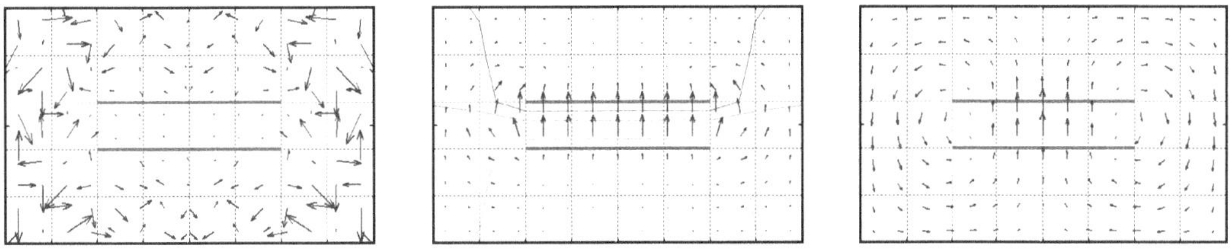

Fig. 3. Model problem: Electric fields of a static eigenmode (gradient of a random discrete potential vector), the plate mode (stationary solution of charging process), and a dynamic mode (oscillating eigenmode) in the plate capacitor.

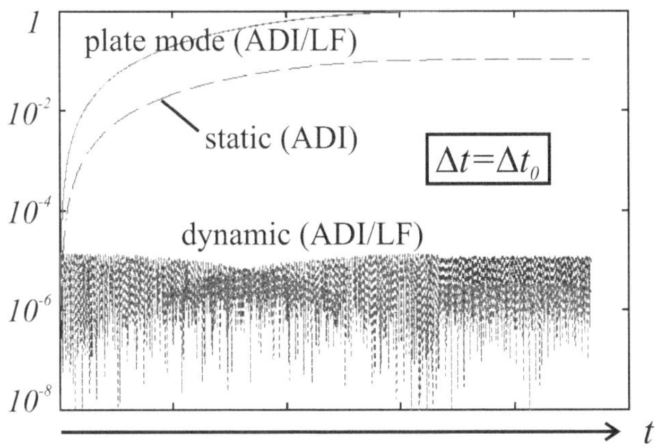

Fig. 4. Transient modal expansion coefficients (logarithmic scale) for the Courant time step $\Delta t = \Delta t_0$: The leapfrog (LF) curves (which serve as a reference result) show the desired plate mode and some dynamic modes excited by the charging process, whereas all static modes are below round off. The ADI curves, however, include a parasitic static mode with a magnitude of about 10% of the total field at steady state.

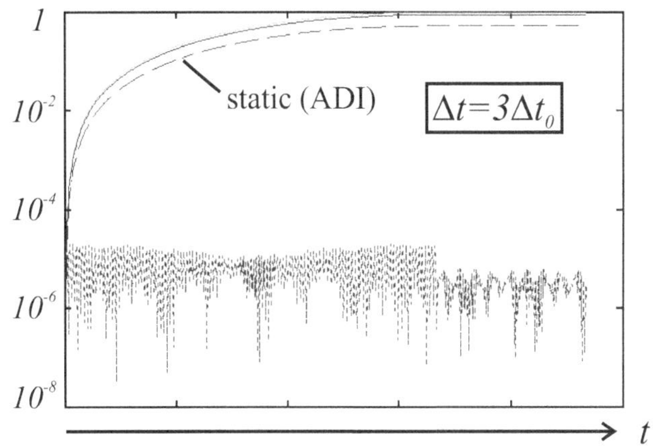

Fig. 5. Transient modal expansion coefficients (logarithmic scale) for enlarged time step $\Delta t = 3 \cdot \Delta t_0$: The magnitude of the parasitic static mode in the ADI solution has considerably increased. No Leapfrog solutions are available abouve the Courant limit (instable time integration).

the order of magnitude of the desired field solution (plate mode). This unphyiscal mode also qualitatively disturbes the stationary field solution, which is not shown here.

As the reason for this behaviour of ADI we postulate here the *loss of orthogonality* between the *eigenvectors* of the iteration matrix: In the leapfrog scheme it can be easily shown that the electric part of the eigenvectors of the iteration matrix $\mathbf{G}_{LF}$ and the system matrix $\mathbf{A}$ are identical, and that they fulfill the same orthogonality condition Eq. (11). In the ADI scheme, however, this property is no longer valid: Although the static modes of $\mathbf{G}_{ADI}$ remain unchanged compared to the static solutions of $\mathbf{A}$ (or $\mathbf{G}_{LF}$) — this can be proven for the 2D TE-case analyzed here — the dynamic modes now include parts of the original static modes. As a consequence, energy can be transferred in each time step between these two classes of eigensolutions (which should be exactly separated in continous theory). As a visualization of this fact, Fig. 6 shows the *orthogonality pattern* of the ADI-matrix for $\Delta t = 3 \cdot \Delta t_0$, exhibiting the parasitic coupling between static and dynamic solutions. As another consequence of this loss of consistency in the discrete model, the field solutions of ADI are no longer source-free, but parasitic electric and magnetic charges arise during the iteration. The intrinsic algebraic reason for these results is the splitting of the operator matrix $\mathbf{C} = \mathbf{C}_1 + \mathbf{C}_2$ in the derivation of the ADI-update equations. Since the spatial curl operator does no longer appear as a whole in the update process, the consistency properties Eqs. (5) and (6) are not applicable any more.

All these effects increase with a growing time step Δt. Although the ADI-method itself shows a 2nd order convergence for $\Delta t \to 0$ (cf. also Garcia et al. (2002)), the results in this analysis show that the parasitic effect can be observed also for moderate time steps, even in the range of the Courant limit of the leapfrog method.

4 Conclusion

A transient modal expansion of the electric field in time domain methods has been used to analyze different time stepping approaches. Using this powerful tool it can be shown that the ADI method, although being unconditionally stable,

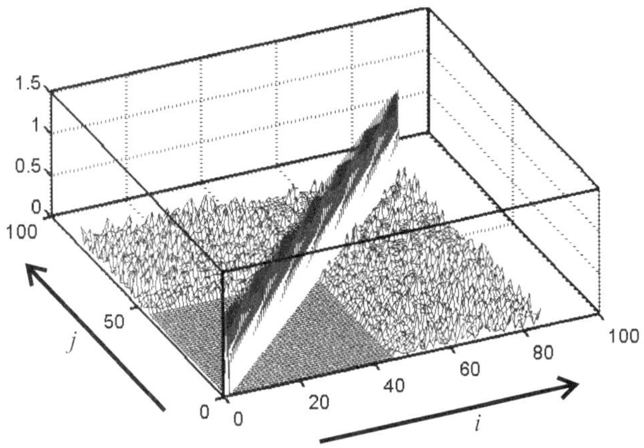

Fig. 6. Orthogonality pattern $\langle \mathbf{x}_i, \mathbf{x}_j \rangle \neq \delta_{ij}$ of the eigenmodes $\mathbf{x}_i$ of the ADI iteration matrix. The (degenerated) static modes can be orthogonalized, but the orthogonality between static and dynamic modes is lost. This unphysical effect increases with the time step width (here: $\Delta t = 3 \cdot \Delta t_0$).

does not conserve the energy of the original dynamic eigenmodes, even for moderate time steps. The main reason for this behaviour is the operator splitting in the construction of the ADI scheme, which leads to a loss of important consistency properties and of the orthogonality of the system's eigensolutions. The same poor results also have to be expected for similar schemes which are based on the idea of splitting the spatial operators to obtain unconditional stability of the time integration.

References

Darms, M.: Analyse des ADI Zeitintegrationsverfahrens für zeitlich veränderliche Feldprobleme, Studienarbeit, Technische Universität Darmstadt, 2001.

Darms, M., Schuhmann, R., Spachmann, H., and Weiland, T.: Dispersion and Asymmetry Effects of ADI–FDTD, IEEE Microwave and Wireless Components Letters, in press, 2003.

Garcia, S. G., Lee, T., and Hagness, S. C.: On the Accuracy of the ADI-FDTD Method, IEEE Antennas and Wireless Propagation Letters, 1, 31–34, 2002.

Kole, J. S., Figge M. T., and De Raedt, H.: Unconditionally Stable Algorithms to Solve the Time-Dependent Maxwell Equations, Physical Review E, 64, 066705, 2001.

Namiki, T.: 3-D ADI–FDTD Method – Unconditionally Stable Time-Domain Algorithm for Solving Full Vector Maxwell's Equations, IEEE Trans. on Microwave Theory and Techniques, 48, 1743–1748, 2000.

Schuhmann, R. and Weiland, T.: Conservation of Discrete Energy and Related Laws in the Finite Integration Technique, Progress In Electromagnetics Research, PIER, 32, 301–316, 2001.

Staker, S. and Piket-May, M.: Algorithm Study of ADI-FDTD, Digest of the 2001 USNC/URSI National Radio Science Meeting, Boston, 255, 2001.

Weiland, T.: A Discretization Method for the Solution of Maxwell's Equations for Six-Component Fields, Electronics and Communication (AEÜ), 31, 116, 1977a.

Weiland, T.: Time Domain Electromagnetic Field Computation with Finite Difference Methods, Int. Journal of Numerical Modelling, 9, 295–319, 1996.

Yee, K. S.: Numerical Solution of Initial Boundary Value Problems Involving Maxwell's Equations in Isotropic Media, IEEE Trans. on Antennas and Propagation, 17, 585–589, 1966.

Simulation of microwave circuits and laser structures including PML by means of FIT

G. Hebermehl[1], J. Schefter[2], R. Schlundt[2], Th. Tischler[3], H. Zscheile[3], and W. Heinrich[3]

[1]Greifswalder Str. 147, 10409 Berlin, Germany
[2]Weierstrass Institute for Applied Analysis and Stochastics (WIAS), Mohrenstr. 39, 10117 Berlin, Germany
[3]Ferdinand-Braun-Institut für Höchstfrequenztechnik (FBH), Albert-Einstein-Str. 11, 12489 Berlin, Germany

Abstract. Field-oriented methods which describe the physical properties of microwave circuits and optical structures are an indispensable tool to avoid costly and time-consuming redesign cycles. Commonly the electromagnetic characteristics of the structures are described by the scattering matrix which is extracted from the orthogonal decomposition of the electric field. The electric field is the solution of an eigenvalue and a boundary value problem for Maxwell's equations in the frequency domain. We discretize the equations with staggered orthogonal grids using the Finite Integration Technique (FIT). Maxwellian grid equations are formulated for staggered nonequidistant rectangular grids and for tetrahedral nets with corresponding dual Voronoi cells. The interesting modes of smallest attenuation are found solving a sequence of eigenvalue problems of modified matrices. To reduce the execution time for high-dimensional problems a coarse and a fine grid is used. The calculations are carried out, using two levels of parallelization. The discretized boundary value problem, a large-scale system of linear algebraic equations with different right-hand sides, is solved by a block Krylov subspace method with various preconditioning techniques. Special attention is paid to the Perfectly Matched Layer boundary condition (PML) which causes non physical modes and a significantly increased number of iterations in the iterative methods.

1 Introduction

The commercial applications of microwave circuits cover the frequency range between 1 GHz and about 100 GHz; special applications, e.g. in radioastronomy, use even higher frequencies up to 1 THz. For optoelectronic devices, frequencies around several hundred THz are common. Higher frequencies lead to decreasing wavelength. Thus they yield increased dimensions of the discretization corresponding to the numerical problem and demand new strategies.

Correspondence to: G. Hebermehl
(hebermehl@wias-berlin.de)

The subject under investigation are three-dimensional structures of arbitrary geometry which are connected to the remaining circuit by transmission lines. Ports are defined at the transmission-lines outer terminations. In order to characterize their electrical behavior the transmission lines are assumed to be infinitely long and longitudinally homogeneous. Short parts of the transmission lines and the passive structure (discontinuity) form the structure under investigation. The entire structure has to be covered with an enclosure.

2 Scattering matrix

The scattering matrix describes the structure in terms of wave modes on the transmission line sections at the ports. The dimension of this matrix is determined by the total number of modes at all ports. The scattering matrix can be extracted from the orthogonal decomposition of the electric field into a sum of mode fields. This has to be done at a pair of neighboring planes z_p and $z_{p+\Delta p}$ on each waveguide (Christ and Hartnagel, 1987). The electric fields at the cross-sectional planes z_p and $z_{p+\Delta p}$ are known from the solution of the eigenvalue problem (see Sect. 5) and from the computation of the boundary value problem (see Sect. 3), respectively.

3 Boundary value problem

A three-dimensional boundary value problem can be formulated using the integral form of Maxwell's equations in the frequency domain in order to compute the electromagnetic field:

$$\oint_{\partial\Omega} \boldsymbol{H} \cdot d\boldsymbol{s} = \int_{\Omega} \jmath\omega[\epsilon]\boldsymbol{E} \cdot d\boldsymbol{\Omega}, \qquad \oint_{\partial\Omega} \boldsymbol{E} \cdot d\boldsymbol{s} = -\int_{\Omega} \jmath\omega[\mu]\boldsymbol{H} \cdot d\boldsymbol{\Omega}, \tag{1}$$

$$\boldsymbol{D} = [\epsilon]\boldsymbol{E}, \quad \boldsymbol{B} = [\mu]\boldsymbol{H}, \tag{2}$$

with

$$[\epsilon] = \epsilon_0 \mathrm{diag}\left(\tilde{\epsilon}_x, \tilde{\epsilon}_y, \tilde{\epsilon}_z\right), \quad [\mu] = \mu_0 \mathrm{diag}\left(\tilde{\mu}_x, \tilde{\mu}_y, \tilde{\mu}_z\right) \tag{3}$$

for rectangular grids and in this paper

$$[\epsilon] = \epsilon_0\epsilon_r, \quad [\mu] = \mu_0\mu_r \tag{4}$$

for tetrahedral grids. The electric field intensity $\boldsymbol{D}$ and the magnetic flux density $\boldsymbol{B}$ are complex functions of the spatial coordinates. ω is the angular frequency of the sinusoidal excitation, and $j^2 = -1$. Ω is an open surface surrounded by a closed contour $\partial\Omega$. The direction of the element ds of the contour $\partial\Omega$ is determined according to a right-hand system.

At the port p the transverse electric field $\boldsymbol{E}_t(z_p)$ is given by superposing the fields of the weighted transmission line modes $\boldsymbol{E}_{t,l}(z_p)$:

$$\boldsymbol{E}_t(z_p) = \sum_{l=1}^{m^{(p)}} w_l(z_p)\boldsymbol{E}_{t,l}(z_p). \tag{5}$$

The transverse electric mode fields $\boldsymbol{E}_{t,l}(z_p)$ have to be computed solving an eigenvalue problem for the transmission lines (see Sect. 5). All other parts of the surface of the computation domain are assumed to be electric or magnetic walls:

$$\boldsymbol{E} \times \boldsymbol{n} = 0 \quad \text{or} \quad \boldsymbol{H} \times \boldsymbol{n} = 0. \tag{6}$$

In order to simulate open structures we apply the PML absorbing boundary condition. The uniaxial formulation according to (Sacks et al., 1995) is implemented. The PML provides absorbing properties for any frequency, polarization and angle of incidence. Its layers are filled with an artificial material with complex anisotropic material properties. Therefore, the complex permittivity $[\epsilon]$ and the complex permeability $[\mu]$ are diagonal tensors (see Eq. 3). ϵ_0 and μ_0 are the permittivity and the permeability in vacuum, respectively.

4 Maxwellian grid equations

Maxwell's equations are discretized on staggered grids using the Finite Integration Technique with lowest order integration formulae

$$\oint_{\partial\Omega} \boldsymbol{f} \cdot d\boldsymbol{s} \approx \sum(\pm f_i s_i), \qquad \int_{\Omega} \boldsymbol{f} \cdot d\boldsymbol{\Omega} \approx f\Omega. \tag{7}$$

In the following we refer to staggered nonequidistant rectangular (Beilenhoff et al., 1992; Weiland, 1977; Hebermehl et al., 1999) grids or tetrahedral nets with corresponding Voronoi cells.

4.1 Staggered nonequidistant rectangular grids

The use of rectangular grids is the standard approach. In general it is very well adapted to planar microwave structures, since most circuits have a rectangular geometry.

Using Eq. (7) for staggered nonequidistant rectangular grids Eqs. (1) are transformed into a set of grid equations:

$$\begin{aligned} \mathbf{A}^T\mathbf{D}_{s/\tilde{\mu}}\boldsymbol{b} &= j\omega\epsilon_0\mu_0\mathbf{D}_{A_{\tilde{\epsilon}}}\boldsymbol{e}, \\ \mathbf{A}\mathbf{D}_s\boldsymbol{e} &= -j\omega\mathbf{D}_A\boldsymbol{b}. \end{aligned} \tag{8}$$

The vectors $\boldsymbol{e}$ and $\boldsymbol{b}$ contain the components of the electric field intensity and of the magnetic flux density of the elementary cells, respectively. The diagonal matrices $\mathbf{D}_{s/\tilde{\mu}}$, $\mathbf{D}_{A_{\tilde{\epsilon}}}$, $\mathbf{D}_s$, and $\mathbf{D}_A$ contain the information on cell dimensions and materials. $\mathbf{A}$ is sparse and contains the values 0, 1, and -1 only.

Eliminating the components of the magnetic flux density from the two Eqs. (8) the number of unknowns can be reduced by a factor of two, and we obtain the system of linear algebraic equations

$$(\mathbf{A}^T\mathbf{D}_{s/\tilde{\mu}}\mathbf{D}_A^{-1}\mathbf{A}\mathbf{D}_s - k_0^2\mathbf{D}_{A_{\tilde{\epsilon}}})\boldsymbol{e} = 0, \quad k_0 = \omega\sqrt{\epsilon_0\mu_0}. \tag{9}$$

Taking into account the boundary conditions, Eq. (9) can be written as $\tilde{\mathbf{A}}\underline{\boldsymbol{e}} = \boldsymbol{r}$, where the vector $\boldsymbol{r}$ contains the boundary values of the electric field according to Eqs. (5) and (6). k_0 is the wavenumber in vacuum, which is proportional to the frequency ω.

4.2 Tetrahedral grids and Voronoi cells

Because of the high spatial resolution CPU time and storage requirements are very high. Using rectangular grids a mesh refinement in one point results in an accumulation of small elementary cells in all coordinate directions even though generally the refinement is needed only in inner regions. In addition, rectangular grids are not well suited for treatment of curved and non-rectangular structures. Thus, an additional finite-volume method was developed, which uses tetrahedral nets with corresponding Voronoi cells for the three-dimensional boundary value problem. This allows to reduce the number of elementary cells by local grid refinement and improves the treatment of curved structures. The primary grid is formed by tetrahedra and the dual grid by the corresponding Voronoi cells (see Fig. 1).

For sake of simplicity, at first, we assume in this paper that the circumcenter of a tetrahedron is located within the tetrahedron. We consider a tetrahedron $ABCD$ with the internal edge AB (see Fig. 2) and the neighbouring elements, which share the edge AB with it (see Fig. 3).

The electric field intensity components (marked with red color in Fig. 1 and in Fig. 2) are located at the centers of the edges of the tetrahedra, and the magnetic flux density components (marked with black color) are normal to the circumcenters of the triangular faces. The Voronoi cells are polytopes. In the special case demonstrated in Fig. 1 the Voronoi cell is a pentadodecahedron which results as dual cell from 20 neighboring regular tetrahedra.

We use the following notations (see Figs. 1, 2, and 3) with $X, Y, Z, W \in \{A, B, C, D\}$, where X, Y, Z, W are different from each other, in order to develop the grid equations for tetrahedral nets:

X, Y, Z, W	nodes,
XY	edge between the nodes X and Y,
XYZ	triangle,
$XYZW$	tetrahedron,
S_{XY}	center of XY,

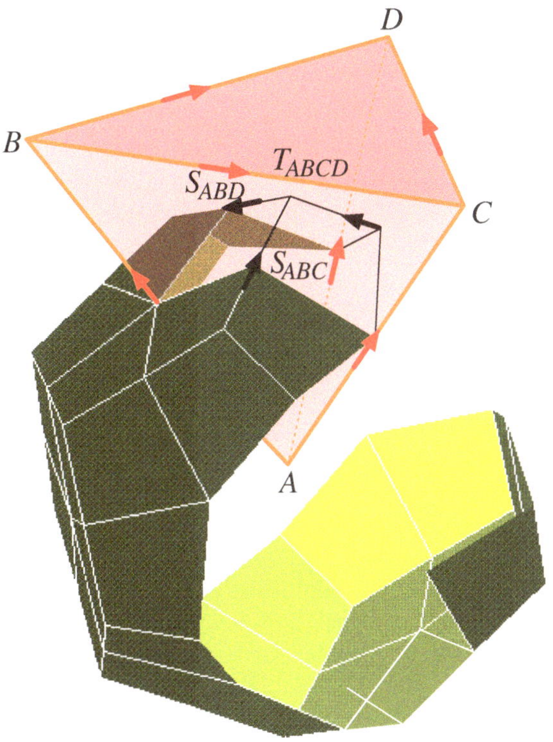

Fig. 1. Voronoi cell and single tetrahedron.

S_{XYZ}	circumcenter of the triangle XYZ,
T_{XYZW}	circumcenter of the tetrahedron $XYZW$,
E_{XY}	magnitude of the electric field on S_{XY} between the nodes X and Y,
B_{XYZ}	magnitude of the magnetic flux density on S_{XYZ}, normal to XYZ,
$\mu_{XYZW} = \mu_0\mu_r$	permeability in $XYZW$,
$\epsilon_{XYZW} = \epsilon_0\epsilon_r$	permittivity in $XYZW$,
l_{XY}	distance of node X to node Y,
l_{XYZ}^{W}	distance of the circumcenter of $XYZW$ to the face XYZ,
d_{XY}^{Z}	distance of the circumcenter of XYZ to the edge XY,
a_{XYZ}	area of triangle XYZ.

E_{XY} and B_{XYZ} satisfy

$$\begin{aligned} E_{XY} &= -\,E_{YX}, \\ B_{XYZ} = \; B_{YZX} &= \; B_{ZXY} = \\ -B_{YXZ} = -B_{XZY} &= -B_{ZYX} \;, \end{aligned} \tag{10}$$

respectively.

Using a finite volume approach with the lowest-order integration formulae (7) Eqs. (1) are transformed into a set of grid equations.

Taking into account the constitutive relations (2) the first equation of (1) is discretized on the dual grid. The internal edge AB is orthogonal to the corresponding Voronoi cell face over which we have to integrate (see Fig. 1). The closed integration path $\partial\Omega$ (see Eqs. 1 and 7) consists of the edges with length $s_i = l_{XYZ}^{W}$ (see Eq. 7 and Fig. 2), and is the

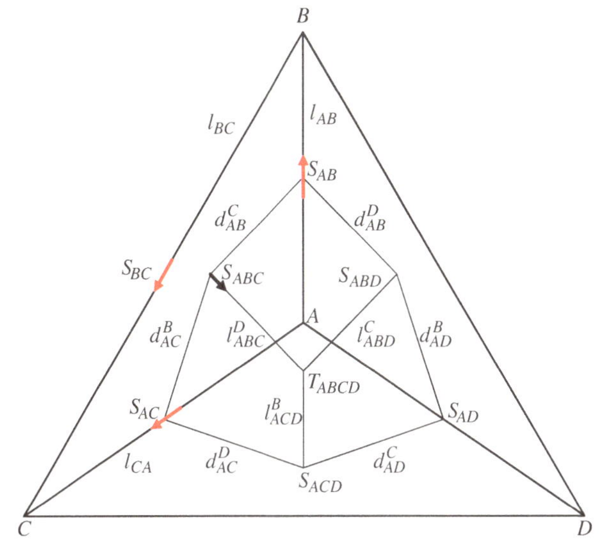

Fig. 2. Tetrahedron with partial areas of the Voronoi cell faces, which correspond to node A.

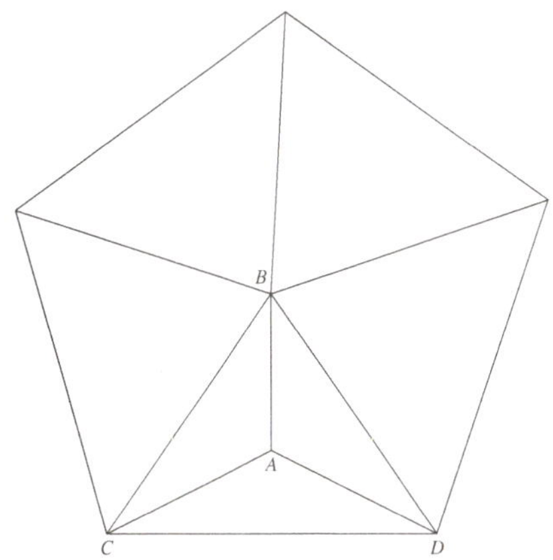

Fig. 3. Tetrahedra which share the edge AB.

polygon around the periphery of the mentioned Voronoi cell face (see Fig. 1, upper pentagon). The vertices of the polygon are the circumcenters of the tetrahedra which share the edge AB with the tetrahedron $ABCD$ (see Fig. 3). $f_i = B_{XYZ}$ denotes the function values on S_{XYZ} (see Fig. 2). Ω is the area of the Voronoi cell face. $f = E_{AB}$ denotes the function value on the center S_{AB}. Thus, the discretized equation takes the form:

$$\begin{aligned} &\sum\nolimits_{CD} \frac{1}{\mu_{ABCD}} \left[l_{ABC}^{D}\, B_{ABC} + l_{ABD}^{C}\, B_{ABD} \right] = \\ &j\omega \left[\sum\nolimits_{CD} \tfrac{1}{2} \epsilon_{ABCD} \left(d_{AB}^{C}\, l_{ABC}^{D} + d_{AB}^{D}\, l_{ABD}^{C} \right) \right] E_{AB} \end{aligned} \tag{11}$$

where the sum is over those tetrahedra $ABCD$, which share the edge AB (see Fig. 3).

The second equation of (1) is discretized using Eq. (7) on the primary grid. We have to integrate over the triangle ABC, i.e. the closed integration path $\partial\Omega$ consists of the edges with length $s_i = l_{XY}$ and is the polygon around the periphery of the mentioned triangle. $f_i = E_{XY}$ denotes the function value on the edges XY. Ω is the area a_{ABC} of the triangle ABC. $f = B_{ABC}$ denotes the function value on the circumcenter S_{ABC} of the triangle ABC. This yields the following form:

$$l_{AB}\, E_{AB} + l_{BC}\, E_{BC} + l_{CA}\, E_{CA} = -\jmath\omega a_{ABC}\, B_{ABC}. \quad (12)$$

Now we address the first of the surface integrals

$$\oint_{\cup\Omega} ([\epsilon]\boldsymbol{E}) \cdot d\boldsymbol{\Omega}{=}0, \quad \oint_{\cup\Omega} ([\mu]\boldsymbol{H}) \cdot d\boldsymbol{\Omega}{=}0, \quad (13)$$

reverting to the dual grid. In Eq. (13) $\cup\Omega$ is a closed surface with an interior volume. We have to integrate over the surface of the Voronoi cell. The surface of the Voronoi cell consists of all partial Voronoi areas, which belong to tetrahedron edges, whose shared corner node is A (see Figs. 1, 2 and 3). A discretization formula, with similar form to the right-hand side of (11) is obtained, i.e.

$$\sum_B \left(\left[\sum_{CD} \frac{1}{2} \overline{\epsilon} \left(d_{AB}^C\, l_{ABC}^D + d_{AB}^D\, l_{ABD}^C \right) \right] E_{AB} \right) = 0, \quad (14)$$

with $\overline{\epsilon} = \epsilon_{ABCD}$, except for the additional outer summation taken over all the nodes B neighboring A (in the primary grid).

For our final integral Eq. (13) the primary grid is used again, but now the integration is over the surface of the tetrahedron $ABCD$. As a consequence, the discretized form

$$\begin{aligned} &-a_{ABC}\, B_{ABC} - a_{ACD}\, B_{ACD} + \\ &+a_{ABD}\, B_{ABD} + a_{BCD}\, B_{BCD} \quad = 0 \end{aligned} \quad (15)$$

can be deduced.

Equations (11) and (12) form a system of linear algebraic equations for the computation of the electromagnetic field. Substituting the components of the magnetic flux density the number of unknowns in this system can be reduced by a factor of two:

$$\begin{aligned} &\textstyle\sum_{CD} \frac{1}{\mu_{ABCD}} \left[\left(\frac{l_{ABC}^D}{a_{ABC}} + \frac{l_{ABD}^C}{a_{ABD}} \right) l_{AB}\, E_{AB} + \right. \\ &+ \frac{l_{ABC}^D\, l_{BC}}{a_{ABC}}\, E_{BC} + \frac{l_{ABC}^D\, l_{CA}}{a_{ABC}}\, E_{CA} + \\ &\left. + \frac{l_{ABD}^C\, l_{BD}}{a_{ABD}}\, E_{BD} + \frac{l_{ABD}^C\, l_{DA}}{a_{ABD}}\, E_{DA} \right] = \\ &= \frac{\omega^2}{2} \left[\textstyle\sum_{CD} \epsilon_{ABCD} \left(d_{AB}^C\, l_{ABC}^D + d_{AB}^D\, l_{ABD}^C \right) \right] E_{AB}. \end{aligned} \quad (16)$$

Here, summation is taken over these tetrahedra $ABCD$, which possess the common edge AB. Equation (16) has to be solved using the boundary conditions (5) and (6).

5 Eigenvalue problem

For the eigenvalue problem, we refer to the rectangular grids (Christ and Hartnagel, 1987).

The transverse electric mode fields (see Eq. 5) at the ports of a transmission line, which is discretized by means of tetrahedral grids, are computed interpolating the results of the rectangular discretization.

The field distribution at the ports is computed assuming longitudinal homogeneity for the transmission line structure. Thus, any field can be expanded into a sum of so-called modal fields which vary exponentially in the longitudinal direction:

$$\boldsymbol{E}(x, y, z \pm 2h) = \boldsymbol{E}(x, y, z)e^{\mp \jmath k_z 2h}. \quad (17)$$

k_z is the propagation constant. $2h$ is the length of an elementary cell in z-direction. We consider the field components in three consecutive elementary cells. The electric field components of the vector $\boldsymbol{e}$ (see Eq. 9) $E_{x_{i,j,k+1}}$, $E_{x_{i,j,k-1}}$, $E_{y_{i,j,k+1}}$, $E_{y_{i,j,k-1}}$, $E_{z_{i,j,k-1}}$, $E_{z_{i+1,j,k-1}}$, and $E_{z_{i,j+1,k-1}}$ are expressed by the values of cell k using ansatz (17). The longitudinal electric field components E_z can be eliminated by means of the electric-field divergence equation (see first equation of (13)). Thus, we obtain an eigenvalue problem for the transverse electric field $\boldsymbol{y}$ on the transmission line region:

$$\mathbf{G}\boldsymbol{y} = \gamma\, \boldsymbol{y}, \quad \gamma = e^{-\jmath k_z 2h} + e^{+\jmath k_z 2h} - 2 = -4\sin^2(hk_z). \quad (18)$$

The sparse matrix $\mathbf{G}$ is in general nonsymmetric and complex. The order of $\mathbf{G}$ is $n = 2n_x n_y - n_b$. $n_x n_y$ is the number of elementary cells at the port. The size n_b depends on the number of cells with perfectly conducting material. The relation between the propagation constants k_z and the eigenvalues γ is nonlinear. The interesting modes of smallest attenuation are found solving a sequence of eigenvalue problems of modified matrices (see (Hebermehl et al., 2003b)) using the invert mode of the Arnoldi iteration (Lehoucq, 1995).

6 PML modes

We use the PML in order to calculate the eigen modes of open waveguide structures as well as for structures that require electrically large computational cross sections. Using only magnetic or electric walls at the outer boundaries causes additional non-physical box modes. Those modes would affect the propagation behavior of the physical waveguide modes. Introducing the PML shifts these box modes within the eigenvalue spectrum, away from the physical ones. The difference, however, is not always large enough to be clearly detectable. Therefore, we need an additional criterion to distinguish these PML modes from the desired physical ones. The PML modes are characterized by their high power concentration in the PML area (Tischler and Heinrich, 2000). Thus, to eliminate the PML modes we calculate the magnitude of the power flow of each computed mode in the PML ($P^{(P)}$), in the waveguide region ($P^{(W)}$), and in the total computational domain (P):

$$\begin{aligned} &P = P^{(P)} + P^{(W)} = \\ &\int_{\Omega^{(P)}} (\boldsymbol{E}_t \times \boldsymbol{H}_{t,m}^*) \cdot d\boldsymbol{\Omega} + \int_{\Omega^{(W)}} (\boldsymbol{E}_t \times \boldsymbol{H}_{t,m}^*) \cdot d\boldsymbol{\Omega}. \end{aligned} \quad (19)$$

A mode is specified as PML-mode if

$$r^{(P)} = \frac{P^{(P)}}{P} > \xi \,, \tag{20}$$

with values $\xi = 0.2, \ldots, 0.6$, found empirically.

7 Laser analysis

The presented method, although developed initially for microwave structures (Hebermehl et al., 2001), is expanded to meet the special requirements of optoelectronic structure calculations. For optoelectronic devices frequencies of about several hundred THz are common. Thus the complex region containing the eigenvalues of potentially propagating modes grows substantially. A significantly higher number of eigenvalue problems have to be solved within our algorithm. Additionally, the maximum cell size of the discretization should be less than $\frac{\lambda}{10}$, where λ denotes the wavelength in the material with the highest $\Re([\epsilon])$. Additional mesh refinements have to be used for structure regions with highly varying fields. All this results in high-dimensional problems which have to be handled.

To reduce the execution times, in a first step the problem is solved using a coarse grid with lower accuracy requirements in order to find the approximate locations of the interesting propagation constants. Anyway, the number of modified eigenvalue problems to be solved is high. Thus, we split the interesting interval into subintervals and compute the corresponding eigenpairs independently and in parallel, for instance on different workstations or shared memory multiprocessors. Finally, the modes of interest are calculated in a second step for an essentially reduced region using a fine grid, that fulfills higher accuracy requirements.

The linear sparse solver PARDISO (Schenk et al., 2000) is applied in order to fulfill the high accuracy requirements of the eigenvalue problem. The parallel CPU mode of PARDISO provides the additional possibility to reduce the computing times for this high-dimensional problem on shared memory multiprocessors without essential additional memory requirements.

A Laser application can be found in (Hebermehl et al., 2003a).

8 Systems of linear algebraic equations

Four kinds of preconditioning and a block quasi-minimal residual algorithm are applied to solve the large scale systems of linear algebraic equations. Details are given with (Schlundt et al., 2001) and (Hebermehl et al., 2003a).

Especially, adding the gradient of the electric field divergence the numerical properties of the system matrix are improved and the equations can be solved faster.

8.1 Rectangular grids

Multiplying (9) by $\mathbf{D}_s^{1/2}$ yields a symmetric form of linear algebraic equations:

$$\bar{\mathbf{A}}\boldsymbol{x} = 0, \quad \bar{\mathbf{A}} = (\mathbf{D}_s^{1/2}\mathbf{A}^T\mathbf{D}_{s/\tilde{\mu}}\mathbf{D}_A^{-1}\mathbf{A}\mathbf{D}_s^{1/2} - k_0^2\mathbf{D}_{A_{\tilde{\epsilon}}}) \tag{21}$$

with $\boldsymbol{x} = \mathbf{D}_s^{1/2}\boldsymbol{e}$. The gradient of the electric field divergence

$$[\epsilon]\nabla([\epsilon]^{-2}\nabla \cdot [\epsilon]\boldsymbol{E}) = 0 \tag{22}$$

is equivalent to the matrix equation

$$\bar{\mathbf{B}}\boldsymbol{x} = 0, \quad \bar{\mathbf{B}} = \mathbf{D}_s^{-1/2}\mathbf{D}_{A_{\tilde{\epsilon}}}\mathbf{B}^T\mathbf{D}_{V_{\tilde{\epsilon}\tilde{\epsilon}}}^{-1}\mathbf{B}\mathbf{D}_{A_{\tilde{\epsilon}}}\mathbf{D}_s^{-1/2}. \tag{23}$$

The rectangular matrix $\mathbf{B}$ is sparse and contains the values 0, 1, and -1 only. The diagonal matrix $\mathbf{D}_{V_{\tilde{\epsilon}\tilde{\epsilon}}}$ is a volume matrix for the 8 partial volumes of the dual elementary cell. Taking into account the boundary conditions (5) and (6) Eqs. (21) and (23) yields the form $\hat{\mathbf{A}}\boldsymbol{x} = \boldsymbol{b}$ and $\hat{\mathbf{B}}\boldsymbol{x} = 0$, respectively, and

$$(\hat{\mathbf{A}} + \hat{\mathbf{B}})\boldsymbol{x} = \boldsymbol{b}, \quad \hat{\mathbf{A}} + \hat{\mathbf{B}} \text{ complex indefinite symmetric}, \tag{24}$$

can be solved faster than Eq. (9).

In comparison to the simple lossy case the number of iterations of Krylov subspace methods increases significantly in the presence of PML. Among others the speed of convergence depends on the relations of the edges in an elementary cell of the nonequidistant rectangular grid in this case. The best results can be obtained using nearly cubic cells. Moreover, overlapping PML conditions at the corner regions of the computational domain lead to an increase of the magnitude of the corresponding off-diagonal elements in comparison to the diagonal of the coefficient matrix. This downgrades the properties of the matrix. Thus, overlapping PML should be avoided.

8.2 Tetrahedral grids

Equations (11), (12), (14) and (15) can be written in matrix form. Besides of the locations and values of the entries, the matrix representations of Eqs. (11) and (12) have the same structure as Eq. (8), and an appropriate (see Eq. 21) symmetric form of linear algebraic equations can be deduced. Adding the gradient of the electric field divergence also gives a system which can be solved faster. Here, the gradient of the divergence at an internal point is obtained considering the partial volumes of the appropriate Voronoi cell. PML are not included.

9 Application

As an example we have simulated a microwave structure with a microstrip changing its width (impedance step, see Fig. 4). The discontinuity of the metallic strip is located at $z = l1$. The metal is assumed to be ideally conducting. The relative permittivity of the substrate below the microstrip line is $\epsilon_r = \tilde{\epsilon}_x = \tilde{\epsilon}_y = \tilde{\epsilon}_z = 11.9$. Above the microstrip line one has air

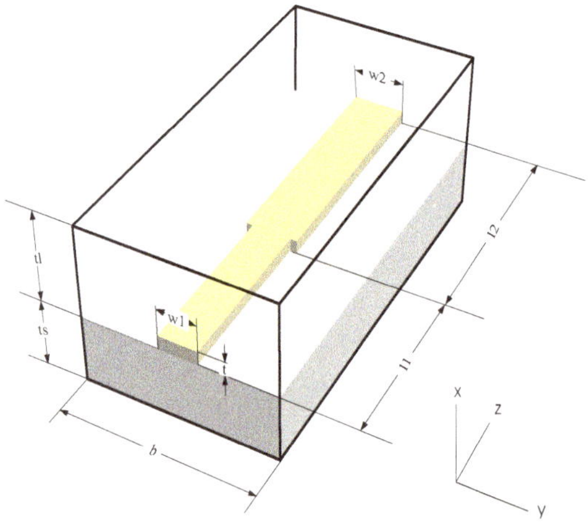

Fig. 4. The microstrip line changes its extension.
Measurements: b = 300 μm, w1 = 60 μm, w2 = 80 μm, ts = 250μm, tl = 250 μm, t = 3 μm, l1 = 200 μm, l2 = 200 μm.

($\epsilon_r = \tilde{\epsilon}_x = \tilde{\epsilon}_y = \tilde{\epsilon}_z = 1$). Ports are located at the planes $z = 0$ and $z = l1 + l2$. The remaining 4 outer surfaces are assumed to be electric walls. The structure is symmetric with respect to the plane $y = b/2$. But, the whole structure is discretized in this test example.

For comparison the structure is subdivided in nonequidistant rectangular three-dimensional elementary cells on the one hand and in tetrahedra on the other hand.

In case of rectangular grids, the order of the system of linear algebraic equations, which corresponds to the boundary value problem (1), is $n = 3n_x n_y n_z = 335\,160$. $n_x n_y n_z$ is the number of cells of the structure which is assumed to be a parallelepiped. We need a high mesh refinement near the microstrip line. Using the rectangular grid the mesh refinement in this region results in an accumulation of elementary cells in all coordinate directions even though the refinement is not necessary in order to approximate the solution with the required accuracy.

The tetrahedral grid consists of $n_n = 29\,615$ nodes, $n_t = 161\,308$ tetrahedra, and $n_p = 16\,100$ peripheral cell faces. The order of the corresponding system of linear algebraic equations is equal to the number of edges and amounts

$$n = n_n + n_t + n_p/2 - 1 = 198\,972. \quad (25)$$

The disadvantage of rectangular grids, the accumulation of elementary cells in all coordinate directions, is avoided here.

10 Conclusions

Finite-difference analysis of microwave circuits including lossy materials and radiation effects leads to an complex eigenvalue problem and large-scale complex indefinite symmetric systems of linear algebraic equations. Staggered nonequidistant rectangular grids and tetrahedral nets with corresponding Voronoi cells are used. The high execution times in the eigenvalue problems for waveguide applications with extremely large mesh sizes, e.g. laser structures, are reduced using a coarse and a fine grid, and two levels of parallelization. The high-dimensional systems of linear algebraic equations are solved using a bock Krylov subspace method.

Acknowledgements. We thank F.-K. Hübner for his support in preparing graphics.

References

Beilenhoff, K., Heinrich, W., and Hartnagel, H. L.: Improved Finite-Difference Formulation in Frequency Domain for Three-Dimensional Scattering Problems, IEEE Transactions on Microwave Theory and Techniques, 40, No. 3, 540–546, 1992.

Christ, A. and Hartnagel, H. L.: Three-Dimensional Finite-Difference Method for the Analysis of Microwave-Device Embedding, IEEE Transactions on Microwave Theory and Techniques, 35, 688–696, 1987.

Hebermehl, G., Schlundt, R., Zscheile, H., and Heinrich, W.: Improved Numerical Methods for the Simulation of Microwave Circuits, Surveys on Mathematics for Industry, 9, 117–129, 1999.

Hebermehl, G., Hübner, F.-K., Schlundt, R., Tischler, T., Zscheile, H., and Heinrich, W.: Numerical Simulation of Lossy Microwave Transmission Lines Including PML, Lecture Notes in Computational Science and Engineering, Springer Verlag, 18, 267–275, 2001.

Hebermehl, G., Hübner, F.-K., Schlundt, R., Tischler, T., Zscheile, H., and Heinrich, W.: Simulation of Microwave and Semiconductor Laser Structures Including Absorbing Boundary Conditions, Lecture Notes in Computational Science and Engineering, Springer Verlag, 35, 131–159, 2003a.

Hebermehl, G., Hübner, F.-K., Schlundt, R., Tischler, T., Zscheile, H., and Heinrich, W.: Perfectly Matched Layers in Transmission Lines, Numerical Mathematics and Advanced Applications, ENUMATH 2001, Springer Verlag Italia, pp. 281–290, 2003b.

Lehoucq, R. B.: Analysis and Implementation of an Implicitly Restarted Arnoldi Iteration, Technical Report **13**, Rice University, Department of Computational and Applied Mathematics, 1–135, 1995.

Sacks, Z. S., Kingsland, D. M., Lee, R., and Lee, J.-F.: A Perfectly Matched Anisotropic Absorber for Use as an Absorbing Boundary Condition., IEEE Transactions on Antennas and Propagation, 43, 1460–1463, 1995.

Schenk, O., Gärtner, K., and Fichtner, W.: Efficient Sparse LU Factorization with Left-Right Looking Strategy on Shared Memory Multiprocessors, BIT, 40, 158–176, 2000.

Schlundt, R., Hebermehl, G., Hübner, F.-K., Heinrich, W., and Zscheile, H.: Iterative Solution of Systems of Linear Equations in Microwave Circuits Using a Block Quasi-Minimal Residual Algorithm, Lecture Notes in Computational Science and Engineering, Springer Verlag, 18, 325–333, 2001.

Tischler, T. and Heinrich, W.: The Perfectly Matched Layer as Lateral Boundary in Finite-Difference Transmission-Line Analysis, IEEE Transactions on Microwave Theory and Techniques, 48, 2249–2253, 2000.

Weiland, T.: A discretization method for the solution of Maxwell's equations for six-component fields, Electronics and Communication (AEÜ), 31, 116–120, 1977.

4

Comparison of errors and stability in FDTD and FVTD

P. Leuchtmann, C. Fumeaux, and D. Baumann

Institut für Feldtheorie und Höchstfrequenztechnik, ETH Zürich, CH-8092 Zürich, Switzerland

Abstract. The methods of finite-difference time-domain (FDTD) and finite-volume time-domain (FVTD) are well known techniques for solving electromagnetic problems. While for FDTD a number of commercial codes is available there are only few based on FVTD. However, FVTD has advantages since it allows completely unstructured meshes. This paper 1. deals with the principle errors made in both approaches and 2. gives a theoretical comparison of the (local) stability of an FVTD scheme as opposed to the conventional Yee-scheme commonly used in FDTD. Finally the principles of a new FVTD program is presented. Its capabilities are demonstrated by the field of an antenna array.

1 From field functions to numbers

In any numerical technique for solving Maxwell's equations the vector-valued $\boldsymbol{r}, t$ (space,time) *functions* $\boldsymbol{E}(\boldsymbol{r}, t)$ (electric field strength), $\boldsymbol{H}(\boldsymbol{r}, t)$ (magnetic field strength), $\boldsymbol{D}(\boldsymbol{r}, t)$ (displacement density) and $\boldsymbol{B}(\boldsymbol{r}, t)$ (magnetic flux density) are mapped to a finite set of *numbers* to be processed by the computer. In the well-known Yee-scheme (Yee, 1966; Taflove and Hagness, 2000) field components are taken at particular locations on a cubic grid such that the component values can be seen as "exact" values at the respective locations and the equations applied to these values can be interpreted as second-order finite-difference approximations of Maxwell's equations. E.g., the z-component of Maxwell's first equation is $-\frac{\partial}{\partial t} B_z(\boldsymbol{r}, t) = \frac{\partial}{\partial x} E_y(\boldsymbol{r}, t) - \frac{\partial}{\partial y} E_x(\boldsymbol{r}, t)$. This can be approximated to second-order accuracy at the location $(\boldsymbol{r}, t)$ by

$$-\frac{B_z(+\delta_t) - B_z(-\delta_t)}{2\delta_t} \approx \frac{E_y(+\delta_x) - E_y(-\delta_x)}{2\delta_x} - \frac{E_x(+\delta_y) - E_x(-\delta_y)}{2\delta_y} \qquad (1)$$

Correspondence to: P. Leuchtmann
(leuchtmann@ifh.ee.ethz.ch)

where $\delta_x, \delta_y, \delta_z, \delta_t$ are the respective half grid constants and say $E_y(+\delta_x)$ is short for $E_y(x + \delta_x, y, z, t)$. This means that the approximation of this particular component of Maxwell's equations at the location $(\boldsymbol{r}, t)$ needs only field values being half space- or time-steps apart from this location. Yee's scheme uses different locations for different field components. Fig. 1a shows the spatial location of both the components and the equations. Note that the material equations ($\boldsymbol{B} = \mu \boldsymbol{H}$ and $\boldsymbol{D} = \varepsilon \boldsymbol{E}$) are easily satisfied for each component separately at its own location. The Yee-scheme consists of set $\mathrm{A_Y}$: six Maxwellian component Eqs. (1) plus set $\mathrm{B_Y}$: six material equations. Set $\mathrm{A_Y}$ is of second-order accuracy, set $\mathrm{B_Y}$ is exactly satisfied.

In the finite integration technique (FIT) (Weiland, 1977) Maxwell's first equation in integral form ($-\frac{\partial}{\partial t} \iint_A \boldsymbol{B} \cdot \boldsymbol{dA} = \oint_{\partial A} \boldsymbol{E} \cdot \boldsymbol{ds}$) is applied to the green rectangle shown in Fig. 1(b).

$$-\frac{\partial}{\partial t} \langle B_z \rangle_{\mathrm{A}} \cdot 2\delta_x \cdot 2\delta_y = \langle E_y \rangle_{\mathrm{L}}^{+\delta_x} \cdot 2\delta_y - \langle E_y \rangle_{\mathrm{L}}^{-\delta_x} \cdot 2\delta_y - \left(\langle E_x \rangle_{\mathrm{L}}^{+\delta_y} \cdot 2\delta_x - \langle E_x \rangle_{\mathrm{L}}^{-\delta_y} \cdot 2\delta_x \right) \qquad (2)$$

Thereby $\langle . \rangle_{\mathrm{L}}^{\delta}$ means a line mean value along the respective side of the rectangle and $\langle . \rangle_{\mathrm{A}}$ means the surface mean value. After integrating the whole Eq. (2) along the time interval from $t - \delta_t$ to $t + \delta_t$ and then dividing it by $8\delta_x\delta_y\delta_t$ we obtain time mean values (denoted by overbars) at the right-hand side and surface mean values evaluated at the boundary of the time interval at the left-hand side:

$$-\frac{\langle B_z \rangle_{\mathrm{A}}(+\delta_t) - \langle B_z \rangle_{\mathrm{A}}(-\delta_t)}{2\delta_t} = \frac{\langle \overline{E_y} \rangle_{\mathrm{L}}^{+\delta_x} - \langle \overline{E_y} \rangle_{\mathrm{L}}^{-\delta_x}}{2\delta_x} - \frac{\langle \overline{E_x} \rangle_{\mathrm{L}}^{+\delta_y} - \langle \overline{E_x} \rangle_{\mathrm{L}}^{-\delta_y}}{2\delta_y} \qquad (3)$$

Though formally identical to the approximative relation (1) this is an *exact* equation. As in Yee's scheme the whole set of Maxwell's eqations can be derived. However, the material equations are now approximatively satisfied since, e.g., the line-time-mean value $\langle \overline{H_z} \rangle_{\mathrm{L}}$ is related to the surface mean

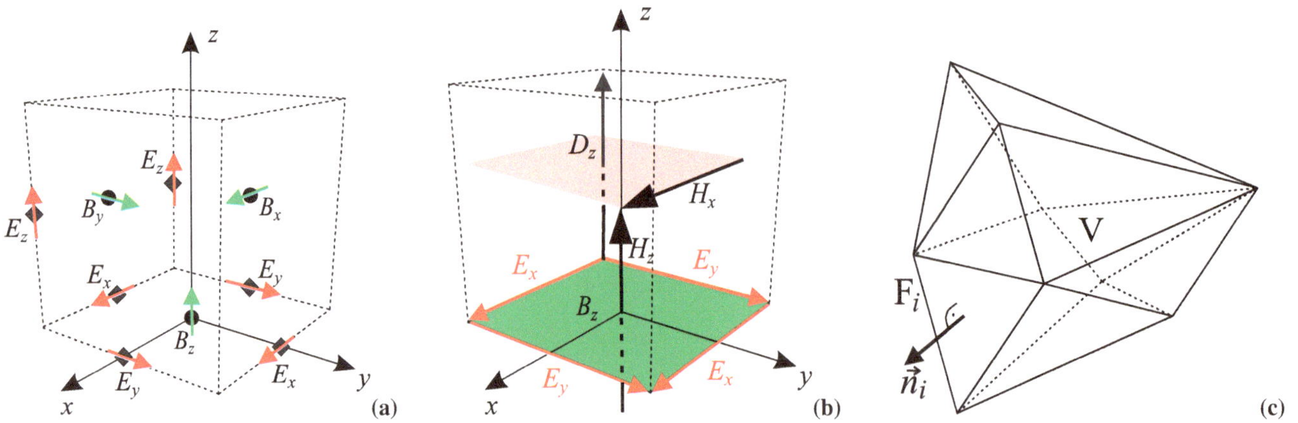

Fig. 1. The well-known Yee-scheme **(a)** places different field components at different locations. We use • for magnetic components and for electric components. The respective values are used to form second-order accurate difference equations. The FIT-approach **(b)** uses mean values (B- and D-components: surface mean values; E- and H-components: line-mean values). In FVTD **(c)** volume-mean values ($\boldsymbol{B}$- and $\boldsymbol{D}$-vectors) and face-mean values ($\boldsymbol{E}$- and $\boldsymbol{H}$-vectors) are used.

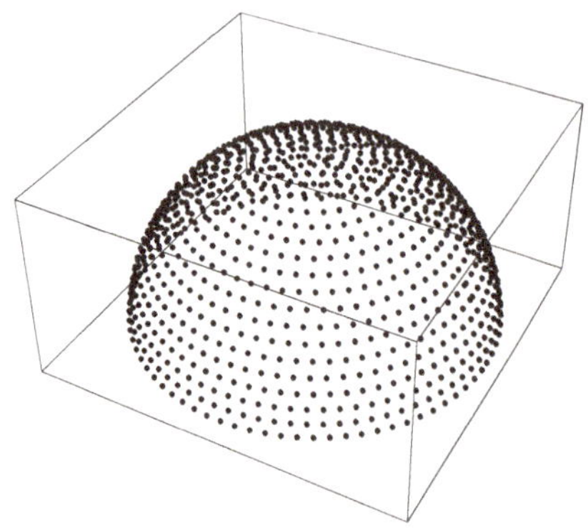

Fig. 2. The picture shows 988 wave vector directions (from the sphere's center). The numerical schemes are tested using plane waves propagating in the respective directions.

value $\langle B_z\rangle_A$ which is a value at a fixed time. The FIT-scheme consists of set A_F: six Maxwellian component equations plus set B_F: six material equations. Set A_F is exactly satisfied while set B_F is approximate.

In the finite-volume time-domain (FVTD) approach Maxwell's equations can also be integrated over a volume V rather than a surface A as done in the FIT-derivation in the previous section. We assume a polyhedron V with N surfaces F_i forming the boundary ∂V of V (see Fig. 1c) and find, e.g.,

$$-\frac{\partial}{\partial t}\underbrace{\iiint_{V}\boldsymbol{B}\,dV}_{V\cdot\langle\boldsymbol{B}\rangle_V}=-\oiint_{\partial V}\boldsymbol{E}\times d\boldsymbol{A}=\sum_{i=1}^{N}\boldsymbol{n}_i\times\underbrace{\iint_{F_i}\boldsymbol{E}\,dA}_{F_i\cdot\langle\boldsymbol{E}\rangle_{F_i}},\quad (4)$$

where $\langle\boldsymbol{E}\rangle_{F_i}$ and $\langle\boldsymbol{B}\rangle_V$ are the time-dependent mean values of the vector functions $\boldsymbol{E}$ (or $\boldsymbol{B}$) on the polyhedrons i-th face F_i (or the volume respectively) and $\boldsymbol{n}_i$ is the respective outer normal unit vector. Hence Eq. (4) is an exact relation between time dependent mean values. If Eq. (4) is integrated over time an exact vector-valued equation somehow similar to Eq. (3) is found:

$$-\frac{\langle\boldsymbol{B}\rangle_V(+\delta_t)-\langle\boldsymbol{B}\rangle_V(-\delta_t)}{2\delta_t}=\frac{1}{V}\sum_{i=1}^{N}F_i\left(\boldsymbol{n}_i\times\langle\overline{\boldsymbol{E}}\rangle_{F_i}\right).\quad (5)$$

Using Maxwell's second equation another exact vector equation involving $\langle\boldsymbol{D}\rangle_V$ and $\langle\overline{\boldsymbol{H}}\rangle_{F_i}$ values is obtained.

Concerning time the face mean values are "older" than the newest volume mean value at the left hand side. "Newer" face mean values might be obtained by the extrapolation

$$\begin{aligned}\langle\overline{\boldsymbol{E}}\rangle_F(t+2\delta_t)&\approx 2\langle\overline{\boldsymbol{E}}\rangle_F(t+\delta_t)-\langle\overline{\boldsymbol{E}}\rangle_F(t),\\ \langle\overline{\boldsymbol{H}}\rangle_F(t+2\delta_t)&\approx 2\langle\overline{\boldsymbol{H}}\rangle_F(t+\delta_t)-\langle\overline{\boldsymbol{H}}\rangle_F(t).\end{aligned}\quad (6)$$

Since each face has two sides one can set up further relations, e.g.

$$\begin{aligned}\langle\overline{\boldsymbol{H}}\rangle_F&\approx\tfrac{w_l}{\mu_l}\langle\boldsymbol{B}\rangle_{V_l}+\tfrac{w_r}{\mu_r}\langle\boldsymbol{B}\rangle_{V_r}\\ \langle\overline{\boldsymbol{E}}\rangle_F&\approx\tfrac{w_l}{\varepsilon_l}\langle\boldsymbol{D}\rangle_{V_l}+\tfrac{w_r}{\varepsilon_r}\langle\boldsymbol{D}\rangle_{V_r}\\ \text{or even}\quad \langle\overline{\boldsymbol{H}}\rangle_F&\approx f\left(\langle\boldsymbol{B}\rangle_{V_l},\langle\boldsymbol{B}\rangle_{V_r},\langle\boldsymbol{D}\rangle_{V_l},\langle\boldsymbol{D}\rangle_{V_r}\right)\\ \langle\overline{\boldsymbol{E}}\rangle_F&\approx g\left(\langle\boldsymbol{B}\rangle_{V_l},\langle\boldsymbol{B}\rangle_{V_r},\langle\boldsymbol{D}\rangle_{V_l},\langle\boldsymbol{D}\rangle_{V_r}\right)\end{aligned}\quad (7)$$

where the indices r and l stand for "left" and "right" (of the face F) respectively. The quantities $w_{r,l}$ are some weights due to geometry. In a simple symmetrical case (two identical cells) it is $w_r = w_l = \frac{1}{2}$. In the more complicated case, f and g are some functions yet to be specified. Regardless of the particular choices both Eqs. (6) and (7) are approximate relations.

FVTD ends up with an exact set A_V of six scalar Eq. (5) and a second set B_V – formed with Eqs. (6) and (7) – of approximate relations.

As it is well known both sets of equations are combined to an update scheme which allows it to compute "new" values (written in n-dimensional vector $[V]_{\text{new}}$) from the "old" values $[V]_{\text{old}}$. The respective scheme can be formally written

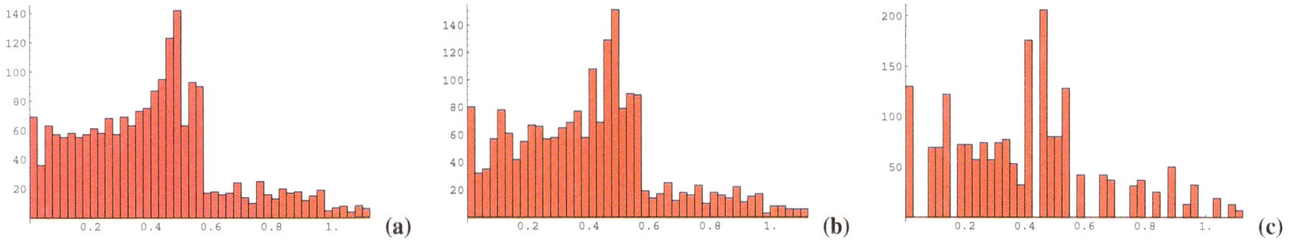

Fig. 3. The histograms of the FIT-errors according to Eq. (10) for $\frac{\lambda}{2\delta_\xi} = 14 \ldots 15$ (cells per wave length). Maximum values are around 1%. From left to right η_{FITx}, η_{FITy} and η_{FITz} are shown.

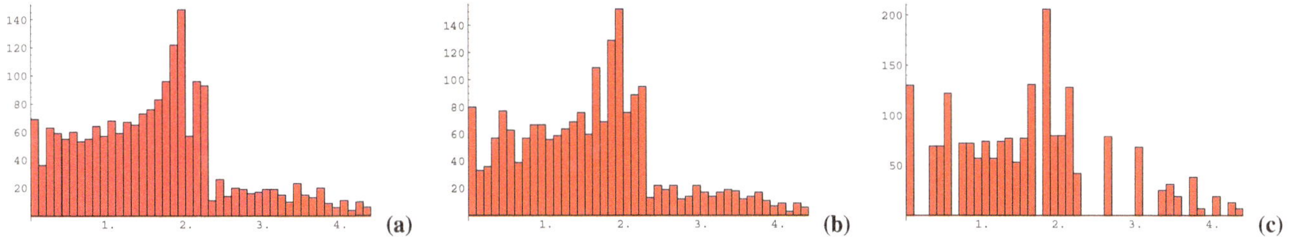

Fig. 4. Histograms of the FVTD-errors according to Eq. (11) for $\frac{\lambda}{2\delta_\xi} = 14 \ldots 15$ (cells per wave length). Maximum values are around 4% which is four times more than in FIT. This higher value is not a principle drawback of FVTD but mainly due to our simple choice in Eqs. (6) and (7): essentially a first order scheme both in space and time. From left to right η_{FVTDx}, η_{FVTDy} and η_{FVTDz} are shown.

with a large matrix $\mathbf{M}_{method}$ where "method" is one out of FDTD, FIT and FVTD:

$$[V]_{\mathrm{new}} = \mathbf{M}_{method} \cdot [V]_{\mathrm{old}}. \quad (8)$$

Two questions can be posed:
Accuracy: "To what degree are the approximate equations wrong in each case?"
Stability: "Under what conditions are the respective schemes numerically stable?"

Since FDTD and FIT differ only in the interpretation of the numbers but lead to formally the same equations we treat only the FIT.

2 The definition of the errors and their values

The accuracy depends on several things such as structure and dimension of the mesh but also on the actual field. We define 988 test cases of the actual field: plane waves propagating in different directions. A single plane wave in a homogeneous medium is described by

$$\boldsymbol{D}(\boldsymbol{r},t)/\varepsilon = \boldsymbol{E}(\boldsymbol{r},t) = \boldsymbol{E}_0 \cdot \cos(\omega t - \boldsymbol{\kappa}\cdot\boldsymbol{r})$$
$$\boldsymbol{B}(\boldsymbol{r},t)/\mu = \boldsymbol{H}(\boldsymbol{r},t) = \boldsymbol{H}_0 \cdot \cos(\omega t - \boldsymbol{\kappa}\cdot\boldsymbol{r})$$
$$\text{with} \quad \boldsymbol{H}_0 = \frac{1}{\omega\mu}\boldsymbol{\kappa}\times\boldsymbol{E}_0, \boldsymbol{\kappa}\cdot\boldsymbol{E}_0 = 0,\ \boldsymbol{\kappa}\cdot\boldsymbol{\kappa} = \omega^2\mu\varepsilon. \quad (9)$$

The directions of $\boldsymbol{\kappa} := (\kappa_x, \kappa_y, \kappa_z)$ and $\boldsymbol{E}_0$ are free beside these restrictions.

Given one of these test fields $[V]_{\mathrm{old}}$ is computed for both methods in the respective way (space and/or time mean values) and also the respective exact $[V]_{\mathrm{new,ex}}$. A slightly different value for $[V]_{\mathrm{new}}$ is obtained by using Eq. (8). A comparison of $[V]_{\mathrm{new,ex}}$ and $[V]_{\mathrm{new}}$ delivers the respective error.

In order to keep the amount of computation within reasonable limits we restrict the error analysis to the approximative equations and use typical discretisation lengths δ_ξ (fractions of the wavelength $\lambda = \frac{2\pi}{|\kappa|}$).

In FIT the material equations are approximate relations, e.g., $\langle D_\xi\rangle_{\mathrm{A}} \approx \varepsilon\langle\overline{E_\xi}\rangle_{\mathrm{L}}$. From this we derive the error $r_{\mathrm{Ex}} := \varepsilon\langle\overline{E_\xi}\rangle_{\mathrm{L}}/\langle D_\xi\rangle_{\mathrm{A}} \overset{!}{=} 1$. Introducing the expressions given in Eq. (9) and evaluating the mean values by analytical integration we find a value different from 1 for the respective exact ratio:

$$\tilde{r}_{\mathrm{Ex}} = \frac{\mathrm{Si}(\kappa_x\delta_x)\cdot\mathrm{Si}(\kappa_y\delta_y)}{\mathrm{Si}(\kappa_z\delta_z)\cdot\mathrm{Si}(\omega\delta_t)} \Rightarrow \eta_{\mathrm{FITx}} := |\tilde{r}_{\mathrm{Ex}} - 1|\cdot 100\% \quad (10)$$

with $\mathrm{Si}(x) := \frac{\sin x}{x}$. The histograms of these error values evaluated for $\delta_x = \delta_y = \delta_z = \sqrt{\frac{3}{\mu\varepsilon}}\delta_t$ and all wave directions shown in Fig. 2 are given in Figs. 3 and 4.

In order to obtain a reasonable basis of comparison we use a regular cubic grid (with $2\delta_x\cdot 2\delta_y\cdot 2\delta_z$-cubes) also in the FVTD case and make the aforementioned simple choices in Eqs. (6) and (7). Moreover we can set $w_1 = w_2 = \frac{1}{2}$. Introducing the expressions from Eq. (9) and performing the integrations according to the definitions in the FVTD-formulae we finally obtain for the update equation related to a face oriented in ξ-direction

$$\eta_{\mathrm{FVTD}\xi} := 2\left|\frac{\mathrm{Si}(2\kappa_\xi\delta_\xi) - \mathrm{Si}(2\omega\delta_t)}{\mathrm{Si}(\omega\delta_t)}\right|\cdot 100\% \quad (11)$$

The respective histograms of these errors are given in Fig. 4.

3 The stability

The time iterative scheme is a repeated application of Eq. (8). This means that the matrix $\mathbf{M}_{method}$ should not have any eigenvalue λ with $|\lambda| > 1$. For a full problem the dimension of $\mathbf{M}_{method}$ is very large and the respective search for all eigenvalues would be extremely expensive. However, there are special numerical schemes (e.g., the Arnoldi scheme) which find the largest eigenvalue within still reasonable time. In this work we do not follow that way but reduce the number of variables (and with it the dimension of $\mathbf{M}_{method}$) by defining a *local stability* by focusing on a single cell. Considering the update scheme for a particular value (e.g., by solving Eq. 3 or 5 for the latest B-value) we find that for computing all "new" values of a single cell the number of the required "old" values is always larger than the number of "new" values. This simply reflects the fact that "old" values from the neighbour cells are also involved. The respective 'local' update equation would have a small but rectangular matrix $\mathbf{M}_{local}$. This matrix can be reduced to a quadratic matrix by applying a spatial Fourier transformation to the "old" values. In this case any "old" value can be written as $v_{\text{old},0} \cdot e^{j(\kappa_x \tilde{x} + \kappa_y \tilde{y} + \kappa_z \tilde{z})}$ where $v_{\text{old},0}$ is the value in the cell's center, $(\tilde{x}, \tilde{y}, \tilde{z})$ denotes the displacement and $\kappa = (\kappa_x, \kappa_y, \kappa_z)$ is the vector of the spatial Fourier frequencies. In particular values required from outside the cell are related to the respective values inside the cell by a simple multiplication with the respective dislocation factor.

The restriction of the stability analysis to a single Fourier term is sufficient if stability is proofed for *any* Fourier term. This can be deduced from Parseval's theorem: the sum of all Fourier terms (which is the true field) remains stable.

Note that in the rectangular grid neighbour values are simply multiplied by $e^{\pm j\kappa_\xi \delta_\xi}$ which remains true even for line-, surface- and volume mean values. Assuming a homogeneous material in and around the cell in FIT/FDTD the (6×6)-matrix $\mathbf{M}$ can be written as

$$\begin{bmatrix} \boldsymbol{H} \\ \boldsymbol{E} \end{bmatrix}_{\text{new}} = \underbrace{\begin{bmatrix} \mathbb{U} + \frac{1}{\mu\varepsilon}\mathbf{M}_E\mathbf{M}_H & \frac{1}{\mu}\mathbf{M}_E \\ \frac{1}{\varepsilon}\mathbf{M}_H & \mathbb{U} \end{bmatrix}}_{\mathbf{M}} \cdot \begin{bmatrix} \boldsymbol{H} \\ \boldsymbol{E} \end{bmatrix}_{\text{old}} \tag{12}$$

with

$$\mathbb{U} = \begin{bmatrix} 1\,0\,0 \\ 0\,1\,0 \\ 0\,0\,1 \end{bmatrix}, \quad \begin{aligned} \mathbf{M}_E &= \begin{bmatrix} 0 & \frac{\delta_t}{\delta_z}(e^{2j\kappa_z\delta_z} - 1) & -\frac{\delta_t}{\delta_y}(e^{2j\kappa_y\delta_y} - 1) \\ -\frac{\delta_t}{\delta_z}(e^{2j\kappa_z\delta_z} - 1) & 0 & \frac{\delta_t}{\delta_x}(e^{2j\kappa_x\delta_x} - 1) \\ \frac{\delta_t}{\delta_y}(e^{2j\kappa_y\delta_y} - 1) & -\frac{\delta_t}{\delta_x}(e^{2j\kappa_x\delta_x} - 1) & 0 \end{bmatrix}, \\ \mathbf{M}_H &= \begin{bmatrix} 0 & -\frac{\delta_t}{\delta_z}(1 - e^{-2j\kappa_z\delta_z}) & \frac{\delta_t}{\delta_y}(1 - e^{-2j\kappa_y\delta_y}) \\ \frac{\delta_t}{\delta_z}(1 - e^{-2j\kappa_z\delta_z}) & 0 & -\frac{\delta_t}{\delta_x}(1 - e^{-2j\kappa_x\delta_x}) \\ -\frac{\delta_t}{\delta_y}(1 - e^{-2j\kappa_y\delta_y}) & \frac{\delta_t}{\delta_x}(1 - e^{-2j\kappa_x\delta_x}) & 0 \end{bmatrix}. \end{aligned} \tag{13}$$

The eigenvalues' amount of $\mathbf{M}$ does not exceed 1 if and only if

$$\frac{\delta_t^2}{\mu\varepsilon} \leq \frac{1}{\left(\frac{\sin\delta_x\kappa_x}{\delta_x}\right)^2 + \left(\frac{\sin\delta_y\kappa_y}{\delta_y}\right)^2 + \left(\frac{\sin\delta_z\kappa_z}{\delta_z}\right)^2} \leq \frac{1}{\left(\frac{1}{\delta_x}\right)^2 + \left(\frac{1}{\delta_y}\right)^2 + \left(\frac{1}{\delta_z}\right)^2} \underset{\delta_x=\delta_y=\delta_z=\delta}{\overset{\uparrow}{=}} \frac{\delta^2}{3}. \tag{14}$$

This is the well-known Courant limit.

In FVTD there are 24 scalar variables per cell: 3 face mean values $\boldsymbol{E}_{1,2,3}$ and $\boldsymbol{H}_{1,2,3}$ plus volume mean values of $\boldsymbol{B}$ and $\boldsymbol{D}$. The scheme can be written as

$$[\boldsymbol{E}_1 \quad \boldsymbol{E}_2 \quad \boldsymbol{E}_3 \quad \boldsymbol{H}_1 \quad \boldsymbol{H}_2 \quad \boldsymbol{H}_3 \quad \boldsymbol{B} \quad \boldsymbol{D}]^{\mathrm{T}}_{\text{new}} = \mathbf{M} \cdot [\boldsymbol{E}_1 \quad \boldsymbol{E}_2 \quad \boldsymbol{E}_3 \quad \boldsymbol{H}_1 \quad \boldsymbol{H}_2 \quad \boldsymbol{H}_3 \quad \boldsymbol{B} \quad \boldsymbol{D}]^{\mathrm{T}}_{\text{old}} \tag{15}$$

with the 24×24-matrix

$$\mathbf{M} = \begin{bmatrix} -\mathbb{U} & \mathbf{0} & \mathbf{0} & \frac{1+\alpha_x}{\varepsilon}\mathbf{A}_x & \frac{1+\alpha_x}{\varepsilon}\mathbf{A}_y & \frac{1+\alpha_x}{\varepsilon}\mathbf{A}_z & \mathbf{0} & \frac{1+\alpha_x}{\varepsilon}\mathbb{U} \\ \mathbf{0} & -\mathbb{U} & \mathbf{0} & \frac{1+\alpha_y}{\varepsilon}\mathbf{A}_x & \frac{1+\alpha_y}{\varepsilon}\mathbf{A}_y & \frac{1+\alpha_y}{\varepsilon}\mathbf{A}_z & \mathbf{0} & \frac{1+\alpha_y}{\varepsilon}\mathbb{U} \\ \mathbf{0} & \mathbf{0} & -\mathbb{U} & \frac{1+\alpha_z}{\varepsilon}\mathbf{A}_x & \frac{1+\alpha_z}{\varepsilon}\mathbf{A}_y & \frac{1+\alpha_z}{\varepsilon}\mathbf{A}_z & \mathbf{0} & \frac{1+\alpha_z}{\varepsilon}\mathbb{U} \\ -\frac{1+\alpha_x}{\mu}\mathbf{A}_x & -\frac{1+\alpha_x}{\mu}\mathbf{A}_y & -\frac{1+\alpha_x}{\mu}\mathbf{A}_z & -\mathbb{U} & \mathbf{0} & \mathbf{0} & \frac{1+\alpha_x}{\mu}\mathbb{U} & \mathbf{0} \\ -\frac{1+\alpha_y}{\mu}\mathbf{A}_x & -\frac{1+\alpha_y}{\mu}\mathbf{A}_y & -\frac{1+\alpha_y}{\mu}\mathbf{A}_z & \mathbf{0} & -\mathbb{U} & \mathbf{0} & \frac{1+\alpha_y}{\mu}\mathbb{U} & \mathbf{0} \\ -\frac{1+\alpha_z}{\mu}\mathbf{A}_x & -\frac{1+\alpha_z}{\mu}\mathbf{A}_y & -\frac{1+\alpha_z}{\mu}\mathbf{A}_z & \mathbf{0} & \mathbf{0} & -\mathbb{U} & \frac{1+\alpha_z}{\mu}\mathbb{U} & \mathbf{0} \\ -\mathbf{A}_x & -\mathbf{A}_y & -\mathbf{A}_z & \mathbf{0} & \mathbf{0} & \mathbf{0} & \mathbb{U} & \mathbf{0} \\ \mathbf{0} & \mathbf{0} & \mathbf{0} & \mathbf{A}_x & \mathbf{A}_y & \mathbf{A}_z & \mathbf{0} & \mathbb{U} \end{bmatrix} \tag{16}$$

where $\mathbf{0}$ stands for a 3-by-3 zero-matrix, $\alpha_x := e^{j\kappa_x\delta_x}$, $\alpha_y := e^{j\kappa_y\delta_y}$, $\alpha_z := e^{j\kappa_z\delta_z}$ and

$$\mathbf{A}_\xi = \frac{\delta_t}{\delta_\xi}(1-1/\alpha_\xi)\mathbf{K}_\xi, \quad \mathbf{K}_x = \begin{bmatrix} 0 & 0 & 0 \\ 0 & 0 & -1 \\ 0 & 1 & 0 \end{bmatrix}, \quad \mathbf{K}_y = \begin{bmatrix} 0 & 0 & 1 \\ 0 & 0 & 0 \\ -1 & 0 & 0 \end{bmatrix}, \quad \mathbf{K}_z = \begin{bmatrix} 0 & -1 & 0 \\ 1 & 0 & 0 \\ 0 & 0 & 0 \end{bmatrix}. \tag{17}$$

Evaluating the eigenvalues of $\mathbf{M}$ yields

$$-1, +1, \pm\sqrt{1-\gamma\frac{\delta_t^2}{\mu\varepsilon} \pm \sqrt{\left(1-\gamma\frac{\delta_t^2}{\mu\varepsilon}\right)^2 - 1}} \quad \text{with} \quad \gamma = \frac{1-\cos 2\kappa_x\delta_x}{\delta_x^2} + \frac{1-\cos 2\kappa_y\delta_y}{\delta_y^2} + \frac{1-\cos 2\kappa_z\delta_z}{\delta_z^2} \tag{18}$$

We find that for $\gamma\delta_t^2/(\mu\varepsilon) \le 2$ all eigenvalues are unimodular complex numbers while for $\gamma\delta_t^2/(\mu\varepsilon) > 2$, there are eigenvalues with an absolute value being larger than one. A stability-criterion is therefore

$$\frac{1}{\delta_x^2} + \frac{1}{\delta_y^2} + \frac{1}{\delta_z^2} \le \frac{\mu\varepsilon}{\delta_t^2} \tag{19}$$

This is exactly the same as Eq. (14) in FIT/FDTD!

---

4 The FVTD program

A FVTD program is developed in parallel to this theoretical study. To take advantage of the geometrical flexibility of the method, the FVTD algorithm is applied in an unstructured tetrahedral mesh. This type of mesh permits a conformal meshing of complicated geometries including, e.g. curved or oblique surfaces.

The basic FVTD Eq. (4) is numerically integrated in each cell of the mesh in a time-stepping iteration. The approximate relations of the type Eq. (7) are implemented using the following approach: For each face of the cell only the field components tangential to the face are considered (plane-wave ansatz) and the fields are split into incoming and outgoing contributions. Second-order accuracy in space is achieved using the MUSCL approach (monotonic upwind scheme for conservation laws (Bonnet et al., 1999)) that interpolates volume values (assumed located in the barycenter of the cell) to face centers using estimated gradients.

When using second-order accurate schemes in space, the first-order time-stepping scheme of the left side of Eq. (5) is advantageously replaced by the second-order predictor-

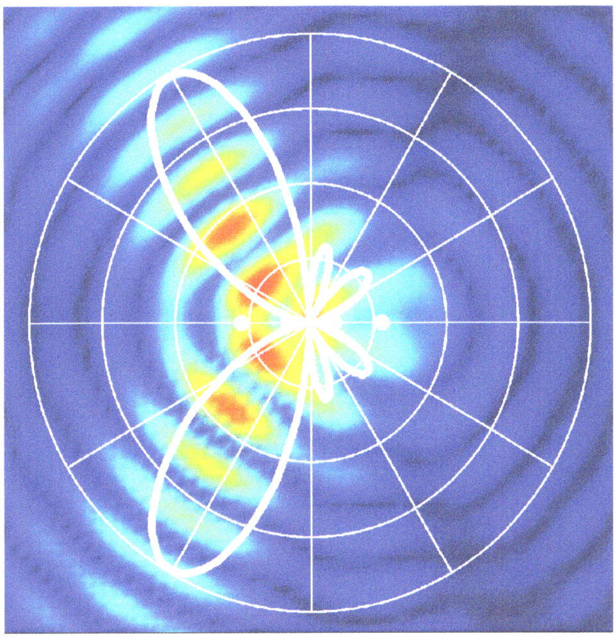

Fig. 5. A four element linear array of Hertzian dipoles (white dots on the horizontal axis). The near-zone E-field (magnitude distribution) is shown. It has been computed with the FVTD method. The array is steered toward 30° off broadside (= vertical direction). The respective far-field pattern is superimposed in white.

corrector Lax-Wendroff scheme (Bonnet et al., 1999). This scheme permits as alternative to Eq. (6) to obtain "newer" face mean values in the numerical estimation of Eq. (5). The resulting implemented algorithm is consequently second-order accurate both in space and time. Absorbing boundary conditions of the Silver-Müller type or Engquist-Majda type are applied to the outer boundary of the computational domain.

Figure 5 shows the near-field distribution of a linear array of four elementary dipoles with half-wave spacing. The plane of observation is perpendicular to the dipoles and located slightly above them. This example validates the FVTD method for EM simulations. The modeling of complicated structures requires only a geometrical definition and meshing of the structure but no change in the FVTD algorithm. Application of the FVTD method to more complicated examples is in progress.

5 Conclusions

Common numerical schemes may involve significant errors in particular equations. The FVTD-schemes have many degrees of freedom. The simple scheme treated here leads to unacceptably high errors. However, it is expected that a more sophisticated scheme delivers much lower errors.

References

Yee, K. S.: Numerical solution of initial boundary value problems involving maxwell's equations in isotropic media, IEEE Transactions on Antennas and Propagation, 14(3): 302–307, 1966.

Taflove, A. and Hagness, S. C.: Computational Electrodynamics – The finite-difference time-domain method, Artech House, Boston, 2nd edition, 2000.

Weiland, T.: A discretization method for the solution of maxwell's equations for six-component fields, Electronics and Communications AEÜ, 31: 116–120, 1977.

Bonnet, P., Ferrieres, X., Michielsen, B. L., Klotz, P., and Roumiguières, J. L.: Time domain electromagnetics (editor: S.M. Rao), chapter Finite-volume time domain method, Recent Research Developments Series, Academic Press, San Diego, 1999.

5

A hybrid method combining the Time-Domain Method of Moments, the Time-Domain Uniform Theory of Diffraction and the FDTD

A. Becker and V. Hansen

Chair of Electromagnetic Theory, University of Wuppertal, Germany

Abstract. In this paper a hybrid method combining the Time-Domain Method of Moments (TD-MoM), the Time-Domain Uniform Theory of Diffraction (TD-UTD) and the Finite-Difference Time-Domain Method (FDTD) is presented. When applying this new hybrid method, thin-wire antennas are modeled with the TD-MoM, inhomogeneous bodies are modelled with the FDTD and large perfectly conducting plates are modelled with the TD-UTD. All inhomogeneous bodies are enclosed in a so-called FDTD-volume and the thin-wire antennas can be embedded into this volume or can lie outside. The latter avoids the simulation of white space between antennas and inhomogeneous bodies. If the antennas are positioned into the FDTD-volume, their discretization does not need to agree with the grid of the FDTD. By using the TD-UTD large perfectly conducting plates can be considered efficiently in the solution-procedure. Thus this hybrid method allows time-domain simulations of problems including very different classes of objects, applying the respective most appropriate numerical techniques to every object.

1 Introduction

Every numerical method has inherent advantages and disadvantages. For example the FDTD is not efficient for modelling electrically large but simple bodies whereas the TD-UTD is well suited for such bodies but cannot be applied to inhomogeneous bodies. Consequently, by combining several techniques hybrid methods can be created to overcome the disadvantages and simultaneously make use of the advantages of the underlying techniques.

In the context of this paper we treat transient problems as shown in Fig. 1: Electrically large perfectly conducting plates are modelled with the TD-UTD, thin-wire antennas are modelled with the TD-MoM and inhomogeneous bodies are modelled with the FDTD. The aim of this paper is the hybridization of all three methods. Therefore, first of all the three methods are shortly presented in Sect. 2–4 without claiming to give a complete overview about state of the art developments of the specific methods, but pointing out the properties, which are important for the hybridization. In Sect. 5 the hybrid method combining all three methods is developed step by step. Finally, numerical examples are presented.

Correspondence to: A. Becker
(abecker@ieee.org)

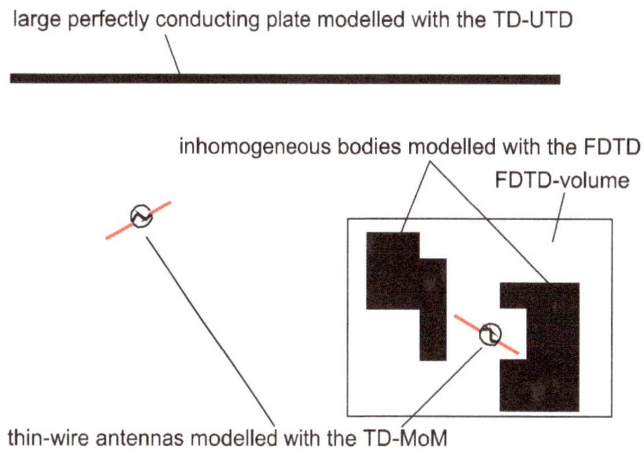

Fig. 1. Schematic representation of the proposed hybrid method.

2 TD-MoM

Considered are perfectly conducting thin-wire antennas (surface S) located in free space. One standard integral equation for the electric field radiated in free space by the current density $\boldsymbol{J}$ flowing on the surface S reads:

$$\boldsymbol{E}^{\text{rad}}(\boldsymbol{r},t) = \int_S \left[-\nabla \frac{\rho(\boldsymbol{r}',t)}{4\pi\varepsilon r} - \frac{\partial}{\partial t} \frac{\mu \boldsymbol{J}(\boldsymbol{r}',t')}{4\pi r} \right] da', \qquad (1)$$

with $r=|\boldsymbol{r}-\boldsymbol{r}'|$, $\rho(\boldsymbol{r}',t)=-\int_0^{t'} \nabla \boldsymbol{J}(\boldsymbol{r}',t')dt$ and $t'=t-\frac{r}{c}$. For the numerical solution the Ansatz

$$\boldsymbol{J}(\boldsymbol{r}',t) = \sum_{i=1}^{N_s} \sum_{j=1}^{N_t} I_{i,j} \boldsymbol{\beta}_i(\boldsymbol{r}') \mathcal{T}_j(t) \qquad (2)$$

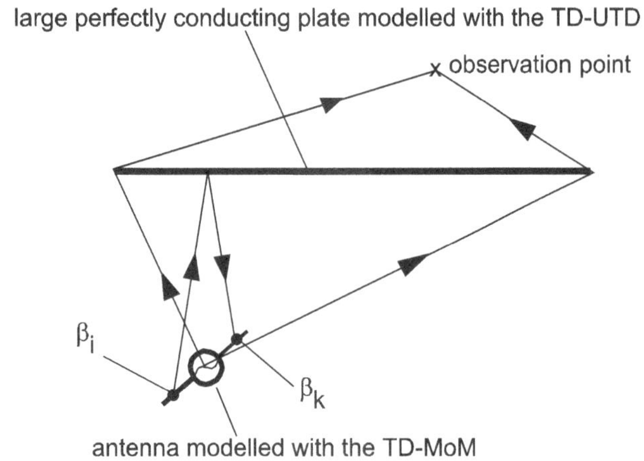

Fig. 2. Schematic representation of the TD-MoM+TD-UTD-Hybrid Method. The diffracted field in an observation point is modelled using one equivalent source point. Additionally shown: One ray reflected from basis function $\boldsymbol{\beta}_i$ to test function $\boldsymbol{\beta}_k$.

is chosen to approximate the spatial and temporal dependency of the electric current density (Gómez et al., 1992). N_t and N_s denote the number of temporal and spatial basis functions, respectively and $I_{i,j}$ are unknown weighting coefficients. Substituting Eq. (2) into Eq. (1) and taking the temporal derivate of the electric field yields

$$\frac{\partial}{\partial t}\boldsymbol{E}^{\mathrm{rad}}(\boldsymbol{r},t) = \sum_{i=1}^{N_s}\sum_{j=1}^{N_t}\frac{I_{i,j}}{4\pi}\int_{S_i}\left[\frac{\nabla\gamma_i(\boldsymbol{r}')\mathcal{T}_j(t')}{\varepsilon r} - \frac{\partial^2}{\partial t^2}\frac{\mu\boldsymbol{\beta}_i(\boldsymbol{r}')\mathcal{T}_j(t')}{r}\right]da', \tag{3}$$

where $\gamma_i(\boldsymbol{r}')=\nabla\boldsymbol{\beta}(\boldsymbol{r}')$. A scalar equation is derived by multiplying Eq. (3) with a test function $\boldsymbol{\beta}_k$ and integrating the result over the domain S_k of the test function. In a next step, the time axis is subdivided into time steps $t=nT$ and it is assumed that

$$\mathcal{T}_j(t) = \begin{cases} 0 & \text{if } |t-jT|\geq T \\ \mathcal{T}(t-jT) & \text{if } |t-jT|<T \end{cases} \tag{4}$$

and that for $t_0 7=n_0 T$ all coefficients $I_{i,j<n_0}$ are known. On the surface of perfectly conducting scatterers the total tangential electric field must vanish, i.e. the sum of the incoming fields $\boldsymbol{E}^{\mathrm{inc}}$ and the radiated fields $\boldsymbol{E}^{\mathrm{rad}}$ must be zero. Fulfilling this boundary condition at the domain of all test functions $\boldsymbol{\beta}_k$ and for t_0 leads to a systems of linear equations (Gómez et al., 1992) to calculate the coefficients I_{i,n_0}. Consequently, in a so-called *marching-on-in-time* procedure all coefficients $I_{i,j}$ can be calculated successively. The potential consider incoming field in the solution procedure is the main point of contact for hybridization.

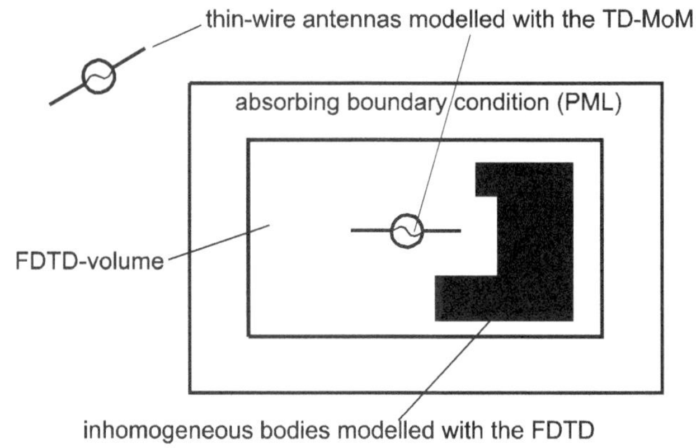

Fig. 3. Schematic representation of the TD-MoM+FDTD-Hybrid Method.

3 Time-domain Geometrical Theory of diffraction

The Geometrical Theory of Diffraction is based on the representation of waves by rays. According to the UTD, the field radiated by a point-source in presence of a perfectly conducting plate can be approximated as the sum of the direct ray, a reflected ray and diffracted rays (Rousseau and Pathak, 1995):

$$\boldsymbol{E} = \boldsymbol{E}_{LOS}U_i + \boldsymbol{E}_R U_R + \boldsymbol{E}_D, \tag{5}$$

with U_i=1 if line of sight (LOS) is given between source and observation point and U_i=0 if no direct ray reaches the observation point. U_R is equal to one, if a reflected ray reaches the observation point and it is equal to zero, otherwise. According to the UTD, the reflected and diffracted fields are found by using the fields in the reflection and diffraction points, respectively. Both U_i and U_R, as well as possible reflection and diffraction points are found by using a ray-tracing algorithm.

The diffracted field is found by convolution of the field $\mathbf{E}_{D_0}$ in the diffraction point with the diffraction coefficient and multiplying the result with the spreading factor $A(s_d)$:

$$\mathbf{E}_D(s_d,t) = \mathbf{E}_{D_0}(t-\frac{s_d}{c}) * \overset{\leftrightarrow}{d}(t-\frac{s+s_d}{c})A(s_d), \tag{6}$$

with

$$A(s_d) = \sqrt{\frac{s'}{s_d(s_d+s')}}. \tag{7}$$

The dyadic diffraction coefficient $\overset{\leftrightarrow}{d}(t)$ can be found in Rousseau and Pathak (1995). In the context of this paper, the fields in the reflection and diffraction points are calculated with the TD-MoM and the FDTD, respectively. Thus these fields act as point of contact for hybridization.

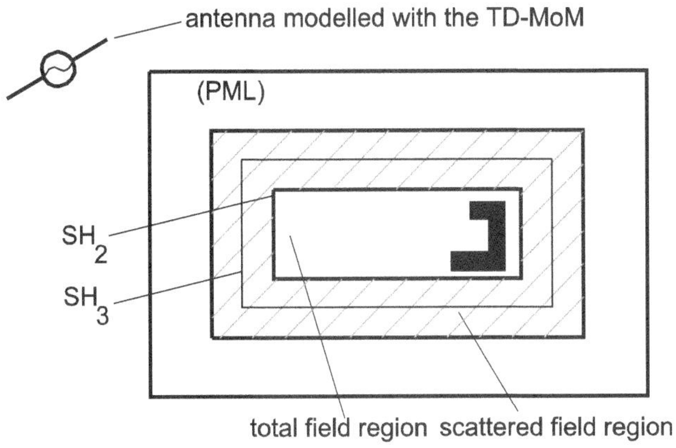

Fig. 4. Subdivision of the FDTD-volume into total field region and scattered field region and the Huygens surfaces used for hybridization.

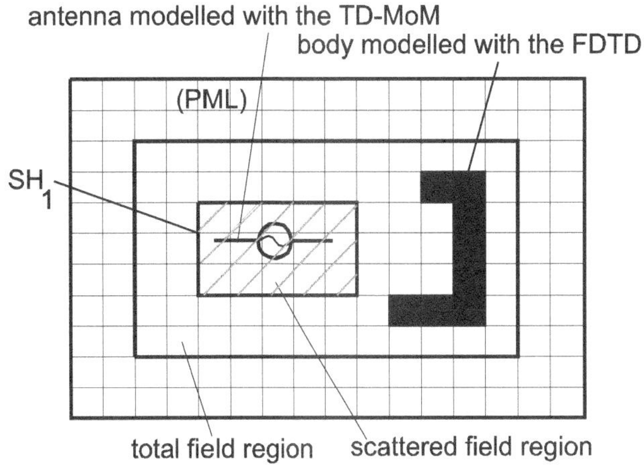

Fig. 5. Subdivision of the FDTD-volume into total field region and scattered field region and the Huygens surfaces used for hybridization.

4 FDTD

For this paper a simple 3D-FDTD with a Perfectly Matched Layer (PML) as absorbing boundary condition is used; for details see e.g. Taflove (2000). The points of contact for hybridization are the total-field/scattered-field formulation (see e.g. Huang et al., 1999) to introduce incoming fields into the FDTD-volume and the calculation of fields outside the FDTD-Volume with a surface integral representation of the electric field (see e.g. Skarlatos et al., 2005).

5 Hybrid methods

The three methods are hybridized by using well known theorems and making use of the superposition principle. In the following – for the sake of simplicity – the process of hybridization is described step by step.

5.1 TD-MoM+TD-UTD-Hybrid Method (Becker and Hansen, 2006)

The schematic representation of this hybrid method is shown in Fig. 2. The hybridization of both methods comprises two steps:

1. The fields in the observation points are calculated with Eq. (5). For this purpose, the fields radiated by the thin-wire antennas are approximated by rays with source points lying on the surface S of the antennas, which are modelled with the TD-MoM. In many cases the rays reaching an observation point can be constructed by using one equivalent source-point in the centre of the antenna (see Fig. 2). The direct fields in the observation points (if line of sight is given, i.e. $U_i \neq 0$) and the fields in the reflection and diffraction points are calculated in the same *marching-on-in-time* procedure as the currents flowing on the antenna. All necessary current coefficients $I_{i,j}$, which excite a field at $t{=}n_0T$ in the observation point are – due to the distance between source and observation point – calculated in the *marching-on-in-time* procedure before.

2. The influence of the surrounding on the currents is considered by treating reflected or diffracted fields, which reach the surface S of the antennas, as incoming fields $\boldsymbol{E}^{\text{inc}}$ in the *marching-on-in-time* procedure. Thus the free space Green's function used for Eq. (1) is implicitly modified. This strategy is well proven as it has been successfully applied for several frequency-domain hybrid methods (Alaydrus et al., 2002). Reflected and scattered fields are constructed by using the centre of any base function $\boldsymbol{\beta}_i$ as origin of the respective rays. For one infinitely large reflecting plate, N_s basis functions and N_s test functions this results into N_s^2 reflected rays. Consequently, this becomes relatively costly.

5.2 TD-MoM+FDTD-Hybrid Method

As shown above, the consideration of the influence of the surrounding on the currents flowing on the thin-wires is relatively costly. Additionally, the TD-MoM+TD-UTD Hybrid Method is restricted to problems involving simply structured scatterers. In the context of this paper, this restriction is overcome by using a TD-MoM+FDTD Hybrid Method. When applying this hybrid method, thin-wire antennas are modelled with the TD-MoM and inhomogeneous bodies are modelled with the FDTD. All inhomogeneous bodies lie inside the so-called FDTD-volume, which is enclosed by a PML (see Fig. 3). The TD-MoM is fully hybridized with the FDTD, if the fields radiated by the thin-wire antennas are considered in the FDTD solution-procedure (at the location of the inhomogeneous bodies) and if the influence of the

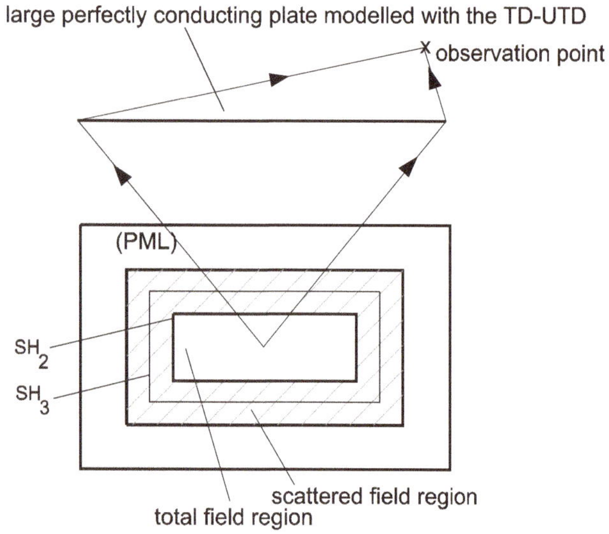

Fig. 6. Schematic representation of the TD-UTD+FDTD-Hybrid Method. Not shown: Rays backscattered to the FDTD-volume.

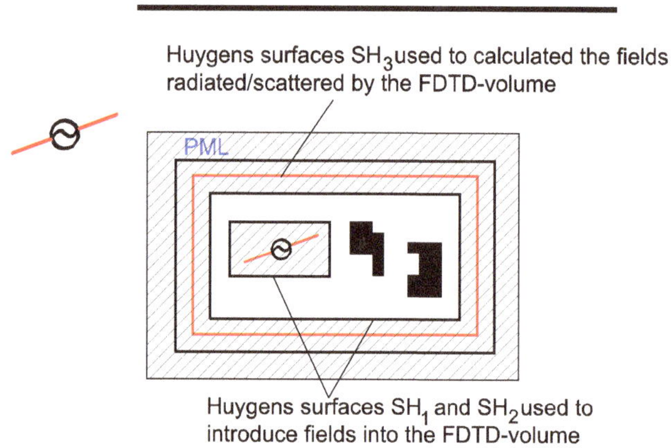

Fig. 7. The three necessary Huygens surfaces for full hybridization of all three methods.

inhomogeneous bodies on the currents flowing on the thin-wire antennas are considered in the TD-MoM solution procedure and vice versa. This is achieved by using the approach described in Huang et al. (1999): The basic idea is to subdivide the FDTD-region into a scattered field region and into a total field region. By this a numeric Huygens source scheme is realised and it is possible to introduce arbitrary incident fields into the FDTD-volume. For the sake of simplicity it is useful to use the following case differentiation:

5.2.1 Thin-wire antennas outside the FDTD-volume

Figure 4 shows the case of an antenna modelled with the TD-MoM, which lies outside the FDTD-Volume. The fields radiated by the thin-wire antennas are introduced into the FDTD solution procedure by using the surface SH_2, which seperates the total field region (inside SH_2) from the scattered field region (outside SH_2). Consequently, at the position of the inhomogeneous bodies, the unknowns of the FDTD include the fields radiated by the thin-wire antennas. The fields radiated or scattered by the FDTD-volume are calculated by using equivalent sources on the Huygens surface SH_3 lying just inside the PML (see Fig. 4) and a surface integral representation of the electric field (see e.g. Taflove, 2000). These fields are used as incoming fields $\boldsymbol{E}^{\text{inc}}$ in the TD-MoM solution procedure.

5.2.2 Thin-wire antennas inside the FDTD-volume

As shown in Fig. 5, the antennas modelled with the TD-MoM now lies inside the FDTD-Volume. The FDTD-volume is seperated by the Huygens surface SH_1 into two regions: A total field region and a scattered field region. The scattered field region needs to be homogeneous and includes the antennas modelled with the TD-MoM. The discretiziation of the latter does not need to agree with the grid of the FDTD. The total field region includes the inhomogenous bodies and can include arbitrary material (see Fig. 5). The scattered field, which is in fact the field back-scattered to the antenna by its surrounding, is used as incoming field in the TD-MoM solution procedure. As the antenna lies inside the FDTD-volume, fields outside the FDTD-volume can be calculated by using SH_3 (not shown in Fig. 5) and the procedure described above. Consequently, for observation points outside the FDTD-volume, the FDTD+TD-MoM Hybrid-Method does not distinguish from a *stand-alone* FDTD method.

5.3 TD-UTD+FDTD-Hybrid Method

The FDTD is hybridized with the TD-UTD by using a strategy very similar to Skarlatos et al. (2005): In the application of the TD-UTD the fields in the reflection and diffraction points, as well as the direct fields, are calculated using once again a surface integral representation and the equivalence principle (surface SH_3). To save memory and CPU-time SH_3 is subdivided into large subdomains and for the ray-representation of the fields one equivalent source point per subdomain is used. Obviously, the farther the reflection and diffraction points are away from the FDTD-volume, the less subdomains are needed. Therefore it is sometimes useful, to use one equivalent source point in the centre of the FDTD-volume as shown in Fig. 6. Fields back-scattered to the FDTD-volume can be taken into account in the FDTD solution procedure by using the same procedure and the same surface SH_2 as described in Sect. 5.2.1.

5.4 TD-MoM+TD-UTD+FDTD-Hybrid Method

When applying this hybrid method thin-wire antennas are modelled with the TD-MoM, large perfectly conducting plates are modelled with the TD-UTD and inhomogeneous bodies are modelled with the FDTD. The hybridization of

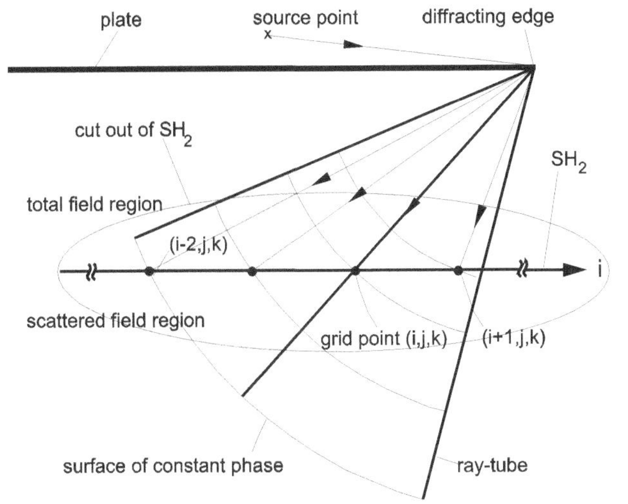

Fig. 8. Rays reaching a cut out of SH_2 and a ray tube enclosing 4 grid points.

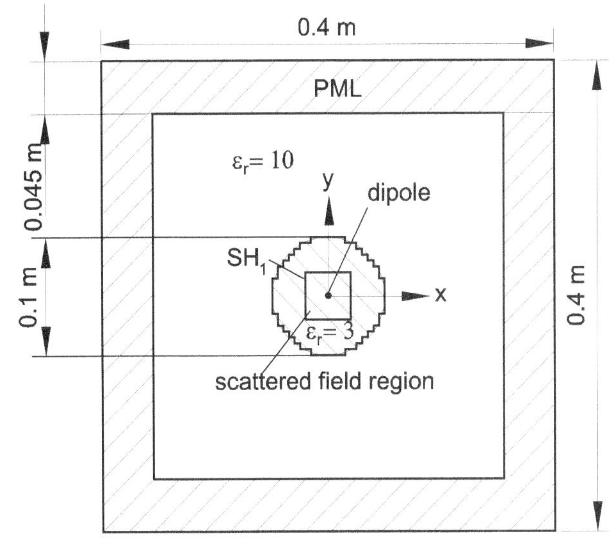

Fig. 9. Dipole embedded into a dielectric cylinder.

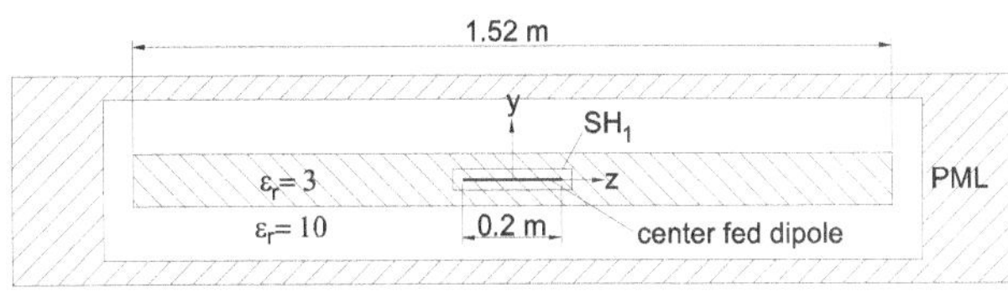

Fig. 10. Dipole embedded into a dielectric cylinder.

all three methods is achieved by starting with the TD-MoM+TD-UTD hybrid method and using the superposition principle. The TD-MoM+TD-UTD hybrid method considers the interactions between the thin-wire antennas and the large plates. Additional FDTD bodies require the use of the Huygens surfaces SH_1 and SH_2 to introduce the fields radiated by the thin-wire antennas into the FDTD-volume. These fields can include fields, which are radiated by the thin-wire antennas and then scattered by the plates. SH_3 is used to calculate the fields radiated or scattered by the FDTD-volume and these fields are considered as incoming fields $\boldsymbol{E}^{\text{inc}}$ in the TD-MoM solution procedure. Additionally, these fields $\boldsymbol{E}^{\text{inc}}$ can include fields radiated by the FDTD-volume and then scattered by the plates. Thus the interactions between the thin-wire antennas and the inhomogeneous bodies are considered including the additional influence of the plates. Consequently, one needs to use all three Huygens surfaces SH_1, SH_2 and SH_3 and by this all three methods are hybridized (see Fig. 7).

5.5 Major drawbacks and improvement opportunities

One of the main drawbacks of the proposed methods are the resource and calculation requirements to calculate the fields radiated by the FDTD-volume. These fields are calculated by using equivalent sources on the Huygens-surface SH_3 and an integral representation of the electric field. This calculation is very costly and we assume that these costs can be reduced by using techniques like the one proposed in Oetting and Klinkenbusch (2005).

Right now, the second main drawback is the fact that the fields scattered by the plates have to be calculated pointwise (see also Sect. 2), in order to be introduced into the FDTD-volume and to be used as incoming field in the TD-MoM solution procedure. That means, that for any grid point on the surface SH_2 individual rays have to be constructed (see Fig. 8). This is extremely costly but in future developments this might be accelerated by making more use of the ray representation of the scattered fields. As indicated in Fig. 8 ray tubes can be constructed, which enclose more than one grid point and by this the cost for constructing the rays becomes much smaller. The drawback of this strategy is that the temporal behaviour of the field in one reference point (grid point (i, j, k) in Fig. 8) has to be stored in memory to be able to calculate the fields on different surfaces of constant phase.

6 Numerical examples

6.1 Dipole embedded into a dielectric cylinder

Considered is the geometry shown in Fig. 9 and Fig. 10, consisting of a center-fed dipole (radius r=0.001 m) embedded into a dielectric cylinder, which is surrounded by a dielectric medium. In Fig. 11 the input impedance of the dipole is shown. The result calculated with the TD-MoM+FDTD-Hybrid Method is compared to the result calculated with a frequency-domain method of moments for cylindrically layered media (Becker and Hansen, 2006).

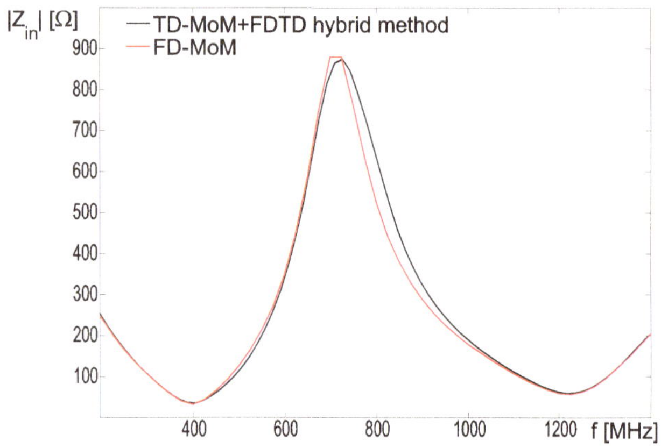

Fig. 11. Input impedance calculated with the TD-MoM+FDTD Hybrid Method and with the frequency domain Method of Moments.

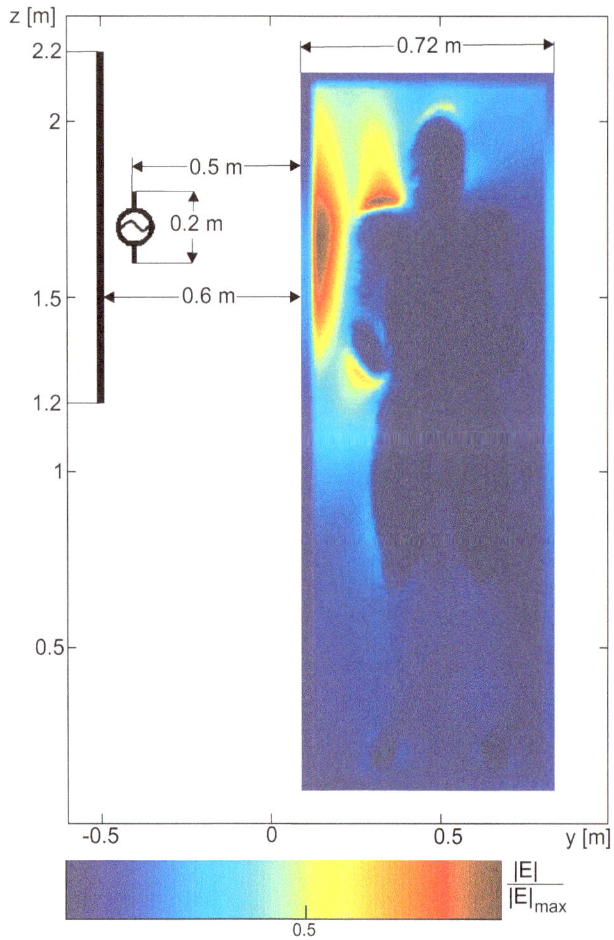

Fig. 12. Human body in front of a center-fed dipole and a perfectly conducting plate.

6.2 Human body in front of a dipole with reflector

As a second example, a center-fed dipole (r=0.001 m) in front of a perfectly conducting plate (infinitely long in ±x-direction) and a model of the human body is considered (based on visible human National Library, 1986; Gabriel, 2006, discretized using $\triangle_x = \triangle_y = \triangle_z = 0.01\,m$). In Fig. 12 the absolute value of the electric field inside the body (cut at x=0.31 m, i.e. the plane including the dipole) is shown. For this calculation diffraction at the edges and corners of the plate and the influence of the human body towards the currents flowing on the dipole are neglected. The fields radiated by the thin-wire antennas and reflected by the plate are calculated by using mirror sources.

7 Conclusions

The presented method combines three techniques to a new hybrid method. Using this it is possible to aply appropriate numerical techniques for modelling different types of bodies: The TD-UTD for large, perfectly conducting bodies, the TD-MoM for small, perfectly conducting bodies and the FDTD for dielectric bodies. The combination of these three approaches yields a powerful tool for handling problems which cannot be treated with any other time-domain technique.

References

Alaydrus, M., Hansen, V., and Eibert, T. F.: Hybrid^2: Combining the three-dimensional hybrid finite element-boundary integral technique for planar multilayered media with the uniform geometrical theory of diffraction, IEEE Trans. Antennas Propagat, vol. 50, pp. 67–74, Jan., 2002.

Becker, A. and Hansen, V.: A Hybrid Method Combining the Multi-Temporal Resolution Time-Domain Method of Moments with the Time-Domain Geometrical Theory of Diffraction for Thin-Wire Antenna Problems, IEEE Trans. Antennas Propagat, vol. 54, pp. 953–960, March, 2006.

Becker, A., Zhou, Y., and Hansen, V.: On the Calculation of the Radiation Properties of Dipole Antennas in Cylindrical Boreholes Considering Large Scatterers, German Microwave Conference, Karlsruhe, 2006.

Gabriel, C.: Brooks Air Force Technical Report AL/OE-TR-1996-0037, http://www.fcc.gov/fcc-bin/dielec.sh.

Gómez, R., Martin, A., and Rubio Bretones, A.: Time-domain integral equation methods for transient analysis, IEEE Magazine Antennas Propagat, vol. 34, pp. 15–23, June, 1992.

Huang, Z., Demarest, K. R., and Plumb, R. G.: An FDTD/MoM hybrid technique for modeling complex antennas in the presence of heterogeneous grounds; Geoscience and Remote Sensing, IEEE Trans. Antennas Propagat, vol. 37, pp. 2692–2698, Nov., 1999.

National Library of Medicine: Visible Human Project: http://www.nlm.nih.gov/research/visible/visible_human.html, 1986.

Oetting, C.-C. and Klinkenbusch, L.: Near-to-Far-Field Transformation by a Time-Domain Spherical-Multipole Analysis, Trans. AP, vol. 53, pp. 2054–2063, 2005.

Rousseau, P. R. and Pathak, P. H.: Time domain geometrical theory of diffraction for a curved wedge, IEEE Trans. Antennas Propagat, vol. 43, pp. 1375–1382, Dec., 1995.

Skarlatos, A., Schuhmann, R., and Weiland, T.: Solution of Radiation and Scattering Problems in Complex Environments Using a Hybrid Finite Integration Technique – Uniform Theory of Diffraction Approach, IEEE Trans. Antennas Propagat, vol. 53, pp. 3347–3357, October, 2005.

Taflove, A.: Computational Electrodynamics, Artech House, Boston, 2000.

6

Correlation properties of dual polarized antennas with finite pattern orthogonality in mobile fading channels

G. Armbrecht, O. Klemp, and I. Rolfes

Leibniz Universität Hannover, Institut für Hochfrequenztechnik und Funksysteme, Appelstr. 9A, 30167 Hannover, Germany

Abstract. Starting from planar broadband log.-per. antenna design, offering the possibility of dual-polarized reception properties, in this article a generalized mathematical approach for rapidly estimating the resulting signal correlation coefficient in a stochastically modeled propagation environment solely based on measured or simulated radiation characteristics of one single antenna element is presented. The obtained results are marking an upper limit and are describing the worst-case scenario according to the signal correlation at the antenna feeding points in terms of line-of-sight (LOS) reception in main beam direction. The knowledge of the derived relationship may be helpful especially for antenna designers to combine antenna performance values with the significant communication system performance parameters, as e.g. in case of Multiple-Input Multiple-Output (MIMO) and diversity configurations.

1 Introduction

Multiple Input Multiple-Output (MIMO) and diversity techniques will be an integral part of future mobile communication systems. With the aim of providing higher data rates or decreased outage probabilities, multi element antenna techniques seek for standardization in cellular systems as well as in the domain of wireless networks. Following Jensen and Wallace (2004), the realistic antenna properties as well as multipath channel propagation characteristics play a role in determining the communication performance. As a consequence from space limitations on the mobile side, the application of multimode operating air interfaces offering the capabilities of polarization- and pattern diverse sensing provide a superior usage in comparison to conventional spatial diversity techniques. Our research work is especially focused on four-arm planar log.-per. antennas with two orthogonal polarized excitation modes, as shown in Klemp et al. (2004, 2005b). In accordance with Klemp and Eul (2005) and Klemp et al. (2005a), systems equipped with such antenna elements show a significantly improved MIMO and diversity system performance. Due to the combination of broadband reception capabilities as well as excellent cross-polarization suppression over a wide frequency range (e.g. 1.7 up to 6.0 GHz), these structures are promising candidates for the integration in mobile devices and hand-held applications. However, finite pattern orthogonality between the two linear polarized operational modes of a multimode antenna yields an unwanted signal cross-coupling, as depicted in Fig. 1, that represents a limiting factor with respect to diversity gain or MIMO channel capacity.

For that reason we intend to derive a generalized mathematical approach for rapidly estimating the resulting signal correlation coefficient just based on measured or simulated radiation characteristics of one single antenna element. Without loss of generality, the presented theory can be applied to any other dual-polarized antenna configuration supporting two independent operational modes, e.g. the cross-dipole. Considering the antenna in a conventional angle-of-arrival (AOA) multipath propagation environment, modeling the AOA dependency in a stochastic formulation according to Pedersen et al. (2000) and Kermoal et al. (2002), we are able to express the modal correlation exclusively in terms of the related field quantities. As a consequence from the symmetry properties of the considered antenna, co- and cross-polar pattern performances can be related to one of the regarded excitation modes. The derived approach will illustrate the relationship of signal-based correlation and finite cross-polarization decoupling between the considered operational modes in a certain line-of-sight (LOS) situation with vanishing angular spreads. This leads to an estimation of a distinct worst-case threshold value for signal-based correlation of dual-polarized operating antennas. Based on these results, antenna designers can easily predict the anticipated upper limit of signal correlation if a distinct antenna design will be applied in a multipath propagation scenario. In this paper, planar four-arm trapezoidal antennas with M=1 and M=5 periods of the trapezoidal unit cell geometry are utilized to

Correspondence to: G. Armbrecht
(armbrecht@hft.uni-hannover.de)

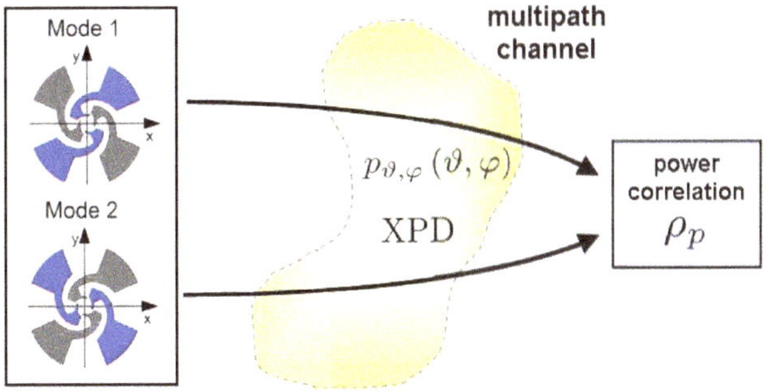

Fig. 1. Finite pattern orthogonality of a dual-polarized trapezoidal antenna (M=1) in a multipath propagation scenario yielding unwanted signal correlation.

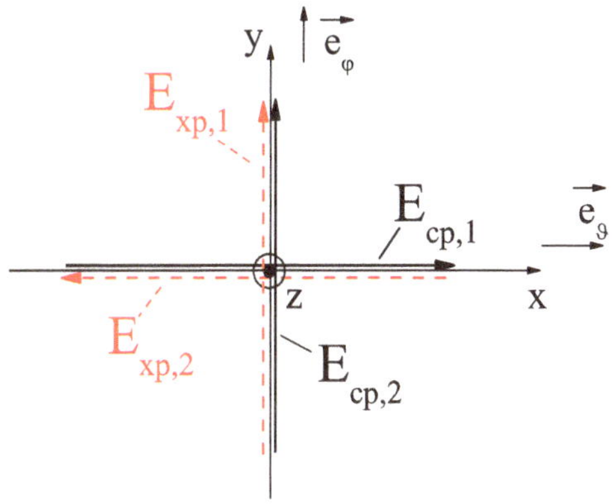

Fig. 2. Far-field dependency in main beam direction (ϑ_0=0°) of a dual-polarized trapezoidal antenna located in the xy coordinate plane.

validate the derived formulation by comparing the theoretical correlation limits with the antenna cross-polarization suppression obtained from the simulated radiation pattern. See Armbrecht et al. (2006) for detailed information about the two trapezoidal antennas.

2 Antenna correlation in mobile fading channels

Multi-port antenna configurations providing two linear excitation modes with finite pattern orthogonality exhibit interelement correlation, which is strongly affected by the properties of the considered multipath propagation environment. A high amount of signal correlation at the antenna feeding point reduces the derived MIMO and diversity system performance significantly. In accordance with Svantesson (2002) the power correlation coefficient ρ_p can be generally computed, as follows:

$$\rho_\mathrm{p} \approx |\rho_{12}|^2 = \left| \frac{R_{12}}{\sqrt{\sigma_1^2 \cdot \sigma_2^2}} \right|^2 , \tag{1}$$

where R_{12} and $\sigma_{1,2}^2$ denote the antenna covariance and variance, respectively. In Taga (1990) the spatial channel properties of a multi element antenna configuration are determined by a two-dimensional probability distribution $p_{\vartheta,\varphi}(\vartheta,\varphi)$ for the AOA behavior at the reception antenna elements in the direction of the elevation angle ϑ and the azimuth angle φ. Using the complete antenna far-field representation given by the complex vector components $E_\vartheta(\vartheta,\varphi)$ and $E_\varphi(\vartheta,\varphi)$ and the proposed two-dimensional probability distribution $p_{\vartheta,\varphi}$, the covariance R_{12} is calculated, as follows:

$$R_{12} = \mathrm{K} \int_0^{2\pi}\int_0^{\pi} \Big[E_{\vartheta,1}(\vartheta,\varphi)\, E_{\vartheta,2}^*(\vartheta,\varphi) + \mathrm{XPD} \cdot E_{\varphi,1}(\vartheta,\varphi)\, E_{\varphi,2}^*(\vartheta,\varphi) \Big] \cdot p_{\vartheta,\varphi}(\vartheta,\varphi) \cdot \sin\vartheta \, d\vartheta \, d\varphi \, . \tag{2}$$

In our case, the indices 1 and 2 represent the excitation modes of the dual-linear polarized antenna structure. K is a proportionality constant and (*) accounts for the complex conjugate operation. Following Taga (1990), the value of XPD in Eq. (2) denotes the cross polarization discrimination which specifies the ratio of the mean incident power in cross-polarized direction at the receiver to the mean incident power in horizontal direction. The respective variances $\sigma_{1,2}^2$ are given by Eq. (3) with ν=1,2.

$$\sigma_\nu^2 = \mathrm{K} \int_0^{2\pi}\int_0^{\pi} \Big[\left|E_{\vartheta,\nu}(\vartheta,\varphi)\right|^2 + \mathrm{XPD} \cdot \left|E_{\varphi,\nu}(\vartheta,\varphi)\right|^2 \Big] \cdot p_{\vartheta,\varphi}(\vartheta,\varphi) \cdot \sin\vartheta \, d\vartheta \, d\varphi \, . \tag{3}$$

Spatial separation between the regarded antenna elements can be neglected due to the fact that the phase centers of all antenna elements coincide in case of polarization diverse sensing.

2.1 Line-of-sight (LOS) propagation scenario with deterministic impinging waves at the receiver

To draw a relationship between the correlation properties and the field-based quantities of the antenna elements, it is necessary to assume a line-of-sight scenario featuring deterministic impinging wave propagation at the reception side. As given by Eq. (4), this behavior can easily be described in compliance with Fujimoto and James (2000) by modifying the stochastic channel description using delta distributions in ϑ- and φ-direction.

$$p_{\vartheta,\varphi}(\vartheta,\varphi) = \frac{\delta(\vartheta-\vartheta_0)\cdot\delta(\varphi-\varphi_0)}{\sin\vartheta} \tag{4}$$

Substituting $p_{\vartheta,\varphi}(\vartheta,\varphi)$ in Eq. (2) with Eq. (4), the delta distributions will directly fade out the complex field quantities of the two adjacent antenna elements in the direction defined by the angles ϑ_0 and φ_0.

In this article, we assume a deterministic wave distribution impinging from the positive z-direction at ϑ_0=φ_0=0°, which

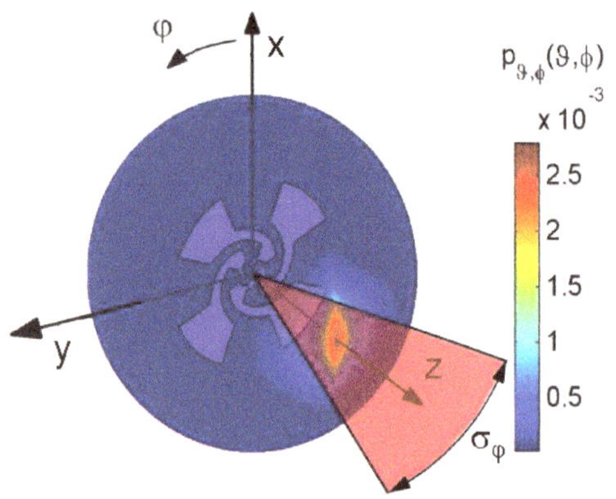

Fig. 3. Variation of the azimuthal angular spread σ_φ in main beam direction of a log.-per. trapezoidal antenna geometry (M=1) located in the xy coordinate plane.

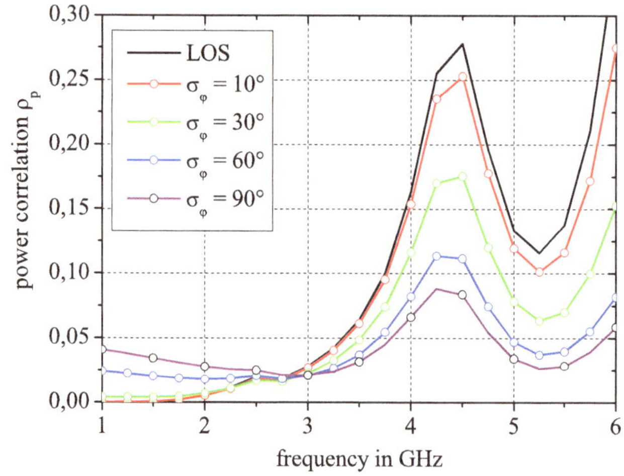

Fig. 4. Validating the proposed analytical correlation threshold value by decreasing the azimuthal angular spread $\sigma_\varphi \rightarrow 0$ for a log.-per. trapezoidal antenna (M=1).

corresponds to the main beam direction of the dual-polarized planar antenna structures located in the xy coordinate plane. In addition, as depicted in Fig. 2, the far-field components of both excitation states can be expressed by considering solely the field quantities of the first mode and representing the fields of the second orthogonal mode by performing a circular permutation. This fact yields to a major simplification according to the mathematical complexity of the aspired formulation. The equivalences accounting for the far-field components of both excitation modes in co- (cp) as well as in cross-polarization (xp) can be formulated, as follows:

$$\begin{aligned} &\text{Mode 1}: \; E_{\mathrm{cp},1} = E_\vartheta \; ; \; E_{\mathrm{xp},1} = E_\varphi \\ &\text{Mode 2}: \; E_{\mathrm{cp},2} = E_\varphi \; ; \; E_{\mathrm{xp},2} = -E_\vartheta \, . \end{aligned} \tag{5}$$

2.2 Analytical computation and validation of signal correlation based on far-field quantities

Applying the LOS channel scenario model in combination with the distinct permutation of field components in main beam direction, as introduced in Sect. 2.1, to Eq. (1), a novel analytical approach for the power correlation coefficient $\rho_{\mathrm{p}}^{\mathrm{LOS}}$ is derived, as follows:

$$\rho_{\mathrm{p}}^{\mathrm{LOS}} = \left| \frac{E_\vartheta E_\varphi^* + \mathrm{XPD}\cdot E_\varphi \left(-E_\vartheta^*\right)}{\sqrt{\left(|E_\vartheta|^2 + \mathrm{XPD}\cdot|E_\varphi|^2\right)\left(|E_\varphi|^2 + \mathrm{XPD}\cdot|E_\vartheta|^2\right)}} \right|^2 . \tag{6}$$

Solely depending on complex vector components E_ϑ and E_φ in main beam direction at $\vartheta_0 = \varphi_0 = 0°$ and the cross polarization discrimination XPD, this equation establishes a direct coupling between field-based and channel-environment-based quantities. Furthermore, it provides a distinct threshold value for the correlation coefficient marking an upper limit due to the vanishing angular spread of the assumed line-of-sight reception scenario. In compliance with Fujimoto and James (2000) increasing angular spreads in the stochastically modeled radio channel leads to a decorrelation of the received signals. This behavior is exploited to validate the derived formulation, as depicted in Fig. 3. Decreasing the azimuthal angular spread σ_φ successively must lead to increasing values of the derived power correlation coefficient ρ_{p} according to Eq. (1) passing into the threshold value $\rho_{\mathrm{p}}^{\mathrm{LOS}}$ for the special case of $\sigma_\varphi \rightarrow 0$, as given by Eq. (6). Without loss of generality, the validation process is conducted using one of the introduced log.-per. trapezoidal antenna structures with M=1 periods of the trapezoidal unit cell geometry. In Fig. 4, the power correlation coefficient ρ_{p} is shown over the frequency sweep from 1.0 to 6.0 GHz for various values of the angular spread σ_φ in a range from 90° down to 10°. In terms of decreasing angular spreads the depicted curves continuously approach the analytical line-of-sight power correlation $\rho_{\mathrm{p}}^{\mathrm{LOS}}$. The obtained approximation verifies the equation given by Eq. (6) yielding a distinct limiting value for vanishing angular spreads in the stochastic channel model for a constant value of XPD=0 dB. Subsequently in Sect. 3, the derived equation given by Eq. (6) will be evaluated under the assumption of mitigating phase influences leading to a straightforward relationship between antenna-based and mobile fading channel effects.

3 Field-based estimation of the correlation performance

If a choice has to be made among a variety of different antenna structures for the application in wireless communication systems exploiting the multiple propagation path using polarization diverse sensing, always the antenna exhibiting the best performance according to the suppression of the cross-polarized field component will be selected.

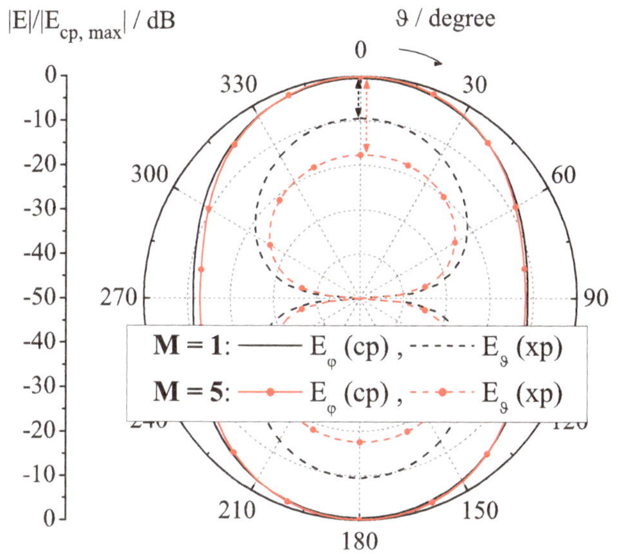

Fig. 5. Simulated far-field radiation pattern of the trapezoidal antennas with $M=1, 5$ in free space at $f=4.5$ GHz exhibiting different value of XP-suppression in main beam direction at $\vartheta=0°$.

Table 1. Upper bound correlation coefficient ρ_p^{max} derived from ratio of cross polarization suppression $|E_{cp}|/|E_{xp}|$ for two trapezoidal antennas at $f=4.5$ GHz compared to correlation values of different scenarios.

	$\|E_{cp}\|/\|E_{xp}\|$	ρ_p^{max}	ρ_p^{LOS}	$\rho_p(\sigma_\varphi=10°)$	$\rho_p(\sigma_\varphi=60°)$
$M=1$	10 dB	0.32	0.278	0.25	0.112
$M=5$	18 dB	0.05	0.014	0.01	0.008

Disregarding the phase information with respect to the complex vector components E_ϑ and E_φ in Eq. (6) will lead to a distinct relationship between cross polarization ratio $|E_{cp}|/|E_{xp}|$ obtained from a radiation pattern representation of the vector field components, as depicted in Fig. 5 for the two considered trapezoidal antennas with $M=1$ and $M=5$ at $f=4.5$ GHz. Accounting for an in-phase superposition of the co- and cross-polarized field quantities yields to an absolute maximum of the power correlation coefficient ρ_p^{max} in line-of-sight wave propagation in main beam direction. Hence, in Fig. 6 these upper bound representation of the power correlation ρ_p^{max} is evaluated in detail concerning various values of the ratio of XP-suppression $|E_{cp}|/|E_{xp}|$ ranging from 0 dB to 20 dB as well as the channel parameter XPD varying from -30 dB up to 0 dB.

As depicted, for low value of XPD the proposed correlation coefficient ρ_p^{max} is close to ≈ 1 due to the power imbalances according to the perceived power in suppressed polarization state. Actually, this kind of correlation behavior is valid for almost any state of cross polarization ratio. The obtained threshold values of ρ_p^{max} will be stronger influenced for increasing XPD ratios.

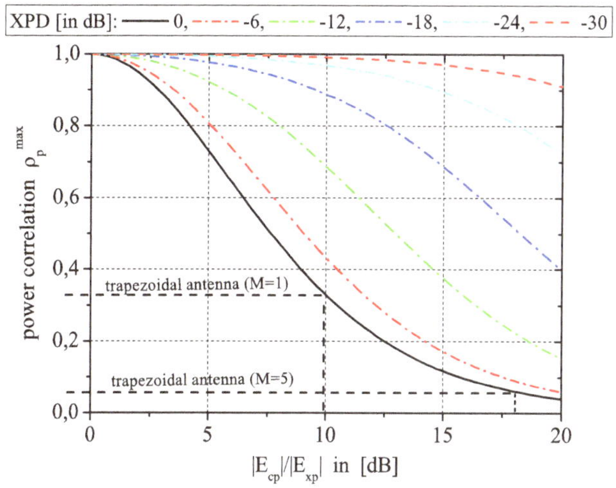

Fig. 6. Maximal correlation behavior ρ_p^{max} solely depending on the ratio of co- to cross-polarization suppression $|E_{cp}|/|E_{xp}|$.

As shown in Table 1 for the particular case of XPD=0 dB, the correlation effects for ρ_p^{max} according to Fig. 6 for two trapezoidal antennas at $f=4.5$ GHz are verified for the cross polarization ratio extracted from Fig. 5. This novel kind of correlation computation is solely depending on simulated radiation pattern for one of the dual-polarized antenna modes. Furthermore, regarding the LOS representation ρ_p^{LOS} and the respective angular spreads of $\sigma_\varphi=10°$ and $60°$ representing typical outdoor and indoor scenarios, the threshold value ρ_p^{max} is proven to be a promising approximation for the worst-case correlation behavior due to the decorrelating impact of wider angular spreads and field-based phase considerations.

4 Conclusions

In this paper, a novel approach to effectively estimate the anticipated correlation properties for dual-linear polarized antenna elements in the presence of finite pattern orthogonality is introduced. Depending on simulated or measured radiation patterns, the phase information of the respective antenna configuration can be neglected in order to perform a worst-case estimation for a line-of-sight spatial propagation scenario. Hence, the proposed computation determines an upper threshold value for the correlation coefficient. According to realistic mobile fading channels, the obtained signal correlation will be permanently smaller due to decorrelation effects of a wider angular spread and actual phase properties of the antenna far-field components. The presented investigation offers communication system designers as well as antenna engineers the possibility to predict rapidly the anticipated system performance by solely knowing the utilized antenna radiation pattern.

References

Armbrecht, G., Klemp, O., and Eul, H.: Spherical mode analysis of planar frequency-independent multi-arm antennas based on its surface current distribution, Adv. Radio Sci., 4, 25–32, 2006, http://www.adv-radio-sci.net/4/25/2006/.

Fujimoto, K. and James, J. R.: Mobile antenna systems handbook, 2nd Edition, Artech House antennas and propagation library, Boston, London, 2000.

Jensen, M. A. and Wallace, J. W.: A Review of Antennas and Propagation in MIMO Wireless Communications, 52, 2810–2824, 2004.

Kermoal, J. P., Schumacher, L., Pedersen, K. I., Morgensen, P. E., and Frederiksen, F.: A Stochastic MIMO Radio Channel Model With Experimental Validation, 20, 1211–1226, 2002.

Klemp, O. and Eul, H.: On the MIMO capacity of planar, log.-per. four-arm antennas, The European Conference on Wireless Technology, ECWT, 2005.

Klemp, O., Schultz, M., and Eul, H.: Miniaturization Techniques For Logarithmically-Periodic Planar Antennas, IEEE 15th International Symposium on Personal, Indoor and Mobile Radio Communications, PIMRC, 2004.

Klemp, O., Schmitz, O., and Eul, H.: Polarization diversity analysis of dual-polarized log.-per. planar antennas, IEEE 16th International Symposium on Personal, Indoor and Mobile Radio Communications, PIMRC, 2005a.

Klemp, O., Schultz, M., and Eul, H.: Novel logarithmically periodic planar antennas for broadband polarization diversity reception, Int. J. Electron. Commun. (AEUE), 59, 268–277, 2005b.

Pedersen, K. I., Andersen, J., Kermoal, J. P., and Morgensen, P.: A Stochastic Multiple-Input-Multiple-Output Radio Channel Model for Evaluation of Space-Time Coding Algorithms, VTC 2000 Boston, 1, 893–897, 2000.

Svantesson, T.: Correlation and Channel Capacity of MIMO Systems Employing Multimode Antennas, 51, 1304–1312, 2002.

Taga, T.: Analysis for Mean Effective Gain of Mobile Antennas in Land Mobile Radio Environments, 39, 117–131, 1990.

7

Link budget comparison of different mobile communication systems based on EIRP and EISL

G. Fischer[1], F. Pivit[2], and W. Wiebeck[2]

[1]Lucent Technologies, Bell Labs Advanced Technologies, Thurn-und-Taxis-Straße 10, 90411 Nuremberg, Germany
[2]Institut für Höchstfrequenztechnik und Elektronik, Universität Karlsruhe, Kaiserstraße 12, 76128 Karlsruhe, Germany

Abstract. The metric EISL (Equivalent Isotropic Sensitivity Level) describing the effective sensitivity level usable at the air interface of a mobile or a basestation is used to compare mobile communication systems either based on time division or code division multiple access in terms of coverage and emission characteristics. It turns out that systems that organize the multiple access by different codes rather than different timeslots run at less emission and offer greater coverage.

1 Introduction

Mobile communication systems are evolving and new ones are added. For instance HSDPA (High Speed Downlink Packet Access) is added to UMTS (Universal Mobile Telecommunication System). It is desirable to have a clear understanding of the pros and cons of each system and the gains obtained in each evolution step. This article mainly focuses on the characteristics of mobile communication systems in terms of emission and coverage capabilities.

Besides scientific interests of understanding those characteristics for each system there are also strong business implications, which are critical for the financial success of mobile operators. Operators often have to run systems in parallel, or they have to make decisions on the selection of a certain system, which is linked to huge investments over a long period. Of course it has to be noted that capacity capabilities and spectral efficiency are further critical characteristics, but this is beyond the scope of this article.

Operators become more and more global players and are working on global standard strategies to offer their customers international roaming. A worldwide standard evolution strategy is linked to even higher investments and higher financial risks.

Correspondence to: G. Fischer
(georgfischer@bell-labs.com)

Figure 1 illustrates the great variety of mobile communication systems and their evolution steps. A major evolution step is a migration from so-called second generation systems (2G) like IS136-TDMA (Time Division Multiple Access) and GSM (Global System for Mobile communication) towards third generation systems (3G, IMT2000 systems) like UMTS and CDMA2000 (Code Division Multiple Access). EDGE (Enhanced Data rates through GSM Evolution) is not considered a 3G system here, because it is not able to offer Quality of Service (QoS) control, what is mandated with 3G systems by the ITU (International Telecommunications Union). A further evolution of EDGE called GERAN (GSM EDGE Radio Access Network) will be able to offer that, but this standard doesn't have a wide acceptance yet.

In this article systems are compared based on their uplink, because the cell radius is mainly determined by the limited transmit power available from the mobile and the basestations sensitivity. Looking at the uplink first is also the usual procedure with RF network planning. The downlink can normally be adjusted to balance the uplink by selecting a power amplifier of appropriate size at the basestation.

In the following a methodology first introduced in Fischer et al. (2002) is used. It allows assessment of the effective usable sensitivity at the air interface of a receiving system, through a metric named EISL (Equivalent Isotropic Sensitivity Level). It follows the same philosophy as EIRP (Equivalent Isotropic Radiated Power, Meinke and Gundlach, 1992). Both metrics EIRP and EISL are power metrics measured in Watt or respectively dBm. EISL is a metric that covers aspects of the receiving systems RF installation like antenna, RF cabling and receiver's noise as well as aspects of its digital signal processing in terms of equalization and decoding. EISL allows to describe the characteristics of RF and signal processing in a joined metric. In that way it allows also to trade off efforts in RF and signal processing against each other.

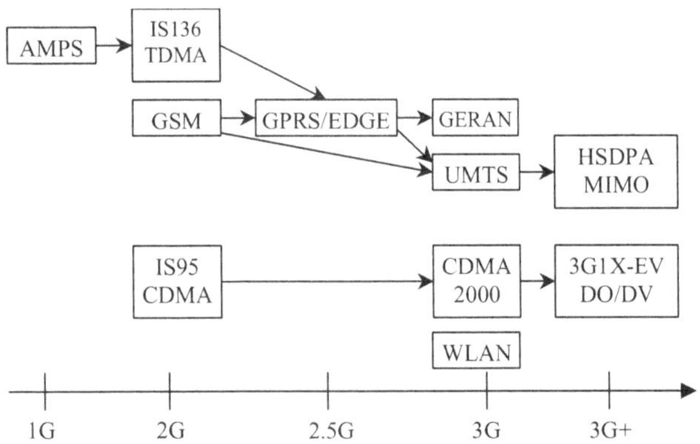

Fig. 1. Evolution of mobile communication standards.

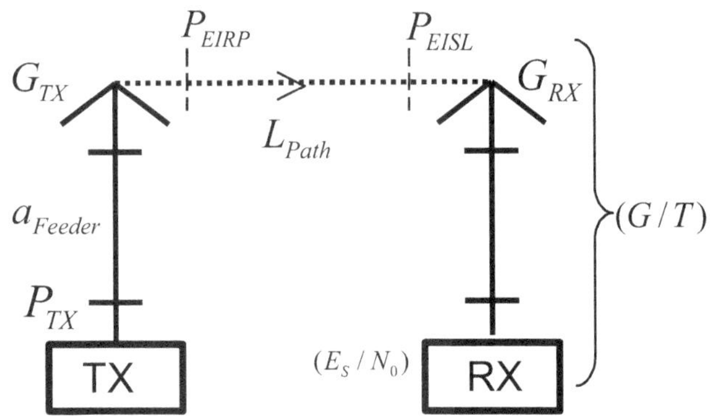

Fig. 2. Radio link metrics.

2 Equivalent isotropic sensitivity level EISL

2.1 Basic approach

As already noted in the introduction, EISL is defined in a similar way as EIRP. Let us first recover the definition of EIRP. It equals the power that would have to be fed into an ideal lossless omnidirectional antenna to create the same receive power as in the main beam direction of a directive antenna. EIRP values are typically much higher than usual power figures known from power amplifiers. This occurs especially for those cases where a high gain antenna is used with the transmitter system. Equivalently EISL is defined as the sensitivity a receiver would need to have if connected to an ideal lossless omnidirectional antenna to provide the same receive quality e.g. measured in terms of bit, frame or block error rate (BER/FER/BLER) as with a receiver connected to a directive antenna with the main beam focusing on the transmitter (Fischer et al., 2002). EISL values are typically much less than sensitivity figures measured at receiver inputs. Like with EIRP this especially happens if a high gain antenna and low noise amplifiers are used with the receiving system. EISL values can even be significantly lower than the physical limit given by the fundamental noise floor determined by the Boltzmann constant.

2.2 Definition of EISL

The following Fig. 2 shows a basic radio link. It defines the essential parameters the metrics, EIRP and EISL are dependent on.

At the transmitter system EIRP is a function of transmit power, the feeder loss and the transmit antenna gain:

$$P_{\mathrm{EIRP}}=f\,(P_{\mathrm{TX}},\ a_{\mathrm{Feeder}},\ G_{\mathrm{TX}})\,. \tag{1}$$

EIRP describes the whole transmit arrangement and thus reflects the effective usable power available at the air interface. Therefore it needs to include all gains and losses in the RF path of the transmit system.

At the receiving system, EISL is dependent on the performance of the digital signal processing stage, E_S/N_0, the quality factor G/T of the RF installation and the receive antenna gain G_{RX}:

$$P_{\mathrm{EISL}}=f\,(E_S/N_0,\ G/T,\ G_{\mathrm{RX}})\,. \tag{2}$$

EISL describes the whole receive arrangement. It reflects the effective usable sensitivity offered at the air interface and therefore includes all gains and losses in the RF and digital path of the receiving system. The quality factor G/T describing RF performance is a metric commonly in use with radio astronomy and satellite antennas (Zinke and Brunswig, 1979; Meinke and Gundlach, 1992). It reflects the gain to noise ratio of the RF installation and therefore includes all noise sources in the receive chain.

2.3 Link budget calculation philosophy

Based on EIRP and EISL the link budget can simply be analysed: The requirement that a link can be established at a certain quality is given by:

$$L_{\mathrm{Path},\,dB}\leq P_{\mathrm{EIRP},\,dBm}-P_{\mathrm{EISL},\,dBm}, \tag{3}$$

whereby L_{Path} denotes the path loss. For equal sign the maximum acceptable path loss is obtained:

$$L_{\mathrm{Path},\,dB,\,\mathrm{max}}=P_{\mathrm{EIRP},\,dBm}-P_{\mathrm{EISL},\,dBm}, \tag{4}$$

which can be converted into a cell radius according to ETSI (1998).

2.4 G/T calculation

The same equation used with dish antennas for satellite reception (Meinke and Gundlach, 1992) can also be applied to a basestation receiving system. Thus we have:

$$\frac{G}{T}=\frac{\eta\cdot D_{\mathrm{RX}}}{\eta\cdot T_{\mathrm{Ant}}+(1-\eta)\cdot T_0+(F-1)\cdot T_0}, \tag{5}$$

whereby G/T reflects the RF figure of merit, D_{RX} the directivity of the receive antenna, η the antenna efficiency, T_{Ant} the equivalent antenna noise temperature, T_0 the environmental temperature and F the cascaded noise figure of the RF chain.

Looking at the denominator in Eq. (5), three sources of noise can be identified. The first term reflects the noise power caused by the effective antenna brightness temperature. It is weighted with the efficiency of the antenna to account for the attenuation inside the antenna. The second term is related to the noise caused by the losses inside the antenna. Therefore the environmental temperature of the surrounding is used. Historically the third term related to the cascaded noise figure dominated the total noise. It reflects the noise of the whole RF processing based on the cascaded noise figure. However given recent advances in low noise amplifiers this term no longer dominates the total noise. It can be observed that nowadays a much more careful modeling of the equivalent antenna noise temperature T_{Ant} is needed (see also Fischer et al., 2002). T_{Ant} has to be obtained by convoluting the three dimensional beam pattern of the antenna with the elevation dependent brightness temperature.

The simple approach to model the antenna as a resistor at environmental temperature like it was done so far with network planning is too pessimistic. It predicts a too high noise floor and therefore implies smaller cell sizes than possible. This means that more cell sites are set-up than really needed. The difference between the exact model based on G/T versus the simple model based on a resistor can easily be 1 or 2 dB, which has strong implications on link budgets and thus cell planning. For instance a 1 dB better link budget means 15% greater cell size and 13% less basestations, see Fischer et al. (2002).

2.5 Example calculation

If some usual parameters are assumed: Gain of receive antenna $G_{\mathrm{RX},dB}=15\,dBi$ ($G_{\mathrm{RX}}=\eta\cdot D_{\mathrm{RX}}$), antenna efficiency $\eta=92\%$, equivalent antenna noise temperature $T_{\mathrm{Ant}}=200\,\mathrm{K}$, environmental temperature $T_0=290\,\mathrm{K}$ and cascaded noise figure of RF chain $F_{dB}=6\,\mathrm{dB}$, we obtain with Eq. (5):

$$\left(\frac{G}{T}\right)_{\mathrm{dB}}=-15.3\,\mathrm{dB/K}. \tag{6}$$

Just for comparison, typical values with radar telescopes are +30...45 dB/K and with home satellite dish antennas are +6...14 dB/K (Meinke and Gundlach, 1992).

2.6 EISL calculation

For the derivation of EISL it is assumed that the signal to noise ratio S/N with a receiving system at the interface between RF and digital signal processing in front of the channel selection filter within a given equivalent noise bandwidth B_{Noise} is identical looked either from the RF or from the digital side.

The derivation of the equivalent noise bandwidth with the channel selection filter is considered uncritical as it drops out of the EISL calculation. The reason for this is that essentially the noise density in terms of W/Hz is the determining factor. Instead of forcing identical SNRs also identical Energy per symbol could have been demanded. SNR was selected due to reason that it has a more practical meaning.

The interface between RF and signal processing is being looked at from both sides. Let us first look from the RF side:

$$\left(\frac{S}{N}\right)_{\mathrm{RF}}=\frac{G}{T}\cdot\frac{P_{\mathrm{EISL}}}{k\cdot B_{\mathrm{Noise}}} \tag{7}$$

with k being the Boltzmann constant:

$$k=1.38\cdot10^{-23}\,\frac{Ws}{K} \tag{8}$$

and then from the digital side:

$$\left(\frac{S}{N}\right)_{\mathrm{Digital}}=\frac{ES}{N_0}\cdot\frac{r_{\mathrm{Symbol}}}{B_{\mathrm{Noise}}}, \tag{9}$$

whereby r_{Symbol} means the symbol rate of the used modulation format. As stated above both have to be equal:

$$\left(\frac{S}{N}\right)_{\mathrm{RF}}=\left(\frac{S}{N}\right)_{\mathrm{Digital}}$$

$$\frac{G}{T}\cdot\frac{P_{\mathrm{EISL}}}{k\cdot B_{\mathrm{Noise}}}=\frac{E_S}{N_0}\cdot\frac{r_{\mathrm{Symbol}}}{B_{\mathrm{Noise}}}$$

$$P_{\mathrm{EISL}}=\frac{E_S}{N_0}\cdot\frac{1}{G/T}\cdot\left(k\cdot r_{\mathrm{Symbol}}\right)$$

$$P_{\mathrm{EISL,\,dBm}}=\left(\frac{E_S}{N_0}\right)_{\mathrm{dB}}-\left(\frac{G}{T}\right)_{\mathrm{dB}}+10\cdot\log_{10}\left(k\cdot r_{\mathrm{Symbol}}\right) \tag{10}$$

Looking at the EISL equation there are 3 terms that are added up. The first term characterizes the digital signal processing section through E_S/N_0, the second one the RF section through G/T and the third one covers constants that are system dependent. From this equation it is apparent that EISL allows merging the performance metrics for RF and digital part into a joined one.

2.7 EISL parameter dependency

Detailed parameter studies can be found in Fischer et al. (2002), however a quick summary is provided here. Given the fact that today's basestations incorporate very low noise RF frontends based on transistors with 0.4 dB noise figure, basestations with 1 dB noise figure got commodity. With typical feeder cables of 3 dB loss and a loss of 2 dB inside the antenna combiner structure, a typical value of 6 dB for the cascaded noise figure is obtained. However if installations with tower mounted low noise amplifiers are considered, the cascaded noise figure may go down to 4.1 dB and with active antenna elements it may come down to even 1.9 dB. Those noise figure variations do not translate 1 to 1 into EISL variations, because with low noise figure the other noise sources get more and more important. EISL may vary by 5 dB if F_{dB} varies by 4 dB. Such an effect occurs especially if the antenna is fairly "cold".

For low noise figures, EISL gets also strongly dependent on seasonal changes and the downtilt of the antenna. As the ground typically is much warmer than the cold sky, the equivalent antenna noise temperature T_{Ant} raises if the beam more and more points towards the "warm" ground.

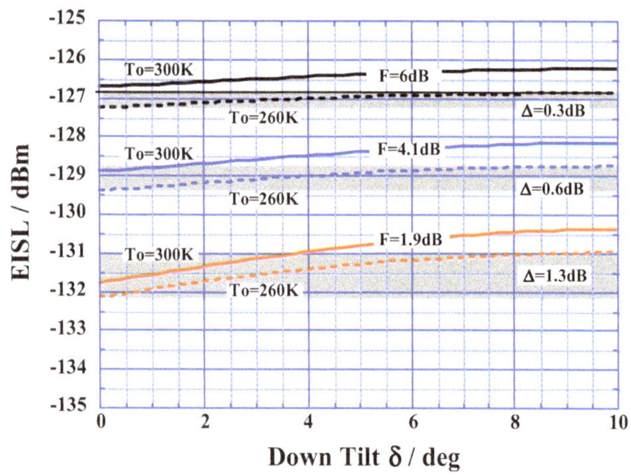

Fig. 3. EISL parameter study (Fischer et al., 2002), environmental temperature.

The variations related to seasonal and downtilt changes in Fig. 3 may be easily up to 1.5 dB (Changes due to downtilt shaded in Fig. 3). These changes have to be related to performance improvements in established signal processing structures for stable standards, which are often limited to less than 0.5 dB.

3 Application to various standards

As stated in the introduction, for coverage and link budget analysis the uplink is the critical path. Further as emission typically is dominated by the handset located next to the head and not by basestations that are farer away and even run higher power, a focus on uplink can be justified.

The comparison considers a basic 12.2 kbit/s voice service in an urban environment. Of course it is desirable also to perform such comparisons for data services, but that is part of ongoing studies.

3.1 GSM-uplink

For the calculation of a GSM mobile's EIRP, the following assumptions are made: Transmit power $P_{\mathrm{TX,\,dBm}}=20\,\mathrm{dBm}$, antenna gain $G_{\mathrm{TX,\,dBi}}=+1\,\mathrm{dBi}$ and body loss $A_{\mathrm{Bodyloss,\,dB}}=3\,\mathrm{dB}$. This results in an EIRP of:

$$\begin{aligned} P_{\mathrm{EIRP,dBm}} &= P_{\mathrm{TX}} + G_{\mathrm{TX,\,dB}} - A_{\mathrm{Bodyloss,\,dB}} \\ &= (29+1-3)\ \mathrm{dBm} \\ &= 27\ \mathrm{dBm}. \end{aligned} \tag{11}$$

The following calculation of the GSM base station's EISL is based on these parameters: Normalized sensitivity $E_S/N_0=5\,\mathrm{dB}$, RF figure of merit $G/T=-15.3\,\mathrm{dB}$, symbol rate $r_{\mathrm{Symbol}}=270\,\mathrm{k_{Sym}/s}$. These parameters are put into Eq. (10) and this leads to:

$$\begin{aligned} P_{\mathrm{EISL,\,dBm}} &= \left(\frac{E_b}{N_0}\right)_{\mathrm{dB}} - \left(\frac{G}{T}\right)_{\mathrm{dB}} + 10\cdot\log_{10}\left(k\cdot r_{\mathrm{Symbol}}\right) \\ &= \left[5-(-15)+10\cdot\log_{10}\left(1.38\cdot 10^{-23}\cdot 270\cdot 10^{3}\right)\right]\ \mathrm{dBW} \\ &= -124.3\,\mathrm{dBm}. \end{aligned} \tag{12}$$

Typical sensitivity figures referred to a basestations antenna connector are around -112 dBm and the physical limit for the fundamental noise floor is around -121 dBm. Once again, EISL is a virtual performance metric convenient for link budget calculation like EIRP.

From Eqs. (11) and (12) the maximum allowable pathloss for the GSM uplink case can be derived:

$$L_{\mathrm{Path,\,dB,\,max}} = 27\,\mathrm{dB}-(-124.3\,\mathrm{dBm})=151.3\,\mathrm{dB}. \tag{13}$$

3.2 IS136-TDMA-uplink

For IS136-TDMA some system dependent parameters of course change compared to the GSM case (Coursey, 1999), but parameters related to the RF processing are kept fixed for a fair comparison.

The derivation of an IS136-TDMA mobile's EIRP is based on these assumptions: $P_{\mathrm{TX,\,dBm}}=28\,\mathrm{dBm}$, $G_{\mathrm{TX,\,dBi}}=+1\,\mathrm{dBi}$, $A_{\mathrm{Bodyloss,\,dB}}=3\,\mathrm{dB}$. These input values lead to:

$$\begin{aligned} P_{\mathrm{EIRP,dBm}} &= P_{\mathrm{TX,\,dBm}} + G_{\mathrm{TX,\,dB}} - A_{\mathrm{Bodyloss,\,dB}} \\ &= (28+1-3)\ \mathrm{dBm} \\ &= 26\ \mathrm{dBm}. \end{aligned} \tag{14}$$

For the determination of an IS136-TDMA base station's EISL, again the parameters related to RF processing are kept identical with the GSM case and only the system dependent ones are modified: $E_S/N_0=19\,\mathrm{dB}$, $G/T=-15.3\,\mathrm{dB}$, $r_{\mathrm{Symbol}}=24.3\,\mathrm{k_{Sym}/s}$. This results in:

$$\begin{aligned} P_{\mathrm{EISL,\,dBm}} &= \left(\frac{E_S}{N_0}\right)_{\mathrm{dB}} - \left(\frac{G}{T}\right)_{\mathrm{dB}} + 10\cdot\log_{10}\left(k\cdot r_{\mathrm{Symbol}}\right) \\ &= \left[19-(-15.3)+10\cdot\log_{10}\left(1.38\cdot 10^{-23}\cdot 24.3\cdot 10^{3}\right)\right]\ \mathrm{dBW} \\ &= -120.7\,\mathrm{dBm}. \end{aligned} \tag{15}$$

3.3 UMTS-uplink

With code multiplexed systems the EISL equation (10) has to be modified to account for processing gain associated with those systems. Process gain is derived from the ratio between the chip rate and bit rate:

$$G_P=\frac{r_{\mathrm{chip}}}{r_{\mathrm{bitrate}}}. \tag{16}$$

The energy per symbol is further replaced by the energy per chip:

$$\left(\frac{S}{N}\right)_{\mathrm{Digital}}=\frac{E_c}{N_0}\cdot\frac{r_{\mathrm{chip}}}{B_{\mathrm{Noise}}}=\frac{E_b}{N_0}\cdot\frac{1}{G_P}\cdot\frac{r_{\mathrm{chip}}}{B_{\mathrm{Noise}}}. \tag{17}$$

The equation for the SNR from the RF side is untouched:

$$\left(\frac{S}{N}\right)_{\mathrm{RF}}=\frac{G}{T}\cdot\frac{P_{\mathrm{EISL}}}{k\cdot B_{\mathrm{Noise}}}. \tag{18}$$

Equations (17) and (18) have to be equal:

$$\left(\frac{S}{N}\right)_{\mathrm{RF}}=\left(\frac{S}{N}\right)_{\mathrm{Digital}}$$

$$\frac{G}{T}\cdot\frac{P_{\mathrm{EISL}}}{k\cdot B_{\mathrm{Noise}}}=\frac{E_b}{N_0}\cdot\frac{1}{G_P}\cdot\frac{r_{\mathrm{chip}}}{B_{\mathrm{Noise}}}$$

$$P_{\mathrm{EISL}}=\frac{E_b}{N_0}\cdot\frac{1}{G_P}\cdot\frac{1}{G/T}\cdot\left(k\cdot r_{\mathrm{chip}}\right)$$

$$P_{\mathrm{EISL,\,dBm}}=\left(\frac{E_b}{N_0}\right)_{\mathrm{dB}}-G_{P,\,\mathrm{DB}}-\left(\frac{G}{T}\right)_{\mathrm{dB}}+10\cdot\log_{10}\left(k\cdot r_{\mathrm{chip}}\right). \quad (19)$$

Looking at above equation it can be seen that the processing gain reduces the EISL and therefore implies an improved sensitivity.

In the following the UMTS mobile's EIRP is determined based on these parameters: Transmit power $P_{\mathrm{TX,\,dBm}}$=21 dBm, antenna gain $G_{\mathrm{TX\,dBi}}$=+1 dBi, body loss $A_{\mathrm{Bodyloss,\,dB}}$=3 dB. This results in:

$$\begin{aligned}P_{\mathrm{EIRP,dBm}} &= P_{\mathrm{TX,\,dBm}}+G_{\mathrm{TX,\,dB}}-A_{\mathrm{Bodyloss,\,dB}}\\ &= (21+1-3)\ \mathrm{dBm}\\ &= 19\,\mathrm{dBm}.\end{aligned} \quad (20)$$

To determine the base station's EISL these values are applied: E_b/N_0=5 dB, r_{chip}=3.84 Mchip/s,

$$G_{P,\,\mathrm{dB}}=10\cdot\log_{10}\left(\frac{3.84\,\mathrm{Mchip/s}}{12.2\,\mathrm{kbit/s}}\right)\ \mathrm{dB}=25\,\mathrm{dB},$$

G/T=−15.3 dB. This results in:

$$P_{\mathrm{EISL,dBW}}=\left(\frac{E_b}{N_0}\right)_{\mathrm{dB}}-G_{P,\mathrm{dB}}-\left(\frac{G}{T}\right)_{\mathrm{dB}}+10\cdot\log_{10}\left(k\cdot r_{\mathrm{chip}}\right)$$
$$=\left[5-25-(-15.3)+10\cdot\log_{10}\left(1.38\cdot10^{-23}\cdot3.84\cdot10^{6}\right)\right]\ \mathrm{dBW}. \quad (21)$$

Comparing Eq. (21) with Eq. (15) a much lower value for the effective sensitivity with UMTS compared to the effective sensitivity with GSM can be observed. However the UMTS related value has to be adjusted to account for effects specific with code multiplex systems.

3.3.1 Further factors to be considered

In the context of code multiplexed system further factors have to be considered (Holma and Toskala, 2000), for instance "cell breathing", which means that the cell size is a function of network load, meaning that capacity influences coverage.

The first factor that has to be quantified into a link budget as a margin is the interference margin. It accounts for inter- and intracell Interference:

$$M_{\mathrm{Interference,\,dB}}=1...3\,\mathrm{dB}\ (20...50\%\ \mathrm{Load}). \quad (22)$$

The second margin that has to be considered is the fast fade margin or sometimes also called power control headroom. It accounts for inaccuracies with the power control algorithm:

$$M_{\mathrm{FastFade,\,dB}}=2...5\,\mathrm{dB}. \quad (23)$$

Besides margins that degrade the link budget there is also a gain that improves it: Soft handover gain. It reflects the gain by fading mitigation through macro diversity:

$$G_{\mathrm{SoftHo,\,dB}}=2...3\,\mathrm{dB}. \quad (24)$$

Now the value for EISL can be adjusted:

$$\begin{aligned}P_{\mathrm{EISL,\,dBm,\,adj}}&=P_{\mathrm{EISL,\,dBm}}+M_{\mathrm{Interference,\,dB}}+M_{\mathrm{FastFade,\,dB}}-G_{\mathrm{SoftHO,\,dB}}\\ &=[-137.5+3+2-3]\,\mathrm{dBm}\\ &=-135.5\,\mathrm{dBm}.\end{aligned} \quad (25)$$

The EISL value gets degraded by high network load. Capacity effects cannot be neglected with analysis of coverage and emission.

3.4 CDMA-uplink

These parameters are assumed for an IS95-CDMA mobile: $P_{\mathrm{TX,\,dBm}}$=23 dBm, $G_{\mathrm{TX,\,dBi}}$=+1 dBi, $A_{\mathrm{Bodyloss,\,dB}}$=380 dBm. They result in an EIRP of:

$$\begin{aligned}P_{\mathrm{EIRP,\,dBm}} &= P_{\mathrm{TX,\,dBm}}+G_{\mathrm{TX,\,dB}}-A_{\mathrm{Bodyloss,\,dB}}\\ &= (23+1-3)\ \mathrm{dBm}=21\,\mathrm{dBm}.\end{aligned} \quad (26)$$

As E_b/N_0 is referenced to the channel decoders input also the bit rate at the channel decoders input which is 14.4 kbit/s and thus slightly higher than the service rate of 12.2 kbit/s has to be selected for derivation of the process gain. The calculation of the IS95-CDMA base station's EISL therefore assumes: G/T=15.3 dB, E_b/N_0=7 dB,

$$G_{P,\,\mathrm{dB}}=10\cdot\log_{10}\left(\frac{1.2288\,\mathrm{Mchip/s}}{14.4\,\mathrm{kbit/s}}\right)\ \mathrm{dB}=10.3\,\mathrm{dB}$$

r_{chip}=1.2288 Mchip/s and results in:

$$P_{\mathrm{EISL}}=(E_bN_0)_{\mathrm{dB}}-G_{P,\,\mathrm{dB}}-\left(\frac{G}{T}\right)_{\mathrm{dB}}+10\cdot\log_{10}\left(k\cdot r_{\mathrm{chip}}\right)$$
$$=\left[7-19.3-(-15.3)+10\cdot\log_{10}\left(1.38\cdot10^{23}\cdot1.2288\cdot10^{6}\right)\right]$$
$$P_{\mathrm{EISL,\,dBm}}=-134.7\,\mathrm{dBm}. \quad (27)$$

3.4.1 Further factors to be considered

Like with the UMTS case further factors accounting for additional margins and gains have to be quantified into the link budget (Yang, 1998):

Interference margin:

$$M_{\mathrm{Interference,\,dB}}=3.4\,\mathrm{dB}$$

Fast fade margin, power control headroom:

$$M_{\mathrm{FastFade,\,dB}}=2...5\,\mathrm{dB}$$

Soft Handover gain:

$$G_{\mathrm{SoftHO,\,dB}}=4\,\mathrm{dB}.$$

The EISL value derived in Eq. (27) can now be adjusted:

$$\begin{aligned}P_{\mathrm{EISL,\,dBm,\,adj}}&=P_{\mathrm{EISL,\,dBm}}+M_{\mathrm{Interference,\,dB}}+M_{\mathrm{FastFade,\,dB}}-G_{\mathrm{SoftHO,\,dB}}\\ &=[-134.7+3.4+2-4]\,\mathrm{dBm}\\ &=-133.3\,\mathrm{dBm}.\end{aligned} \quad (28)$$

Table 1. Typical EIRP and EISL values with wireless uplink and their implications on RF network planning.

	$P_{\mathrm{EIRP,\,dBm}}$	$P_{\mathrm{EIRP,\,dBm,adj}}$	$L_{\mathrm{Path,\,dB}}$	Relative cell radius	Relative cell area	Change in no. of sites
GSM	27	−124.3	151.3	1	1	0
UMTS	19	−135.5	154.5	1.24	1.54	−35%
IS136-TDMA	26	−120.7	146.7	0.73	0.54	+86%
IS95-CDMA	21	−133.3	154.3	1.23	1.5	−33%

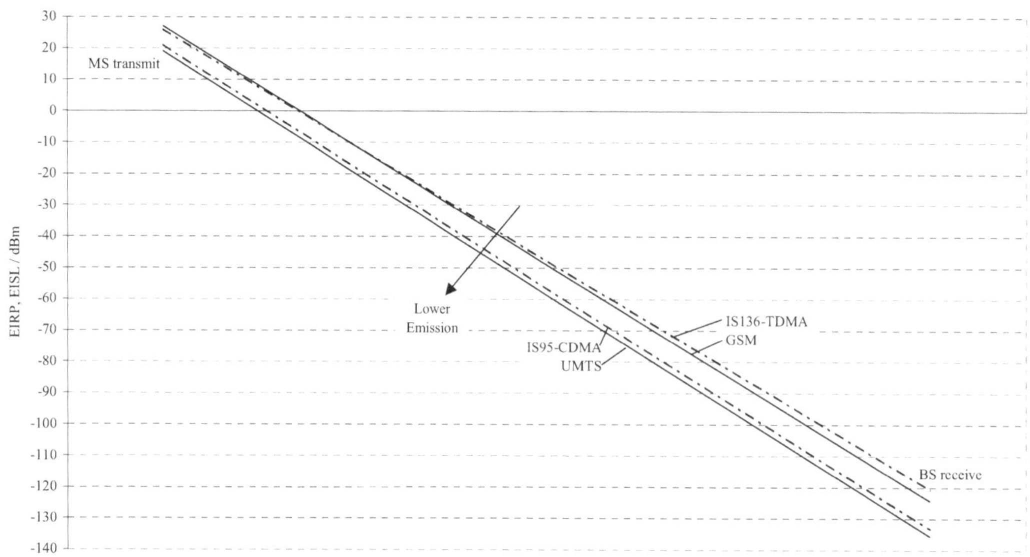

Fig. 4. Link budget comparison of mobile communication standards.

3.5 Comparison and consequences

The results obtained in Sects. 3.1 to 3.4 are summarized in Table 1 and Fig. 4. Table 1 also transfers the differences in link budget into relative changes of cell radius, cell area and number of basestations sites needed. Hereby the GSM link budget is used as a reference point. The relative change in cell radius assumes that the receive power is decaying with distance weighted by an exponent of 3.4, instead of 2 normal with free space.

From Table 1 it can be observed that code multiplexed systems are ahead of time multiplexed ones. They offer a larger cell radius. For covering a larger area around 33% less basestations are needed. As total site costs are about 3 times the cost of a basestation, network operators can obtain huge savings with network deployment. However the figure of −33% assumes a regular cell grid, which is not the case in real scenarios. Therefore practical savings are less. Furthermore it has to be considered that above conclusions are only true if an operator has the choice to deploy either of the standards in the same frequency band. CDMA and UMTS do better by around 3 dB compared to GSM, but if their operating frequency is twice compared to GSM this gain is no longer present as an increase of operating frequency by factor 2 implies about 9 dB more path loss for the same cell radius (see ETSI, 1998). Statements like that UMTS is an inefficient system as it requires much more sites than GSM are therefore questionable. The increase in number of sites is mainly a consequence of the high frequency band not a consequence of the standard. Instead if UMTS would be deployed at 900 MHz, in the same band as GSM, instead of 2100 MHz, sites could be saved. Benefits of low frequency bands in terms of emission and coverage have e.g. been analyzed in a research project funded by the German federal ministry on research and education BMBF (BMBF, 2002).

Figure 4 provides a quick insight into the emission characteristics. The lines shown are just straight connections between data points for mobile EIRP and basestation EISL, with no mapping to the abscissa. The lines are drawn to visualize the slope as the slope of the lines can be interpreted as the maximum path loss supported. IS136-TDMA has a more flat slope than the others and this indicates that it supports less path loss as already shown by Table 1. Furthermore the lower the lines the lower the emission. Systems can offer the same maximum path loss but one may operate at generally lower power levels. This is the case for the code multiplexed systems compared to the time multiplexed ones. They operate generally at 10 dB less power and therefore run at 10 dB less emission. However it has to be stressed that with time multiplexed systems the transmitter isn't on air all the time. With GSM due to 8 timeslots the average emission drops by

9 dB and in IS136-TDMA 3 timeslots are used resulting in a drop by around 5 dB.

In this context it has to be mentioned that the biological effects of pulsed signals are discussed very controversy and that a mathematical rule for weighting the variations of the RF envelope doesn't exist yet.

4 Conclusion

It has been shown that the concept of analyzing the effective usable sensitivity of a receiving system through the metric EISL, describing the effective usable sensitivity at the air interface aside the well known metric EIRP for the effective radiated power, is a powerful tool for doing link budget comparisons between systems of different nature.

EISL is derived from the metric G/T describing the performance of the RF processing part including antenna and RF cabling and $E_{\mathrm{S\,or\,chip}}/N_0$ describing the performance of the digital signal processing part. EISL merges both metrics into a single number together with some system related constants.

EISL shows strong dependence on architectural changes, beam downtilt and seasonal changes.

From the link budget comparisons it is derived that code multiplexed systems run at generally lower emission levels and offer greater cell radius than time multiplexed ones.

It is also observed that the frequency band of operation has a much higher impact on cell size than the differences in the link budgets between the various standards.

The comparisons presented in this article focused on a basic voice service at 12.2 kbit/s. Of course performing comparisons also for data services would be of high interest and that is the scope of ongoing activities in this area.

Acknowledgements. The authors would like to thank A. Dekorsy, P. Gunreben, and F. Obernosterer for their support in the area of performance figures for signal processing with GSM, CDMA and UMTS.

References

BMBF: Final Report on research project MiniWatt, study of opportunities to reduce emission, German Federal Ministry on Research and Education, 2002.

Coursey, K. C.: Understanding Digital PCS – The TDMA Standard, Artech, 1999.

ETSI: Specification GSM-03.30, Radio network planning aspects, 1998.

Fischer, G., Pivit, F., and Wiesbeck, W.: EISL, the Pendant to EIRP: A Measure for the Receive Performance of Base Stations at the Air Interface, EUMW, Milan, 2002.

Holma, H. and Toskala, A.: W-CDMA for UMTS, Wiley, 2000.

Meinke, H. H. and Gundlach, F.-W.: Taschenbuch der Hochfrequenztechnik, 5. Auflage, Springer, 1992.

Yang, S. C.: CDMA RF System Engineering, Artech, 1998.

Zinke, O. and Brunswig, H.: High Frequency Techniques II, 4th Edition, Springer, Berlin–Heidelberg–New York, 2. Auflage, 228, 1979.

Geometry-based channel modelling of MIMO channels in comparison with channel sounder measurements

G. Del Galdo, M. Haardt, and C. Schneider

Communications Research Laboratory, Ilmenau University of Technology, P.O. Box 100565, 98684 Ilmenau, Germany

Abstract. In this paper we propose a flexible geometry-based propagation model for wireless communications developed at Ilmenau University of Technology. The IlmProp comprises a geometrical representation of the environment surrounding the experiment and a precise representation of the transmitting and receiving antennas. The IlmProp is capable of simulating Multi-User MIMO scenarios and includes a complete collection of tools to analyze the synthetic channels. In order to assess the potentials as well as the limits of our channel simulator we reconstruct the scenario encountered in a recent measurement campaign at Ilmenau University of Technology leading to synthetic data sets similar to the ones actually measured. The measurements have been collected with the RUSK MIMO multi-dimensional channel sounder. From the comparisons of the two channel matrices it is possible to derive useful information to improve the model itself and to better understand the physical origins of small-scale fading. In particular the effects of the different parameters on the synthetic channel have been studied in order to assess the sensibility of the model. This analysis shows that the correct positioning of a small number of scatterers is enough to achieve frequency selectiveness as well as specific traits of the channel statistics. The size of the scattering clusters, the number of scatterers per cluster, and the Rician K-factor can be modified in order to tune the channel statistics at will. To obtain higher levels of time variance, moving scatterers or time dependent reflection coefficients must be introduced.

1 Introduction

The high complexity of the propagation phenomenon sets a vast challenge in the effort of modelling the wireless channel. The most accurate approach would be to solve the Maxwell equations. However the computational effort required would be prohibitive, especially if we considered multiple antennas at both ends of the link, i.e. a MIMO system (Multiple Input Multiple Output). In fact, if M_R and M_T are the number of antennas at the receiver and at the transmitter, respectively, the number of channels to be calculated grows to $M_R \cdot M_T$. Furthermore a precise physical and geometrical description of the objects would be necessary.

In order to reduce the complexity of the model the so called GTD (Geometric Theory of Diffraction) can be employed (Correia, 2001). This approach postulates the existence of direct, diffracted and reflected rays only. As the wavelength λ approaches zero, this approximation becomes increasingly accurate.

With a ray-tracing engine and a full 3D description of the environment it is then possible to calculate all possible rays linking the antennas. However with a GTD approach a complete geometrical representation of the scenario is still required. A further approximation is to simplify the geometry surrounding the antennas with a discrete number of reflectors, called *scatterers*; when only direct and reflected rays are considered the computational complexity drops significantly. In this paper we present the IlmProp, a simplified model which generates channels which mimic most characteristics of measured channels. To this aim it is extremely important to determine which features of the synthetic channel are affected by the different parameters present in the model. A relatively low computational complexity allows us to generate channels with all possible combinations of values for the parameters. Then we analyze the data by finding strong traits which link a particular parameter to a particular feature. At the same time we are able to assess whether a parameter has no effect on a specific feature.

This extensive search and its results are reported in Sect. 3. Then we modeled the scenario found in one of the measurement campaigns undertaken at Ilmenau University of Technology. Section 4 reports the direct comparison between the synthetic and measured channels which led to a better understanding on the origins of the fast and slow fading processes. Section 5 deals in particular with the comparison of the phase responses. Lastly, Sect. 6 concludes this paper.

Correspondence to: G. Del Galdo
(giovanni.delgaldo@tu-ilmenau.de)

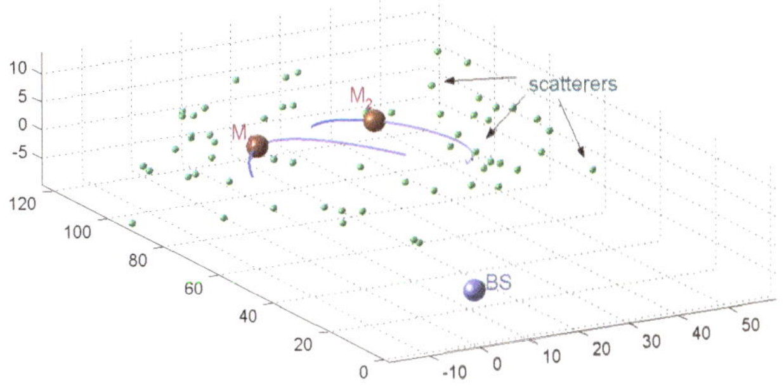

Fig. 1. In this example 2 Mobiles (M's) surrounded by scatterers move in the proximity of one fixed Base Station (BS).

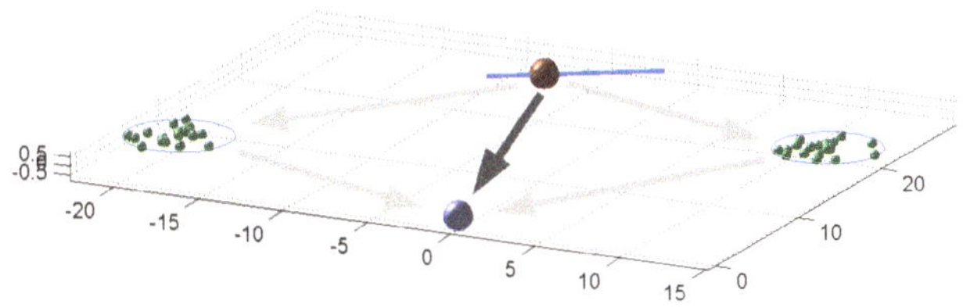

Fig. 2. The geometrical representation of the model $\mathcal{M}_\alpha$. Two scattering clusters and a strong LOS are visible. The transmitter (at the top) moves on a linear path.

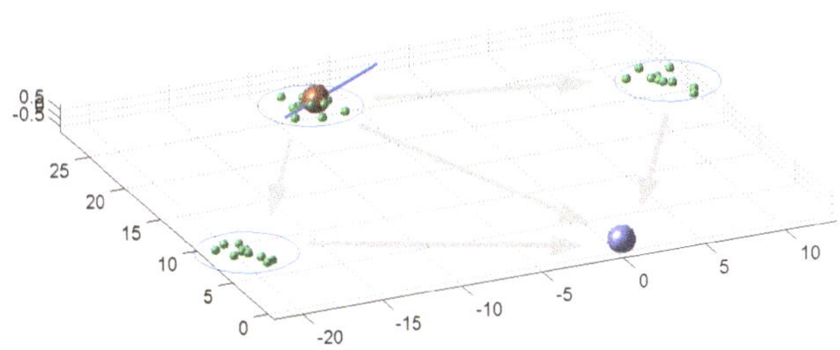

Fig. 3. The geometrical representation of the model $\mathcal{M}_\beta$. The model is characterized by three scattering clusters; one of them surrounds the transmitter.

2 The IlmProp: a geometry-based channel model

The IlmProp developed at TU Ilmenau relies entirely on a 3D geometric representation and has been designed to simulate a Multi-User-MIMO scenario. It has been simplified to single-bounce reflections in order to reduce the computational complexity. Clusters of scatterers can be arbitrarily positioned in the 3D reconstruction of the environment. Each scatterer corresponds to a single ray and is characterized by a complex coefficient which, for simplicity, is independent of the Direction of Arrival (DoA) and the Direction of Departure (DoD); but may vary in time with an arbitrary law. The coefficient determines the phase shift and power attenuation introduced by the scatterer.

The receiver thus receives a number of rays equal to the number of scatterers plus the Line Of Sight (LOS), which links the transmitter to the receiver directly. Figure 1 shows a simple example of this modelling approach. Such a simplified model still requires a fairly high computational complexity because the total number of rays to be computed is: $M_R \cdot M_T \cdot N_{scatt}$.

The advantage of a geometry-based model with respect to a stochastic one is indeed the possibility to easily generate a frequency selective time variant channel. In fact the frequency selectiveness is caused by the multi-path effect while the time variance is set off by the movement of the antennas and of the scatterers as well as the varying scattering coefficients with respect to time. In other words it well models a typical frequency selective time variant radio channel. In order to characterize such a channel we need to describe it in one of the four possible two-dimensional domains named the Bello domains. (Bello, 1963). Either $\{t, f\}$, $\{t, \tau\}$, $\{f_D, f\}$ or $\{f_D, \tau\}$ where t, f_D, f, and τ denote time, Doppler frequency, frequency, and delay time, respectively. The most convenient domain in which to compute the channel is certainly $\{t, f\}$. Having a geometrical representation of all objects in time makes the choice of t over f_D quite obvious. The choice of f over τ requires more explanation. Thanks to the geometrical representation it is simple to calculate the path lengths for each ray present in the scenario. Assuming that the velocity of light, c, is equal to the velocity of all rays, we can calculate the delay time associated with every ray, i.e. the time needed to reach the receiver. The delay times found will not necessarily match the sampling grid chosen. For this reason more processing is required. On the contrary, in the frequency domain we can calculate the channel response for the specific frequency bins that we are interested in. The choice of a specific Bello domain is at the end a matter of convenience since it is possible to derive the other representations via Fourier transforms. For instance, having $H(f, t)$, i.e. the channel expressed in delay time and time, we can calculate $h(\tau, t)$, the channel in frequency and time, as:

$$h(\tau, t) = \int H(f, t)e^{j2\pi f\tau}df. \tag{1}$$

The amplitude and phase of each ray are calculated separately at the receiver. The phase of the i-th ray at a certain time snapshot t_o and frequency f_o are calculated as follows:

$$\varphi_i(t_o, f_o) = \angle(e^{-j2\pi\frac{f_o}{c}L_i}) + \theta_i, \tag{2}$$

where L_i is the ray's path-length and θ_i is the phase shift introduced by the scatterer. The amplitude depends simply on the scattering coefficient, antenna gain, and path-loss (Van Trees, 2002).

3 An exhaustive investigation

The first step in understanding how sensible the model is to the different parameters and which features of the synthetic channel are affected, is to model some typical scenarios and to generate channels with all possible values for the parameters. We implemented two models, to which we will refer as $\mathcal{M}_\alpha$ and $\mathcal{M}_\beta$, seen in Figs. 2 and 3. $\mathcal{M}_\alpha$ models a typical scenario in which the areas around the transmitter and the receiver are free of scatterers while some are present in the middle of the link. $\mathcal{M}_\beta$, on the other hand, models a scattering cluster in the proximity of the transmitter. We generated each channel for the following sampling grid:

- A bandwidth B of 400 MHz at a center frequency of 1 GHz sampled with a $\Delta f = 1.3$ MHz for a total of 301 samples.
- A total of 200 time snapshots were taken, one every 10 ms for a total duration of 2 s.

The choices of these parameters determine the sampling grids in the coupled domains. In fact if the complex baseband signal is sampled in the range $[-\frac{B}{2}, \frac{B}{2}]$, then $\Delta\tau$, the resolution in delay time, is equal to $\frac{1}{B}$. In the same way the bandwidth in the Doppler frequency, $2 \cdot f_{D,max}$ can be calculated from the resolution in time Δt as:

$$f_{D,max} = \frac{1}{2 \cdot \Delta t}. \tag{3}$$

- The resolution in delay time is $\Delta\tau = 2.5$ ns, which corresponds to a distance of less than a meter (75 cm). We can successfully resolve paths (i.e. without aliasing), which are up to 220 meters long.
- Doppler shifts up to $f_{D,max} = \pm 50$ Hz are visible without aliasing.

The transmitter moves with a uniform velocity of about 5 m/s. Both transmitter and receiver employ a ULA (Uniform Linear Array) with $M_R = M_T = 4$ antennas spaced by $\frac{\lambda}{2}$.

After several experiments with different geometries and different configurations we narrowed down our extensive search to three dominant parameters:

- Cluster size
- Number of scatterers per cluster
- Rician K-factor

A fourth dominant parameter is certainly the modelling of the reflection coefficients, and more importantly their trend in time. This leaves, however, too many degrees of freedom since we could implement any arbitrary function to describe it in time. For this reason we left this important parameter to future and more specific investigations. In our simulations the reflection coefficients are kept constant along all dimensions, i.e. frequency and time. The Rician K-factor is defined as

$$K = \frac{P_{\text{LOS}}}{P_{\text{scatt}}}, \tag{4}$$

i.e. the power attributed to the Line Of Sight divided by the power of the scatterers. Clearly it is a function of the pathlosses which on their side are determined by the geometry, the transmit power, and the reflection coefficients. We scale the amplitudes of the reflection coefficients to obtain a desired K-factor.

We then generate many channels with different values for these three parameters and then analyze the data to find any strong dependency.

3.1 K-factor

The K-factor affects several features of the channel. Firstly it influences its variance with respect to all four domains. It is thus very interesting to assess this information quantitatively computing the RMS delay spread τ_{RMS}, the coherence bandwidth $(\Delta f)_{\text{c}}$, the RMS Doppler spread $f_{\text{D,RMS}}$ and the coherence time $(\Delta t)_{\text{c}}$.

The RMS delay spread of the channel, τ_{RMS}, is derived from the multipath intensity profile $\psi_{De}(\tau)$ which represents the average power of the channel output as a function of delay time (Paulraj et al., 2003; Stüber, 1996).

$$\tau_{\text{RMS}} = \sqrt{\frac{\int_0^{\tau_{\max}} (\tau - \bar{\tau})^2 \psi_{De}(\tau) d\tau}{\int_0^{\tau_{\max}} \psi_{De}(\tau) d\tau}}. \tag{5}$$

$\tau_{\max}$ is the maximum delay spread while $\bar{\tau}$ is the average delay spread given by

$$\bar{\tau} = \frac{\int_0^{\tau_{\max}} \tau \psi_{De}(\tau) d\tau}{\int_0^{\tau_{\max}} \psi_{De}(\tau) d\tau}. \tag{6}$$

The coherence bandwidth $(\Delta f)_{\text{c}}$ corresponds to the frequency lag in which the channel's autocorrelation function reduces to 0.7 (Paulraj et al., 2003). As previously described, the channel representations in the four domains are coupled through Fourier transforms. For this reasons $(\Delta f)_{\text{c}}$ and τ_{RMS} are connected as well. In fact the coherence bandwidth is inversely proportional to the RMS delay spread so that

$$(\Delta f)_{\text{c}} = \frac{const_1}{\tau_{\text{RMS}}}. \tag{7}$$

The same concepts are applied on the $\{f_D, t\}$ pair. The RMS Doppler spread, $f_{\text{D,RMS}}$, is derived in a similar way from ψ_{Do}, the Doppler power spectrum, which corresponds to the average power as a function of Doppler frequency f_D.

$$\begin{aligned} f_{\text{D,RMS}} &= \sqrt{\frac{\int_{\mathcal{F}} (f_{\text{D}} - \bar{f}_{\text{D}})^2 \psi_{Do}(f_{\text{D}}) df_{\text{D}}}{\int_{\mathcal{F}} \psi_{Do}(f_{\text{D}}) df_{\text{D}}}} \\ \bar{f}_{\text{D}} &= \frac{\int_{\mathcal{F}} f_{\text{D}} \psi_{Do}(f_{\text{D}}) df_{\text{D}}}{\int_{\mathcal{F}} \psi_{Do}(f_{\text{D}}) df_{\text{D}}} \end{aligned} \tag{8}$$

The integration domain, $\mathcal{F}$, corresponds to the Doppler frequency interval where the Doppler power spectrum is non zero. The coherence time $(\Delta t)_{\text{c}}$ is calculated in the same way as the coherence bandwidth. It corresponds to the time lag at which the autocorrelation function drops to 0.7 (Paulraj et al., 2003). $(\Delta t)_{\text{c}}$ indicates how time variant the channel is and similarly to the $\{f, \tau\}$ pair

$$(\Delta t)_{\text{c}} = \frac{const_2}{f_{\text{D,RMS}}} \tag{9}$$

Figure 4 shows typical trends for τ_{RMS}, $f_{\text{D,RMS}}$, $(\Delta f)_{\text{c}}$, and $(\Delta t)_{\text{c}}$ for different K-factors. Similar plots can be observed for both models $\mathcal{M}_\alpha$ and $\mathcal{M}_\beta$ for almost any cluster size and number of scatterers. As K grows τ_{RMS} decreases, meaning that the channel becomes rapidly less frequency selective. The same observation can be derived from the trend of $(\Delta f)_{\text{c}}$. $f_{\text{D,RMS}}$ and $(\Delta t)_{\text{c}}$ show the same trends. For bigger K's the channel becomes increasingly less time variant. The

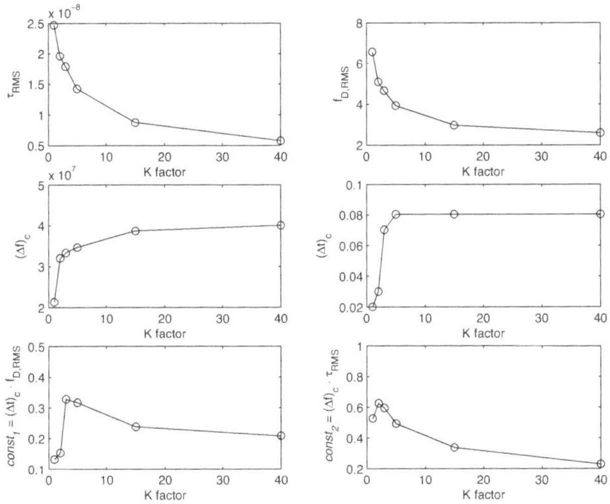

Fig. 4. τ_{RMS}, $f_{\mathrm{D,RMS}}$, $(\Delta f)_c$ and $(\Delta t)_c$ plotted for different K-factors.

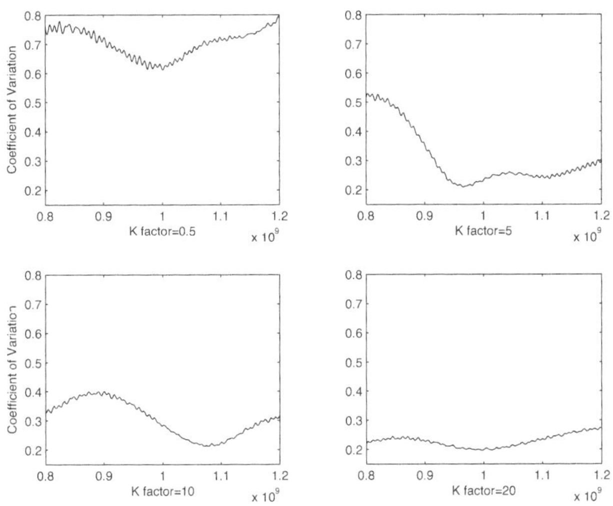

Fig. 5. The *coefficient of variation* ξ plotted against frequency f for different K-factors.

two constants $const_1$ and $const_2$ are plotted as well to show that the inverse proportionality expressed in Eqs. (7) and (9) is a good approximation.

Another interesting measure of the variation of the channel is to look at the fluctuation of $\|\boldsymbol{H}\|_F^2$. A good index is the so called *coefficient of variation* (Nabar, 2003) which is commonly used as a quantitative measure of the spread (fluctuation) of a random variable. The coefficient of variation ξ is defined as:

$$\xi = \frac{\sqrt{E\{(\|\boldsymbol{H}\|_F^2)^2\} - (E\{\|\boldsymbol{H}\|_F^2\})^2}}{E\{\|\boldsymbol{H}\|_F^2\}}. \tag{10}$$

We calculated ξ for all frequency bins to discover that for growing K's the coefficient of variation becomes increasingly smaller (Fig. 5). This is actually expected if we associate

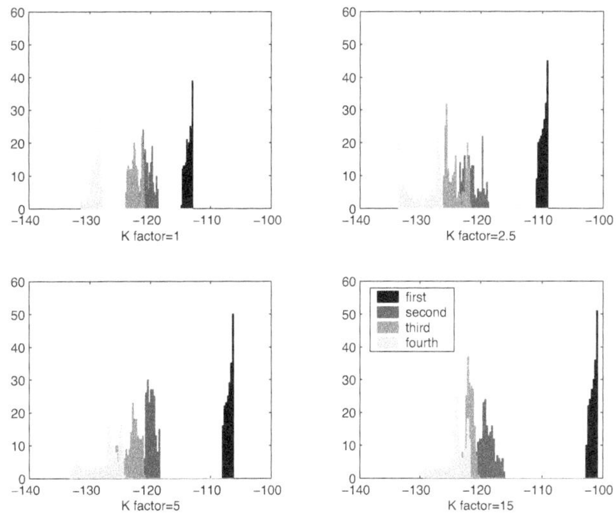

Fig. 6. Eigenvalue distributions in dB as K grows for $\mathcal{M}_\alpha$.

ξ with the *effective diversity order*, as described in (Nabar, 2003).

In fact the effective diversity order, N_{div}, can be written as:

$$N_{div} = \frac{1}{\xi^2}, \tag{11}$$

and it approaches infinity for a channel with only the LOS component (AWGN channel) while $N_{div} = M_R \cdot M_T$ for Rayleigh channels. The trend in ξ and consequently in N_{div} can be observed for both models proposed and independently from other parameters.

Another interesting analysis can be performed on the statistics of the eigenvalues. Figure 6 shows their distributions for different K-factors for the model $\mathcal{M}_\alpha$. Eigenvalue decompositions have been applied on the spatial correlation matrices at the receiver $\boldsymbol{R}_R$ calculated for each time snapshot. Then we generated the histograms to derive the amplitudes distribution on a dB scale. The spatial correlation matrix $\boldsymbol{R}_R$ at the receiver for the specific time t_o is:

$$\boldsymbol{R}_R(t_o) = \frac{1}{F}\sum_{i=1}^{F} \boldsymbol{H}(:,:,i,t_o) \cdot \boldsymbol{H}(:,:,i,t_o)^H, \tag{12}$$

where F is the total number of frequency bins, $\boldsymbol{H}^H$ is the Hermitian transpose of $\boldsymbol{H}$ and the channel $\boldsymbol{H}$ has dimensions $M_R \times M_T \times F \times T$. The MATLAB like notation $\boldsymbol{H}(:,:,i,t_o)$ denotes the two-dimensional channel matrix $\boldsymbol{H} \in \mathbb{C}^{M_R \times M_T}$ for a specific frequency $f = i$ and time $t = t_o$. Figure 6 shows the distributions of the eigenvalues plotted in dB. The channels have been generated without noise and without any normalization, i.e. the absolute amplitudes of the eigenvalues appear very low (around -120 dB). This has no physical meaning and only the relative distances between the eigenvalues should be taken into account. As the K-factor grows, the Line Of Sight becomes increasingly stronger and the channel displays low rank traits. In other words the strongest eigenvalue separates itself from the weaker ones as K increases.

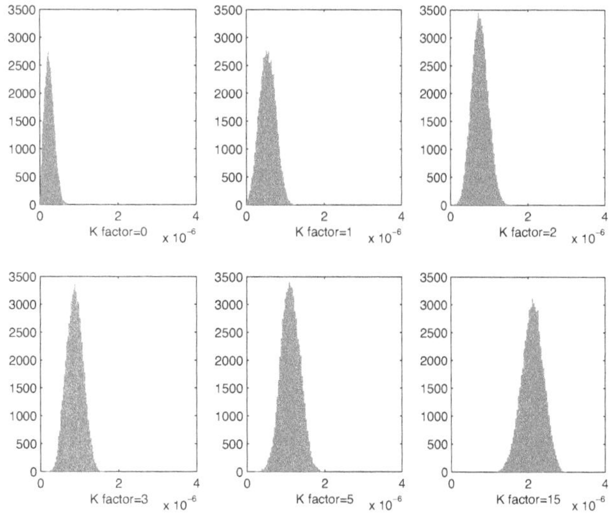

Fig. 7. Amplitude distributions of channels considering all frequencies and time snapshots for different Rician K-factors for $\mathcal{M}_\alpha$.

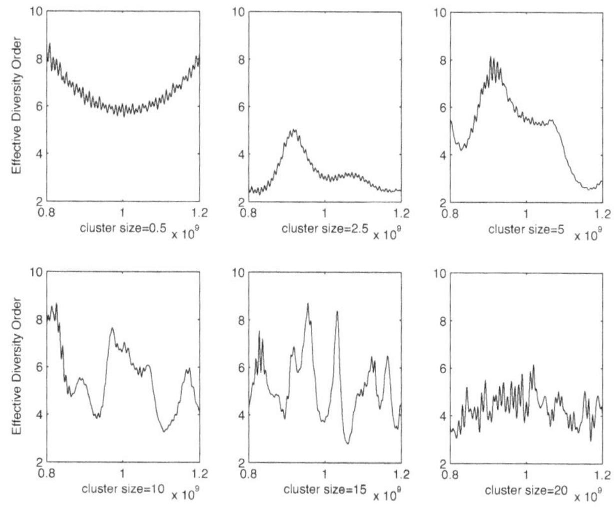

Fig. 8. The *effective diversity order* N_{div} plotted against frequency f for different cluster sizes (expressed in λ's) for $\mathcal{M}_\beta$.

A higher K-factor inhibits the effects of the fast-fading. This can be easily seen by plotting the distributions of the amplitudes of the channel considering all frequencies and time snapshots (Fig. 7).

3.2 Cluster size and number of scatterers

While there exists a strong relationship between the different channel statistics and the Rician K-factor, there is no strict relation with the parameters which characterize a scattering cluster, namely the number of scatterers, the cluster size and the type of distribution of the scatterers inside the cluster. In the first simulations we investigated whether the geometrical distribution affected the channel. We generated clusters with scatterers distributed uniformly on a sphere, on a circle on the horizontal plane (as seen in Figs. 2 and 3), and on the surface of a cylinder or a sphere. All channels showed the same trends and for this reason we continued with more detailed investigations fixing the distribution of the scatterers on a circle. We generated models characterized by cluster sizes with diameters ranging from $\frac{\lambda}{2}$ to 20λ and with numbers of scatterers ranging from 1 up to 100 reflection points per cluster.

We then investigated trends in the different channel features fixing all parameters as either the cluster size or the number of scatterers changed. All features but the effective diversity order did not show any strong dependency even though they were all affected. For instance, increasing the size of the scattering clusters affects the frequency selectiveness of the channel, even though there is neither a direct nor an inverse proportionality between the two. Figure 8 shows the effective diversity order N_{div} plotted for different frequencies f and for different cluster diameters expressed in λ's for the first model, $\mathcal{M}_\alpha$. Analyzing N_{div} along different frequencies we can recognize a slow-moving trend corrupted by faster variations. For instance, in the first plot of Fig. 8 (cluster size = $0.5\ \lambda$), we see a slow moving trend which looks like the sum of a concave semicircle and a zero mean noise-like process which introduces fast variations. After many simulations we observed that the magnitude of the fast moving N_{div} changed at every realization while the variance of the slow moving N_{div} changed with the cluster diameter with an undoubtedly direct proportionality. In other words, for bigger cluster sizes, N_{div} changes more rapidly along frequency while this does not necessarily correspond to higher frequency selectiveness nor to greater time variance. The number of scatterers affects the channel in a similar way but with a much weaker dependency.

4 A comparison with measurements

In order to assess the results derived in Sect. 3 with an even more realistic scenario we modeled the surroundings of a measurement campaign undertaken at TU Ilmenau. Additionally the simulation allowed us to investigate the importance of other critical parameters of the model: the positions of the scatterers and the modelling of the reflection coefficients. The first step was to determine whether a rough positioning of the scatterers was enough to recreate the main features which characterize the measured channel. To this end, we chose, among the many measurements gathered, a channel with a strong LOS and characterized by a simple geometry so that the position of the scatterers could easily be estimated. Figure 9 shows the bird's eye view of the reconstructed scenario. The receiver was a 4×8 polarization sensitive patch-array URA (Uniform Rectangular Antenna) and it was positioned by a window on the second floor of a building. The transmitter was a single 2 GHz omnidirectional antenna mounted on a wheeled cart pulled by hand up a road in the direction of a second building (Fig. 10). The transmitter path is represented by a continuous line in Fig. 9.

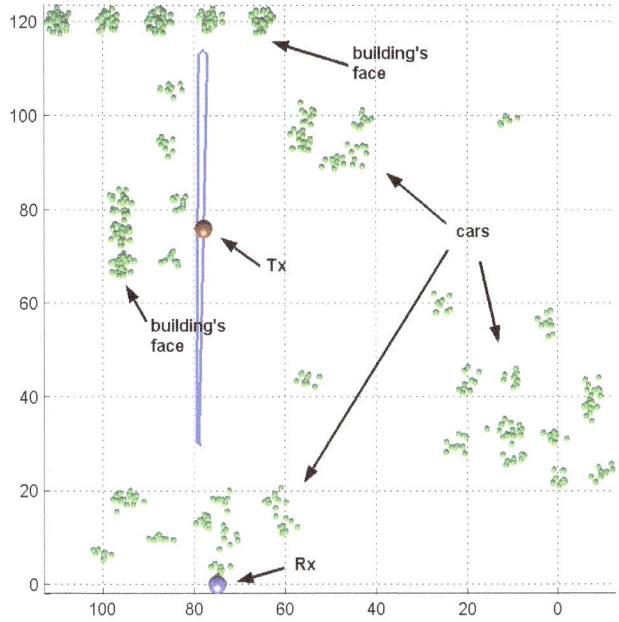

Fig. 9. Bird's eye view of the synthetic reconstruction of the scenario where the measurements were taken.

Fig. 10. The receiver (top circle) and the transmitting antenna (bottom circle) mounted on a cart.

In the proximity of the building the cart was turned around and pulled in the opposite direction (towards the receiver) approximately on the same path. The whole measurement took approximately 1 min at a speed of 3 m/s. The measurement device employed was the broadband real-time channel sounder RUSK MIMO from MEDAV (Thomä et al., 2001). The following parameters characterize the measurements:

- A bandwidth B of 120 MHz at a center frequency of 1.95 GHz sampled with $\Delta f = 300$ kHz for a total of 385 samples.

- A total of 6313 time snapshots were taken, one every 9.2 ms for a total of 58 s.

The same parameters were chosen for the synthetic IlmProp model so that the two data sets had same dimensions and sampling grids. However, in order to compare the measurements with the synthetic channels, adding the noise and performing a proper normalization must be taken into account.

The need for a normalization is due to the fact that the absolute powers sent by the antennas have not been considered. Only the relative difference in power between the LOS and the scattering components have been modeled through the Rician K-factor which can be set at will. This simplified the modelling effort even though the results need post-processing in order to be finally compared with the measurements. The normalization is performed in such a way that the total power received for the synthetic channel matches the one received in the measurements. Clearly both data sets must have the same number of samples, dimensions and more in general, the same sampling grids. If $\boldsymbol{H}_M$ denotes the channel for the measurements and $\boldsymbol{H}_I$ the one generated with the IlmProp then the whole channel matrix must be scaled so that

$$\sum_{\Theta} \|\boldsymbol{H}_M\|^2 = \sum_{\Theta} \|\boldsymbol{H}_I\|^2 \tag{13}$$

where Θ spans the complete domain of all dimensions. In our simulations $\boldsymbol{H}_I$ has been generated without noise. The normalization and the adding of the noise must then be done concurrently. First we have to estimate the power of the noise from the measurements. We assume constant AWGN (Additive White Gaussian Noise) over all dimensions (delay time, time, antennas, etc.). An effective way to measure the noise floor in the measurements is to observe the channel matrix in the time and delay time domain. If the path lengths are short enough, so that the last echoes extinguish before the maximum delay resolvable, we have a measurement of the noise without signal. If the number of samples is sufficient we can use this data to estimate the power of the noise. The noisy synthetic channel will then be

$$\boldsymbol{H}_{I,noisy} = \boldsymbol{H}_{I,noiseless} + \boldsymbol{H}_w, \tag{14}$$

where $\boldsymbol{H}_w$ is a matrix whose elements are ZMCSCG (Zero Mean Circular Symmetric Complex Gaussian) random numbers with variance σ^2 (Paulraj et al., 2003).

Equation (13) then becomes:

$$\sum_{\Theta} \|\boldsymbol{H}_M\|^2 - \gamma \cdot \sigma^2 = \sum_{\Theta} \|\boldsymbol{H}_I\|^2, \tag{15}$$

where γ is the total number of samples present in $\boldsymbol{H}$, i.e. $\gamma = M_R \cdot M_T \cdot T \cdot F$, where T and F are the number of time snapshots and frequency bins, respectively.

The geometry has been roughly reconstructed with the help of blueprints of the area and through visual inspection. Cars, buildings, light poles, and trash bins have been identified as the probable sources of scattering.

Figure 11 shows the channel impulse responses in time t and delay time τ. The channel has been gathered originally in the $\{t, f\}$ domain; the CIRs in delay time have been computed via the Fourier transform described in Eq. (1). The

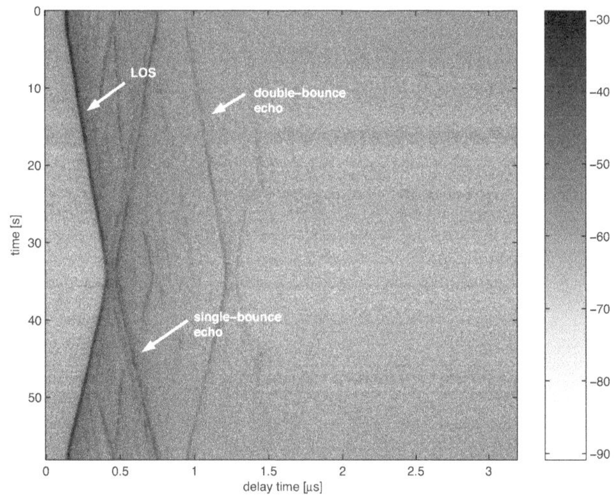

Fig. 11. Channel Impulse Responses from the measurements in time t and delay time τ. The Line Of Sight LOS and echos are clearly visible. The grayscale represents amplitudes in a dB scale

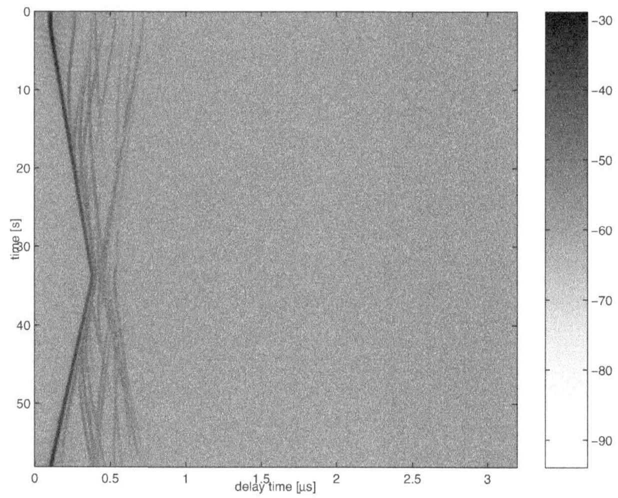

Fig. 12. The synthetically generated Channel Impulse Responses in time t and delay time τ.

LOS is clearly visible and as expected its delay lag grows as the transmitter moves further away from the receiver. When the cart is turned around and pulled back towards the receiver, the arrival delay time decreases. Many echoes are visible as well. Note that as the transmitter moves most echoes arrive earlier, meaning that their path length becomes smaller. When the curve corresponding to one echo changes in time with the same slope characterizing the Line Of Sight, then we can deduce that the path length of the echo changes with the same rate of the path length of the LOS. The strong echo labeled *double-bounce echo* has the same trend as the LOS just delayed by approximately 0.8 μs, which corresponds to 240 m, the distance between the top and bottom buildings doubled. For this reason it is very likely that this arrival corresponds to the signal reflected by the bottom building first, and then by the top building, thus a doble-bounce echo. Figure 12 shows the channel generated by our model. The trend of the LOS is extremely well reproduced. This was, however, expected since the LOS is the only component easily predictable once the position of the transmitter in known. Most of the scattering contributions can be observed in the synthetic data set as well, with the exception of the double-bounce paths which have been omitted. Analyzing the two data sets in the $\{t, \tau\}$ domain suggests improvements in the modelling approach: in the measurements we can observe that certain echo traces are stronger than others. In the modeled channel this does not happen. This is a result of the modelling of the reflection coefficients. In fact while their phase has been generated uniformly distributed in the $[0, 2\pi]$ range, the amplitude has been taken equal for all. The comparison suggests that the reflection coefficients need a more sophisticated modelling. Furthermore, a closer look on the measurements reveals that some echoes exist only in certain time periods. This suggests that the amplitude of the reflection coefficients (perhaps their phases too) should be

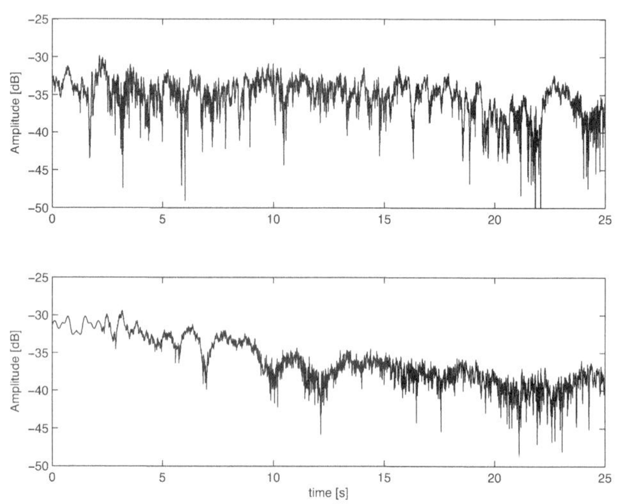

Fig. 13. Amplitudes of the CIRs for a specific frequency bin plotted in dB against time for the measurements (top) and the IlmProp model (bottom).

dependent on the Direction of Arrival (DoA) and Direction of Departure (DoD). Figure 13 shows the fluctuations of the amplitudes of the channel for a specific frequency f_o plotted against time for the first 25 s. Fast-fading characterizes both signals even though the measurements appear somehow more variant. In order to assess this feature quantitatively we compute the multipath intensity profile $\psi_{De}(\tau)$, the Doppler power spectrum $\psi_{Do}(f_D)$ and the autocorrelation functions in frequency acf(Δf) and time acf(Δt) for both data sets. From their analysis we can obtain the RMS delay spread τ_{RMS}, the coherence bandwidth $(\Delta f)_c$, the RMS Doppler spread $f_{D,\mathrm{RMS}}$ and the coherence time $(\Delta t)_c$ (Figs. 14 and 15). The RMS delay spread shows that the energy is similarly distributed in delay time τ. In the measurements two

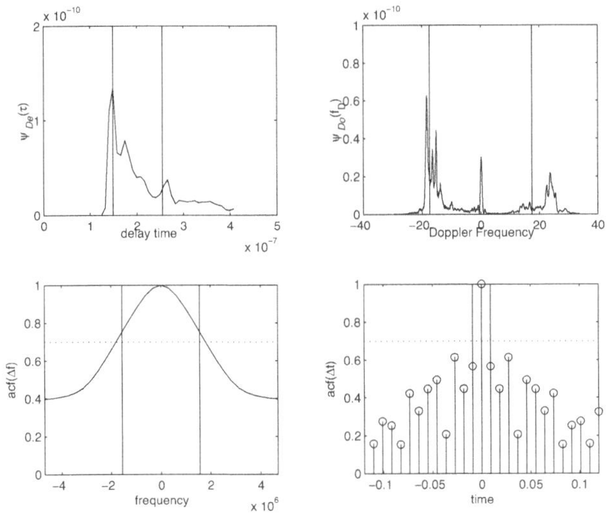

Fig. 14. The multipath intensity profile $\psi_{De}(\tau)$, Doppler power spectrum $\psi_{Do}(f_D)$ and the autocorrelation functions in frequency $\mathrm{acf}(\Delta f)$ and time $\mathrm{acf}(\Delta t)$ for the measurements

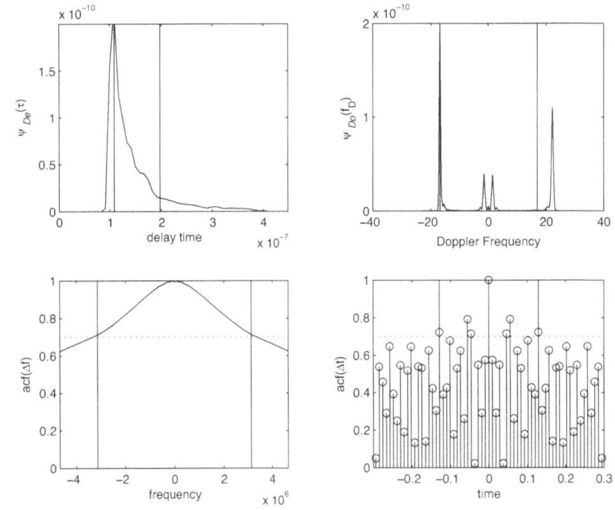

Fig. 15. The multipath intensity profile $\psi_{De}(\tau)$, Doppler power spectrum $\psi_{Do}(f_D)$ and the autocorrelation functions in frequency $\mathrm{acf}(\Delta f)$ and time $\mathrm{acf}(\Delta t)$ for the IlmProp model

more spikes can be observed. This is due to the fact that two specific echoes have a significantly stronger magnitude compared to other echoes. The RMS delay spread τ_{RMS} is anyhow very similar suggesting that the two channels will have a similar frequency selectiveness. The RMS delay spread corresponds to the distance between the two vertical lines plotted in Figs. 14 and 15. From the autocorrelation function in the frequency domain $\mathrm{acf}(\Delta f)$ we can derive the coherence bandwidth $(\Delta f)_c$ which confirms that the two channels have approximately the same coherence bandwidth. In the $\{t, f_D\}$ pair the two channels show more differences. While the Doppler power spectrum $\psi_{Do}(f_D)$ has very similar trends, the autocorrelation function in the time domain shows that the measured channel is much more time variant than the synthetic one. Furthermore this analysis proves that Eq. (9) only represents a good empirical law, valid in most but not all cases. In Fig. 13 we can in fact recognize that the measurements have a faster moving small-fading effect, especially during the first 10 s. We then investigated how we could generate more time variance changing only positions, the size of the scattering clusters, and the number of scatterers per cluster, discovering that this was not possible. Our simulations showed that in order to achieve higher order of time variance it is absolutely necessary to model time variant reflection coefficients or to introduce moving scatterers along the time dimension.

5 A deeper look at the phase response

Looking more closely at the phase of the received signal leads to interesting comparisons and remarks. Figure 16 shows how the phase of the received signal (in its complex base-band representation) is constructed in a flat-fading frequency selective channel. The resulting phase φ_R will be

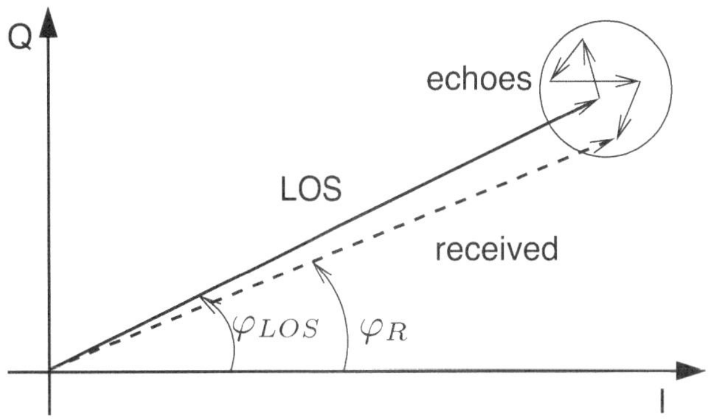

Fig. 16. When the amplitude of the echoes is small the phase of the received signal φ_R is a good approximation of the phase of the LOS component only φ_{LOS}.

the sum of the phase introduced by the LOS φ_{LOS} plus the additional phase introduced by the echoes. The smaller the amplitude of the echoes is compared to the LOS component, the smaller will be the difference $\varphi_R - \varphi_{LOS}$. As the radial distance between transmitter and receiver changes, the LOS component, φ_{LOS}, will change accordingly. If the movement is slow enough and the phase-noise introduced by the echoes is negligible, it is possible to derive the displacement from the phase trend of the received signal. In fact, a movement of λ in the radial direction corresponds to a full rotation of 2π radians in the phase. If this wavelength has been sampled with more than two samples per wavelength, thus fulfilling Shannon's theorem, the phase can be efficiently 'unwrapped', or in other words, absolute jumps greater than π are converted to their 2π complement. The spatial displacement can be then derived from the phase simply multiplying by the factor: $\frac{\lambda}{2\pi}$. Figure 17 shows the unwrapped phase for every frequency bin derived directly from the measured

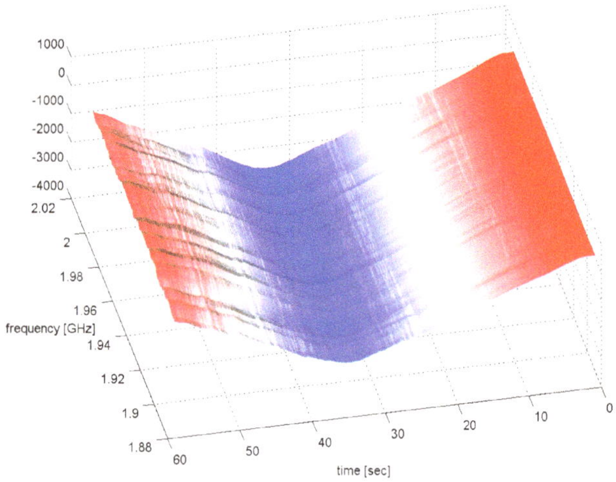

Fig. 17. Unwrapped phase trend derived from measurements plotted in radians against frequency f and time t.

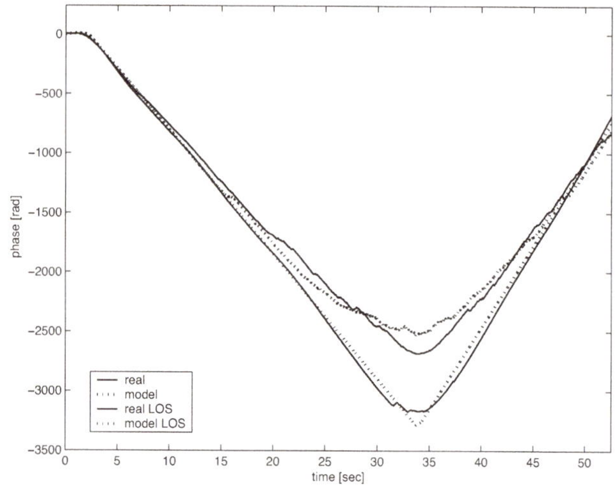

Fig. 18. Phase comparison between the measurements and its synthetic reconstruction. The upper curves are obtained simply unwrapping the signal received at one specific frequency. The lower curves represent the LOS component only.

channel analyzed in Sect. 4. Even though a different fluctuation perturbs each frequency bin, the overall trend is constant and shows a descending phase for half of the time and a symmetrical ascending phase for the other half, in accordance with the experiment's geometry. The different slope is due to the different velocity. In fact the transmitter moved faster in the second half of the experiment. If the channel were flat-fading we would see the slow moving LOS component only, absolutely equal for every frequency. It is interesting to compare the overall phase trend with the LOS component only. In case of the synthetic channel this is very easy to obtain. More complicated is the case of the measurements. Once the channel has been transformed in the $\{t, \tau\}$ domain, it is possible to identify the samples in delay time τ with highest energy for every time snapshot. The changing phase of these samples against time approximates the phase trend introduced by the LOS component only. The comparison of the unwrapped LOS phase with the overall phase is shown in Fig. 18. The model and the measurements show very similar traits. The upper curves show the phase trend for one frequency bin derived directly from the unwrapping of the phase. The lower curves are derived, respectively, with the LOS approximation for the measurements, and from a direct calculation of the displacement for the model. In fact, for the synthetic data, the exact path of the transmitter is known. From these coordinates it is possible to derive the radial displacement and consequentially the phase. The phase trends match for the first 10–15 s, while they start to diverge significantly around 25 s. This can be simply explained considering that the Rician K-factor drops in the middle part of the experiment. When the amplitude of the echoes is comparable with the LOS, then the received vector will have a phase uniformly distributed distributed in the interval $[0, 2\pi]$. The unwrapper will not find any trend and will jump from ascending to descending phases randomly. The resulting unwrapped signal will have a slope close to 0, that is horizontal. For this reason, as we pass from a higher K-factor to a smaller one we find increasingly less steep slopes. In other words the phase-noise introduced by the echoes is uniformly distributed, but due to the non-linearity of the unwrapper, the resulting error leads always to more moderate slopes.

6 Conclusions

Our simulations show that for certain modelling applications a simple geometry-based single-bounce model can mimic most channel features with sufficient precision, among which the frequency selectiveness and the time variance are the most characterizing. A rough positioning of the scattering clusters allows us to achieve the desired frequency selectiveness as well as other specific traits of the channel statistics by tuning three basic parameters, i.e, the size of the scattering clusters, the number of scattering coefficients, and the Rician K-factor. To obtain higher levels of time variance and more realistic synthetic channels, moving scatterers and time dependent reflection coefficients must be modeled as well.

Acknowledgements. The authors gratefully acknowledge the support of MEDAV (www.channelsounder.de) and of the Electronic Measurements Engineering Laboratory at Ilmenau University of Technology in performing the channel measurements.

References

Bello, P.: Characterization of randomly time-variant linear channels, IEEE Trans. Comm. Syst., 11, 360–393, 1963.

Correia, L. M.: Wireless Flexible Personalised Communications, Wiley, 2001.

Nabar, R. U.: Performance Analysis and Transmit Optimization for General MIMO Channels, Doctoral Dissertation, Stanford, 2003.

Paulraj, A., Nabar, R., and Gore, D.: Introduction to space-time wireless communications, Cambridge University Press, Cambridge, 2003.

Stüber, G.: Principles of Mobile Communication, Kluwer, Norwell, MA, 1996.

Thomä, R., Hampicke, D., Richter, A., Sommerkorn, G., and Trautwein, U.: MIMO vector channel sounder measurement for smart antenna system evaluation, European Transaction on Telecommunications, Special issue on smart antennas, 12, 2001.

Van Trees, H. L.: Optimum Array Processing: Part IV of Detection, Estimation, and Modulation Theory, John Wiley & Sons Inc, New York, 2002.

On the optimum antenna pattern for widebeam radar reflectivity estimation

F. Gerbl and E. M. Biebl

Fachgebiet Höchstfrequenztechnik, Technische Universität München, Munich, Germany

Abstract. Noise is a limiting factor in radar systems. The power received from a target depends on the target reflectivity varying with the aspect angle, the target range and the antenna pattern, amongst others. Inevitably, an estimation algorithm for the mean target reflectivity weights noise contributions from greater aspect angles stronger than such from smaller aspect angles due to the target range increasing with increasing aspect angle. However, as an opposed effect, due to the supposed equidistant rather than equiangular sampling along the linear aperture, noise contributions for greater aspect angles have lower influence than those for smaller aspect angles. In the paper, the signal-to-noise ratio for a linear, equidistantly sampled aperture and an arbitrary antenna pattern is determined. Moreover, an upper limit for the achievable signal-to-noise ratio is given, and the antenna pattern maximizing the signal-to-noise ratio is derived.

1 Introduction

Noise is a limiting factor in radar systems. The power received from a target depends on the target reflectivity varying with the aspect angle, the target range and the antenna pattern, amongst others. Directing power only into a narrow region of space, i.e. using a high gain antenna, yields higher received power and higher azimuth resolution than transmitting power to a broader region of space. However, using a narrow antenna beam to linearly scan a target scene limits the aspect angles, from which targets with generally unknown orientation are illuminated. Widening the antenna beam and processing the received signal in an appropriate manner, an image containing estimates of the mean target reflectivities over the aspect angle can be obtained. For wide antenna beams, the range of a target as measured from the radar system, and therefore the power received from even an omnidirectionally scattering target, varies as the radar system passes by the target.

Correspondence to: F. Gerbl
(gerbl@tum.de)

2 Setup

The radar system is moved along a straight path, passing the scene consisting of targets to be imaged. The orientation of the antenna remains unchanged throughout the passing. The system transmits a signal which is reflected by the objects in the scene and in turn received by the system. The amplitude and phase of the received signal are determined and sampled equidistantly along the straight path. The radar system is assumed to be monostatic. Figure 1 illustrates the setup and indicates the changes the signal undergoes during the passing. Distance and aspect angle and therefore level of free space attenuation, phase shift and the influence of the antenna pattern vary as the antenna passes by the target.

3 Coordinate system

The basis for subsequent considerations are coordinate systems as depicted in Fig. 2. The coordinate systems are antenna-centered. The trajectory of the target in the antenna-centered coordinate systems is a straight line in case the antenna is moved in a certain distance to the target on a straight line. It is assumed that the relative motion of the target occurs with respect to the x coordinate, i.e. y and z remain constant. Since antenna patterns are usually given in spherical coordinates, a spherical coordinate system is attached to the Cartesian one. The spherical coordinate system is oriented in such a way that its meridians intersect on the x axis. This orientation allows to describe the position of the target by only one varying angle coordinate, α, while the other an-

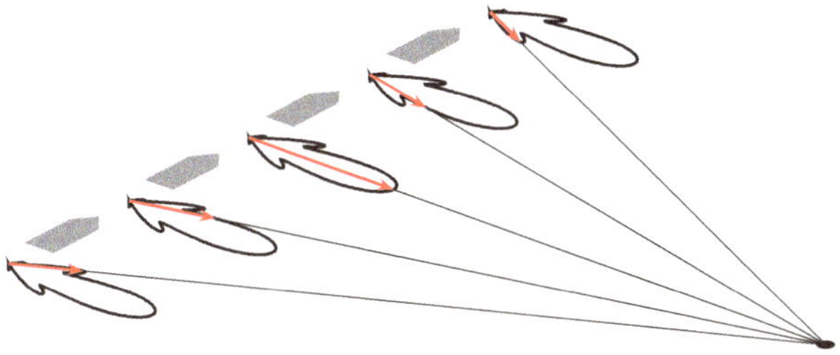

Fig. 1. Sketch of the setup.

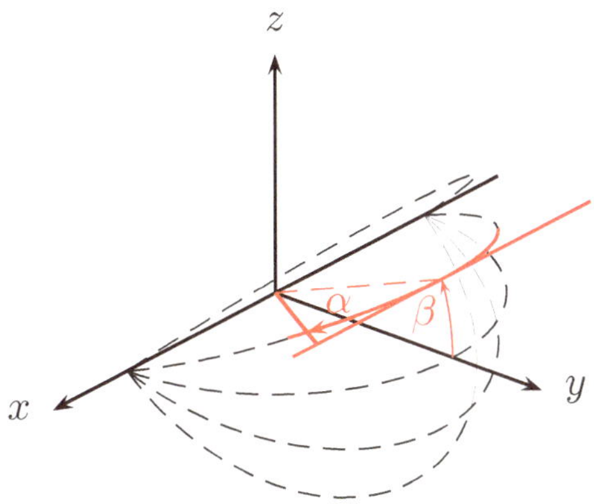

Fig. 2. Antenna-centered coordinate systems and trajectory of a target.

gle coordinate, β, remains constant at the value

$$\beta_t = \arctan \frac{z_t}{y_t}, \tag{1}$$

where y_t and z_t are the y and z coordinates of the target. In Fig. 2, a possible trajectory of a target is indicated with a red line parallel to the x axis.

4 Signal processing

The received signal is a function of the aspect angle α and time t and can be written as

$$s_{RX}(\alpha, t) = a \cdot \rho(\alpha) \cdot \frac{1}{r^2(\alpha)} \cdot D_A^2(\alpha) \cdot \exp\{-j2kr(\alpha)\} + n_{\Re}(t) + jn_{\Im}(t) \tag{2}$$

in the presence of noise, assuming there is only one single point-like target. a is a constant determined by system parameters like radiated power, mixer losses, and others. ρ is the reflectivity of the target, r the distance between antenna and target, D_A the antenna amplitude pattern, k the wave number, and $n_{\Re}(t)$ and $n_{\Im}(t)$ are the noise contributions in real and imaginary part of the received signal, respectively. Actually, the antenna pattern is a function of both α and β. However, since the antenna pattern is assumed to be constant over β except for a rectangular window that will be accounted for by a proper choice of integration limits, the pattern will be denoted as a function of only α instead of α and β in the majority of cases in order to avoid unnecessarily long expressions.

Compensating the aforementioned effects and integrating (in reality summarizing) the compensated signals yields a noisy estimate $\hat{\overline{\rho}}_{\alpha_0}$ of the mean reflectivity over the angle α within $\pm\alpha_0$

$$a \cdot 2\alpha_0 \cdot \hat{\overline{\rho}}_{\alpha_0} = \int_{-\alpha_0}^{\alpha_0} a \cdot \rho(\alpha) d\alpha + \int_{-\alpha_0}^{\alpha_0} n'(\alpha, t) d\alpha \tag{3}$$

with

$$n'(\alpha, t) = (n_{\Re}(t) + jn_{\Im}(t)) \cdot r^2(\alpha) \cdot \frac{1}{D_A^2(\alpha)} \cdot \exp\{j2kr(\alpha)\}, \tag{4}$$

where the estimate consists of the desired signal and a noise contribution. It is desirable to minimize the noise contribution in order to maximize the signal-to-noise ratio.

Throughout the paper, integrals are used for the discussion where possible rather than sums. There remains an uncertainty in the estimation of the mean reflectivity due to the non-zero distance between two samples. However, the spacing has to be chosen relatively small in order to avoid aliasing in case of wide antenna beams. It is assumed that the spacing is so small that sum and integral can be regarded as equal.

For that pixel of the reflectivity map that corresponds to the true target location, the signal-to-noise ratio of the processing result is

$$\mathrm{SNR} = \frac{S}{N}, \tag{5}$$

where

$$S = \left| \int_{-\alpha_0}^{\alpha_0} a \cdot \rho(\alpha) d\alpha \right|^2 = \left| a \cdot 2\alpha_0 \cdot \overline{\rho}_{\alpha_0} \right|^2 \tag{6}$$

is the power of the desired signal and

$$N = \mathrm{E}\left\{ \left| \int_{-\alpha_0}^{\alpha_0} n'(\alpha, t) d\alpha \right|^2 \right\}. \tag{7}$$

is the mean power of the noise contribution after processing. $\mathrm{E}\{\cdot\}$ denotes the expected value. Note that N is not the power of the noise contributions $n_{\Re}(t)$ and $n_{\Im}(t)$ but that of the noise in the processing result. It is shown in Appendix A that the SNR can be written as

$$\mathrm{SNR} = \frac{4\alpha_0^2 \left| a \cdot \overline{\rho}_{\alpha_0} \right|^2}{\Delta x \cdot r_0^3 \cdot P_n' \cdot \int_{-\alpha_0}^{\alpha_0} \frac{1}{D^2(\alpha)\cos^2(\alpha)} d\alpha}, \tag{8}$$

where $\overline{\rho}_{\alpha 0}$ is the true mean value of the target reflectivity within $\pm\alpha_0$, Δx the spatial distance between two samplings, and r_0 the minimum distance of the target to the x-axis. The power of the noise term $n_{\Re}(t)$ is assumed to be equal to that of $n_{\Im}(t)$. The sum of those noise powers is proportional to P'_{n}. $D(\alpha)=D_{\text{A}}^2(\alpha)$ is the antenna power pattern.

5 Optimization

From Eq. (8) it is obvious that the SNR is influenced by the antenna pattern. The antenna pattern is to be optimized in that sense that the SNR after the signal processing is maximized, i.e. the expression

$$\int_{-\alpha_0}^{\alpha_0} \frac{1}{D^2(\alpha)\cos^2(\alpha)} d\alpha \tag{9}$$

in the denominator of Eq. (8) is to be minimized. The boundary condition for the optimization is generally

$$\iint D(\Omega) d\Omega = 4\pi, \tag{10}$$

and particularly for the coordinate system used here

$$\int_{-\pi}^{\pi} \int_{-\pi/2}^{\pi/2} D(\alpha, \beta)\cos(\alpha) d\alpha d\beta = 4\pi. \tag{11}$$

i.e. the integral of the antenna pattern over all solid angle elements of a sphere has to be equal to 4π.

The first step to maximizing the SNR is to restrict the radiation to only those regions of space from where signals will be processed. For a given maximum processing angle α_0, the antenna pattern has to vanish for $|\alpha| > \alpha_0$. For the following considerations it is assumed that the antenna pattern does not vary with the angle β except for the fact that the pattern vanishes for $|\beta| > \beta_0$ resulting in a limited field of view, within which the SNR is independent of the target's location with respect to the related angle β. Under the prerequisites of a pattern vanishing for $|\alpha| > \alpha_0$ and $|\beta| > \beta_0$ and being constant with respect to β inside its support region, the boundary condition can be written as

$$2\beta_0 \int_{-\alpha_0}^{\alpha_0} D(\alpha)\cos(\alpha) d\alpha = 4\pi. \tag{12}$$

It is shown in Appendix B that the optimum antenna pattern, D_{opt,α_0}, for a given maximum processing angle α_0 that maximizes the SNR is

$$D_{\text{opt},\alpha_0}(\alpha) = \begin{cases} \frac{\pi}{\alpha_0\beta_0} \cdot \frac{1}{\cos(\alpha)} & \text{for } |\alpha| \le \alpha_0 \text{ and } |\beta| \le \beta_0 \\ 0 & \text{else} \end{cases} \tag{13}$$

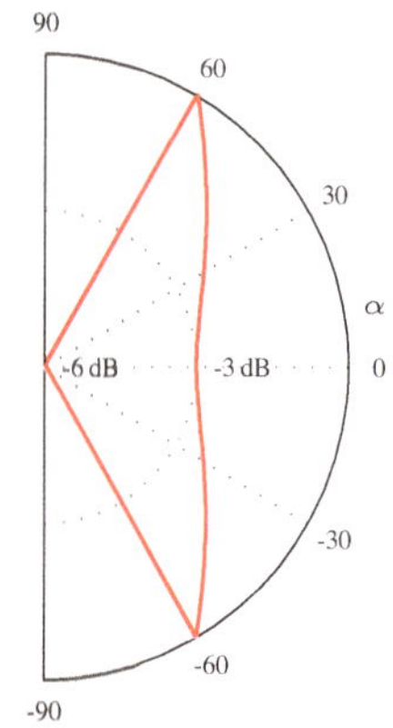

Fig. 3. Normalized optimum pattern $D_{\text{opt},\alpha_0}(\alpha)$ for $\alpha_0 = 60°$.

For every maximum processing angle α_0, a different antenna pattern is optimum. Using the respective pattern, the maximum achievable signal-to-noise ratio SNR_{max} can be determined with Eqs. (8) and (13) to be

$$\text{SNR}_{\text{max}} = \frac{2\pi^2 \left|a\overline{\rho}_{\alpha_0}\right|^2}{\Delta x \cdot r_0^3 \cdot P'_{\text{n}} \cdot \alpha_0 \cdot \beta_0^2}. \tag{14}$$

The maximum achievable signal-to-noise ratio decreases with increasing antenna beam width. Nevertheless, there might be reasons (e.g. decreasing the probability of missing a target that reflects strongly anisotropically, increasing azimuth resolution) to use wide antenna beams. The signal-to-noise ratio is only one figure of merit amongst others. Using the appropriate optimum pattern yields the maximum achievable SNR for a given α_0. Figure 3 shows exemplarily the optimum pattern for α_0=60°.

6 Benefit of the findings

Equation (8) allows the determination of the achievable SNR for a given antenna pattern and a certain maximum processing angle α_0 that might be dictated by special needs of a certain application. Equation (14) provides a benchmark for the antenna under consideration. The Figs. 4, 5 and 6 show three different antenna patterns and the resulting normalized signal-to-noise ratios for different maximum processing angles α_0 along with the normalized maximum achievable signal-to-noise ratios using the appropriate optimum patterns matched to the respective maximum processing angles. Reference for the normalization is the signal-to-noise ratio obtainable for α_0=90° with the respective optimum pattern.

Figure 4 shows the normalized pattern (red) of an antenna that radiates power uniformly into all directions of the hemisphere. Nominally, this antenna has a beam width of ±90°. However, its optimum maximum processing angle is $\alpha_0 \approx 54°$. The maximum signal-to-noise ratio achievable with

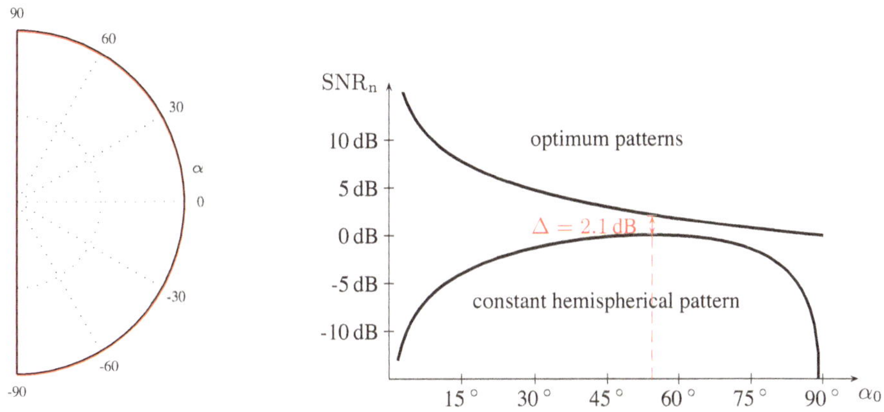

Fig. 4. Plot of a constant hemispherical pattern (red) and the resulting normalized signal-to-noise ratios compared to the optimum pattern.

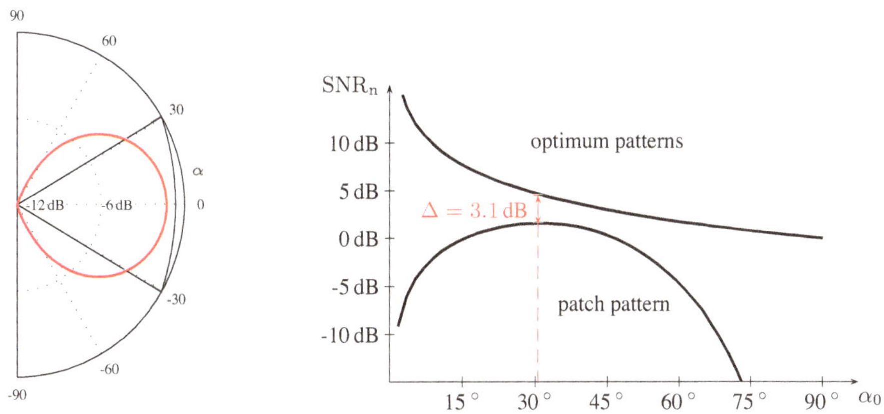

Fig. 5. Plot of the pattern of a patch antenna (red) and the resulting normalized signal-to-noise ratios.

this pattern is about 2.1 dB below that achievable with the respective optimum pattern.

Figure 5 shows the normalized pattern (red) of a patch antenna. Its optimum maximum processing angle is $\alpha_0 \approx 31°$. The pattern plot also shows the pattern (black) of the optimum pattern for $\alpha_0 = 31°$. The maximum signal-to-noise ratio achievable with the patch pattern is about 3.1 dB below that achievable with the respective optimum pattern.

Figure 6 shows a fictitious pattern (red) as it might emerge during the design process of an antenna. Its optimum maximum processing angle is $\alpha_0 \approx 31°$ as for the patch antenna considered before. The pattern plot also shows the pattern (black) of the optimum pattern for $\alpha_0 = 31°$. The maximum signal-to-noise ratio achievable with this fictitious pattern is about 1 dB below that achievable with the respective optimum pattern. The reason for the drop-off of the signal-to-noise ratio for angles above the optimum angle is that only a very small fraction of the energy is radiated into the respective directions and the compensation of signals received from those directions causes strong noise contributions due to a multiplication with the reciprocal of the almost vanishing antenna pattern.

The curve of the maximum achievable signal-to-noise ratio given by Eq. (14) can be used advantageously to judge the quality of the approximation of a desired antenna pattern by the current design.

7 Conclusions

In this paper, the signal-to-noise ratio of radar reflectivity estimation by sampling a linear aperture equidistantly and processing the received signals properly has been derived. It has been shown that the signal-to-noise ratio depends on the antenna pattern, amongst others. The optimum patterns, in that sense that they maximize the signal-to-noise ratio, as a function of the aspect angle and the maximum processing angle have been determined. The maximum achievable signal-to-noise ratio using the optimum patterns is stated. The findings

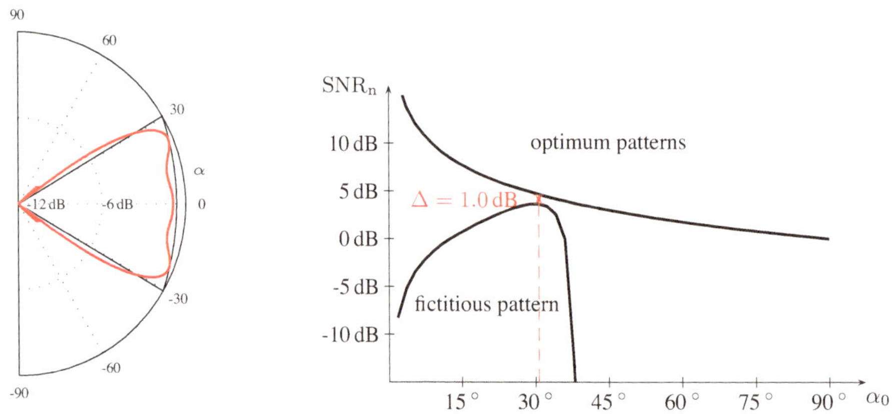

Fig. 6. Plot of a fictitious pattern (red) and the resulting normalized signal-to-noise ratios.

can be used to evaluate the performance of existing antennas for the task of estimating the radar reflectivity within different maximum processing angles. The results show that the signal-to-noise ratio achievable with a certain pattern as it might emerge in one of the iterations of an antenna design process does not depend on single values of the pattern but on an integral over a function of the antenna pattern. Thus, the difference of the signal-to-noise ratio achievable with the pattern under consideration and that achievable with the appropriate optimum pattern can be used as a figure of merit for the evaluation of an antenna pattern.

Appendix A Determination of the signal-to-noise ratio

Content of this section is the derivation of Eq. (8). Samples of the received signal are taken equidistantly at discrete positions along a straight line. Therefore, the number of samples per angle element is greater for greater angles α than for smaller ones. To account for this fact, the statistical evaluation of the noise contribution has to be done after the substitution of α by x. With

$$\alpha = \arctan\frac{x}{r_0}, \quad \frac{d\alpha}{dx} = \frac{r_0}{r_0^2 + x^2} \tag{A1}$$

Equation (7) can be written as

$$N = \mathrm{E}\left\{\left|\int_{-x_0}^{x_0} n'(\alpha(x), t)\frac{r_0}{r_0^2 + x^2}dx\right|^2\right\}. \tag{A2}$$

Using

$$r_0^2 + x^2 \equiv r^2(\alpha(x)) \tag{A3}$$

and

$$n''(\alpha(x), t) = \frac{n'(\alpha(x), t)}{r^2(\alpha(x))}, \tag{A4}$$

the noise power after processing reads as

$$N = r_0^2\,\mathrm{E}\left\{\left|\int_{-x_0}^{x_0} n''(\alpha(x), t)dx\right|^2\right\} \tag{A5}$$

and consequently

$$\begin{aligned} N &= r_0^2\,\mathrm{E}\left\{\left(\int_{-x_0}^{x_0} \Re\left\{n''(\alpha(x), t)\right\}dx\right)^2 \right. \\ &\qquad \left. + \left(\int_{-x_0}^{x_0} \Im\left\{n''(\alpha(x), t)\right\}dx\right)^2\right\} \\ &= r_0^2\left(\mathrm{E}\left\{X_1^2\right\} + \mathrm{E}\left\{X_2^2\right\}\right) \end{aligned} \tag{A6}$$

with

$$X_1 = \int_{-x_0}^{x_0} \Re\left\{n''(\alpha(x), t)\right\}dx \tag{A7}$$

and

$$X_2 = \int_{-x_0}^{x_0} \Im\left\{n''(\alpha(x), t)\right\}dx. \tag{A8}$$

The integrands are

$$\Re\left\{n''(\alpha(x), t)\right\} = \frac{n_\Re(t)\cos(2kr(\alpha(x))) - n_\Im(t)\sin(2kr(\alpha(x)))}{D(\alpha(x))} \tag{A9}$$

and

$$\Im\left\{n''(\alpha(x), t)\right\} = \frac{n_\Re(t)\sin(2kr(\alpha(x))) + n_\Im(t)\cos(2kr(\alpha(x)))}{D(\alpha(x))}. \tag{A10}$$

$n_{\Re}(t)$ and $n_{\Im}(t)$ are zero-mean. Therefore, the sum of several samples of those noise processes is zero-mean:

$$\mathrm{E}\{X_1\} = \mathrm{E}\{X_2\} = 0. \tag{A11}$$

Since generally

$$\mathrm{E}\left\{X^2\right\} = \mathrm{Var}\{X\} + \mathrm{E}^2\{X\} \tag{A12}$$

and under the prerequisite that $n_{\Re}(t)$ and $n_{\Im}(t)$ have the same properties, it can be stated that

$$\mathrm{E}\left\{X_1^2\right\} = \mathrm{E}\left\{X_2^2\right\} = \mathrm{Var}\{X_1\} = \mathrm{Var}\{X_2\}. \tag{A13}$$

Therefore,

$$N = 2r_0^2\,\mathrm{Var}\{X_1\}. \tag{A14}$$

For further evaluation, the integrals are temporarily approximately replaced by sums. The reason is that in the continuous integral two "samples" of the noise process are directly adjacent to each other and therefore correlated. In reality, samples are taken with non-zero intervals between them. Assuming that those intervals are great enough, which is satisfied in many cases in reality, two samples of the noise process can be treated as uncorrelated. With the substitution

$$f(x) = 2kr(\alpha(x)), \tag{A15}$$

the variance of X_1 can be evaluated as

$$\begin{aligned}
&\mathrm{Var}\{X_1\}\\
&= \mathrm{Var}\left\{\int_{-x_0}^{x_0} \frac{n_{\Re}(t)\cos(f(x)) - n_{\Im}(t)\sin(f(x))}{D(\alpha(x))}dx\right\}\\
&\approx \mathrm{Var}\left\{\sum_{l=-l_0}^{l_0} \frac{n_{\Re}(t)\cos(f(l\Delta x))}{D(\alpha(l\Delta x))}\Delta x - \sum_{l=-l_0}^{l_0} \frac{n_{\Im}(t)\sin(f(l\Delta x))}{D(\alpha(l\Delta x))}\Delta x\right\}\\
&= \Delta x^2\,\mathrm{Var}\left\{\sum_{l=-l_0}^{l_0} \frac{n_{\Re}(t)\cos(f(l\Delta x))}{D(\alpha(l\Delta x))}\right\} + \Delta x^2\,\mathrm{Var}\left\{\sum_{l=-l_0}^{l_0} \frac{-n_{\Im}(t)\cos(f(l\Delta x))}{D(\alpha(l\Delta x))}\right\}\\
&= \Delta x^2 \sum_{l=-l_0}^{l_0} \mathrm{Var}\left\{\frac{n_{\Re}(t)\cos(f(l\Delta x))}{D(\alpha(l\Delta x))}\right\} + \Delta x^2 \sum_{l=-l_0}^{l_0} \mathrm{Var}\left\{\frac{-n_{\Im}(t)\cos(f(l\Delta x))}{D(\alpha(l\Delta x))}\right\}\\
&= \Delta x^2\,\mathrm{Var}\{n_{\Re}(t)\} \sum_{l=-l_0}^{l_0} \left(\frac{\cos(f(l\Delta x))}{D(\alpha(l\Delta x))}\right)^2 + \Delta x^2\,\mathrm{Var}\{n_{\Im}(t)\} \sum_{l=-l_0}^{l_0} \left(-\frac{\sin(f(l\Delta x))}{D(\alpha(l\Delta x))}\right)^2\\
&= \frac{1}{2}\cdot\Delta x\cdot P_n'\cdot \sum_{l=-l_0}^{l_0} \frac{1}{D^2(\alpha(l\Delta x))}\Delta x\\
&\approx \frac{1}{2}\cdot\Delta x\cdot P_n'\cdot \int_{-x_0}^{x_0} \frac{1}{D^2(\alpha(x))}dx.
\end{aligned} \tag{A16}$$

With the substitution

$$x = r_0\tan\alpha, \quad dx = \frac{r_0}{\cos^2(\alpha)}d\alpha \tag{A17}$$

and Eqs. (8), (6), (7), (A16), the the maximum achievable signal-to-noise ratio can be written as in Eq. (14).

Appendix B Optimization

Content of this section is the minimization of expression (9) with the boundary condition (12) by means of Lagrange multipliers and the calculus of variations. With the substitution

$$y(\alpha) = D(\alpha)\cos(\alpha), \tag{B1}$$

the Lagrange functional

$$\begin{aligned}
F(\alpha) &= \int_{-\alpha_0}^{\alpha_0} \left(\frac{1}{y^2(\alpha)} + \lambda y(\alpha)\right)d\alpha - \lambda\frac{2\pi}{\beta_0}\\
&= \int_{-\alpha_0}^{\alpha_0} f(y(\alpha))d\alpha - \lambda\frac{2\pi}{\beta_0}
\end{aligned} \tag{B2}$$

with

$$f(y(\alpha)) = \frac{1}{y^2(\alpha)} + \lambda y(\alpha) \tag{B3}$$

can be formed. A necessary condition for D being an extremal is

$$f_y = 0. \tag{B4}$$

This condition is fulfilled by

$$D(\alpha) = \left(\frac{2}{\lambda}\right)^{\frac{1}{3}}\cdot\frac{1}{\cos(\alpha)}. \tag{B5}$$

Using the boundary condition (12), λ can be determined to be

$$\lambda = 2\left(\frac{\beta_0\alpha_0}{\pi}\right)^3. \tag{B6}$$

With Eq. (B5), the optimum pattern as given in Eq. (13) follows.

Acknowledgement. The authors would like to thank K. Frick, Institute of Computer Science, University of Innsbruck, Austria, for the fruitful correspondence regarding the optimization method.

References

Ansorge, R. and Oberle, H. J.: Mathematik für Ingenieure, Band 2, Wiley, Berlin, 2000.

Gerbl, F., Biebl, E. M.: SNR Considerations for Widebeam, Short-Range Synthetic Aperture Radar Processing, GeMiC, 2006.

Orfanidis, S. J.: www.ece.rutgers.edu/~orfanidi/ewa/ch14.pdf, 2004.

10

Advances in hybrid finite element – boundary integral – multilevel fast multipole – uniform geometrical theory of diffraction method

A. Tzoulis[1] **and T. F. Eibert**[2]

[1]FGAN-FHR, Neuenahrer Str. 20, 53343 Wachtberg, Germany
[2]Institute of Radio Frequency Technology, Universität Stuttgart, 70550 Stuttgart, Germany

Abstract. Numerical modeling of problems including composite metallic/dielectric objects with arbitrary shapes and electrically large conducting objects within a common environment is performed in an optimum way with the recently developed powerful hybrid numerical method, which combines the Finite Element Boundary Integral (FEBI) method and the Multilevel Fast Multipole Method (MLFMM) with the Uniform Geometrical Theory of Diffraction (UTD), giving full electromagnetic coupling between all involved objects. In this contribution, the hybrid FEBI-MLFMM-UTD method is extended to double diffracted fields on pairs of straight metallic edges, formulated with the hard and soft scalar diffraction coefficients of UTD. The diffraction points on each pair of edges are determined by an iterative three-dimensional parametric realization of the generalized Fermat's principle. The divergence factor of the double diffracted field is computed by multiplying the appropriate divergence factors of the single diffracted UTD fields on each edge for the particular case. Thereby, the ray caustic distance of the diffracted field at the second edge is determined by linear interpolation between the radii of curvature in the two principal planes of the incident astigmatic ray tube. Further, fast near-field computation in the postprocessing stage of the hybrid method is extended in each translation domain to ray optical contributions due to the presence of electrically large objects, according to the hybridization of MLFMM with UTD. Formulations and numerical results will be presented.

Correspondence to: A. Tzoulis
(tzoulis@fgan.de)

1 Introduction

Recently, a powerful hybrid numerical method was introduced, which combines the Finite Element Boundary Integral (FEBI) method and the Multilevel Fast Multipole Method (MLFMM) with the Uniform Geometrical Theory of Diffraction (UTD), giving full electromagnetic coupling between composite metallic/dielectric objects with arbitrary shapes and electrically large conducting objects within a common environment (Tzoulis, Oct. 2005). In that work, single and multiple reflections on flat metallic objects were considered, whereas diffractions were taken into account only by single mechanisms on straight metallic edges.

In this contribution, the hybrid FEBI-MLFMM-UTD method is extended to double diffracted fields on pairs of straight metallic edges, formulated with the scalar hard and soft diffraction coefficients of UTD according to (Kouyoumjian, 1974). For configurations, in which the second diffraction point is in the transition region of the first, the scalar diffraction coefficients of UTD are not valid, because the incident field on the second edge is not purely ray optical, by means of rapid spatial variations due to the Fresnel transition functions in the diffraction coefficients of the first edge. Double diffracted field formulations, which are valid in transition regions of the first edge, have been developed in (Tiberio et al., 1989)–(Albani, 2005) using several approaches, such as a spectral extension of the UTD or spherical spectral synthesis.

In the present implementation of standard double diffracted UTD fields, the diffraction points on the pair of edges are determined by an iterative three-dimensional parametric realization of the generalized Fermat's principle, allowing treatment of non-coplanar and skewed edges. According to this, the position vectors of the diffraction points are expressed in terms of two parameters defined on the edges. On each edge, the diffraction point is computed in closed form for a given set of source and observation point,

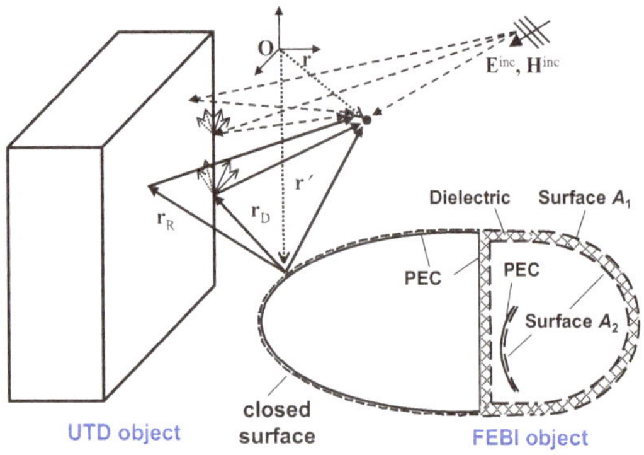

Fig. 1. Hybrid FEBI-MLFMM-UTD configuration.

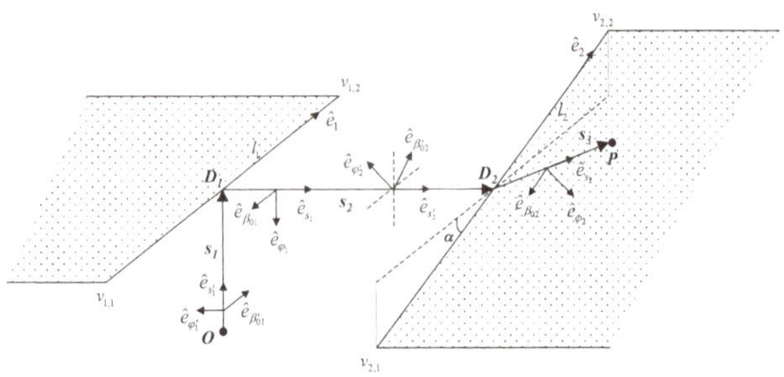

Fig. 2. Double diffraction edge configuration.

where the corresponding parameter is determined by a three-dimensional realization of the generalized Fermat's principle. For double diffraction, the above procedure is repeated iteratively for each edge, using each time a point on the other edge as source or observation point, respectively, until the law of diffraction is satisfied on both edges with desired accuracy (Ivrissimtzis, 1991).

The divergence factor of the double diffracted field is determined by multiplying the appropriate divergence factors of the single diffracted UTD fields on each edge for the particular case. Thereby, the ray caustic distance of each single diffracted field is needed and for straight metallic edges it is given by the radius of curvature of the incident wavefront in the plane of incidence. For the diffracted field at the second edge, the incident wave is an astigmatic ray tube and the ray caustic distance is determined by linear interpolation between the radii of curvature in the two principal planes of the incident astigmatic ray tube.

Further, near-field computations in the postprocessing stage of the hybrid method are accelerated by MLFMM according to (Tzoulis, 2005), in which high-frequency contributions due to the presence of electrically large objects are taken into account within each translation domain according to the hybridization of MLFMM with UTD introduced in (Tzoulis, Oct. 2005). In the following, formulations will be given and numerical results will be shown.

2 Formulation

The configuration for the hybrid FEBI-MLFMM-UTD concept can be seen in Fig. 1. Composite metallic/dielectric objects are treated with the FEBI technique, in which fast integral equation solution is achieved by MLFMM. Electrically large metallic objects, which are in the same environment, are treated with UTD and full coupling with the surface currents of the Boundary Integral (BI) part is given according to the hybridization introduced in (Tzoulis, Oct. 2005). In that work, single and multiple reflections on flat metallic objects were considered, whereas diffractions were taken into account only by single mechanisms on straight metallic edges.

In the hybrid method, double diffracted UTD field contributions on pairs of edges are taken into account according to the configuration shown in Fig. 2. The edges of the pair can be in general non-coplanar and skewed by an angle α. The diffraction points D_1 on edge 1 and D_2 on edge 2 are numerically determined by an iterative three-dimensional parametric realization of the generalized Fermat's principle. According to this, the position vectors of the diffraction points are expressed in terms of two parameters defined on the edges as

$$\mathbf{r}_{D_j} = \mathbf{r}_{v_{j,1}} + p_j l_j \hat{e}_j, \qquad j = 1,2 \tag{1}$$

where $\mathbf{r}_{v_{j,1}}$ and l_j are the position vector of the first vertex and the length of edge j, respectively. The parameter p_j is defined on edge j and takes values between [0, 1]. Also, $\hat{e}_j$ is the unit vector along edge j defined from the first to the second vertex of the edge. On each edge j, the diffraction point $\mathbf{r}_{D_j}$ is computed in closed form for a given set of source and observation point, by a three-dimensional realization of the generalized Fermat's principle, as shown in Fig. 3. According to this, the parameter p_j is given by

$$p_j = \frac{x_d + x_1}{l_j}, \tag{2}$$

with

$$x_d = \frac{h_1}{h_1 + h_2}(x_2 - x_1). \tag{3}$$

h_1 and h_2 are the distances of the source and observation point from the edge, respectively. Also, x_1 and x_2 are the distances of the projection of the source and observation point from the first vertex of the edge, respectively. For double diffraction, the above procedure is repeated iteratively for each edge, using each time a point on the other edge as source or observation point, respectively, until the law of diffraction is satisfied on both edges with desired accuracy (Ivrissimtzis, 1991). This is typically achieved after only few iterations.

The double diffracted field at the observation point P has the form

$$\mathbf{E}^{dd}(P) = \bar{\mathbf{D}}_2 \cdot \bar{\mathbf{D}}_1 \cdot \mathbf{E}^i(D_1) A_1 A_2 \mathrm{e}^{-jk(s_2+s_3)}, \tag{4}$$

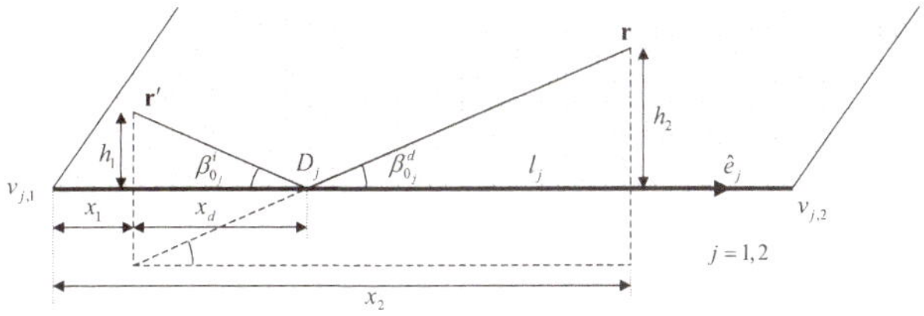

Fig. 3. Computation of diffraction point on each edge.

where $\mathbf{E}^i(D_1)$ is the incident field at D_1 and

$$\bar{\mathbf{D}}_j = -D_{\parallel_j}\hat{e}_{\beta_{0j}}\hat{e}_{\beta'_{0j}} - D_{\perp_j}\hat{e}_{\varphi_j}\hat{e}_{\varphi'_j}, \qquad j = 1, 2 \tag{5}$$

are the diffraction tensors, in which $\hat{e}_{\beta'_{01,2}}$,$\hat{e}_{\varphi'_{1,2}}$ and $\hat{e}_{\beta_{01,2}}$,$\hat{e}_{\varphi_{1,2}}$ are the transversal components in the edge-fixed coordinate system of incidence and diffraction on edge 1 or edge 2, respectively, as shown in Fig. 2. The double diffracted field is formulated with the scalar diffraction coefficients $D_{\parallel}$,$D_{\perp}$ of UTD (Kouyoumjian, 1974). In addition,

$$A_1 = \sqrt{\frac{s_1}{s_2(s_2 + s_1)}} \tag{6}$$

is the divergence factor of the diffracted field at edge 1. In that, the ray caustic distance ρ_{e_1} is assumed to be equal to s_1, which stands for the case of incident spherical wavefront on a straight edge. Further,

$$A_2 = \sqrt{\frac{\rho_{e_2}}{s_3(s_3 + \rho_{e_2})}} \tag{7}$$

is the divergence factor of the diffracted field at edge 2. The incident field on edge 2 is the diffracted field at edge 1, which is an astigmatic ray field, as shown in Fig. 4.

The ray caustic distance ρ_{e_2} of the diffracted field at edge 2 is determined from the radius of curvature of the incident field in the plane of incidence, defined by $\hat{e}_{s'_2}$ and $\hat{e}_2$. In case of non-coplanar and skewed edges, the plane of incidence at edge 2 is in general somewhere in between the two principal planes of the incident astigmatic ray tube. Consequently, the ray caustic distance ρ_{e_2} depends on the radii of curvature $\rho^i_{e_{21}}$, $\rho^i_{e_{22}}$ of the incident wavefront in both principal planes, which correspond to the first and second caustic of the incident ray field at edge 2, respectively. It is obvious, that for coplanar edges (α=0), the plane of incidence at edge 2 coincides with the principal plane corresponding to the second caustic of the incident ray tube and for perpendicular edges ($\alpha=\frac{\pi}{2}$), the plane of incidence at edge 2 coincides with the principal plane corresponding to the first caustic of the incident ray tube, which lies on edge 1. The ray caustic distance ρ_{e_2} for values of α between 0 and $\frac{\pi}{2}$ is determined by linear interpolation between these two principal values of ρ_{e_2}, resulting into

$$\rho_{e_2} = s_1 + s_2 - \frac{2\alpha}{\pi}s_1, \tag{8}$$

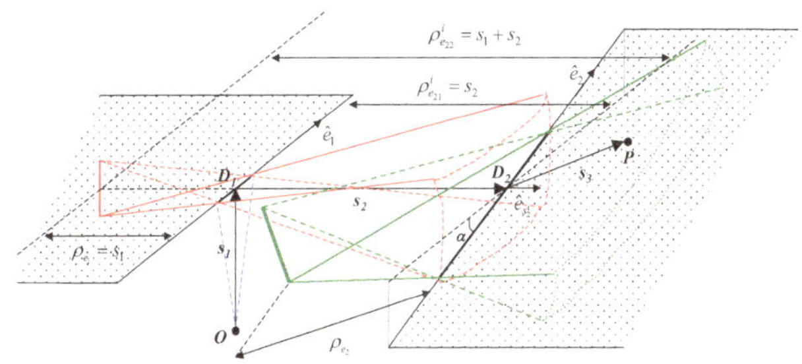

Fig. 4. Ray tubes of double diffraction.

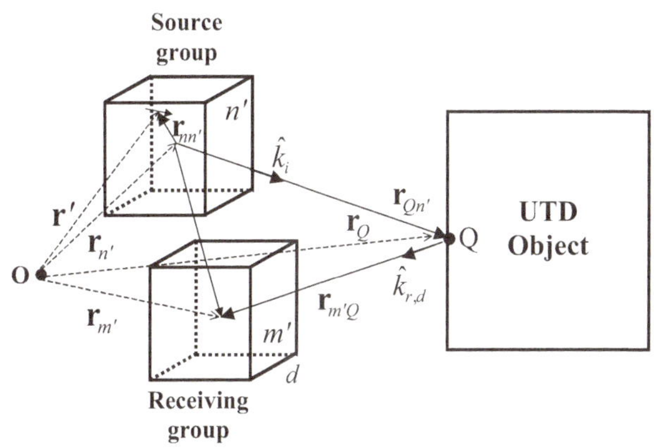

Fig. 5. Hybrid MLFMM-UTD configuration.

where α is given in radians. Consequently, the divergence factor of the diffracted field at edge 2 is given by

$$A_2 = \sqrt{\frac{s_1 + s_2 - \frac{2\alpha}{\pi}s_1}{s_3(s_3 + s_1 + s_2 - \frac{2\alpha}{\pi}s_1)}}. \tag{9}$$

Finally, near-field computations in the postprocessing stage of the hybrid method are accelerated by the MLFMM approach presented in (Tzoulis, 2005), where additional ray optical contributions due to the presence of electrically large objects are taken into account in both translation domains according to the hybridization of MLFMM with UTD introduced in (Tzoulis, Oct. 2005). The total electric field at an observation point $\mathbf{r}$ in the near-field region of a perfectly conducting arbitrarily shaped object, which is in the same environment with electrically large objects, is computed in the postprocessing stage by

$$\mathbf{E}_{tot}(\mathbf{r}) = \mathbf{E}(\mathbf{r}) + \mathbf{E}_{UTD}(\mathbf{r}), \tag{10}$$

where $\mathbf{E}(\mathbf{r})$ is the direct field contribution given in (Tzoulis, 2005) and $\mathbf{E}_{UTD}(\mathbf{r})$ are the ray optical field contributions received at the observation point due to the presence of the electrically large objects. For field points located inside the domain covered by MLFMM groups, the received ray optical contributions due to the presence of the UTD objects are

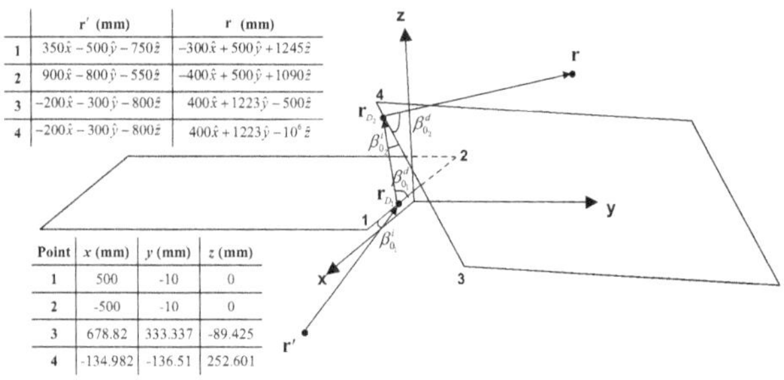

Fig. 6. General example of edges skewed in all directions.

Table 1. Double diffraction points for edge configuration and source/observation points shown in Fig. 6.

	$\mathbf{r}_{D_1}$ (mm)	$\mathbf{r}_{D_2}$ (mm)
$\mathbf{r}_1', \mathbf{r}_1$	$165.0\hat{x}-10\hat{y}+0.0\hat{z}$	$135.7\hat{x}+19.8\hat{y}+138.8\hat{z}$
$\mathbf{r}_2', \mathbf{r}_2$	$279.8\hat{x}-10\hat{y}+0.0\hat{z}$	$195.6\hat{x}+54.4\hat{y}+113.7\hat{z}$
$\mathbf{r}_3', \mathbf{r}_3$	$241.1\hat{x}-10\hat{y}+0.0\hat{z}$	$318.2\hat{x}+125.1\hat{y}+62.1\hat{z}$
$\mathbf{r}_4', \mathbf{r}_4$	$167.2\hat{x}-10\hat{y}+0.0\hat{z}$	$223.3\hat{x}+70.4\hat{y}+10.2\hat{z}$

given by

$$
\begin{aligned}
&\mathbf{E}_{UTD}^{NF}(\mathbf{r}) = \\
&-j\frac{\omega\mu}{4\pi} e^{-j\mathbf{k}_r\cdot\mathbf{r}_{mm'}} \sum_s A_{R_s} \bar{\mathbf{R}}_s^E T_L^{UTD}(kr_{Qn'}) \\
&\qquad \cdot \left(\bar{\mathbf{I}} - \hat{k}_r\hat{k}_i\right) \cdot \sum_n J_n \tilde{\boldsymbol{\beta}}_n(\hat{k}_i) \\
&-j\frac{\omega\mu}{4\pi} e^{-j\mathbf{k}_d\cdot\mathbf{r}_{mm'}} \sum_v A_{D_v} \bar{\mathbf{D}}_v^E T_L^{UTD}(kr_{Qn'}) \\
&\qquad \cdot \left(\bar{\mathbf{I}} - \hat{k}_d\hat{k}_i\right) \cdot \sum_n J_n \tilde{\boldsymbol{\beta}}_n(\hat{k}_i) - \cdots \\
&+ \mathbf{E}_{UTD}^{inc}(\mathbf{r}),
\end{aligned}
\tag{11}
$$

where

$$
\tilde{\mathbf{J}}_n(\hat{k}) = J_n \tilde{\boldsymbol{\beta}}_n(\hat{k}) = J_n \iint_A \boldsymbol{\beta}_n(\mathbf{r}_n) e^{j\mathbf{k}\cdot\mathbf{r}_{nn'}} da' \tag{12}
$$

is the $\hat{k}$-space representation of the electric current densities on the surface elements with $\mathbf{k}=k\hat{k}$. The distance vectors $\mathbf{r}_{nn'}$ and $\mathbf{r}_{mm'}$, as well as the distance $r_{Qn'}$ can be seen in Fig. 5. $T_L^{UTD}(kr_{Qn'})$ is the far-field MLFMM translation operator used to translate outgoing ray optical fields from source groups to incoming ray optical fields at receiving groups (Tzoulis, Oct. 2005). A_{R_s} and A_{D_v} are the divergence and phase factors for reflection and diffraction, respectively, and $\bar{\mathbf{R}}_s^E$ and $\bar{\mathbf{D}}_v^E$ the dyadic reflection and diffraction coefficients for electric field defined by basic GO and UTD concepts (Kouyoumjian, 1974). For observation points outside the MLFMM grouping domain, the ray optical field contributions become

$$
\begin{aligned}
&\mathbf{E}_{UTD}^{FF}(\mathbf{r}) = \\
&-j\frac{\omega\mu}{4\pi} \sum_s A_{R_s} \bar{\mathbf{R}}_s^E T_L^{UTD}(kr_{Qn'}) \\
&\qquad \cdot \left(\bar{\mathbf{I}} - \hat{k}_r\hat{k}_i\right) \cdot \sum_n J_n \tilde{\boldsymbol{\beta}}_n(\hat{k}_i) \\
&-j\frac{\omega\mu}{4\pi} \sum_v A_{D_v} \bar{\mathbf{D}}_v^E T_L^{UTD}(kr_{Qn'}) \\
&\qquad \cdot \left(\bar{\mathbf{I}} - \hat{k}_d\hat{k}_i\right) \cdot \sum_n J_n \tilde{\boldsymbol{\beta}}_n(\hat{k}_i) - \cdots \\
&+ \mathbf{E}_{UTD}^{inc}(\mathbf{r}),
\end{aligned}
\tag{13}
$$

where

$$
\begin{aligned}
\mathbf{E}_{UTD}^{inc}(\mathbf{r}) &= \sum_s A_{R_s} \bar{\mathbf{R}}_s^E \cdot \mathbf{E}^{inc}(\mathbf{r}_{R_s}) \\
&\quad + \sum_v A_{D_v} \bar{\mathbf{D}}_v^E \cdot \mathbf{E}^{inc}(\mathbf{r}_{D_v}) + \cdots
\end{aligned}
\tag{14}
$$

are the ray optical contributions of the incident electric field received at the observation point $\mathbf{r}$. The field contributions for magnetic currents, as well as the formulations for the magnetic field in the postprocessing stage can be given in the same way.

Table 2. Angles of incidence and diffraction for configuration shown in Fig. 6, using the computed diffraction points of Table 1.

	$\beta_{0_1}^i$ (°)	$\beta_{0_1}^d$ (°)	$\beta_{0_2}^i$ (°)	$\beta_{0_2}^d$ (°)
$\mathbf{r}_1', \mathbf{r}_1$	78.33	78.33	66.69	66.69
$\mathbf{r}_2', \mathbf{r}_2$	57.20	57.20	60.24	60.24
$\mathbf{r}_3', \mathbf{r}_3$	117.40	117.40	128.80	128.80
$\mathbf{r}_4', \mathbf{r}_4$	113.35	113.35	110.04	110.04

3 Numerical examples

First, the computation of double diffraction points is tested with the general skewed edge configuration shown in Fig. 6. In the same figure, the Cartesian coordinates of the vertices of the edges, as well as the sets of source and observation points for the determination of the double diffraction points can be seen. The computed diffraction points are shown in Table 1 and the particular angles of incidence and diffraction are shown in Table 2. It can be seen, that the double diffraction points are determined correctly, since in each case the law of diffraction is satisfied.

Second example is the problem of the double diffraction of a plane wave on a pair of skewed edges, formed by two plates placed with a relative angle of 5°, as shown in Fig. 7. As a

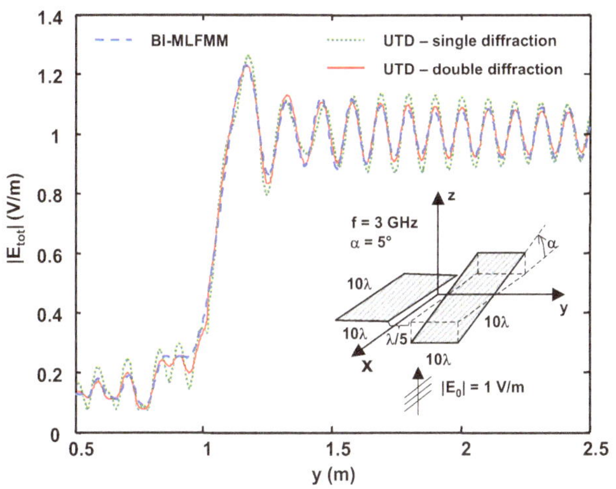

Fig. 7. Double diffraction of plane wave on pair of skewed edges.

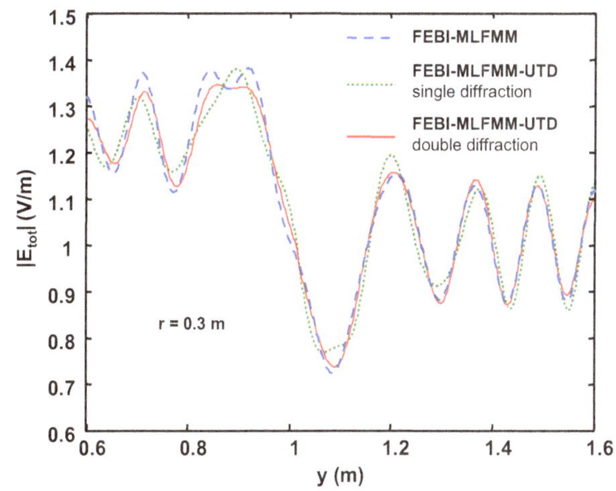

Fig. 9. Electric near-field distribution along direction of observation for configuration shown in Fig. 8.

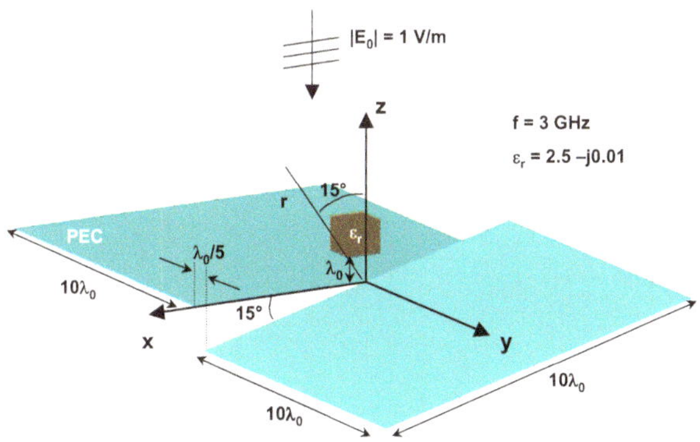

Fig. 8. Dielectric cube over pair of skewed edges.

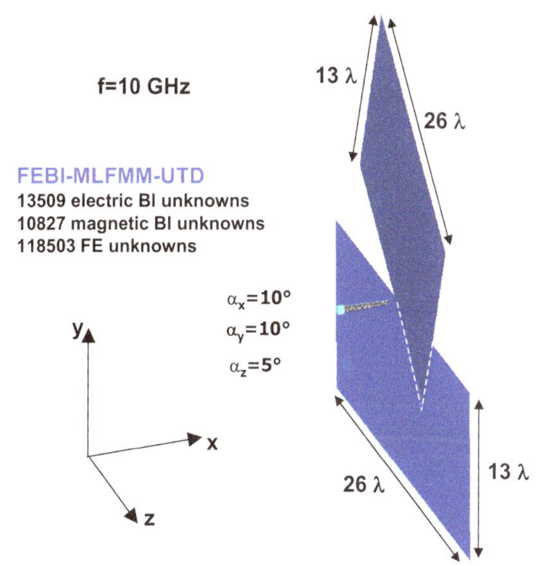

Fig. 10. Dielectric rod antenna radiating in front of skewed flat plates.

reference, the full BI solution is used and the ray optical solution was achieved with the hybrid method by activating only the UTD part. In the same figure, the electric near-field distribution along the y-axis and 5° of the z-axis is shown, for a distance of 0.3 m from the y-axis. This direction corresponds to the Keller cone of the right edge. It can be seen, that considering double diffracted field contributions on the pair of edges, the UTD result shows excellent agreement compared to the reference solution.

Next example is shown in Fig. 8. It consists of a dielectric cube placed over a pair of edges with relative angle of 15°. The dielectric constant of the cube was $\varepsilon_r = 2.5 - j0.01$ and the frequency $f = 3$ GHz. The problem was excited by a -z-traveling plane wave. In this case, the electric field is observed in a direction along the y-axis and 15° of the z-axis for a distance of $r = 0.3$ m. This direction corresponds to the Keller cone of the right edge of the pair. That way, the field first diffracted on the left edge and then on the right edge of the pair is observed. In Fig. 9, the electric near-field distribution along this direction can be seen. A significant improvement of the result in the case of considering the double diffracted contribution, compared to the full FEBI reference solution, can be observed.

Last example is a dielectric rod antenna radiating in front of two skewed flat plates, placed as shown in Fig. 10. The operating frequency of the antenna was 10 GHz and a delta-gap voltage source was used for excitation. The length of the rod was $3\lambda_0$ with $\varepsilon_r = 2.5$, where λ_0 is the free-space wavelength, and a metallic mounting was used to hold the rod. The small radius of the rod was $0.24\lambda_0$ and the large radius at the connection point with the metallic mounting was $0.45\lambda_0$. As a reference, the full FEBI solution is used and with the hybrid approach two simulations where performed

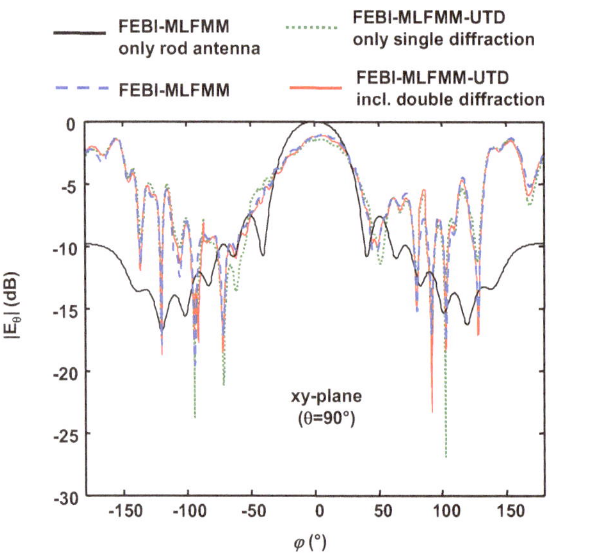

Fig. 11. Radiation pattern in xy-plane of the antenna shown in Fig. 10.

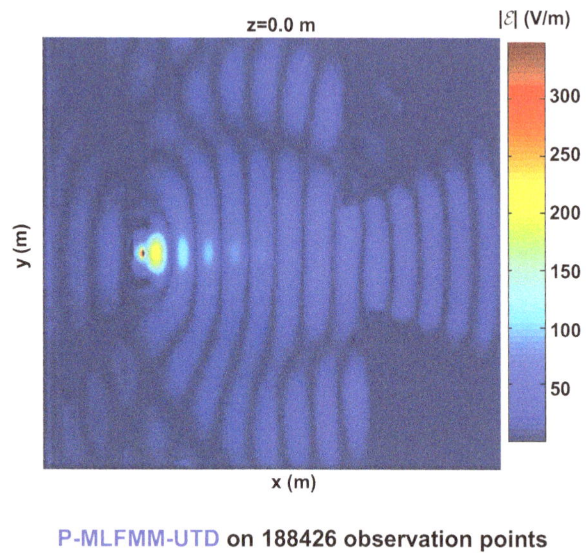

Fig. 12. Instantaneous electric field for t=0 sec in xy-plane for the antenna shown in Fig. 10.

for both, with and without double diffracted contributions on the nearby edges. In Fig. 11, the radiation pattern of the antenna in the xy-plane, cutting through the antenna in the middle, is shown. It can be seen, that the hybrid result including double diffraction shows excellent agreement with the reference solution. The simulations are also compared to the radiation pattern of the antenna radiating in free space. The influence of the edges on the radiation of the antenna is obvious. Deformation of the main lobe can be seen, as well as significant increase of the radiation in backward direction. A disturbance of the symmetric radiation can also be observed. The same result can be seen in Fig. 12, where the instantaneous electric field for $t = 0$ sec is shown in the xy-plane. The influence of the edges in the near-field distribution can be clearly seen.

4 Conclusions

In this contribution, advances in hybrid FEBI-MLFMM-UTD method were presented. In particular, the hybrid technique was extended to double diffracted fields on pairs of straight metallic edges, formulated with the hard and soft scalar diffraction coefficients of UTD. The diffraction points are determined by an iterative three-dimensional parametric realization of the generalized Fermat's principle, allowing treatment of non-coplanar and skewed edges. For the computation of the double diffraction divergence factor, the ray caustic distance of the diffracted field at the second edge is determined by linear interpolation between the radii of curvature in the two principal planes of the incident astigmatic ray tube. Acceleration of the near-field computations in the postprocessing stage of the hybrid method is extended in each translation domain to ray optical contributions due to the presence of electrically large objects, according to the hybridization of MLFMM with UTD.

References

Tzoulis, A. and Eibert, T. F.: A hybrid FEBI-MLFMM-UTD method for numerical solutions of electromagnetic problems including arbitrarily shaped and electrically large objects, IEEE Trans. Antennas Propagat, 53(10), 3358–3366, 2005.

Kouyoumjian, R. and Pathak, P. H.: A uniform geometrical theory of diffraction for an edge in a perfectly conducting surface, Proc. IEEE, 2(11), 1448–1461, 1974.

Tiberio, R., et al.: High-frequency electromagnetic scattering of plane waves from double wedges, IEEE Trans. Antennas Propagat., 37(9), 1172–1180, 1989.

Schneider, M. and Luebbers, R.: A general, uniform double wedge diffraction coefficient, IEEE Trans. Antennas Propagat., 39(1), 8–14, 1991.

Ivrissimtzis, L. P. and Marhefka, R. J.: Double diffraction at a coplanar skewed edge configuration, Radio Sci., 26(4), 821–830, 1991.

Capolino, F., et al.: Double diffraction at a pair of coplanar skew edges, IEEE Trans. Antennas Propagat., 45(8), 1219–1226, 1997.

Albani, M.: A uniform double diffraction coefficient for a pair of wedges in arbitrary configuration, IEEE Trans. Antennas Propagat., 53(2), 702–710, 2005.

Tzoulis, A. and Eibert, T. F.: Efficient electromagnetic near-field computation by the multilevel fast multipole method employing mixed near-field/far-field translations, IEEE Antennas Wireless Propag. Lett., 4, 449–452, 2005.

11

Integral equation analysis of complex (M)MIC-structures with optimized system matrix decomposition and novel quadrature techniques

T. Vaupel[1] **and V. Hansen**[2]

[1]FGAN-FHR, Neuenahrer Str. 20, 53343 Wachtberg, Germany
[2]Department of Electromagnetic Theory, University of Wuppertal, Germany

Abstract. Using integral equation methods for the analysis of complex (M)MIC structures, the computation and storage effort for the solution of the linear systems of equations with their fully populated matrices still forms the main bottleneck. In the last years, remarkable improvements could be achieved by means of diakoptic methods and related preconditiners. In this contribution, we present a method based on the optimized decomposition of the system matrix depending on the circuit topology. The system matrix is splitted in a densely populated matrix and a mainly blockdiagonal matrix with overlapping submatrices. The latter matrix is used for the generation of high performance preconditioners within Krylov subspace methods using sparsified matrix storage methods, adaptive Cholesky decompositions and optimized forward/backward substitutions. Furthermore, we present an integration technique using a complete analytical treatment for the strongly oscillating parts of the spectral domain integrands allowing the analysis of very large structures as compared to the wavelength.

1 Introduction

For the analysis and design process of (M)MIC structures, integral equation techniques based on the method of moments (MoM) are well suited and widely used. Many new techniques have been introduced for the efficient generation of the system matrix as well as for the computational effort reduction during the solution of the linear systems of equations. However, problems may still occur for large structures as compared to the wavelength and with complex topology. Based on the techniques in (Vaupel and Hansen, 2001)(Vaupel and Hansen, 2002), an optimized system matrix decomposition is now performed leading to high performance preconditioners whereas the application of a Legendre-Filon quadrature allows the versatile treatment of highly oscillatory spectral domain integrands. The basic idea for the optimized system matrix decomposition stems from the experiences with the diakoptics-based MoM implementations given in (Ooms and De Zutter, 1998). The generation of macro basis functions (MBs) is extended to entire domain diakoptic basis functions using block Gauss-Seidel procedures, but we could prove in (Vaupel and Hansen, 2002), that this approach still behaves $\sim O(N^3)$. On the other hand, a direct iterative improvement of the current distribution derived by the MBs using block relaxation processes did not converge (Ooms and De Zutter, 1998). We can show now, that this diverging behavior can be traced back on an insufficient system matrix decomposition used during the relaxation process. We have derived rules leading to an extended system matrix decomposition which guarantees fast convergence within a block Jacobi process and is especially well-suited for high performance preconditioners.

The creation of the system matrix of large structures requires the evaluation of spectral domain integrals with highly oscillatory integrands. To overcome this, several strategies (e.g. Pilz and Menzel, 1999) based on Filon integration principles (Filon, 1928) have been proposed, but they mainly suffer from a low integration order. Thus, we have extended the Gauss-Legendre quadrature by solving small linear systems of equations with modified right hand sides providing integration weights already containing the oscillatory parts of the integrands.

2 Formulation

A typical section of a (M)MIC structure is given in Fig. 1. As already outlined in Vaupel and Hansen (2002), additional ports are introduced, causing a block decomposition of the structure, in this example the ports P1...P4 lead to the blocks B1...B6. Each port is allocated two blocks which are connected with the port basis function. The macro basis functions (MBs) are created by the subsequent excitation of the ports and their allocated blocks in absence of all other blocks.

Correspondence to: T. Vaupel
(vaupel@fgan.de)

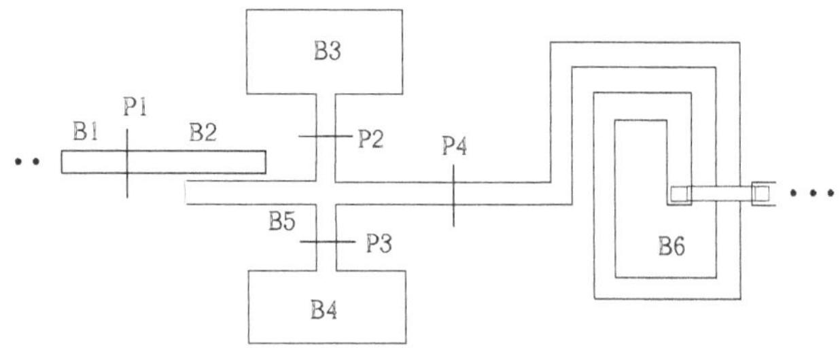

Fig. 1. Section of a typical (M)MIC-structure with port and block numbering.

For example port 3 excites block 3 and block 5. The current profiles on the corresponding blocks (computed with standard MoM) form the MBs. In the next step, we perform an upper level MoM by computing the couplings between the MBs, for example the coupling between the MBs "k" and "l" is computed by

$$Z_{kl} = \sum_{m \in \mathrm{MB}_k} \sum_{n \in \mathrm{MB}_l} I_{mk} Z_{nm} I_{nl} \tag{1}$$

with Z_{nm} the entries of the lower level MoM, $m \in \mathrm{MB}_k$ is the index m of all basis functions allocated to the MB number k, I_{mk} the current profile of the kth MB etc. The Z_{kl} are the entries of the upper level MoM system matrix. Solving for the amplitudes of the MBs and the subsequent superposition of the weighted current profiles of the MBs lead to a first approximation of the overall current distribution, but often the error exceeds 10%. To diminish this error, a block Gauss-Seidel process was proposed (Ooms and De Zutter, 1998), applied to each MB, transforming the original subdomain MB into an entire domain diakoptic basis function within a few iterations. The weighted superposition of the diakoptic basis function provides again the overall current. The error remains below 1% compared to a direct solution, but we could already show (Vaupel and Hansen, 2002), that the overall process is still $\sim O(N^3)$, but is 10–30 times faster than the direct solution.

To reduce this high numerical complexity, a direct improvement of the current distribution $\boldsymbol{I}_0$ provided by the subdomain MBs was proposed, performing the block Jacobi process

$$\mathbf{Z}^{\mathbf{i}} \cdot \boldsymbol{I}_{n+1} = \boldsymbol{U} - \mathbf{Z}^{\mathbf{c}} \cdot \boldsymbol{I}_n, \quad \mathbf{Z} = \mathbf{Z}^{\mathbf{i}} + \mathbf{Z}^{\mathbf{c}} \tag{2}$$

using $\boldsymbol{I}_0$ as start vector.

$\mathbf{Z}^{\mathbf{i}}$ is proposed a sparse block diagonal matrix containing the self and mutual couplings of each block forming diagonal submatrices, whereas $\mathbf{Z}^{\mathbf{c}}$ comprises all remaining mutual couplings.

Unfortunately, this process does not converge in most cases since the convergence condition

$$\rho(\mathbf{Z}^{\mathbf{i}^{-1}}\mathbf{Z}^{\mathbf{c}}) < 1$$

cannot be fulfilled with the chosen matrix decomposition where $\rho(\mathbf{A})$ denotes the spectral radius of a matrix $\mathbf{A}$.

But we get a convergent Jacobi process by using an appropriate extension of the matrix $\mathbf{Z}^{\mathbf{i}}$ employing the following rule: If block s and block t form together a MB, i.e. it exists a current coupling between the two blocks, then the submatrix $\mathbf{Z}_{\mathbf{st}}$ must be added to the matrix $\mathbf{Z}^{\mathbf{i}}$.

For the structure in Fig. 1 we create the following matrix $\mathbf{Z}^{\mathbf{i}}$:

$$\mathbf{Z}^{\mathbf{i}} = \begin{bmatrix} \mathbf{Z}_{11} & \mathbf{Z}_{12} & 0 & 0 & 0 & 0 \\ \mathbf{Z}_{21} & \mathbf{Z}_{22} & 0 & 0 & 0 & 0 \\ 0 & 0 & \mathbf{Z}_{33} & 0 & \mathbf{Z}_{35} & 0 \\ 0 & 0 & 0 & \mathbf{Z}_{44} & \mathbf{Z}_{45} & 0 \\ 0 & 0 & \mathbf{Z}_{53} & \mathbf{Z}_{54} & \mathbf{Z}_{55} & \mathbf{Z}_{56} \\ 0 & 0 & 0 & 0 & \mathbf{Z}_{65} & \mathbf{Z}_{66} \end{bmatrix} \tag{3}$$

We can recognize an additional subdivision of this matrix into three further quadratic submatrices $\mathbf{Z}_1 \ldots \mathbf{Z}_3$ where $\mathbf{Z}_1$ comprises the submatrices from $\mathbf{Z}_{11}$ to $\mathbf{Z}_{22}$, $\mathbf{Z}_2$ the submatrices from $\mathbf{Z}_{33}$ to $\mathbf{Z}_{55}$ and $\mathbf{Z}_3$ the submatrices from $\mathbf{Z}_{55}$ to $\mathbf{Z}_{66}$, thus $\mathbf{Z}_3$ overlaps with $\mathbf{Z}_2$ in their common submatrix $\mathbf{Z}_{55}$. We can observe that the topology of the (M)MIC-circuitry is completely transferred to the submatrix decomposition of $\mathbf{Z}^{\mathbf{i}}$. As next step, we perform an adapted Cholesky decomposition of $\mathbf{Z}^{\mathbf{i}}$ using the described submatrix decomposition thus minimizing the number of computational steps. In this context, we can also perform an incomplete Cholesky decomposition to avoid possible fill-ins of matrix entries. The Cholesky decomposition, denoted with $\mathbf{C}$, can be used to perform Eq. (2), but it is much better to use it as left and right preconditioner within Krylov subspace methods, employing the equivalent linear system of equations

$$\mathbf{C}^{-1}\mathbf{Z}\,\mathbf{C}^{-\mathbf{T}}\mathbf{C}^{\mathbf{T}} \cdot \boldsymbol{I} = \mathbf{C}^{-1} \cdot \boldsymbol{U}. \tag{4}$$

This results in decisive benefits in contrast to Eq. (2):

- about the same computational effort per iteration as Eq. (2)
- excellent convergence behavior nearly independent on start vector
- No explicit generation of MBs with additional port excitation necessary

⇒ block decomposition also at broad transmission lines, arbitrary structure excitation

- fast convergence also with incomplete Cholesky decomposition.

Another problem arising with large structures is combined with strongly oscillating coupling integrands.

Regarding e.g. the coupling integral of two x-directed basis functions we get

$$Z_{nm} = \int_{k_x=0}^{k_{x,a}} \int_{k_y=0}^{k_{y,a}} G^E_{Jxx} \mathrm{Re}\{F_{m0}(k_x) F_{n0}(k_x) e^{jk_x \Delta x_{nm}}\} \mathrm{Re}\{F_{m0}(k_y) F_{n0}(k_y) e^{jk_y \Delta y_{nm}}\} dk_x dk_y. \tag{5}$$

Here, F_{m0} ,F_{n0} denote the Fourier transforms of the nonshifted basis functions. The integrand becomes highly oscillatory with large lateral distances Δx_{nm}, Δy_{nm} thus a very high number of sampling points is necessary using standard quadrature techniques. To overcome this constraint, we employ the following procedure, e.g. outlined for the integration with regard to k_x.

The aim is to integrate

$$\int_a^b f(k_x)e^{jck_x}dk_x \approx C_{sc}\sum_{j=1}^{N}\tilde{w}_j f(x_j) \tag{6}$$

with $f(k_x)$ typically a smooth function of k_x, $\tilde{w}_j$ integration weights adapted to e^{jck_x} and x_j appropriate sample points. Using the substitution $k_x(t) = \frac{b-a}{2}t + \frac{a+b}{2}$, $\frac{dk_x}{dt} = \frac{b-a}{2}$, we can introduce the Legendre expansion

$$f(k_x(t)) = \sum_{n=1}^{N-1} a_n p_n(t), t \in [-1, 1] \tag{7}$$

with a_n the expansion coefficients. We can write

$$\int_a^b f(k_x)e^{jc\cdot k_x}dk_x = \int_{-1}^{1} f(k_x(t))e^{jc\cdot k_x(t)}\frac{dk_x}{dt}dt \approx$$

$$\frac{b-a}{2}e^{j\cdot e}\sum_{n=1}^{N-1} a_n \int_{-1}^{1} p_n(t)e^{jd\cdot t}dt \tag{8}$$

with $d = \frac{c\cdot(b-a)}{2}$, $e = \frac{c\cdot(b+a)}{2}$ and $C_{sc} = \frac{b-a}{2}e^{j\cdot e}$. The best approximation for the integral with the given Legendre expansion is obtained if we achieve

$$\sum_{n=1}^{N-1} a_n \int_{-1}^{1} p_n(t)e^{jd\cdot t}dt = \sum_{n=1}^{N-1} a_n \sum_{j=1}^{N}\tilde{w}_j p_n(t_j),\ n = 0, .., N-1. \tag{9}$$

For this goal, we have to fulfill

$$\int_{-1}^{1} e^{jd\cdot t}p_n(t)dt = \sum_{j=1}^{N}\tilde{w}_j p_n(t_j),\ n = 0, .., N-1. \tag{10}$$

This leads to the generation of the linear system of equations

$$\begin{Bmatrix} p_0(t_1) & p_0(t_2) & \dots & p_0(t_N) \\ p_1(t_1) & p_1(t_2) & \dots & p_1(t_N) \\ \vdots & \vdots & \vdots & \vdots \\ p_{N-1}(t_1) & p_{N-1}(t_2) & \dots & p_{N-1}(t_N) \end{Bmatrix} \cdot \begin{Bmatrix} \tilde{w}_1 \\ \tilde{w}_2 \\ \vdots \\ \tilde{w}_N \end{Bmatrix} =$$

$$= \begin{Bmatrix} \int_{-1}^{1} p_0(t)e^{jd\cdot t}dt \\ \int_{-1}^{1} p_1(t)e^{jd\cdot t}dt \\ \vdots \\ \int_{-1}^{1} p_{N-1}(t)e^{jd\cdot t}dt, \end{Bmatrix}. \tag{11}$$

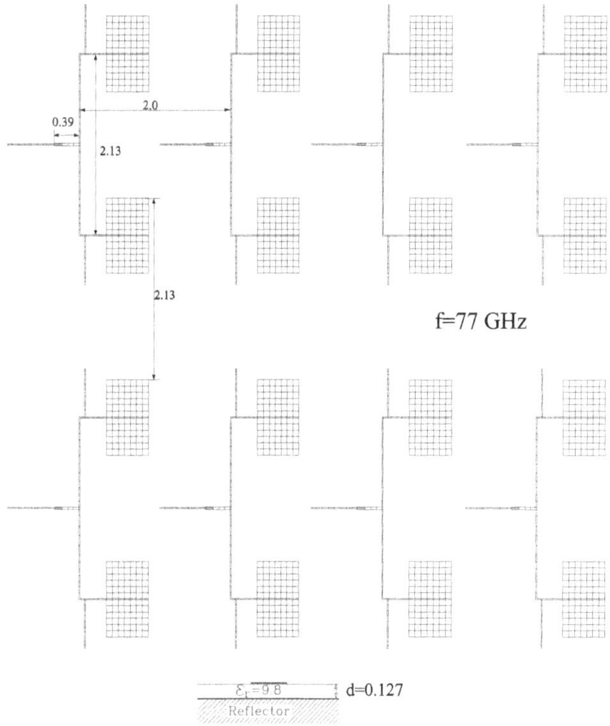

Fig. 2. Discretization of a 16-element patch array.

whose solution provides the integration weights $\tilde{w}_i$. The matrix has only to be computed once using the well known recurrence formula for the Legendre polynomials. Afterwards, we apply a LU-decomposition of this matrix with partial pivotisation and additional storage of a pivot index vector.

Thus the efficient computation of the right hand side of Eq. (11) forms the crucial point for the weight determination.

Here the main task is to solve

$$T_n(d) = \int_{-1}^{1} t^n e^{jd\cdot t}dt = \frac{1}{jd}t^n e^{jd\cdot t}\Big|_{-1}^{1} - \frac{n}{jd}\int_{-1}^{1} t^{n-1}e^{jd\cdot t}dt \tag{12}$$

by a recurrence relation. For the first term right hand side in Eq. (12), we get

$$\frac{1}{jd}t^n e^{jd\cdot t}\Big|_{-1}^{1} = \begin{cases} \frac{2}{d}\sin(d), & n \text{ even} \\ \frac{2}{jd}\cos(d), & n \text{ odd} \end{cases}. \tag{13}$$

Nevertheless, for $d < 0.5$ the computation with Eqs. (12) becomes unstable due to cancellation effects. In these cases, we compute the $T_n(d)$ with an iterative procedure using the power series representation of the complex exponential function, leading to compact series representations with very fast convergence. Using the weights $\tilde{w}_i$ allows to integrate the function $f(k_x)$ exactly up to polynomials of order $N-1$ providing a much higher accuracy than with classical Filon integration rules. Since we use a similar basic principle, our method may be called Legendre-Filon quadrature.

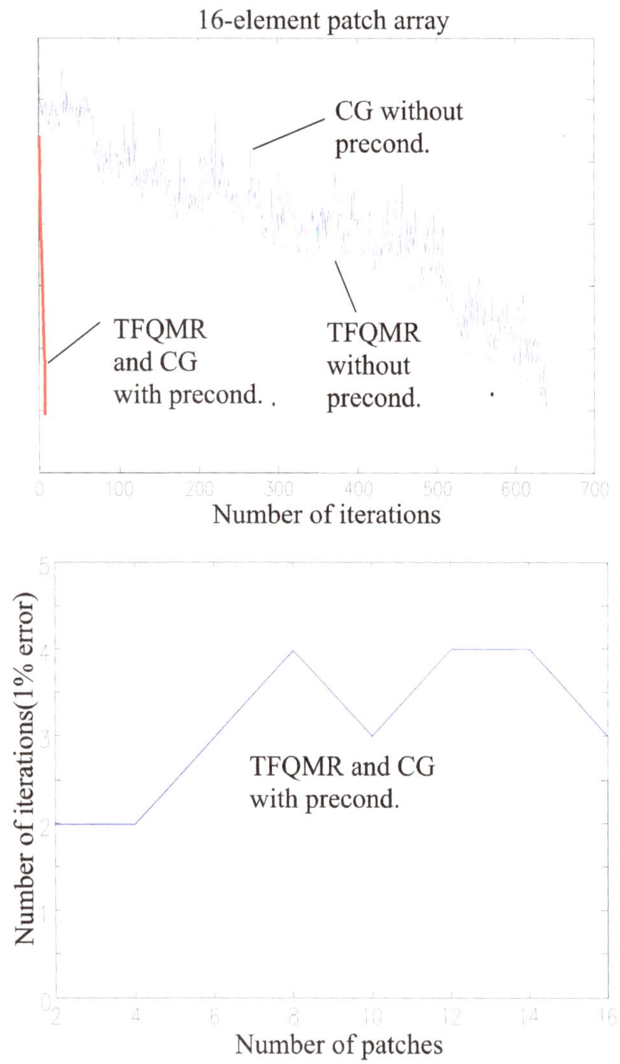

Fig. 3. Solution behavior of 16-element patch array. Top: Convergence history. Bottom: Iteration number in dependence on the number of patches.

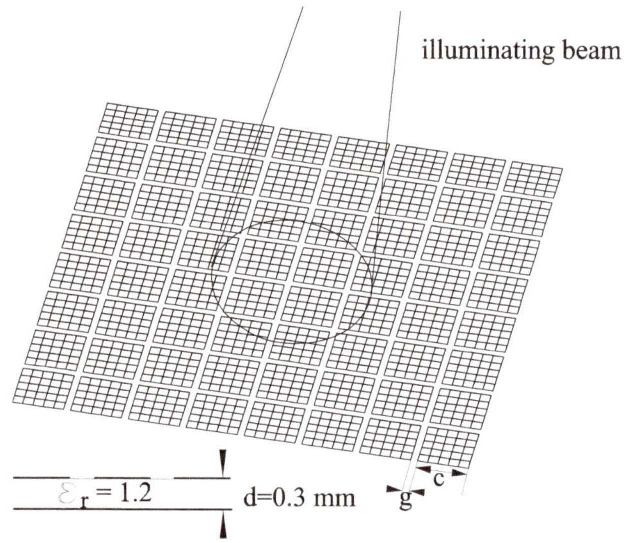

Fig. 4. Multibeam antenna array as part of an astronomical telescope. g=0.15 mm, c=1.1 mm.

Fig. 5. Current distribution (A/mm) of a multibeam antenna excited by an incident plane wave at f=250 GHz.

The effort for computing the weights can be further reduced by using a composite Legendre-Filon integration. This means that the sampling points are subdivided into several intervals whereas the weights are only computed for the first intervall. The weights for the other intervalls are simply derived from the weights of the first intervall by multiplying them with a fixed exponential term.

3 Applications

As first application we present the analysis of a 16-element patch array consisting of 2-element subarrays. Figure 2 shows the dimensions and discretization of the structure.

The structure was discretited with 2400 basis functions. For the matrix decomposition, we introduced 40 ports leading to 48 blocks.

The solution behavior is given in Fig. 3. Figure 3 top shows the convergence history with a Transpose Quasi Minimum Residual (TFQMR) method and a standard CG-method. Without preconditioning the CG-method shows a strong oscillatory residual error whereas the TFQMR-method exhibits a smooth decay. With preconditioning, both methods show a very fast decay with only three necessary iterations for a residual error of 1%. Figure 3 bottom shows the necessary number of iterations (1% error) in dependence on the number of patches. We observe a nearly constant number of necessary iterations, thus the solution performance of our preconditioner is nearly independent on the number of unknowns.

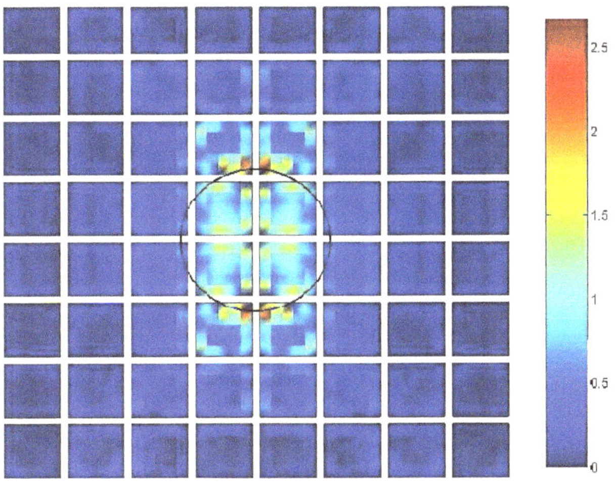

Fig. 6. Current distribution of a multibeam antenna excited by an incident plane wave only illuminating a circular region with diameter 3.05 mm.

A so-called multibeam antenna we present as a second application. It is intended as part of astronomical telescopes for image processing purposes in the submm-wave region. As given in Fig. 4, it consists of 64 patch elements mounted on a foam layer ($\epsilon_r \approx 1.2$) with reflector. We applied a discretization with 2560 unknowns. The lateral dimension of the structure amounts to about 12 wavelength at 250 GHz. For the preconditioning, each antenna element was assigned to one block at first, but we also assigned two elements to one block. The two-element blocks can be considered as isolated or we can introduce so-called virtual ports between the elements. With virtual ports between successive elements e.g. in each row, the elements of a row are considered as connected by current coupling although only field coupling exists between the elements. This leads to overlapping matrices similar as in in Eq. (3). Nevertheless, the performance of this preconditioning strategies is nearly the same and leads to about 11 iteration using a perpendicular plane wave excitation (E-field in x-direction with 1000 V/cm) over the whole array (Fig. 5). If we only illuminate a circular region of the array (Fig. 6), we need up to 24 iterations which can be traced back on the inhomogeneous excitation situation.

4 Conclusions

An optimized system matrix decomposition adapted to the topology of complex (M)MIC-structures was presented, leading to high performance preconditioners within Krylov subspace solvers. Using adaptive Cholesky decompositions and forward/backward substitutions, the computational and storage effort for the solution process of the linear systems of equations could be minimized. For the very accurate evaluation of the system matrix entries a Legendre-Filon integration technique was applied to the often highly oscillatory spectral domain integrands formulated in Cartesian wavenumbers. The methods were applied to microstrip array antennas and multibeam reflectarrays and have demonstrated excellent performance for both the matrix fill and the linear equation solution process.

References

Ooms, S. and De Zutter, D.: A New Iterative Diakoptics-Based Multilevel Moments Method for Planar Circuits, IEEE Trans. on Microwave Theory and Techn., 46, 3, 280–291, 1998.

Vaupel, T., Hansen, V., and Schäfer, F.: Radiation Efficiency Analysis of Submm-Wave Receivers Based on a Modified Spectral Domain Technique, Radio Science, 38, 4, 2003.

Vaupel, T. and Hansen, V.: Integralgleichungsanalyse von (M)MIC-Strukturen mit modifizierter Spektralbereichsintegration und diakoptischen Strategien, Kleinheubacher Berichte, 2001.

Vaupel, T. and Hansen, V.: Anwendungen einer modifizierten Spektralbereichsintegration für implizite Multilevelstrategien und detaillierte Abstrahlungsanalysen, Kleinheubacher Berichte, 2002.

Filon, L. N. G.: On a quadrature formula for trigonometric integrals, Proc. Roy. Soc. Edinburgh, 49, 38–47, 1928.

Pilz, D. and Menzel, W.: Mixed Integration Method for the Evaluation of the Reaction Integrals Using the Spectral Domain Method, IEE Proc. Microw. Antennas Propagation, 146, 3, 1999.

12

Concept for accelerated ray-based monostatic RCS simulations using bistatic approximations

H. Buddendick[1] **and T. F. Eibert**[2]

[1]Institute of Radio Frequency Technology, Universität Stuttgart, 70550 Stuttgart, Germany
[2]Lehrstuhl für Hochfrequenztechnik, Technische Universität München, 80290 München, Germany

Abstract. An approach enabling accelerated shooting and bouncing rays (SBR) simulations to determine the backscattering properties of electrically large and complex objects is presented. Instead of performing independent simulations for all required aspect angles, the concept is based on the idea of additionally exploiting bistatic information for some neighboring aspect angles. Therefore the results of the geometrical ray tracing, which consumes a large part of the computational resources in case of complex shaped objects can be reused multiple times with only low additional computational resources.

This method works well for objects with a sufficiently smooth shape and if a large number of aspect angles is to be simulated. A simple generic simulation example is used to show the general applicability of the method and to examine the degradation of the results depending on the applied bistatic angle. Furthermore, the acceleration that can be expected by the presented approach is determined and verified with the simulation example.

1 Introduction

The Radar Cross Section (RCS) is an important characterization for the electromagnetic scattering properties of complex objects. It is common practice to use this quantity in the fields of radar system design and dimensioning, as well as for radar algorithm development and validation. In many areas it is desired to have a good estimation for it for a wide range of aspect angles. Unfortunately, it is computationally very demanding to perform the required simulations for complex shaped objects like vehicles or airplanes at very high frequencies, even if suitable high frequency approximations like the widely used Geometrical Optics (GO) and Physical Optics (PO) are used.

Based on a well known relationship for the scattering properties at different aspect angles (sometimes referred to as bistatic monostatic equivalence principle) we present an approach which enables us to accelerate considerably these ray-based multi-aspect RCS simulations.

Basically, in SBR rays representing the incident fields are used to determine the resulting surface currents and the resulting field contributions at the given observation points according to the Physical Optics approach. This is explained in Sect. 2 with special focus on those aspects that are interesting for the application of the bistatic monostatic equivalence principle which is introduced in the first part of the following Sect. 3. The second part of Sect. 3 demonstrates how, with a proper phase correction, contributions related to deviating illumination and observation angles can be utilized for an approximation of the desired monostatic RCS. In this way from each determined ray, contributions to several neighboring monostatic receivers can be obtained, and therefore the total number of rays required for a proper sampling of the object can be reduced significantly.

In Sect. 4 some remarks about the angular sampling are given, as it has a very important influence on the efficiency of the presented method. In Sect. 5 simulation results are presented to demonstrate the applicability and performance of the approach and, finally, Sect. 6 summarizes the results we have obtained up to now.

2 The applied shooting and bouncing rays technique

To analyze the scattering properties of electrically large and complex objects asymptotic high frequency methods are to be applied. In this paper we use an approach which is based on a shooting and bouncing rays (SBR) algorithm to determine the local incident fields on the scatterer and the Physical

Correspondence to: H. Buddendick
(buddendick@ihf.uni-stuttgart.de)

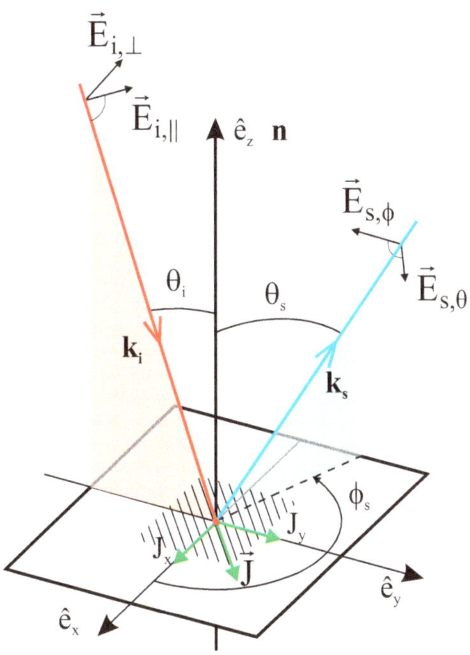

Fig. 1. Local coordinate system used for the PO integration around the interaction point of a ray on a plane surface element.

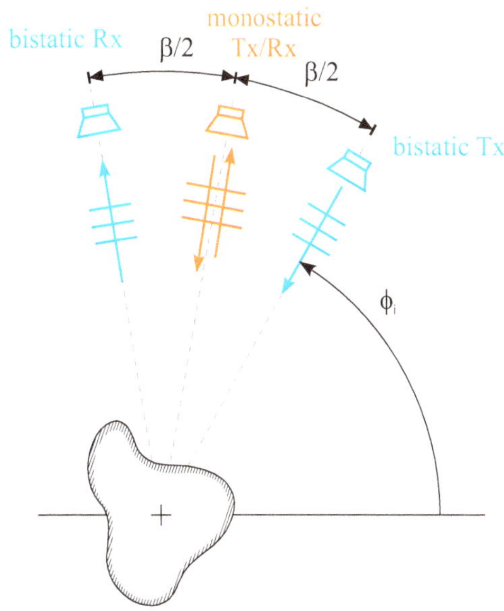

Fig. 2. Equivalence of bistatic and monostatic RCS.

Optics (PO) to determine the resulting field contributions. In the SBR approach, see for example Ling et al. (1989), a number of rays is launched towards the scatterer and the rays are traced with one or more interactions until they exit the scenario, so that the fields at the last interaction point can be used to determine the surface currents. Nowadays, the concept of SBR is widely applied to compute the scattering fields of large objects in scientific as well as in commercial codes. Some examples can be found in Weinmann (2006); Galloway and Welsh (2005); Andersh et al. (2000).

Basically, our algorithm splits into two parts:

- Ray tracing is used to find the fields on the objects surface according to the GO principle.
- PO integration determines the scattered fields based on the surface currents.

An important observation is that usually the ray tracing part of the algorithm consumes most of the computational resources if detailed object models are considered. For typical applications the complexity of the geometric model can be in the order of a few 100 000 triangles. This clearly indicates the need for a very efficient implementation, which, in our case, is based on a uniform spatial subdivision (Fujimoto, 1986).

The number of rays that are directed towards the object must be large enough to capture all relevant object details sufficiently for the coherent superposition of the field contributions. In our case a small random component is added to the ray directions to prevent aliasing effects at surface borders or wedges. Due to this stochastic ray launching process the results of two independent simulation runs (with identical parameter settings) will differ from each other and include a stochastic component as well. With an increasing number of rays this deviation will decrease and the simulations will converge. Obviously, the stochastic uncertainty will be largest for low scattering levels and the expected accuracy can be used to adjust the required number of rays.

The scattering contributions are determined from the last interaction point of the ray on the object. According to Fig. 1 a local coordinate system adapted to the current ray is used and the surface currents can be computed in the following form:

$$J_y = 2\hat{n} \times \boldsymbol{H}_{i,\|} = \frac{2}{Z_{F0}} E_{i,\|}\, \hat{e}_y, \tag{1}$$

$$J_x = 2\hat{n} \times \boldsymbol{H}_{i,\perp} = \frac{-2\cos\theta_i}{Z_{F0}} E_{i,\perp}\, \hat{e}_x. \tag{2}$$

Finally, the far-field scattering contributions caused by these currents can be derived from the wave equation in two orthogonal polarization directions at the given observation points. For details the reader may refer to Balanis (1989), from where we have adopted the notation.

3 Description of the acceleration approach

3.1 Monostatic bistatic equivalence

Considering the scattering properties of an object and a geometrical setup as shown in Fig. 2, a simple and well known

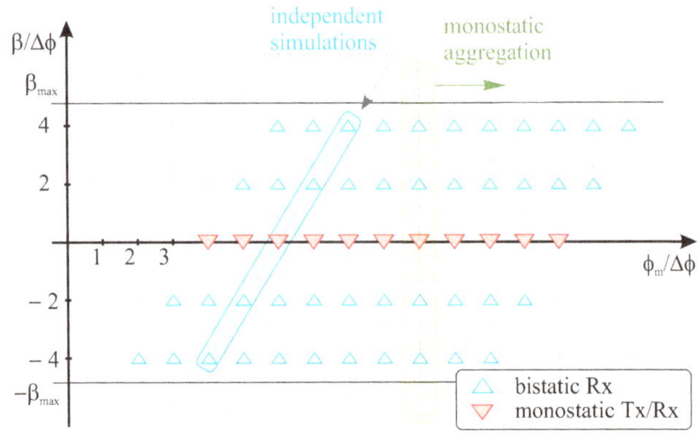

Fig. 3. Symmetric distribution of temporary bistatic receivers around the monostatic receivers and final monostatic aggregation.

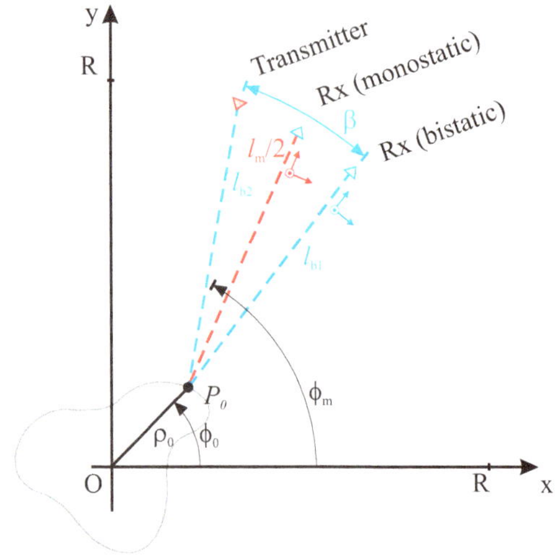

Fig. 4. The actual bistatic path differs from the monostatic path for which the contribution is to be considered and the phase has to be corrected accordingly.

relationship between bistatic and monostatic RCS values exists and can be expressed in the following form:

$$\sigma_m\left(\phi_i + \beta/2\right) = \sigma_b\left(\phi_i, \beta\right), \tag{3}$$

with σ_m and σ_b being the monostatic and bistatic RCS, respectively, and ϕ_i the angle of incidence and β the bistatic angle. This means that the bistatic RCS equals the monostatic RCS observed at the bisecting angle (see Fig. 2). The underlying assumptions for this identity are the consideration of very high frequencies and a sufficiently smooth surface of the scattering object. Details on the monostatic bistatic equivalence including an analytical discussion can be found in Kell (1965) and Siegel et al. (1968). Several interesting applications of the aforementioned principle can be found. Some applications are in the context of RCS measurements, e.g. monostatic RCS values derived from bistatic measurement setups, or, in the reverse direction, bistatic RCS derived from monostatic measurements (Kell, 1965). In the field of simulation techniques the equivalence has been successfully used to accelerate FVTD, FDTD, or MoM codes in a post-processing manner, see for example Schuh et al. (1994). According to the authors knowledge no publication on the use of the bistatic monostatic equivalence on a direct ray-by-ray basis exists.

Another interesting use of this principle is related to the possibility of generating multi-frequency data from multi-aspect (bistatic) data. This can be advantageously utilized in the field of Synthetic Aperture Radar (SAR) imaging as described for example by Mensa (1991). As these methods are based on a coherent signal processing, a simple power based formulation as given in Eq. (3) is not sufficient, and a coherent scattering model has to be used instead. Similarly, in the presented SBR approach rays are added up coherently and consequently a coherent scatterer model is used to analyze the monostatic and bistatic relation. This results in a simple but important phase correction factor to be considered for bistatic contribution as derived in the following section.

3.2 Bistatic acceleration of the SBR algorithm

Figure 3 shows the basic transmitter (Tx) and receiver (Rx) arrangement that is used to capture the bistatic information during the SBR algorithm. For each desired observation direction a conventional monostatic Tx/Rx pair is placed equally spaced by $\Delta\phi$ in the angular domain. Additionally, for each simulation a number of temporary bistatic receivers is created symmetrically around the desired observation points, each separated by $2\Delta\phi$. The bistatic reuse factor N_{bi} equals the number of monostatic observation angles within a sector $\beta_{\max}$:

$$N_{bi} = 2\left\lfloor \frac{\beta_{\max}}{2\Delta\phi} \right\rfloor + 1. \tag{4}$$

After all observation directions have been simulated, the contributions captured by the bistatic receiver can be converted to the bisecting monostatic receiver (this is called monostatic aggregation here). In this way the rays launched from each monostatic direction are reused for several neighboring directions as well. Consequently, for the same accuracy as discussed in Sect. 2, the number of rays launched for each observation direction can be reduced significantly.

During the SBR algorithm a bistatic field contribution is determined for all rays at the location of all temporary bistatic receiver within the bistatic sector defined by $\beta_{\max}$ symmetrically around the transmitter.

Now let's assume for a coherent consideration a point scatterer on the surface of the object as depicted in Fig. 4. The length of the actual bistatic path $l_b = l_{b1} + l_{b2}$ of the ray differs from the length l_m of the path to the bisecting monostatic observation point to which the ray finally shall contribute according to the bistatic equivalence principle. Due to this path

Table 1. Examples for the required angular sampling interval.

f [GHz]	$\rho_{\max}$ [m]	$\Delta\phi_{\max}$ [°]	N_ϕ ϕ=0..180°	N_{bi} $\beta_{\max}$=10°
10	0.09	4.77	38	3
10	0.9	0.477	377	21
24	2.5	0.0716	2514	139
80	2.5	0.0215	8378	465

length difference a phase correction is required. For a far field observation distance with $R \gg \rho_0$ the path length difference depending on the coordinates of P_0 and the observation point can be determined to be

$$\Delta l = l_m - l_b = -2\rho_0 \cos(\phi_m - \phi_0)\,(1 - \cos(\beta/2)), \quad (5)$$

or, alternatively depending only on the quantities directly accessible in the simulation:

$$\Delta l = (l_b - 2R)\left(\frac{1}{\cos(\beta/2)} - 1\right). \quad (6)$$

Following this model a simple phase correction corresponding to

$$\Delta\psi = k_0 \Delta l, \quad (7)$$

with the free-space wave number $k_0 = 2\pi/\lambda_0$ is applied to all field contributions on a ray-by-ray basis. Furthermore it can be seen from Fig. 4 that the coordinate system of the (temporary) bistatic receiver is rotated by $\beta/2$. This should not be a problem as it must be done only once at the end, when the coherent sum of the temporary bistatic receiver is finally added to the corresponding monostatic receiver during the monostatic aggregation.

3.3 Expected performance and limitations

Obviously, the use of the bistatic monostatic equivalence principle limits the presented approach to high frequency applications and to sufficiently smooth objects. The first restriction is not very stringent as this is the main focus of this acceleration approach anyway, and it scales well with its necessity and efficiency, as it will be shown later. The phase correction as explained above degrades in case of multiple bounces (double reflection). This is also closely related to the smoothness of the object and therefore not a new restriction. Here, the smoothness of the object is partly related to a similar illumination of the object for the bistatic and the monostatic case. This may be effected by exposed geometrical features causing a self-shadowing on the object itself. The example in Sect. 5 shows how the smoothness of the object effects the results.

The acceleration that can be expected could be in the range of the bistatic reuse factor N_{bi} as this is the factor by which

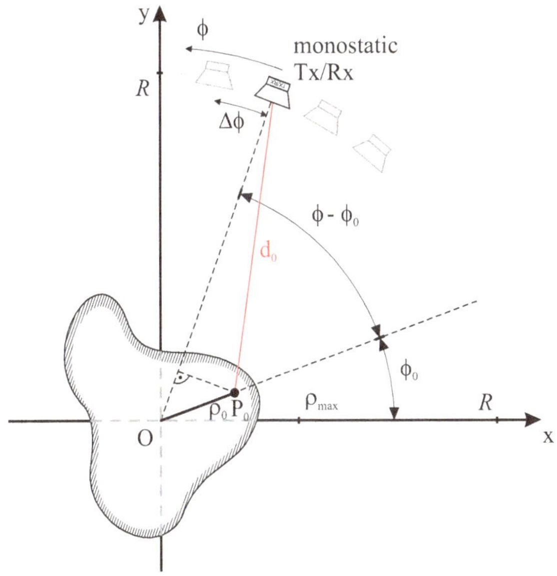

Fig. 5. Geometry for the determination of the minimum angular sampling rate. A point source P_0 on the objects surface is assumed.

the number of effectively evaluated rays exceeds the number of distinct rays. Consequently, the number of rays required to achieve results comparable with pure monostatic simulations could be lowered by N_{bi} and for realistic scenarios the required CPU time is proportional to the number of rays.

Unfortunately, the improvement that can be realized will be somewhat lower. On the one hand this is due to a degradation in the quality of the contribution of the rays with larger bistatic angles. This will lead to systematic error contributions. On the other hand an additional overhead in both, memory and computational cost, due the temporary bistatic receiver will decrease the theoretical performance. Hence, the resulting acceleration that can be achieved will be below the bistatic reuse factor N_{bi}. Furthermore, as each observation point contains (bistatic) information originating from neighboring transmitters, the simulated observation range has to be extended by $\beta_{\max}/2$ at both ends to capture the bistatic information symmetrically. But for many practical scenarios with large observation ranges this has only minor impact.

4 Angular sampling

From the preceding sections we can conclude that the efficiency that can be expected from the bistatic acceleration depends on the bistatic reuse factor N_{bi} and as the maximum bistatic angle $\beta_{\max}$ is determined only by the geometry of the object and is therefore fixed, the resulting efficiency depends on the applied (monostatic) sampling rate. Often it is desired to have a full representation of the backscattering fields

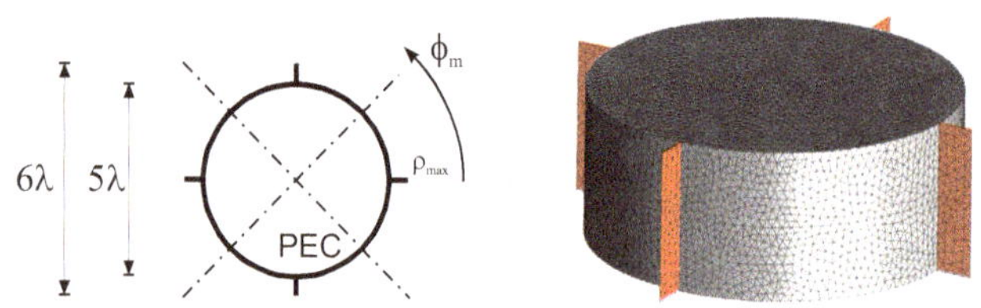

Fig. 6. Generic object representing a smooth body (cylinder) with blades causing bistatic shadowing and double reflections.

of an object for a certain aspect range, e.g. in the azimuthal plane. In this case the required observation angles are to be determined according to the rules of the sampling theorem. Assuming far field conditions an angular sampling interval of

$$\Delta\phi_{\max} = \frac{\lambda_0}{4\rho_{\max}} \tag{8}$$

is required, with λ_0 being the carrier wavelength and $\rho_{\max}$ the maximum object radius in the considered observation plane. Some sampling intervals and the resulting N_{bi} for an bistatic sector of $\beta_{\max}=10°$ are given in Table 1. Note that at frequencies as high as 80 GHz objects of the size of a passenger car require a large amount of observation directions according to the sampling theorem. On the other hand the increasing bistatic reuse factor N_{bi} offers a compensation for this enormous computational burden of such problems at very high frequencies. For the derivation of Eq. (8) we assume a scattering model with a geometrical setup as shown in Fig. 5 with a point scatterer P_0 on the surface of an object. For large observation distances R the amplitude of the field contributions at the observation point is assumed to be constant and the phase is proportional to the path length

$$\psi(d_0) = k_0 2 d_0(R,\phi), \tag{9}$$

with d_0 being the distance between P_0 and the Tx/Rx location. Similar to time dependent signals, a variability f_ϕ in the ϕ-domain can be obtained from the derivative of the phase according to the ϕ angle:

$$2\pi f_\phi = \frac{\partial}{\partial\phi}\psi(d_0) = 2k_0\frac{\partial}{\partial\phi}d_0. \tag{10}$$

Here, the geometrical path length is given by

$$d_0(R,\phi) = \sqrt{R^2+{\rho_0}^2-2\rho_0 R\cos(\phi-\phi_0)}, \tag{11}$$

$$\frac{\partial}{\partial\phi}d_0(R,\phi) = \frac{-\rho_0 R\sin(\phi-\phi_0)}{\sqrt{R^2+{\rho_0}^2-2\rho_0 R\cos(\phi-\phi_0)}}. \tag{12}$$

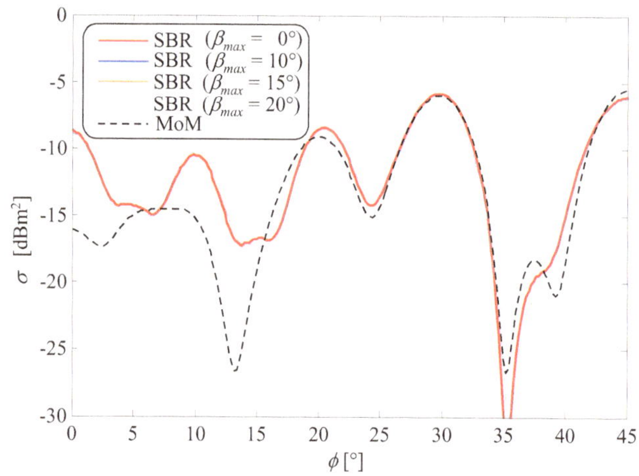

Fig. 7. RCS of the generic cylinder model with different levels of bistatic information.

According to the Nyquist sampling theorem a complete signal representation requires observation angles with a resolution below

$$\Delta\phi_{\max} = \frac{1}{2f_{\phi,\max}} = \frac{\lambda_0}{4\max\left[\left|\frac{\partial}{\partial\phi}d_0\right|\right]}. \tag{13}$$

Obviously, the maximum in Eq. (13) is achieved for $|\phi-\phi_0|=\pi/2$, i.e. for P_0 having the largest lateral distance to the central observation line:

$$\max_{\phi}\left[\left|\frac{\partial}{\partial\phi}d_0(R,\phi)\right|\right] = \frac{R\rho_{\max}}{\sqrt{R^2+\rho_{\max}^2}} \tag{14}$$

This leads to

$$\Delta\phi_{\max} = \frac{\lambda_0\sqrt{R^2+\rho_{\max}^2}}{4R\rho_{\max}}, \tag{15}$$

which, for the far field with $R\to\infty$, directly results in Eq. (8).

5 Simulation results for a generic test object

To determine the effect of the bistatic approximation in case of a non-smooth object including self-shadowing features, a small metal cylinder with four radial blades is considered as depicted in Fig. 6. With the given relative dimensions of the blades, the resulting shadows on the object will be much more penalizing for our approach than what is expected from a realistic object like a passenger car. Besides these shadowing effects multiple reflections will occur between the blades and the cylinder. Therefore, we conclude that this object, qualifies well for a test case and to derive an upper bound for $\beta_{\max}$.

Due to symmetry only a 45° sector must be considered for the monostatic RCS. Figure 7 shows in red the RCS obtained with the straightforward (not accelerated) SBR for a

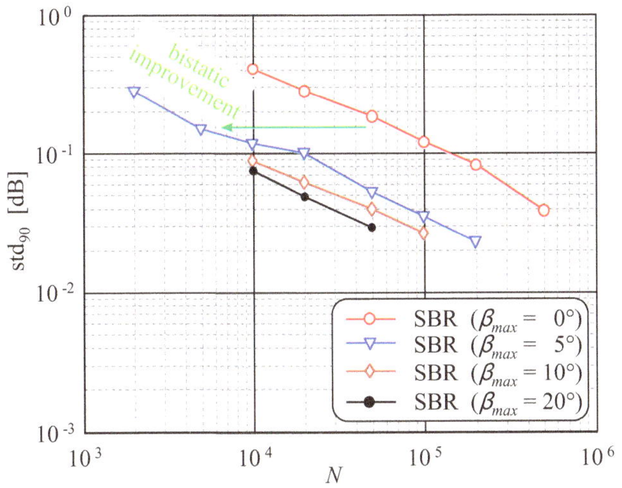

Fig. 8. Influence of the bistatic information on SBR convergence. Standard deviations vs. number of rays for different $\beta_{\max}$.

very high number of rays. This can be considered as reference of what can be achieved with the SBR approach. Additionally, the results from a Method of Moments (MoM) implementation is shown as dashed line. A good agreement for the basic scattering characteristic can be observed. Nevertheless, the deviations for $\phi<15°$ are obvious and remind us that the SBR itself is an approximation with certain restrictions. This should be kept in mind if additional deviations due to the bistatic approximation are considered. It should be added that the object is fairly small for the SBR and that no diffraction effects are included here.

Concentrating on the effect of the bistatic approximation, it can be seen that the degradation for a bistatic angle $\beta_{\max}=10°$ is still very limited. Only a bistatic angle $\beta_{\max}\geq 15°$ will cause a considerable impact due to bistatic shadowing.

It should be noted that Fig. 7 is obtained with a high degree of oversampling (see Sect. 4) for visualization purposes. Of course, this is not a sufficient setup for a comparison of the computational efficiency as the improvement that can be observed using the bistatic approximation could also partly be based on averaging the oversampling redundancy. To avoid this, a simulation with the generic object being scaled by a factor of 10 compared to Fig. 6 has been studied. This results in faster oscillations and an increased required sampling rate. For different $\beta_{\max}$ and with varying numbers of rays the deviations of the RCS from the SBR reference, i.e. with a very high number of rays are determined. By comparing the required number of rays for a given level of deviation from the reference the bistatic acceleration can be estimated (see Fig. 8). For $\beta_{\max}=10°$ an achievable acceleration factor in the range of 10–20 can be concluded.

6 Conclusions

Based on the principle of monostatic bistatic equivalence, an accelerated ray-based RCS simulation algorithm is proposed and some important implementation details are given. Additional bistatic information, which is collected with very low additional computational cost, can be reused to improve the convergence of the stochastic shooting and bouncing rays algorithm, which means that less rays have to be used for a required accuracy. In this way the monostatic sampling of the object of interest is partly replaced by the usage of bistatic side information. Simulation results are presented for a simple canonical object to demonstrate on the one hand the unavoidable level of degradation that comes along with the usage of a certain amount of bistatic information and how it is related to $\beta_{\max}$. On the other hand the possible speed-up of a complete multi-aspect simulation run is shown by comparing the scattering fields for different number of rays with and without using bistatic information.

From the simulations using the canonical object it can be concluded that, even though this object is shaped quite unfavorable for the application of the bistatic acceleration, the impairment of the monostatic scattering fields is still acceptable for bistatic angles up to 10 or 15°. Therefore a bistatic angle of 10° can be proposed for the simulation of a passenger car model which will be done in the next future.

It should be noted that for the presented ray-based approach the inaccuracies introduced by the applied assumptions can be expected to be distributed over all observation points, as they are all based on the same ratio of monostatic to bistatic information. Therefore the resulting quality and accuracy of all aspect angles is affected in a similar way. This is in clear contrast to other simulation techniques, where the equivalence principle is used in a post-processing manner to interpolate observation points leading to a reduced accuracy of the interpolated points compared to the directly simulated ones.

Finally, the presented acceleration technique has an advantageous scaling property as it is expected to be most efficient for very high frequencies, where, at the same time, it is needed the most to reduce computational complexity.

References

Andersh, D., Moore, J., Bhalla, R., Hughes, J., et al.: Xpatch 4: the next generation in high frequency electromagnetic modeling and simulation software, in: Proc. IEEE International Radar Conference 2000, 844–849, 2000.

Balanis, C. A.: Advanced Engineering Electromagnetics, New York, Wiley, 1989.

Fujimoto, A., Tanaka, T., and Iwata, K.: ARTS: Accelerated Ray Tracing System, IEEE Comput. Graph. Appl., 6(4), April 1986.

Galloway, P. and Welsh, T.: Extensions to the Shooting Bouncing Ray algorithm for scattering from diffuse and grating structures, in: Proc. International Radar Symposium, IRS, Berlin, September 2005.

Kell, R. E.: On the Derivation of Bistatic RCS from Monostatic Measurements, in: Proc. IEEE, vol. 53, August 1965.

Ling, H., Chou, R. C., and Lee, S. W.: Shooting and Bouncing Rays: Calculating the RCS of an Arbitrarily Shaped Cavity, IEEE Trans. Antennas Propag., 37(2), 194–205, 1989.

Mensa, D. L.: High Resolution Radar Cross Section Imaging, Artech House, Boston, 1991.

Schuh, J., Woo, A. C., and Simon, M. P.: The Monostatic/Bistatic Approximation, IEEE Antenn. Propag. M., 36(4), 76–78, 1994.

Siegel, K. M., Crispin, J. W., Newman, J. R., et al.: Methods of Radar Cross-Section Analysis, Academic Press, 1968.

Weinmann, F.: Ray Tracing with PO/PTD for RCS Modeling of Large Complex Objects, IEEE Trans. Antennas Propag., 54(6), 1797–1806, June 2006.

Hybrid multilevel plane wave based near-field far-field transformation utilising combined near- and far-field translations

C. H. Schmidt[1] **and T. F. Eibert**[2]

[1]Institute of Radio Frequency Technology, Universität Stuttgart, Germany
[2]Lehrstuhl für Hochfrequenztechnik, Technische Universität München, Germany

Abstract. The radiation of large antennas and those operating at low frequencies can be determined efficiently by near-field measurement techniques and a subsequent near-field far-field transformation. Various approaches and algorithms have been researched but for electrically large antennas and irregular measurement contours advanced algorithms with low computation complexity are required. In this paper an algorithm employing plane waves as equivalent sources and utilising efficient diagonal translation operators is presented. The efficiency is further enhanced using simple far-field translations in combination with the expensive near-field translations. In this way a low complexity near-field transformation is achieved, which works for arbitrary sample point distributions and incorporates a full probe correction without increasing the complexity.

1 Introduction

Far-field antenna radiation pattern measurements of electrically large antennas or antennas at low operating frequencies are often complicated due to the large far-field distance. Near-field measurements can be performed in indoor test ranges of limited size and efficient near-field far-field transformation algorithms are known, which compute the desired radiation characteristic of the antenna under test (AUT) with high accuracy. Thereby the radiation behaviour of the AUT is modelled by equivalent sources. Various types of equivalent sources are feasible resulting in different near-field transformation algorithms all having their own benefits and drawbacks. Popular approaches are equivalent current methods (Sarkar and Taaghol, 1999), which are well suited for irregular and non-closed measurement grids as well as near-field samples close to the AUT. These methods assume an equivalent current distribution either on a fictitious surface or the radiating structure itself and relate the currents to the near-field values using an electric field integral equation. Another approach, implemented in various commercial tools, is based on eigenmode expansions (Hansen, 1988; Kerns, 1981) of the radiated AUT fields. The AUT fields are expanded in the eigenmodes of the employed measurement contour, namely plane waves for Cartesian grids, and cylindrical and spherical waves for cylindrical and spherical measurement contours. The orthogonality relations of the eigenmodes are used to relate the wave amplitudes to the measured near-field values. A detailed description of the different techniques is beyond the scope of this paper and an overview can be found in Yaghjian (1986).

Correspondence to: C. H. Schmidt
(schmidt@ihf.uni-stuttgart.de)

The algorithm presented in this paper works with plane waves as equivalent sources, which are assumed for all directions on the Ewald sphere (Schmidt et al., 2008). One of the main advantages are the diagonal translation operators known from fast multipole methods (FMM) (Coifman et al., 1993; Chew et al., 2001), which are used to translate the plane waves from the AUT to the field probe position in an efficient way, where all waves are superimposed. Using a multilevel grouping scheme of the measurement points (Schmidt and Eibert, 2008), translations are carried out to the box centres on the highest level only and the plane wave contributions are processed towards the measurement points using a disaggregation and anterpolation scheme. Thereby a complexity of $\mathcal{O}(N \log N)$, N being the number of measurement points can be achieved. The unknown wave amplitudes are obtained in an inverse process. In this paper a hybrid approach is presented, which combines the efficient far-field translations with the standard, more expensive near-field translations. Far-field translations can be utilised for measurement points fulfilling the far-field condition. The far-field criterion can be relaxed by subdividing the AUT in several source boxes each with a decreased far-field distance. This approach is not based on the evaluation of orthogonal basis functions and is thus able to deal with arbitrary and also irregular sample point distributions.

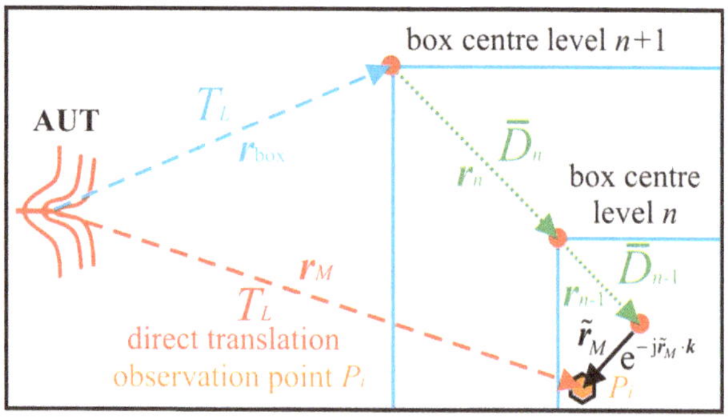

Fig. 1. Multilevel approach with disaggregation and anterpolation.

For practical measurements electric near-field values can be measured indirectly via the output voltage of a field probe only. The finite geometric extent and the receiving characteristic of the probe have a more or less strong impact on the measurement results, depending on the order of the probe. Therefore it is necessary to consider and compensate this influence in the near-field transformation to obtain accurate transformation results. This is called probe correction. For the well-known spherical multipole techniques different probe correction algorithms are available depending on the order of the probe and also the desired accuracy. A full correction of arbitrary probes becomes time consuming since either long transformation or measurement times are required. In the presented approach a full probe correction, similar to the classical plane wave approaches, is already included without increasing the complexity of the algorithm.

In Sect. 2 the multilevel near-field transformation algorithm based on plane wave expansion and diagonal translation operators is described. Section 3 introduces the hybrid approach utilising combined near- and far-field translations. Section 4 shows a performance evaluation of the presented method together with some results and Sect. 5 concludes the paper.

2 Multilevel near-field far-field transformation

The near-field transformation algorithm presented in this paper utilises plane waves $\tilde{\boldsymbol{J}}(\hat{k})$ on the entire Ewald sphere as equivalent sources. The output voltage (Schmidt et al., 2008)

$$U(\boldsymbol{r}_M) = -\mathrm{j}\frac{\omega\mu}{4\pi} \oiint T_L(\hat{k}, \hat{r}_M)\boldsymbol{P}(\hat{k}, \boldsymbol{r}_M) \cdot \left(\bar{\boldsymbol{I}} - \hat{k}\hat{k}\right) \cdot \tilde{\boldsymbol{J}}(\hat{k})d\hat{k}^2 \qquad (1)$$

of the field probe measuring the radiated AUT field distribution is obtained by translating the plane waves to the field probe position $\boldsymbol{r}_M$, where they are superimposed in a weighted manner. ω is the angular frequency, $T_L(\hat{k}, \hat{r}_M)$ is the diagonal FMM translation operator (Coifman et al., 1993; Chew et al., 2001), and

$$\boldsymbol{P}(\hat{k}, \boldsymbol{r}_M) = \begin{pmatrix} \nu_{\mathrm{Pol1}}\mathrm{FF}_{\mathrm{Pol1}} \\ \nu_{\mathrm{Pol2}}\mathrm{FF}_{\mathrm{Pol2}} \end{pmatrix} \qquad (2)$$

is the probe correction coefficient, which is used to weight the incident plane waves at the field probe prior to superposition. Due to the far-field characteristic of the plane waves used as equivalent sources, the probe correction coefficient is simply the far-field characteristic of the probe FF multiplied with the antenna factor ν converting the measured electric field to the probe output voltage.

In the approach presented so far translations are carried out to every measurement point explicitly, which becomes numerically intensive for large antennas and a high number of measurement points. A multilevel version of the algorithm (Schmidt and Eibert, 2008), related to the multilevel fast multipole method (MLFMM) (Chew et al., 2001), organises the measurement scenario in a multilevel box structure, where translations are carried out to the box centres on the highest level N only, resulting in the spectrum

$$\tilde{\boldsymbol{J}}_N^{i_N}(\hat{k}) = T_L(\hat{k}, \hat{r}_{\mathrm{box}})\left(\bar{\boldsymbol{I}} - \hat{k}\hat{k}\right) \cdot \tilde{\boldsymbol{J}}(\hat{k}) \qquad (3)$$

at the i_N-th box. The plane waves

$$\tilde{\boldsymbol{J}}_n^{i_n}(\hat{k}) = \bar{\boldsymbol{D}}_n^{i_n}(\boldsymbol{k}, \boldsymbol{r}_n^{i_n}) \cdot \left(\bar{\boldsymbol{I}} - \hat{k}\hat{k}\right) \cdot \tilde{\boldsymbol{J}}_{n+1}^{i_{n+1}}(\hat{k}) \qquad (4)$$

are processed through the different levels towards the measurement points recursively using a disaggregation and anterpolation procedure as illustrated in Fig. 1.

Disaggregation is a simple phase shift from the box centre on the higher level to the box centre on the lower level and finally the measurement points. Anterpolation is the adjoint operation to interpolation and reduces the sampling rate (Schmidt and Eibert, 2008; Chew et al., 2001). This is possible since the spectral content of the plane wave spectrum is proportional to the box size on a given level. This reduces the complexity of the algorithm from $\mathcal{O}(N^2)$ to $\mathcal{O}(N\log N)$, N being the number of measurement points. The phase shift and the anterpolation dyad $\bar{\boldsymbol{V}}_n(\hat{k})$ are combined in

$$\bar{\boldsymbol{D}}_n^{i_n}(\boldsymbol{k}, \boldsymbol{r}_n^{i_n}) = \bar{\boldsymbol{V}}_n(\hat{k})\mathrm{e}^{-\mathrm{j}\boldsymbol{r}_n^{i_n}\cdot\boldsymbol{k}}. \qquad (5)$$

Arrived on the lowest level, the final phase shift from the box centre on the lowest level to the measurement point is performed and also the probe correction is carried out for a minimum number of plane wave samples, resulting in the probe output voltage

$$U(\boldsymbol{r}_M) = -\mathrm{j}\frac{\omega\mu}{4\pi}\sum_{\hat{k}} W(\hat{k})\mathrm{e}^{-\mathrm{j}\tilde{\boldsymbol{r}}_M^{i_n}\cdot\boldsymbol{k}}\boldsymbol{P}(\hat{k}, \boldsymbol{r}_M) \cdot \left(\bar{\boldsymbol{I}} - \hat{k}\hat{k}\right) \cdot \tilde{\boldsymbol{J}}_0^{i_0}(\hat{k}). \qquad (6)$$

The integrals are evaluated numerically using Gauss-Legendre quadrature (Abramowitz and Stegun, 1972) with the weights $W(\hat{k})$.

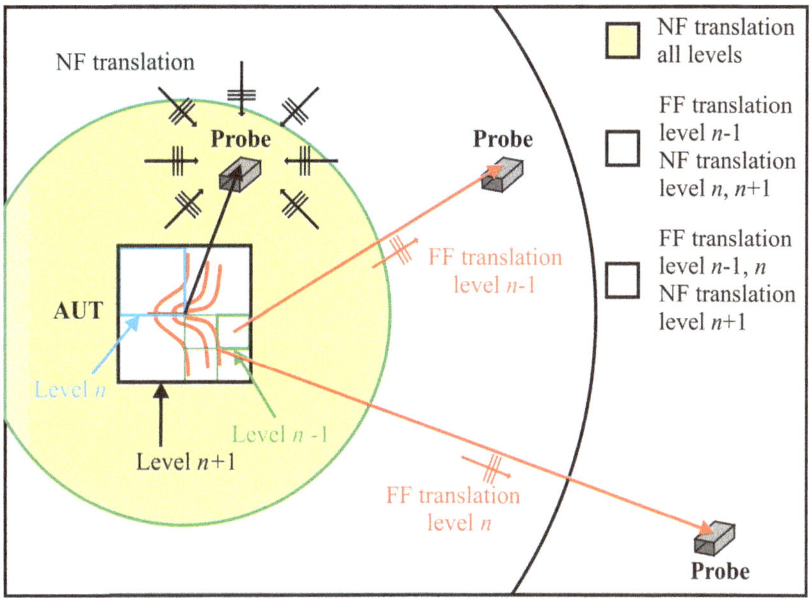

Fig. 2. Near-field and far-field translation ranges.

Electrically large antennas require a huge number of plane wave samples and so iterative solver techniques like the generalised minimum residual method (GMRES) (Saad, 2003) are employed to solve the normal equation on the fly. On the fly means, the probe output voltages are computed from a given set of plane waves for the forward operator without forming a coupling matrix explicitly. The adjoint operator is applied to the result of the forward operator in order to solve the normal system of equations.

3 Hybrid approach

For the derivation of the near-field transformation algorithm presented in the previous section, a multipole expansion and a subsequent plane wave expansion have been applied. In order for the plane wave representation to converge, the minimum spheres enclosing AUT and probe must not overlap. For any larger distance the diagonal FMM translation operators can be applied to translate the plane waves to the field probe position. These operators work with a complete set of plane waves and the translations are referred to as near-field translations. For measurement points with a distance larger than the far-field distance

$$r_{\mathrm{far}} = \frac{\pi(0.87d)}{\lambda} \tag{7}$$

from the AUT, it is possible to use efficient far-field translations (Chew et al., 2002), which require the translation of a single plane wave in the direction towards the measurement point only. d is the side length of the minimum box enclosing the AUT. Nevertheless the far-field condition is normally not fulfilled for a near-field measurement and only for some measurement points at the border of a planar scan surface for example, far-field translations might be applicable. In order to facilitate far-field translations, it is possible to recursively subdivide the AUT in smaller source boxes, each of them fulfilling Eq. (7). The different translation ranges for near- and far-field translations resulting thereby are shown in Fig. 2.

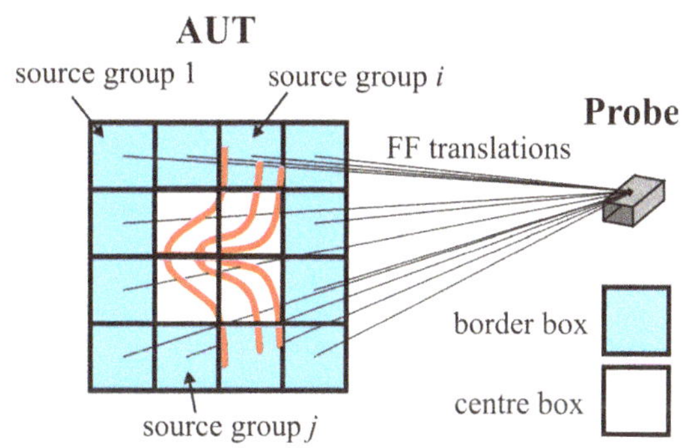

Fig. 3. Source box grouping and far-field translations.

Each source box contributes with a plane wave to the total probe voltage at a measurement point $\boldsymbol{r}_M$ resulting in a partial voltage

$$\boldsymbol{U}_{\mathrm{sb}}(\boldsymbol{r}_M) = -\mathrm{j}\frac{\omega\mu}{4\pi}\frac{\mathrm{e}^{-\mathrm{j}k(r_M - r_{\mathrm{sb}})}}{r_M - r_{\mathrm{sb}}}\boldsymbol{P}(\hat{k}, \boldsymbol{r}_M)\cdot\left(\bar{\boldsymbol{I}} - \hat{k}_0\hat{k}_0\right) \cdot\tilde{\boldsymbol{J}}(\hat{k}_0)\,, \tag{8}$$

with $\boldsymbol{r}_{\mathrm{sb}}$ the source box centre. Therefore a separate set of plane waves is assumed in every border box (see Fig. 3)

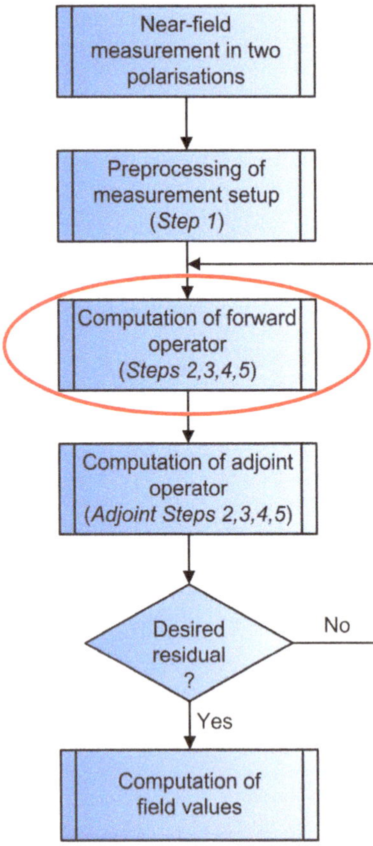

Fig. 4. Flow chart near-field transformation.

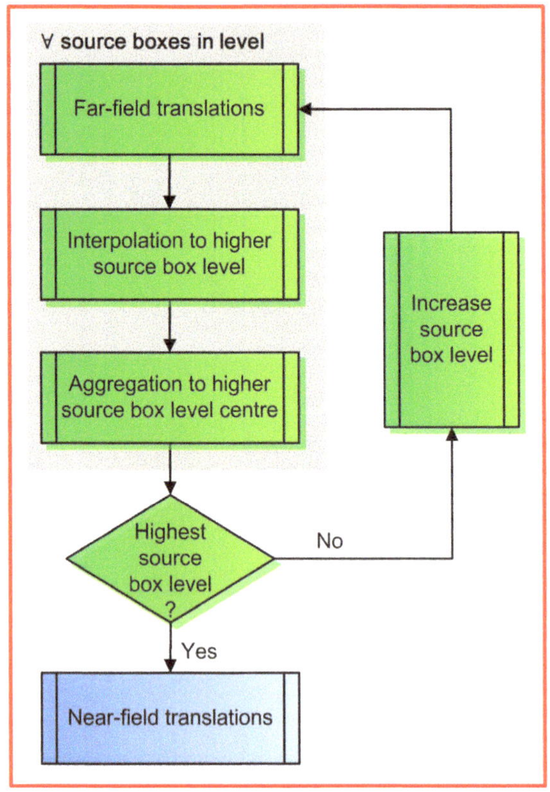

Fig. 5. Flow chart hybrid forward operator.

as equivalent sources and the plane wave in the direction $\hat{k}_0 = (r_M - r_{sb})/|r_M - r_{sb}|$ is interpolated. Since it is sufficient to have equivalent sources enclosing the AUT it is not required to have additional sources in the centre boxes. The total probe output voltage

$$U(r_M) = \sum_i U_{\mathrm{sb},i}(\boldsymbol{r}_M, \hat{k}_{0,i}) \tag{9}$$

is obtained as superposition of all source box contributions.

A typical near-field measurement with subsequent near-field transformation is described in the flow chart in Fig. 4. First the near-field measurement is carried out for two (usually orthogonal) field polarisations. In the second step the measurement setup is pre-processed, e.g. the multilevel box structure is established and translation and disaggregation operators are pre-computed for an efficient computation of the forward operator. The desired amplitudes of the equivalent sources are determined by the iterative GMRES solver. The matrix-vector-products to determine the base of the Krylov subspace are carried out on the fly by applying the adjoint operator to the result of the forward operator. After each iteration the stop criterion is checked and if the desired residuum is reached field values can be computed from the known equivalent sources. Otherwise further iterations are performed.

To emphasise the novel hybrid approach, the forward operator is described in the flow chart in Fig. 5 in detail. The green boxes show steps of the hybrid approach while blue boxes belong to the standard approach. The formulation starts at the lowest level with the highest number of source boxes. Far-field translations to the measurement points fulfilling the far-field criterion for the source boxes on the given level are performed first. For every measurement point the highest level for which far-field translations may be used is determined in the pre-processing step and translations are preferably carried out on the highest possible level (Tzoulis and Eibert, 2005), since the overall number of translations is smaller for a lower number of source boxes. In the next step the plane wave spectra of the boxes on the given level are interpolated to the sampling rate on the next higher level and aggregated to the centre of the parent box. This is done recursively until arrived on the highest level, where only one box remains enclosing the complete AUT. The plane wave samples of this spectrum are used as equivalent sources in the standard approach and they are used to carry out the near-field translations in the conventional way, for measurement points not fulfilling the far-field condition with the given number of source box levels. For an increasing number of

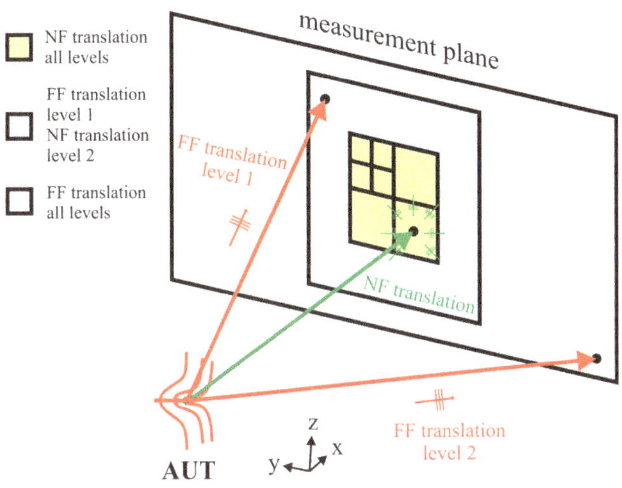

Fig. 6. Planar setup for performance evaluation.

Table 1. Performance of different translation types.

Translation type	Time	# NF points	# FF points
NF	19.5 s	22 801	0
NF/FF (1 level)	8.0 s	7377	15 424
NF/FF (2 levels)	6.0 s	1597	15 424/5780
FF	2.0 s	0	22 801

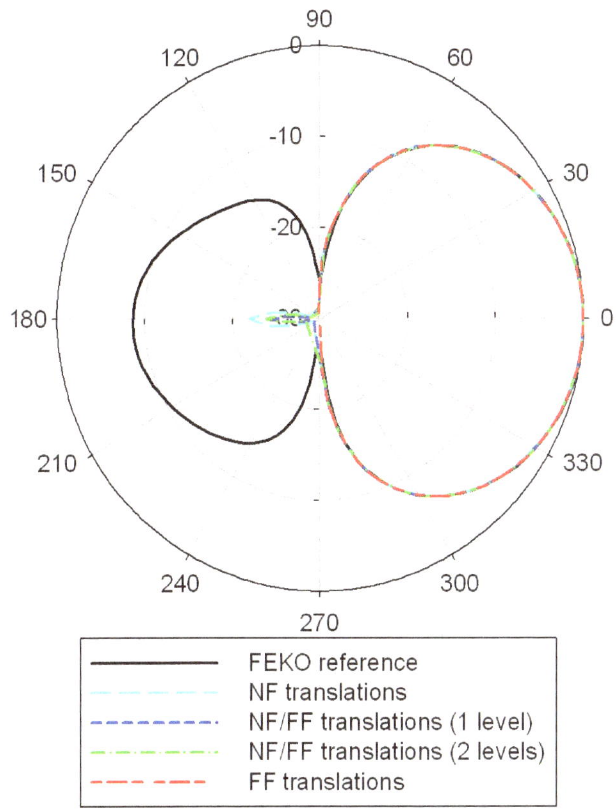

Fig. 7. Comparison of far-field patterns for different translation types (horizontal cut).

source boxes, the computational effort for the far-field translations is increasing and therefore the number of source box levels has to be chosen carefully in order to find an optimal combination of near- and far-field translations.

The adjoint operator is determined in a similar way to the forward operator utilising aggregation and interpolation as adjoint operations to disaggregation and anterpolation.

4 Results and performance evaluation

In this section the functionality and the performance of the presented hybrid near-field transformation algorithm should be shown. Therefore a simple near-field measurement scenario is considered, which was obtained by the electromagnetic field simulator FEKO (http://www.feko.info). Considered is a two-element Yagi-Uda antenna operating at 1.8 GHz. A large planar scan surface of 12 m times 12 m was chosen allowing to investigate the different translation types in a comfortable way. The principal measurement setup with different translation ranges is shown in Fig. 6.

In order to compare the performance of the different translation types, the measurement plane is chosen such that it is in far-field distance to the AUT. Two transformations, one utilising near-field (NF), the other utilising far-field (FF) translations only, have been carried out to get the minimum and maximum transformation times. The results are shown in Table 1 and the time in the second column is given for a single solver iteration. For the first hybrid transformation (1 level), NF translations have been used for the centre points of the measurement plane and FF translations for the points at the borders. Around two thirds of the measurement points are treated with FF translations and the computation time could be reduced significantly.

By comparing the times for NF translations only and FF translations only with the hybrid approach it becomes obvious, that the main part of the computation time is spent for NF translations. For the second hybrid transformation (2 levels), a second source box level is introduced further decreasing the number of NF translations. The transformation time is further decreasing, but it can be seen from the decrease of 2 s compared to the first hybrid transformation, that FF translations with an increasing number of source boxes also increase the transformation time. In Figs. 7 and 8 the results of the different transformations are shown together with the FEKO reference far-field pattern in a horizontal and vertical cut. Due to the planar measurement surface, the far-field pattern is of course only valid in the forward hemisphere or slightly beyond. The results are all comparable to the reference solution showing no degradation of the transformation accuracy by the hybrid approach.

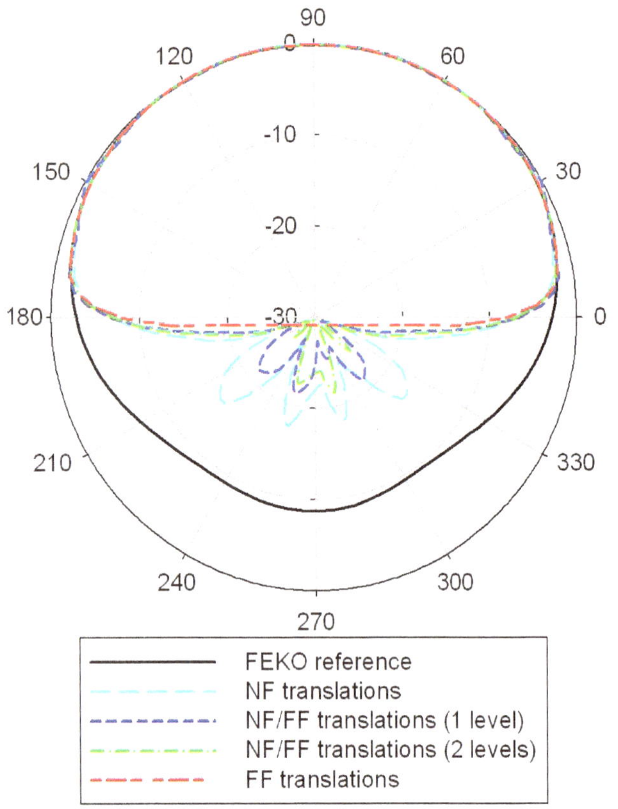

Fig. 8. Comparison of far-field patterns for different translation types (vertical cut).

5 Conclusions

A hybrid approach for the near-field far-field transformation employing plane wave expansion and diagonal translation operators has been presented. The algorithm works with a multilevel plane wave representation resulting in an overall complexity of $\mathcal{O}(N \log N)$, N being the number of measurement points. Utilising very efficient far-field translations in combination with the standard near-field translations, the efficiency is further enhanced. With a source box grouping scheme, the required far-field condition can be achieved also for larger antennas. The algorithm has a full probe correction included and is suited for electrically large antennas due to the low complexity. No special requirements for the measurement grid do apply and also irregular sample point distributions can be handled.

References

Abramowitz, M. and Stegun, I. E.: Handbook of Mathematical Functions, Dover Publications, Inc., New York, 9th edn., 1972.

Chew, W. C., Jin, J. M., Michielssen, E., and Song, J. M.: Fast and Efficient Algorithms in Computational Electromagnetics, Artech House, Inc., 2001.

Chew, W. C., Cui, T. J., and Song, J. M.: A FAFFA-MLFMA Algorithm for Electromagnetic Scattering, IEEE Trans. Antennas Propag., 50, 1641–1649, 2002.

Coifman, R., Rokhlin, V., and Wandzura, S.: The Fast Multipole Method for the Wave Equation: A Pedestrian Prescription, IEEE Antennas and Propag. Mag., 35, 7–12, 1993.

FEKO Suite 5.4, http://www.feko.info, access: 24 February 2009.

Hansen, J. E.: Spherical Near-Field Antenna Measurements, IEE Electromagnetic Wave Series 26, Exeter, UK, 1988.

Kerns, D. M.: Plane-Wave Scattering-Matrix Theory of Antennas and Antenna-Antenna Interactions, National Bureau of Standards, Boulder CO, 1981.

Saad, Y.: Iterative Methods for Sparse Linear Systems, Society for Industrial and Applied Mathematics, 2nd edn., 2003.

Sarkar, T. K. and Taaghol, A.: Near-Field to Near/Far-Field Transformation for Arbitrary Near-Field Geometry Utilizing an Equivalent Current and MoM, IEEE Trans. Antennas Propag., 47, 566–573, 1999.

Schmidt, C. H. and Eibert, T. F.: Multilevel Fast Near-Field Far-Field Transformation for Electrically Large Antennas, IEEE AP-S International Symposium, San Diego, USA, 2008.

Schmidt, C. H., Leibfritz, M. M., and Eibert, T. F.: Fully Probe-Corrected Near-Field Far-Field Transformation Employing Plane Wave Expansion and Diagonal Translation Operators, IEEE Trans. Antennas Propag., 56, 737–746, 2008.

Tzoulis, A. and Eibert, T. F.: Efficient Electromagnetic Near-Field Computation by the Multilevel Fast Multipole Method Employing Mixed Near-Field/Far-Field Translations, IEEE Antennas Wireless Propag. Lett., 4, 449–452, 2005.

Yaghjian, A. D.: An Overview of Near-Field Antenna Measurements, IEEE Trans. Antennas Propag., 34, 30–45, 1986.

14

Micro- and macroscopic simulation of periodic metamaterials

R. Schuhmann[1], **B. Bandlow**[1], **G. Lubkowski**[2], **and T. Weiland**[2]

[1]Univ. Paderborn, EIM-E, FG Theoretische Elektrotechnik, Warburger Straße 100, 33098 Paderborn, Germany
[2]TU Darmstadt, Inst. für Theorie Elektromagnetischer Felder, Schloßgartenstraße 8, 64289 Darmstadt, Germany

Abstract. In order to characterize three-dimensional, left-handed metamaterials (LHM) we use electromagnetic field simulations of unit cells. For waves traveling in one of the main directions of the periodic LHM-arrays, the analysis is concentrated on the calculation of global quantities of the unit cells, such as scattering parameters or dispersion diagrams, and a careful interpretation of the results. We show that the concept of equivalent material values – which may be negative in a narrow frequency range – can be validated by large "global" simulations of a wedge structure. We also discuss the limitations of this concept, since in some cases the macroscopic behavior of an LHM cannot be accurately described by equivalent material values.

1 Introduction

In his triggering publication in 1968, Veselago proposed the existence of left-handed waves for materials with double negative material parameters, i.e. a permittivity and permeability below zero (Veselago, 1968). During the last 15 years, a number of realizations of such materials have been developed as *composites*: periodic arrays of unit cells with metallic or dielectric substructures on a microscopic (sub-wavelength) scale. The reversal of many concepts of wave propagation, such as phase velocity or the angle of refraction, and the possibility to *engineer* the material parameters within a large and unusual range, open many prospectives for the design of new components in microwave techniques and optics.

Being composites with local fields on a microscopic scale, the negative material parameters of LHM realizations can only be understood in terms of *effective material values*, and their determination for a given setup has become a central issue of the research on this field. Clearly, these effective material values will show dispersion, and very often the well-known Lorentz- and Drude formulas from atomic material models can be successfully applied.

The extraction methods used today either start from a microscopic point of view, i.e. local fields within the unit cells, or from global quantities such as reflection and transmission coefficients of single cells or of larger arrays. The raw data can be obtained by analytical considerations (Pendry et al., 1999) or measurements (Shelby et al., 2001), or – with increasing importance – by field simulations (Smith et al., 2000; Weiland et al., 2001). Typical results are the complex effective permittivity and permeability values as functions of frequency.

The effective material values which are obtained by such techniques are an important result for the synthesis of the composites. However, their main purpose is to be used in the design process of large-scale components, where the novel properties of metamaterials shall be profitably used. Thus, the extraction results should of course be as accurate as possible, in the sense that the equivalent model can describe the behavior of the metamaterial within the interesting frequency band. However, the model should also be *physical*, i.e. it should be passive (as long as no active materials are present on the microscopic scale) and last but not least it should be causal. As a critical review of published results (such as, e.g., in Simovski and Tretyakov, 2007) shows, these properties are sometimes not fulfilled, and it seems as if this is not a result of shortcomings in the respective implementations.

The rest of the paper is organized as follows: first, we shortly introduce the simulation model for a typical metamaterial unit cell, which is used as an example to discuss various extraction methods for effective material values. In Sect. 2.4 we describe a validation approach for this extraction process and show first results of this validation for the so-called PFDM approach. Finally, in Sect. 3 we propose to characterize metamaterials by the more general (and of

Correspondence to: R. Schuhmann
(schuhmann@tet.upb.de)

course well-known) approach of dispersion diagrams in order to avoid the problems with non-physical effective material parameters. Using such dispersion diagrams as reference solution it is shown that the characterization of metamaterials based on scattering matrices can be enhanced, if higher order modes at the ports are considered.

2 Equivalent material models for metamaterials

2.1 Numerical model

We concentrate on 3-D metamaterial concepts such as the well-known split ring resonator (SRR) and similar structures, which can be analyzed by electromagnetic field simulation. All simulations are performed using commercial as well as laboratory implementations of the Finite Integration Technique, FIT (Weiland, 1977, 1996), a full-wave simulations method based on a three-dimensional computational grid. If used on Cartesian grids and in time domain, in combination with the leapfrog integration scheme, the FIT is computationally equivalent to the FDTD method (Yee, 1966). However, the more general formulation also allows for various other solvers, e.g. for the eigenvalue formulation in frequency domain with periodic boundary conditions, as discussed below. The implementations include the described extraction methods for effective materials as well as all excitation techniques and boundary conditions needed for the simulation models. Further, for the so-called macroscopic simulations advanced material models are available to treat dispersive permittivity and permeability parameters in time domain (Gutschling et al., 2000). The structure modeling and parts of the postprocessing and visualization are performed by the commercial tool CST Microwave Studio (MWS, by Computer Simulation Technology GmbH, 2007).

2.2 Extraction methods

One of the first ideas to define effective material values in metamaterials goes back to Pendry et al. (1999). The local field distribution in three-dimensional unit cells of SRRs and similar structures is estimated by quasi-static approximations, and the effective material values are defined as the ratio of averaged field quantities. Following this technique, the dispersion of the effective permeability is found to be of Lorentz type. For the effective permittivity of an array of conducting wires a similar investigation leads to a Drude-type dispersion curve, and both together may constitute a double-negative material.

In order to consider also more general geometric shapes of the microscopic elements and their cross-coupling effects, several implementations of the field averaging approach *based on simulated field data* have been published, e.g. in Smith et al. (2000); Weiland et al. (2001). However, this type of methods have turned out to be not very robust, e.g. the desired dispersion types can not always be confirmed.

The second (and most often used) class of methods are oriented on classical measurement techniques and try to define the effective medium based on scattering matrices (SM). Typically these are the reflection and transmission coefficients of plane waves illuminating a metamaterial array with finite length. The effective material parameters can be determined by inverting the dependency of typical quantities (like S-parameters S_{11}, S_{21}, or the wave impedance Z) on ϵ and μ. This step is performed at each frequency to gain the desired $\epsilon(\omega)$ and $\mu(\omega)$-curves, respectively. Again, the expected Lorentz- and Drude-dispersion models are often not revealed.

This last problem can be healed by the so-called parameter fitting of dispersive models (PFDM) approach in Lubkowski et al. (2006), where these dispersion models are assumed to hold a-priori, and their parameters are fitted to the measured or simulated SM data. Thus, the resulting material values are physical (i.e. passive and causal) by definition, whereas the achievable accuracy and the valid bandwidth are not easy to determine a-priori. The example given below shows, however, that the PFDM approach can be a very useful tool, which opens a straight-forward way for a validation of the effective LHM model.

2.3 Properties and limitations of equivalent material models

A main issue of effective material values is the question whether they are simultaneously negative in order to support left-handed waves. However, like all dispersive models, the permittivity and permeability are complex quantities, making the question of sign a bit more involved. A nice overview on this topic has recently been published in Tsukerman (2007).

The next issue to be discussed is the passivity and causality of the effective material, representing important constraints in the calculation of the complex permittivity and permeability over frequency. From above mentioned extraction methods, only the analytical (quasi-static) estimation and the PFDM method yield provably physical results. On the other side, the standard SM-based techniques typically face a couple of problems: First, they often suffer from numerical relicts within the extraction process such as numerical noise in the raw data, ill-conditioned inversion formulas, or improperly chosen branches of complex roots. Next, such methods only consider one principle wave traveling through the metamaterial (in most cases a plane wave). This issue is discussed in Sect. 3.2.

As a result, the material curves may show imaginary parts (i.e. losses) in lossless simulation models, the wrong sign of such imaginary parts (i.e. an active effective material), or a violation of the Kramers-Kronig relations and thus of the principle of causality.

Finally, all scattering matrix based results (including those from PFDM) are only valid for one single direction of propagation of the waves.

Fig. 1. Validation of effective material models: Simulations setup of an LHM wedge with microscopic (left) and effective (right) simulation. The microscopic model contains all geometric details of a 16×9 array of SRR-type unit cells.

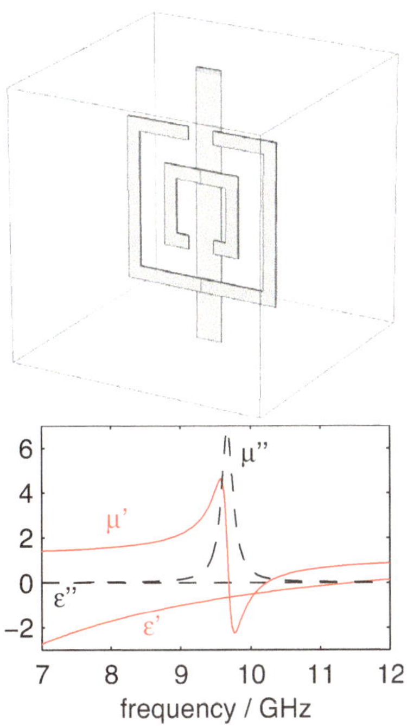

Fig. 2. Unit cell of the LHM in the validation setup (left) and extracted effective material parameters $\underline{\epsilon}=\epsilon'-j\epsilon''$ and $\underline{\mu}=\mu'-j\mu''$. From the PFDM approach we obtain exact Drude and Lorentz models, respectively. The double-negative frequency band is between 9.7 GHz and 10.2 GHz.

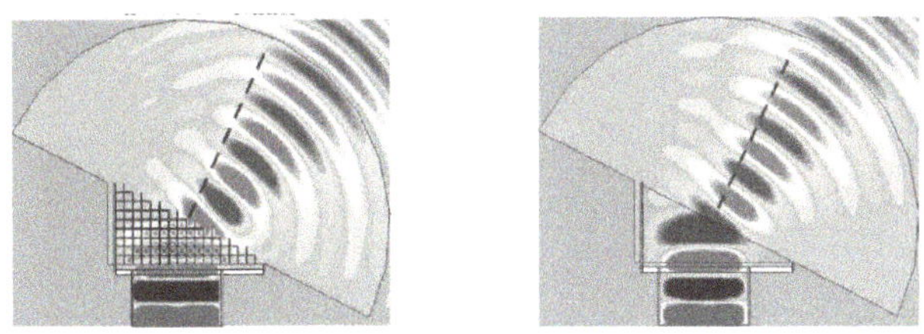

Fig. 3. Results of the LHM-wedge simulations: Both models (microscopic, left, and macroscopic, right) show a negative angle of refraction and thus the desired left-handed behavior.

2.4 Validation of equivalent material models

In spite of the large number of publications on effective material values in metamaterials, not many of them include a thorough validation of the results. Concerning the extraction of the material values from local fields, a validation should compare the behavior of the full (microscopic) LHM model with the effective one. Such a setup is shown in Fig. 1 (full geometric data are given in Lubkowski et al., 2006, 2007): Following the idea of the famous *wedge experiment* published by Shelby et al. (2001), the propagation of a wave through a wedge-shaped LHM and its refraction at the interface to the surrounding air are studied. In a first simulation, all details of the LHM, i.e. the microscopic structure of an array of split ring resonators and wires are considered in a single, large numerical model. For the given wedge we need several millions of grid points (with six unknowns per point), and it is obvious that this procedure is not feasible for arbitrary components. The results are compared to a second simulation, where the LHM is represented by a homogeneous block of material with the dispersive material properties gained by the PFDM-extraction process. Note that this extraction is performed in advance, based on a separate (and computationally cheap) FIT simulation of one single unit cell of the array. The results from the PFDM extraction are the parameters of exact Drude- and Lorentz-type dispersion models for the permittivity and the permeability, respectively (cf. Fig. 2). They are fed into the advanced material models within FIT for the second wedge simulation. Thus, the effective model has the potential to save a considerable amount of computation resources, since its realization in the simulation tools is by far less expensive compared to the microscopic model. In our case, we are able to downscale the computational effort by approximately two orders of magnitude.

As an exemplary result, Fig. 3 shows the fields at one frequency in the left-handed frequency band ($f=10$ GHz): Obviously, a negative angle of refraction is supported in both simulations. Moreover, the setup also allows for a quantitative analysis of the wave propagation and thus of the extracted material parameters. A more detailed investigation will be presented in a subsequent publication. However, these preliminary results already clearly indicate that the PFDM extraction process can be sufficiently accurate, at least in some important parts of the desired frequency spectrum. It

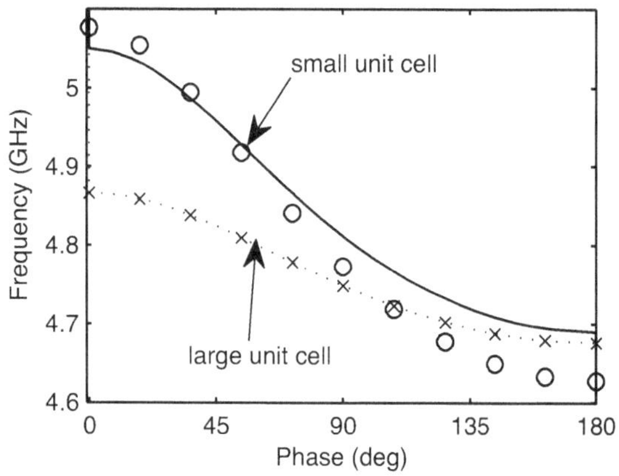

Fig. 4. Dispersion diagram of a metamaterial unit cell. Dots: eigenmode solution (reference), lines: extracted from S-matrix.

is also important to note, that for this type of validation the dispersion data of the effective materials are needed, e.g. as a direct result of the PFDM approach. In case of non-physical material values, violating the passivity or causality rules, the effective simulation will most probably become unstable.

3 Dispersion diagram-based characterization

3.1 Dispersion diagrams as reference solutions

One of the principle weaknesses of the standard SM-based approach is that it only considers plane waves both in the excitation of the unit cells and in the analysis of reflected waves. The accuracy of the material characterization can be improved, if also higher modes (in terms of the aperture of a unit cell) are taken into account. However, this idea hinders the definition of effective materials, since a simple homogeneous (equivalent) medium will not be able to represent all effects of a scattering matrix in higher dimensions.

In order to establish a reference solution for this type of analysis, the characterization of metamaterials using dispersion curves is used. Starting again with the field simulation of single unit cells of the arrays, such dispersion curves can be obtained in frequency domain using periodic boundary conditions with a varying phase angle φ (see also Smith et al., 2000). The corresponding phase factor $\exp(j\varphi)$ is imposed in the two longitudinal boundary conditions (the ports of the SM approach), and from the arising eigenvalue problem we obtain one point of each $\omega(\varphi)$-curve per simulation run.

3.2 Higher order transfer matrices

An alternative way is to calculate the higher-order scattering matrices (S_{ij}), which relates the wave amplitudes a_i, b_i at the "ports" to each other, i.e. the normalized amplitudes of the plane wave and the higher order mode fields at the aperture of the unit cells. By imposing the Bloch condition

$$b_{1i}(\omega) = a_{2i}(\omega)e^{-j\varphi(\omega)}, \quad b_{2i}(\omega) = a_{1i}(\omega)e^{j\varphi(\omega)} \tag{1}$$

we can solve for the dispersion relation $\varphi(\omega)$. Note that for single modes at the ports (plane waves only) and the correlated 2×2-S-matrices, the expressions

$$\mu_{r,\mathrm{eff}} = \frac{Z_c}{\mu_0 d_z \omega}\varphi(\omega), \quad \epsilon_{r,\mathrm{eff}} = \frac{1}{\epsilon_0 d_z \omega Z_c}\varphi(\omega) \tag{2}$$

(with the vacuum material parameters ϵ_0 and μ_0, the unit cell size in propagation direction d_z and the characteristic impedance Z_c) convert this quantity back into the effective material parameters (Simovski and Tretyakov, 2007).

3.3 Quantitative impact of higher-order modes

But we concentrate here on a more detailed analysis of the dispersion relation $\varphi(\omega)$. Figure 4 shows the dispersion diagram for two typical SRR-wire unit cells with different lengths (array parameters, cf. Fig. 5) and left-handed frequency bands around 4.8 GHz. Obviously there exists a discrepancy between the dispersion relation obtained by a fundamental mode scattering matrix and a 3-D full-wave eigenmode simulation with periodic boundary conditions in propagation direction, *if the unit cell size d_z is small*. The field distribution of 3-D full-wave eigenmode simulation at the port plane in Fig. 5 reveals, that the Bloch wave existing in the lattice can not be modeled as a simple plane wave since the field in the aperture is contaminated by higher order modes.

To improve the agreement between the dispersion relations obtained by these two approaches, it is necessary to take into account higher order modes. The extraction of the desired $\varphi(\omega)$-curve out of the simulated $2n\times2n$ S-matrix (for n modes at each port) is performed by the intermediate step of the *transfer scattering matrix* formulation

$$\begin{pmatrix} \mathbf{b}_1(\omega) \\ \mathbf{a}_1(\omega) \end{pmatrix} = \begin{pmatrix} \mathbf{T}_{11} & \mathbf{T}_{12} \\ \mathbf{T}_{21} & \mathbf{T}_{22} \end{pmatrix} \cdot \begin{pmatrix} \mathbf{a}_2(\omega) \\ \mathbf{b}_2(\omega) \end{pmatrix}. \tag{3}$$

Note that all the internal quantities are matrices and vectors now. The formulas for the transformation of **S**- into **T**-matrices can be found in standard textbooks.

We now solve for eigenvalues of the **T**-matrix, which describe the desired Bloch waves. From the phase angles of the propagating ones (those which have a magnitude of 1), we arrive at an improved dispersion relation which includes now the higher order modes at the ports. A typical convergence behavior is shown in Fig. 6. As the number of considered modes increases, the deviation to the 3-D full-wave eigenmode simulation decreases.

As expected, the need for higher-order modes in the scattering matrix approach depends on the size of the metamaterial unit-cells. The fundamental Bloch wave can only be modeled as a plane wave at the port plane, if the ports are sufficiently far away from the SRR and wire elements, and

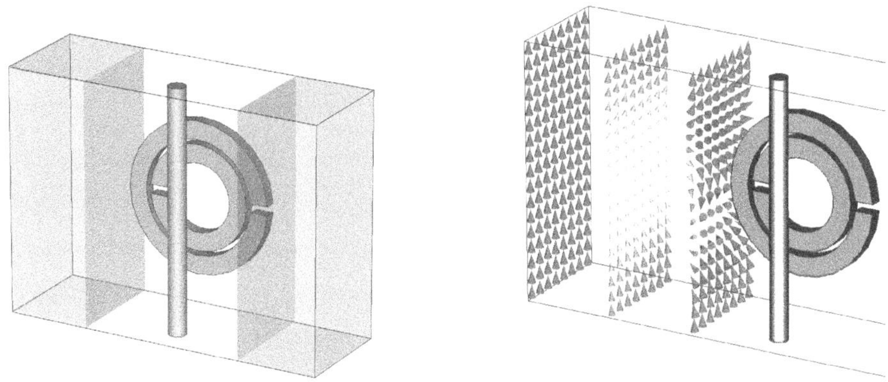

Fig. 5. Electrical fields in port planes for three different lengths of the unit cell. (All two-dimensional field plots of three different structures are projected in one single figure.)

the internal substructures are not too densely packed. Yet a densely packed substructure may be desirable, since the complete bandwidth of the LHM band is increased (cf. Fig. 4).

The analysis also shows, how the impact of higher-order modes can be quantitatively assessed by a comparison with the frequency-domain reference solution with periodic boundaries. In cases where no or only a small number of higher-order modes are necessary for accurate results, the standard time-domain SM approach (with broadband excitation signals) keeps its important advantage to be computationally more efficient than the multiple solution of eigenvalue problems.

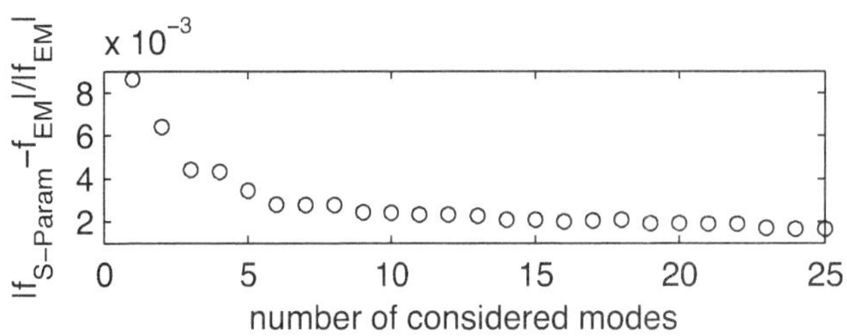

Fig. 6. Convergence behavior: Accuracy of the dispersion diagram depending on the number of considered modes in the higher-order scattering matrix approach.

4 Conclusions

The concept of effective material values for electromagnetic metamaterials is very popular, since there seems to be a simple relation between such parameters and the phenomenon of left-handed waves. We have shortly discussed the various extraction methods for effective materials, and we have shown, that in some cases their results can be validated by large scale field simulations.

However, many difficulties which have been reported in the extraction of effective material values from measured or simulated data seem to be not only due to insufficient implementations. The concept itself is rather limited in its practical application: First, such parameters are only valid for one specific polarization, one specific frequency and one specific angle of incidence. Second, as we have shown in our analysis, the situation gets even more complicated if the structure size gets too small. Some more limitations which we have not discussed here may appear if the dependence between electric and magnetic effects gets too strong, or if interfaces between such materials have to be properly modeled.

A more robust way to characterize such materials is the well-established technique of dispersion diagrams of periodic structures, where all such effects are inherently included. Nevertheless, the concept of effective material parameter may be beneficial in some applications, if its limitations are kept in mind.

Acknowledgements. The work of G. Lubkowski is generously supported by Deutsche Forschungsgemeinschaft under contract GRK 1037/1-04.

References

Computer Simulation Technology GmbH: CST Studio Suite 2006, http://www.cst.com, 2007.

Gutschling, S., Krüger, H., and Weiland, T.: Time Domain Simulation of Dispersive Media with the Finite Integration Technique, Int. J. Num. Modelling: Electronic Networks, Devices and Fields, 13, 329–348, 2000.

Lubkowski, G., Schuhmann, R., and Weiland, T.: Extraction of effective metamaterial parameters by parameter fitting of dispersive models, Microwave Opt. Technol. Lett., 49, 285–288, 2006.

Lubkowski, G., Hirtenfelder, F., Schuhmann, R., and Weiland, T.: 3D Full-wave Field Simulations of Double Negative Metamaterial Macrostructures, in: Proc. Metamaterials 2007, Rome, 22–24 October 2007, pp. 731–734, 2007.

Pendry, J. B., Holden, A. J., Robbins, D. J., and Stewart, W. J.: Magnetism from conductors and enhanced nonlinear phenomena, IEEE T-MTT, 47, 2075, 1999.

Shelby, R. A., Smith, D. R., and Schultz, S.: Experimental Verification of a Negative Index of Refraction, Science, 292, 77–79, 2001.

Simovski, C. R. and Tretyakov, S. A.: Local constitutive parameters of metamaterials from an effective-medium perspective, Phys. Rev. B (Condensed Matter and Materials Physics), 75, 195111, doi:10.1103/PhysRevB.75.195111, http://link.aps.org/abstract/PRB/v75/e195111, 2007.

Smith, D. R., Vier, D. C., Kroll, N., and Schultz, S.: Direct calculation of permeability and permittivity for a left-handed metamaterial, Appl. Phys. Lett., 77, 2246–2248, 2000.

Tsukerman, I.: The Anatomy of Negative Refraction, ICS Newsletter, 14, 3, 2007.

Veselago, V.: The electrodynamics of substances with simultaneously negative values of ϵ and μ, Soviet Physics Uspekhi, 10, 509–514, 1968.

Weiland, T.: Eine Methode zur Lösung der Maxwellschen Gleichungen für sechskomponentige Felder auf diskreter Basis, Archiv für Elektronik und Übertragungstechnik, 31, 116–120, 1977.

Weiland, T.: Time domain electromagnetic field computation with finite difference methods, Int. J. Num. Modeling, 9, 295–319, 1996.

Weiland, T., Schuhmann, R., Greegor, R., Parazzoli, C., Vetter, A., Smith, D., Vier, D., and Schultz, S.: Ab initio numerical simulation of left-handed metamaterials: Comparison of calculations and experiment, J. Appl. Phys., 90, 5419–5424, 2001.

Yee, K. S.: Numerical Solution of Initial Boundary Value Problems Involving Maxwell's Equations in Isotropic Media, IEEE Trans. Antennas and Propagation, 17, 585–589, 1966.

3-D eigenmode calculation of metallic nano-structures

B. Bandlow and R. Schuhmann

University Paderborn, EIM-E, FG Theoretische Elektrotechnik, Warburger Straße 100, 33098 Paderborn, Germany

Abstract. In the calculation of eigenfrequencies of 3-D metallic nanostructures occurs the challenge that the material parameters depend on the desired eigenfrequency. We propose a formulation where this leads to a polynomial eigenvalue problem which can be tackled by different solving strategies. A comparison between a Newton-type method and a Jacobi-Davidson algorithm is given.

1 Introduction

The focus of our analysis is on nanostructures which include a metallic substructure. In the microwave spectrum metals can often be treated as perfect conductors without a significant loss of accuracy. If, however, the nanostructures are supposed to operate at optical frequencies, the finite conductivity of the metallic parts and their frequency dependence must not be neglected. Since we are interested in the computation of eigensolutions of such nanostructures, there is the challenge that the operating frequency (the eigenvalue) is not a-priori known. Thus, the material dispersion leads to a non-linear eigenvalue formulation.

The rest of the paper is organized as follows: Sect. 2 briefly reviews the material behavior of metals at optical frequencies. In Sect. 3 we derive in the first part a continuous eigenvalue representation which is able to take the dispersion into account. In the second part of Sect. 3 this representation is discretized. Section 4 reviews several solving strategies for polynomial eigenvalue problems. Finally a numerical example is presented in Sect. 5.

2 Metals at optical frequencies

There are several publications which investigate experimentally the optical properties of noble metals such as gold and silver (Johnson and Christy, 1972; Ordal et al., 1983; Palik, 1997; Rakic et al., 1998). It turns out that the material values obtained by measurements can be fairly approximated by Drude or a Drude-Lorentz models. These are rational functions which are able to depict one or more resonance effects in a specific frequency range. In general the dependency of the permittivity on the frequency $\epsilon(\omega)$ shall be approximated by a general 2nd order model

$$\epsilon(\omega) = \epsilon_0 \left(\epsilon_\infty + \frac{\beta_0 + j\omega\beta_1}{\alpha_0 + j\omega\alpha_1 - \omega^2} \right). \tag{1}$$

Here ϵ_0 is the permittivity of free-space, the parameters ϵ_∞, α_0, α_1, β_0 and β_1 are real-valued, and an $e^{j\omega t}$ time dependency is implied. Figure 1 exemplarily shows the real and imaginary part of the relative permittivity of silver in the infrared range. The data from (Johnson and Christy, 1972) can be fitted by the parameter set ϵ_∞=5, α_0=0, α_1=3.22e13, β_0=1.96e32, β_1=0 of a Drude approximation.

The imaginary part of the complex permittivity can be reinterpreted as a conductivity, and from the conductivity the skin-depth can easily be calculated. For instance, at a wavelength of 1.5 microns we obtain a skin-depth of 140 nm, which implies that we can not model thin sheets of a material like silver as a good or even perfect electrical conductor at optical frequencies.

3 Eigenmode formulation

3.1 Non-linear and polynomial eigenvalue problem

In the following, we derive the non-linear eigenvalue formulation for structures with dispersive materials. For the sake of simplicity, we assume a homogeneous medium and begin in the continuous rather than the discrete regime. The eigenmode formulation for the electrical field strength $\boldsymbol{E}$ follows from Maxwell's equations in frequency domain and reads

$$\text{curl}\ \frac{1}{\mu}\text{curl}\ \boldsymbol{E} = \omega^2\ \epsilon(\omega)\ \boldsymbol{E}. \tag{2}$$

Correspondence to: B. Bandlow
(bandlow@tet.upb.de)

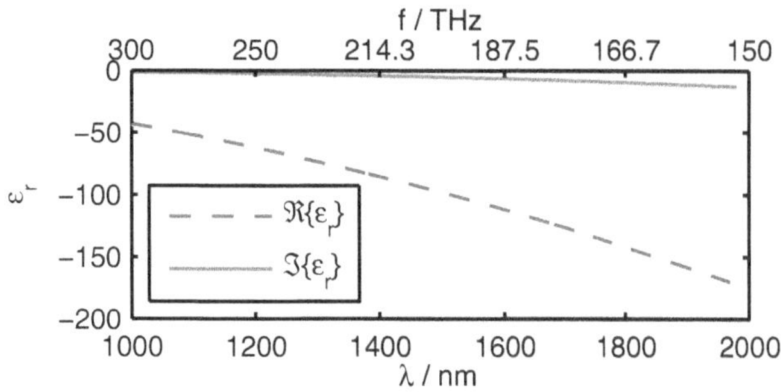

Fig. 1. Real and imaginary part of the relative permittivity of silver in the infrared range.

The dispersion of the permittivity $\epsilon(\omega)$ shall be given by a Drude model as motivated in Sect. 2:

$$\epsilon(\omega) = \epsilon_0 \left(\epsilon_\infty + \frac{\beta_0}{j\omega\alpha_1 - \omega^2} \right). \tag{3}$$

We insert Eq. (3) into Eq. (2) and obtain the complex, non-hermitian, polynomial eigenvalue problem (PEP) in ω

$$(\omega^3 a_3 + \omega^2 a_2 + \omega a_1 + a_0)\boldsymbol{E} = 0. \tag{4}$$

The coefficients a_i are given by

$$a_0 = -j\alpha_1 A_{cc}, \quad a_1 = \epsilon_0\beta_0 + A_{cc},$$

$$a_2 = j\epsilon_0\epsilon_\infty\alpha_1, \quad a_3 = -\epsilon_0\epsilon_\infty, \quad A_{cc} = \text{curl}\,\frac{1}{\mu}\text{curl}.$$

3.2 Discrete formulation

The eigenvalue formulation is discretized using the finite integration technique (FIT) (Weiland, 1977, 1996). In this very general framework Maxwell's equations are transformed into algebraic equations – the Maxwell's Grid Equations – which can be used further to formulate e.g. a discrete wave equation. The degrees of freedom of the FIT approach are the so-called grid voltages $\widehat{\boldsymbol{e}}$, which are defined on the edges of a three-dimensional Cartesian grid. Finally, the discrete representation of the non-linear eigenvalue problem Eq. (2) reads

$$\mathbf{A}_{cc}\widehat{\boldsymbol{e}} = \omega^2 \mathbf{M}_\varepsilon(\omega)\widehat{\boldsymbol{e}}. \tag{5}$$

Here, the large and sparse matrix $\mathbf{A}_{cc}$ is the curl-curl system operator and includes the double curl operation as well as the permeability distribution of the structure. The diagonal matrix $\mathbf{M}_\varepsilon$ is the generalized permittivity operator. The searched eigenvalue is the squared angular frequency ω^2, and the field distribution is defined by the eigenvector $\widehat{\boldsymbol{e}}$. The dimension of the problem is $N_e \times N_e$, with N_e the number of grid edges.

In a straight forward manner, we can also find a discrete representation of the PEP Eq. (4) using FIT

$$\Psi(\omega)\widehat{\boldsymbol{e}} = (\omega^3\mathbf{A}_3 + \omega^2\mathbf{A}_2 + \omega\mathbf{A}_1 + \mathbf{A}_0)\widehat{\boldsymbol{e}} = 0, \tag{6}$$

where $\mathbf{A}_i$ are the spatially discretized coefficient matrices corresponding to the a_i of Eq. (4). This representation also supports an arbitrary inhomogeneous material distribution (including the dispersive permittivity), and it turns out that the usual facet-weighted averaging procedure at material interfaces does not need any special treatment. In FIT with Cartesian grids the matrices $\mathbf{A}_3$ and $\mathbf{A}_2$ are diagonal, and $\mathbf{A}_1$ and $\mathbf{A}_0$ are sparse. The resulting PEP is complex and non-Hermitian.

4 Solver for polynomial eigenvalue problems

The PEP from Sect. 3.2 can be solved in different ways, and we briefly discuss four variants.

4.1 Fixed-point iteration

A first idea is to evaluate $\mathbf{M}_\varepsilon(\omega)$ at a certain frequency ω_i and to solve a standard linear eigenvalue problem from Eq. (5) for the eigenfrequency ω_{i+1}:

$$\mathbf{A}_{cc}\widehat{\boldsymbol{e}} = \omega_{i+1}^2 \mathbf{M}_\varepsilon(\omega_i)\widehat{\boldsymbol{e}}. \tag{7}$$

This approach defines a fixed-point iteration process $\omega_{i+1}=\Phi(\omega_i)$, where the operator Φ includes a solving step of standard linear eigenvalue problem.

This scheme works well (but not very fast) in many cases. However, the proof of its convergence for general cases – e.g. using Banach's fixed-point theorem – is quite challenging due to the complex nature of the eigenvalue problem involved.

4.2 Linearization via companion matrix

A direct way to solve any PEP is to use the so-called companion matrix of the PEP. A PEP of order i can be recast into a generalized linear eigenvalue problem of the form $\mathbf{Ax}=\lambda\mathbf{Bx}$, where

$$\mathbf{A} = \begin{bmatrix} 0 & \mathbf{I} & \cdots & 0 \\ \vdots & \vdots & \ddots & \vdots \\ \vdots & \vdots & \vdots & \mathbf{I} \\ -\mathbf{A}_0 & -\mathbf{A}_1 & \cdots & -\mathbf{A}_{i-1} \end{bmatrix}, \tag{8}$$

$$\mathbf{B} = \begin{bmatrix} \mathbf{I} & & \\ & \ddots & \\ & & \mathbf{I} \\ & & & \mathbf{A}_i \end{bmatrix}, \quad \mathbf{x} = \begin{bmatrix} \widehat{\boldsymbol{e}} \\ \omega\widehat{\boldsymbol{e}} \\ \vdots \\ \omega^{i-1}\widehat{\boldsymbol{e}} \end{bmatrix}.$$

For a problem of polynomial order i with $n \times n$ matrices, this approach leads to $(i-1)\times n$ additional eigenvalues, which are not necessarily all solutions of the original PEP and therefore have to be dropped. Since this approach leads to large and typically ill-conditioned matrices, it is only feasible for small PEPs.

4.3 Newton-type methods for PEPs

The Newton method for polynomial eigenvalue problems may be regarded as a generalization of the method of inverse iteration (Schreiber, 2008). The algorithm yields one eigenpair at a time and is based on a function f of the eigenvector $\boldsymbol{u}$ and the eigenvalue ω

$$f\left(\begin{bmatrix}\boldsymbol{u}\\ \omega\end{bmatrix}\right)=\left(\begin{bmatrix}\Psi(\omega)\boldsymbol{u}\\ \boldsymbol{w}^H\boldsymbol{u}-1\end{bmatrix}\right). \tag{9}$$

It includes the matrix polynomial $\Psi(\omega)$ from Eq. (6) and a normalization vector $\boldsymbol{w}$, which has to be chosen such that $\boldsymbol{w}^H\boldsymbol{u}=1$ holds throughout the iteration. The first derivative of Eq. (9) is given by the Jacobian

$$J\left(\begin{bmatrix}\boldsymbol{u}\\ \omega\end{bmatrix}\right)=\begin{bmatrix}\Psi(\omega) & \Psi'(\omega)\boldsymbol{u}\\ \boldsymbol{w}^H & 0\end{bmatrix} \tag{10}$$

where $\Psi'(\omega)$ denotes the derivation of Eq. (6). For a given starting guess of the eigenpair $(\boldsymbol{u}_0, \omega_0)$, the Newton correction at step i is defined by

$$J\left(\begin{bmatrix}\boldsymbol{u}_i\\ \omega_i\end{bmatrix}\right)\begin{bmatrix}\Delta\boldsymbol{u}_{i+1}\\ \Delta\omega_{i+1}\end{bmatrix}=-f\left(\begin{bmatrix}\boldsymbol{u}_i\\ \omega_i\end{bmatrix}\right) \tag{11}$$

and

$$\boldsymbol{u}_{i+1}=\boldsymbol{u}_i+\Delta\boldsymbol{u}_{i+1}, \qquad \omega_{i+1}=\omega_i+\Delta\omega_{i+1}.$$

Therefore, the linear system Eq. (11) has to be solved in each iteration step. The Newton algorithm is terminated if the norm of the residual

$$\boldsymbol{r}=\Psi(\omega)\boldsymbol{u} \tag{12}$$

is sufficiently small.

4.4 Jacobi-Davidson algorithm for PEPs

The PEP in the form Eq. (6) can also be solved using a Jacobi-Davidson algorithm (JD) (Sleijpen et al., 1996; Bai et al., 2000). This method is intended to find one or more interior eigenvalues of the spectrum near a given target ω_0.

The main idea within the JD method is to project the PEP on a low-dimensional orthogonal subspace $\mathbf{V}$, which leads to a low-dimensional PEP with coefficient matrices $\mathbf{M}_i=\mathbf{V}^*\mathbf{A}_i\mathbf{V}$. The low-dimensional PEP can be solved by the companion matrix approach from Sect. 4.2 and any method for generalized eigenvalue problems. The low-dimensional eigenvector $\boldsymbol{s}$ is expanded to full size again, $\boldsymbol{u}=\mathbf{V}\boldsymbol{s}$, which leads to the current approximative eigenvector $\boldsymbol{u}$. Since $\boldsymbol{u}$ is an approximation, the residual Eq. (12) has to be calculated, and the process stops if the norm of $\boldsymbol{r}$ is sufficiently small. Since this typically does not occur after the first iteration step, a so-called correction equation is formulated and solved, which produces an additional vector, which extends the subspace.

This procedure is repeated until convergence. The computationally most expensive task inside the JD iteration is the solution of the correction equation, which reads

$$\left(\mathbf{I}-\frac{\boldsymbol{p}\boldsymbol{u}^*}{\boldsymbol{u}^*\boldsymbol{p}}\right)\Psi(\theta)(\mathbf{I}-\boldsymbol{u}\boldsymbol{u}^*)\boldsymbol{t}=-\boldsymbol{r}. \tag{13}$$

Here, Ψ is the polynomial from Eq. (6) evaluated at the last, best estimation θ (namely the Ritz value), $\boldsymbol{t}$ is the correction vector, $\boldsymbol{r}$ the residual, $\boldsymbol{u}=\mathbf{V}\boldsymbol{s}$ the best actual approximation of the searched eigenvector, and $\boldsymbol{p}=\Psi'(\theta)$. In the beginning of the JD process the user-defined target ω_0 may be more accurate than an extracted Ritz value. Therefore we set $\boldsymbol{p}=\Psi'(\omega_0)$ until the residual is below a predefined tolerance, in order to get a better correction vector $\boldsymbol{t}$. (This was proposed in (Hochstenbach and Sleijpen, 2008).) For the solution of the correction Eq. (13) we use a preconditioned bicgstab(l) method (Sleijpen and Fokkema, 1993). As preconditioner we use an LU decomposition of the polynomial evaluated at the target value ω_0. Therefore the LU decomposition has to be established only once per JD run.

4.5 Validation

A simple way to validate a specific implementation from the Newton method from Sect. 4.3 or the JD method from Sect. 4.4 is to go back to the fixed-point formulation in Eq. (7) and to execute one single step of the iteration. In all numerical tests our results yield accuracies in the range of numerical noise.

5 Numerical example

As an example we take a bow tie slot antenna which can be regarded as a resonator at optical wavelengths. A potential application is in spectroscopy where field confinement and focusing is desirable. An extensive study can be found e.g. in Guo et al., 2008.

5.1 Setup

The structure consists of a layer of silver and the dimensions are given in Fig. 2. The material parameters of silver are taken from Sect. 2, where Eq. (1) returns a Drude model Eq. (3) and the resulting PEP is of order three. The permeability is chosen to be $\mu=\mu_0$. For the geometric modeling we use the commercial tool CST MICROWAVE STUDIO (Computer Simulation Technology AG (CST)). The discretization of one quarter of the whole structure (due to symmetry) uses a rather coarse mesh with only $15\times16\times22$ grid lines, leading to 13 657 complex degrees of freedom. (Note that in this presentation we concentrate on solving strategies rather than the convergence of our grid model towards the "real" solution.) In order to model a whole array of bow tie

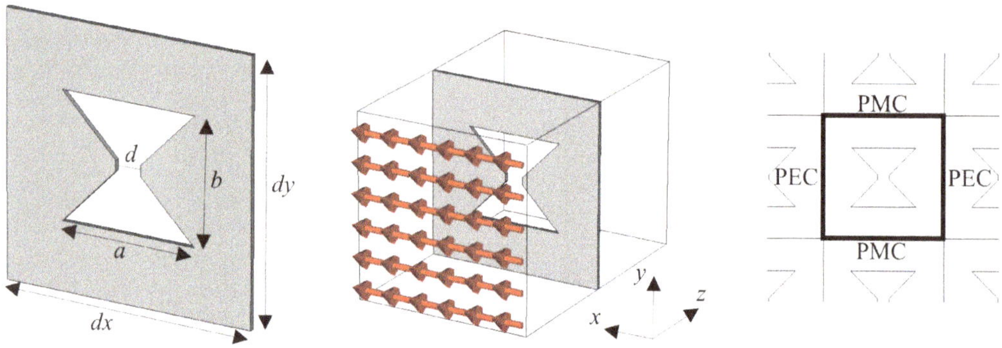

Fig. 2. Bow tie antenna structure with a=270 nm, b=240 nm, d=50 nm and thickness of t=10 nm. The transversal dimensions are dx=dy=500 nm. The polarization of the incident TEM wave and the resulting transversal boundary conditions are shown at the center and at the right.

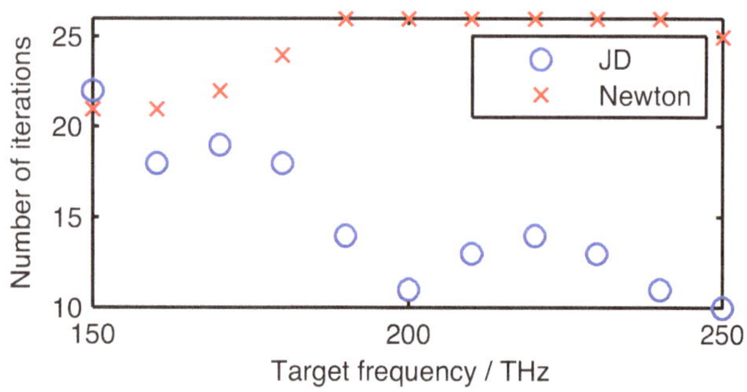

Fig. 3. Number of iterations of the JD and Newton method to converge for different target frequencies ω_0.

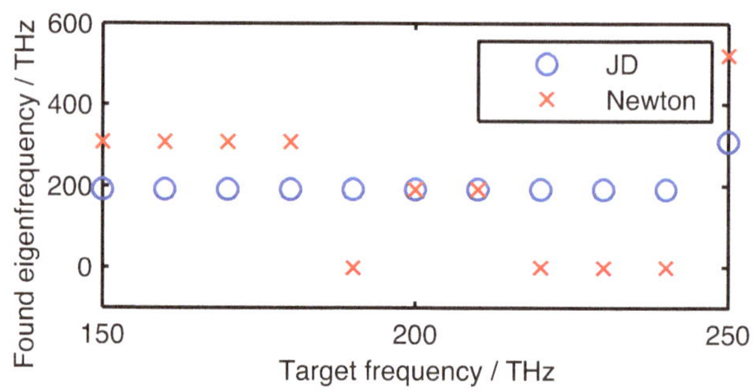

Fig. 4. Resulting eigenvalues of the JD and Newton method for different target frequencies ω_0.

antennas illuminated by a plane wave, the transversal boundary conditions are chosen to be perfectly electric and perfectly magnetic conducting (cf. Fig. 2). In propagation direction we apply Berenger's perfectly matched layer (PML) (Berenger, 1994) to truncate the computational domain appropriately. The PML itself consists of frequency dependent materials, but here we evaluate the PML once at our starting frequency ω_0 instead of including its frequency dependence into the eigenvalue formulation. Nevertheless, to determine the eigenfrequency of the bow tie antenna, a complex non-Hermitian polynomial eigenvalue problem has to be solved, and we compare the results of the Newton method from Sect. 4.3 to those of the JD method from Sect. 4.4, both of them implemented in MATLAB. Convergence is supposed to be reached when the norm of the residual Eq. (12) is less than 1e-9.

5.2 Results

An estimation of the eigenfrequency can be obtained, e.g., by measurements or a scattering parameter simulation, see (Guo et al., 2008) for details. In our case the searched eigenfrequency is at 192.4 THz.

Figure 3 shows the number of iterations needed by the Jacobi-Davidson and Newton method to generate an eigenpair for different target frequencies ω_0. In almost all cases the Jacobi-Davidson method needs less iterations than the Newton method for our specific setup. However, it is questionable *which eigenpairs* are actually found

Figure 4 shows the eigenfrequencies generated by the Jacobi-Davidson and Newton method for different target frequencies ω_0. It turns out that the result of the Newton method strongly depends on the chosen target frequency ω_0 for the first step of Eq. (11). On the contrary the Jacobi-Davidson method reliably generates the same eigenvalue also for less accurately chosen target frequencies.

Figure 5 shows the time needed for the computation of the eigenfrequency (including only those runs which led to the desired eigenvalue of 192.4 THz). Again the JD method outperforms the Newton approach in all cases considered here.

Finally Fig. 6 shows the distribution of electrical field strength of the eigenfrequency at 192.4 THz. It is mainly

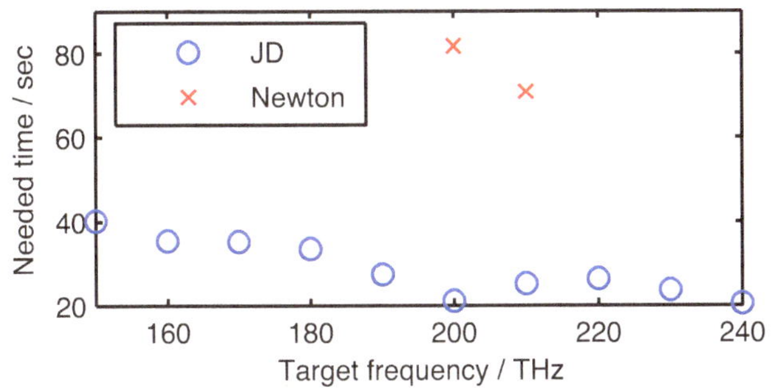

Fig. 5. Computation times of JD and Newton method for target frequencies, which lead to the searched eigenvalue.

orientated in x-direction (like the incident plane wave) and is concentrated within the slot of the bow tie antenna as proposed in the literature.

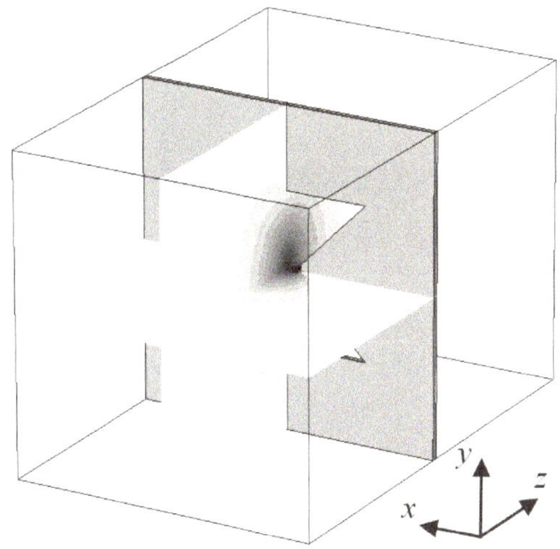

Fig. 6. Contour plot of the field distribution of the eigensolution at 192.4 THz: The x-component of the electrical field strength is confined within the slot of the bow tie antenna.

6 Conclusions

After a review of the behavior of metals at optical frequencies, an eigenfrequency formulation which is capable to deal with frequency dispersive materials has been presented. The formulation is based on the finite integration technique and leads to a polynomial eigenvalue problems (PEP). Four strategies to solve this PEP have been presented and tested. As a numerical example we have used a bow tie slot antenna (Guo et al., 2008) which consists of silver and operates at a wavelengths around 1.5 microns. We have computed its resonance frequency and compared the Newton method and the Jacobi-Davidson method. It turns out that the Jacobi-Davidson method is more robust in terms of poorly chosen target frequencies and also faster at least in our implementation.

If the match of measurement data of material properties requires higher order polynomial functions, our formulation of Sect. 3 can easily be extended and will lead to higher order matrix polynomials.

References

Bai, Z., Demmel, J., Dongarra, J., Ruhe, A., and van der Vorst, H.: Templates for the Solution of Algebraic Eigenvalue Problems: A Practical Guide, SIAM, Philadelphia, 2000.

Berenger, J.-P.: A perfectly matched layer for the absorption of electromagnetic waves, J. Comput. Phys., 114, 185–200, 1994.

Computer Simulation Technology AG (CST): CST Studio Suite 2008, http://www.cst.com, 2008.

Guo, H., Meyrath, T. P., Zentgraf, T., Liu, N., Fu, L., Schweizer, H., and Giessen, H.: Optical resonances of bowtie slot antennas and their geometry and material dependence, Opt. Express, 16, 7756–7766, http://www.opticsexpress.org/abstract.cfm?URI=oe-16-11-7756, 2008.

Hochstenbach, M. E. and Sleijpen, G. L. G.: Harmonic and refined Rayleigh-Ritz for the polynomial eigenvalue problem, Numerical Linear Algebra with Applications, 15, 35–54, 2008.

Johnson, P. B. and Christy, R. W.: Optical Constants of the Noble Metals, Phys. Rev. B., 6, 4370–4379, 1972.

Ordal, M. A., Long, L. L., Bell, R. J., Bell, S. E., Bell, R. R., R. W. Alexander, J., and Ward, C. A.: Optical properties of the metals Al, Co, Cu, Au, Fe, Pb, Ni, Pd, Pt, Ag, Ti, and W in the infrared and far infrared, Appl. Opt., 22, 1099–1119, http://ao.osa.org/abstract.cfm?URI=ao-22-7-1099, 1983.

Palik, E. D.: Handbook of Optical Constants of Solids , 1997.

Rakic, A. D., Djurišic, A. B., Elazar, J. M., and Majewski, M. L.: Optical Properties of Metallic Films for Vertical-Cavity Optoelectronic Devices, Appl. Opt., 37, 5271–5283, http://ao.osa.org/abstract.cfm?URI=ao-37-22-5271, 1998.

Schreiber, K.: Nonlinear Eigenvalue Problems: Newton-type Methods and Nonlinear Rayleigh Functionals, Ph.D. thesis, TU Berlin, http://nbn-resolving.de/urn:nbn:de:kobv:83-opus-18754, 2008.

Sleijpen, G. L. G. and Fokkema, D. R.: Bi-CGSTAB(l) for linear equations involving unsymmetric matrices with complex spectrum, Elec. Trans. Numer. Anal., 1, 11–32, 1993.

Sleijpen, G. L. G., Booten, A. G. L., Fokkema, D. R., and der Vorst, H. A. V.: Jacobi-Davidson type methods for generalized and polynomial eigenproblems, BIT, 36, 595–633, 1996.

Weiland, T.: Eine Methode zur Lösung der Maxwellschen Gleichungen für sechskomponentige Felder auf diskreter Basis, Archiv für Elektronik und Übertragungstechnik, 31, 116–120, 1977.

Weiland, T.: Time Domain Electromagnetic Field Computation with Finite Difference Methods, International Journal of Numerical Modelling, 9, 295–319, 1996.

Efficient extrapolation methods for electro- and magneto-quasistatic field simulations

M. Clemens, M. Wilke, and T. Weiland

Technische Universität Darmstadt, Dept. of Electrical Engineering and Information Technology, Computational Electromagnetics Laboratory (TEMF), Schloßgartenstr. 8, 64289 Darmstadt, Germany

Abstract. In magneto- and electroquasi-static time domain simulations with implicit time stepping schemes the iterative solvers applied to the large sparse (non-)linear systems of equations are observed to converge faster if more accurate start solutions are available. Different extrapolation techniques for such new time step solutions are compared in combination with the preconditioned conjugate gradient algorithm. Simple extrapolation schemes based on Taylor series expansion are used as well as schemes derived especially for multi-stage implicit Runge-Kutta time stepping methods. With several initial guesses available, a new subspace projection extrapolation technique is proven to produce an optimal initial value vector. Numerical tests show the resulting improvements in terms of computational efficiency for several test problems.

In quasistatischen elektromagnetischen Zeitbereichsimulationen mit impliziten Zeitschrittverfahren zeigt sich, dass die iterativen Lösungsverfahren für die großen dünnbesetzten (nicht-)linearen Gleichungssysteme schneller konvergieren, wenn genauere Startlösungen vorgegeben werden. Verschiedene Extrapolationstechniken werden für jeweils neue Zeitschrittlösungen in Verbindung mit dem präkonditionierten Konjugierte Gradientenverfahren vorgestellt. Einfache Extrapolationsverfahren basierend auf Taylorreihenentwicklungen werden ebenso benutzt wie speziell für mehrstufige implizite Runge-Kutta-Verfahren entwickelte Verfahren. Sind verschiedene Startlösungen verfügbar, so erlaubt ein neues Unterraum-Projektion--Extrapolationsverfahren die Konstruktion eines optimalen neuen Startvektors. Numerische Tests zeigen die aus diesen Verfahren resultierenden Verbesserungen der numerischen Effizienz.

Correspondence to: M. Clemens
(clemens/wilke/weiland@temf.tu-darmstadt.de)

1 Introduction

Spatial discretizations of transient quasistatic electromagnetic field formulations with e.g. the Finite Element Method (FEM) or the Finite Integration Technique (FIT), commonly result either in large stiff ordinary differential or index 1 differential-algebraic systems of equations of the form

$$\mathbf{D}\frac{d}{dt}\mathbf{x}(t) + \mathbf{K}[\mathbf{x}(t)]\,\mathbf{x}(t) = \mathbf{b}(t), \tag{1}$$

with the $N-$component vector $\mathbf{x}(t) \in \mathbb{R}^N$ whose evolution in time has to be considered over a given time interval. For magneto-quasistatic formulations see e.g. Kameari (1990) or Clemens and Weiland (1999) and for electro-quasistatic systems Preis et al. (2002) or Clemens et al. (2002a).

The solution of (1) requires to use suitable implicit time integration schemes, e.g. one-step schemes such as the simple $\theta-$methods, the multi-stage embedded (singly diagonal) implicit Runge-Kutta ((SD)IRK) methods in Cameron (1999) and linear-implicit time integration schemes of Rosenbrock-Wanner-type in Hairer and Wanner (1996) or Lang (2001), which were just recently introduced to quasistatic electromagnetic field simulations in Clemens et al. (2002c), or the multi-step backward differentiation schemes described in Hairer and Wanner (1996). In these methods in each time step one or several large (non)linear algebraic systems of equations have to be solved. For the iterative solution methods of these systems *extrapolation schemes* can provide suitable initial start values. Such first approximations of the solutions at the respective new time steps should be computationally cheap to attain and allow to effectively reduce the number of subsequent iteration steps of the preconditioned conjugate gradient solvers. This was shown e.g. in Clemens et al. (2002d), where several extrapolation techniques are compared to produce an initial vector $\mathbf{x}_0^{(n+1)}$ for the iterative solution $\mathbf{x}^{(n+1)} \approx \mathbf{x}(t^{n+1})$ of the algebraic system at time $t^{(n+1)}$

$$\mathbf{M}\mathbf{x}^{(n+1)} = \mathbf{b}^{(n+1)}, \tag{2}$$

with $\mathbf{M} = [\alpha\mathbf{D}+\mathbf{K}]$, where α is a scalar parameter depending on the chosen time step length Δt.

This paper compares several techniques to improve the numerical performance of these algorithms, i.e., to reduce the required number of iterations in the solution methods for the linear algebraic systems of equations as e.g. in the preconditioned conjugate gradient (PCG) method. These scheme are here mainly used in the context of FIT-based magnetic field formulations, but extend also to other simulations where implicit time integration schemes are used with iterative solution methods. No benefits will arise from the presented methods if direct Gauss elimination schemes are applied which do not require start values.

2 Simple extrapolation techniques in implicit time integration schemes

For each new time step solution $\mathbf{x}^{(n+1)} \approx \mathbf{x}(t^{(n+1)})$ of the linear(-ized) system $\mathbf{M}\mathbf{x}^{(n+1)} = \mathbf{b}^{(n+1)}$ suitable initial solutions $\mathbf{x}_0^{(n+1)}$ for the iterative solution to start on have to be provided. The iterative conjugate gradient solution process is based on a Lanczos orthogonalization where a spectral approximation of the system matrix with tridiagonal matrices is performed as shown in van der Sluis and van der Vorst (1986). Thus the complete reduction of the spectral error components in the approximate solution should not benefit from an available good start solution $\mathbf{x}_0$. However, since the PCG process will be terminated when a user-specified finite accuracy is attained, small error components of the start vectors may not require additional iterations. As simple strategies for the choice of the start vectors we consider the following options that arise with **Taylor series expansion extrapolations** of zero-th order

$$\mathbf{x}_0^{(n+1)} := 0, \tag{3}$$

with first order

$$\mathbf{x}_0^{(n+1)} := \mathbf{x}^{(n)} \tag{4}$$

or with second order

$$\mathbf{x}_0^{(n+1)} := \mathbf{x}^{(n)} + \Delta t^{(n+1)} \frac{d}{dt}\mathbf{x}^{(n)}. \tag{5}$$

Starting the iterations using a homogeneous start vector $\mathbf{x}_0^{(n+1)} := 0$ may be expected to require the most iterative work, although it will best allow the conjugate gradient scheme to maintain its weak divergence property for non-gauged, i.e., singular, magneto-quasistatic vector potential formulations (Cp. Clemens and Weiland (1999)). A more commonly used choice of an initial vector for the iterations at time t^{n+1} consists in the solution of the previous time step $\mathbf{x}^{(n)}$ under the assumption that provided a short enough time step $\Delta t = t^{n+1} - t^n$ the solutions will not differ too much, i.e., using a first order Taylor approximation. An extension of equation (4) is to choose a second order Taylor expansion which involves the evaluation of the vector $\frac{d}{dt}\mathbf{x}^{(n)}$. It should be noted that in magnetodynamic simulations the calculation of the eddy current vector $j_{ec}^{(n+1)} = -\mathbf{M}kap\frac{d}{dt}\mathbf{x}^{(n+1)}$ requires to evaluate these time derivative vectors anyway. If the chosen implicit time integration scheme does not provide an approximation also for $\frac{d}{dt}\mathbf{x}^{(n+1)}$ it can be calculated separately, for instance by using a second order one-step update scheme

$$\frac{d}{dt}\mathbf{x}^{(n+1)} \approx \frac{2}{\Delta t}\left(\mathbf{x}^{(n+1)} - \mathbf{x}^{(n)}\right) - \frac{d}{dt}\mathbf{x}^{(n)} \tag{6}$$

or using a BDF-formula, as e.g. the multi-step BDF2-approximation

$$\frac{d}{dt}\mathbf{x}^{(n+1)} \approx \frac{1}{\Delta t}\left(\frac{3}{2}\mathbf{x}^{(n+1)} - 2\mathbf{x}^{(n)} + \frac{1}{2}\mathbf{x}^{(n-1)}\right). \tag{7}$$

These simple and cost effective to implement start value selection strategies based on the Taylor series expansion can be applied to any suitable implicit time integration scheme.

A similarly simple and seemingly straightforward approach for an extrapolation strategy arises with

$$\mathbf{x}_0^{(n+1)} := \widetilde{\mathbf{M}}^{-1}\mathbf{b}^{(n+1)} \tag{8}$$

which involves the use of an easily invertible preconditioning matrix $\widetilde{\mathbf{M}}$ approximating the system matrix $\mathbf{M}$. A simple choice is e.g. to take the SSOR-preconditioning matrix $\widetilde{\mathbf{M}}_{\text{SSOR}} = (\mathbf{D}-\mathbf{L})\mathbf{D}^{-1}(\mathbf{D}-\mathbf{U})$, with $\mathbf{M} = \mathbf{D}-\mathbf{L}-\mathbf{U}$.

Since the first step within a PCG method actually consists in the application of (8), no considerable improvements of the process convergence are to be expected from this strategy. Numerical tests in Clemens et al. (2002d) confirm this. These simulation results also indicate that using only the start strategy in (4) may be advantageous for constant or nearly constant current excitations whereas the second order Taylor extrapolation in (5) was tested with better results for current excitations with strong variations in the considered time interval. Thus it is not initially clear which strategy to follow.

3 Extrapolation techniques for multi-stage SDiRK methods

In the past years, higher order singly diagonal implicit Runge-Kutta (SDiRK) methods were established to be suitable methods for transient magnetic field simulations in Nicolet and Delincé (1996). The increased interest in this family of time integration methods for transient magnetic fields arises from the fact that these one-step methods provide stiffly accurate schemes of almost arbitrarily high order well suited for numerical time integration of differential-algebraic systems of equations of Index 1 of magnetodynamic formulations. An s-stage SDiRK scheme requires to solve s (non)-linear algebraic systems of equations to yield the *stage variables* $\mathbf{Y}_i$, i.e., intermediate solution vectors, to be evaluated at times $t_i = t^{(n)} + c_i\Delta t$ with

$$\mathbf{Y}_i^{(n+1)} = \mathbf{x}^{(n)} + \Delta t \sum_{j=1}^{i} a_{ij}\mathbf{Y}_i', \quad i = 1, \ldots, s, \tag{9}$$

involving the previous time step solution $\mathbf{x}^{(n)}$, the *stage derivatives* $\mathbf{Y}'_i$, the SDiRK-coefficient matrix $\mathbf{A} = \{a_{ij}\}_{1\le i,j\le s}$, ($a_{ij} = 0$ for all $i < j$). The new time step solution of the SDiRK-scheme is given by

$$\mathbf{x}^{(n+1)} = \mathbf{x}^{(n)} + \Delta t \sum_{j=1}^{s} b_j \mathbf{Y}'_i. \tag{10}$$

The method coefficient vectors $\mathbf{b} = \{b_j\}_{1\le j\le s}$, $\mathbf{c} = \{c_j\}_{1\le j\le s}$, and the method matrix $\mathbf{A}$ completely specify the chosen SDiRK method. The internal multi-stage construction principle in some cases also allows to construct embedded schemes corresponding to a second coefficient vector $\hat{\mathbf{b}}$ to provide an additional solution of a lower order. In Cameron et al. (1998) a stiffly accurate, L-stable 4-stage SDiRK3(2) scheme of third (embedded scheme: second) order was proposed and its use for the purposes of an error-controlled variable step length time integration of the discrete magnetodynamic systems based on Finite-Element and Finite Integration method formulations was further refined in Wang et al. (2001) and Clemens et al. (2002b).

For the SDiRK schemes we consider the following strategies to choose a suitable start value. The most simple extrapolation formulation is given with a first order **Taylor series expansion extrapolation**

$$\mathbf{Y}_{i,0}^{(n+1)} := \mathbf{x}^{(n)}, \tag{11}$$

where the last time step solution $\mathbf{x}^{(n)}$ is taken as start value for each stage variable vector to be calculated for the next time step.

It is, however, also possible to use the intermediate solutions of the internal SDiRK-stages to construct extrapolated start vectors $\mathbf{Y}_{i,0}^{(n+1)}$ for the solution at the new time step $t^{(n+1)}$. An extensive mathematical treatment of these techniques is found in Cameron (1999), where a **stage extension extrapolation** method is described with

$$\mathbf{Y}_{i,0}^{(n+1)} := \begin{cases} \mathbf{x}^{(n)} : \ i = 1, \\ \mathbf{Y}_l^{(n+1)}: \text{for } c_l = \max_{1\le k<i}\{c_k | c_k \le c_i\} \\ \frac{c_i - c_j}{c_k - c_j}\mathbf{Y}_j^{(n+1)} + \frac{c_k - c_i}{c_k - c_j}\mathbf{Y}_k^{(n+1)} : \\ \quad c_i \text{ for } c_l \le c_j \le c_i \le c_k \le c_m \\ \quad \text{for all } l \le j, k < i; k \le m \end{cases}, \tag{12}$$

where a start solution for the corresponding stage value vector is constructed using already available stage values of the current time step $\Delta t^{(n+1)} = t^{(n+1)} - t^{(n)}$. Not for every SDiRK method the coefficients c_i of the intermediate times $t^{(n)} + c_i \Delta t^{(n)}$ increase monotonically in size such that $c_i \le c_j, i > j$, holds. Thus, either an already available previous stage value is used as start solution or an interpolation is performed using the nearest in time stage values $\mathbf{Y}_j$ and $\mathbf{Y}_k$ with $c_j \le c_i \le c_k$ following a method described in Cameron (1999). Note, that certain methods allow to consider two stage variable vectors $\mathbf{Y}_i, \mathbf{Y}_j, i \ne j$, at the same time point in the integration interval, i.e., $c_i = c_j$.

A variant of this approach is the **continuous extension extrapolation** scheme. The solution vectors $\bar{\mathbf{x}}$ at times $t^\star = t^{(n)} + \Delta t^{(n)} + c_i \Delta t^{(n+1)} := t^{(n)} + \sigma_i \Delta t^{(n)}$, can be extrapolated from the stage derivatives of the last time step by

$$\mathbf{Y}_{i,0}^{(n+1)} := \bar{\mathbf{x}}(t^\star) = \mathbf{x}^{(n)} + \sigma_i \Delta t \sum_{j=1}^{s} \bar{b}_j(\sigma_i)\mathbf{Y}'_i, \tag{13}$$

which involves a coefficient vector $\bar{\mathbf{b}} = \{\bar{b}_j\}_{1\le j\le s}$, to achieve the extrapolated stage variable $\mathbf{Y}_{i,0}^{(n+1)}$ of the new time step.

For the SDiRK3(2) scheme the following extrapolator coefficient set is available from Cameron (1999) with

$$\bar{\mathbf{b}}^T(\sigma) := \frac{1}{30}\begin{bmatrix} 1 & \sigma & \sigma^2 \end{bmatrix} \begin{bmatrix} 22\sqrt{2}+15 & 236-159\sqrt{2} & 1-7\sqrt{2} & 135\sqrt{2}-222 \\ -14\sqrt{2}-15 & 135\sqrt{2}-262 & 13+14\sqrt{2} & 264-135\sqrt{2} \\ 2\sqrt{2}+5 & 16+20\sqrt{2} & -9-7\sqrt{2} & -12-15\sqrt{2} \end{bmatrix}.$$

Note, that one-stage SDiRK schemes are closely related to the standard one-step θ−methods described e.g. in Hairer and Wanner (1996). For these time integration methods the strategies in Eqs. (11) and (12) will coincide with method (4) and strategy (13) is identical to method (5). Another specific advantage of stiffly accurate SDiRK arises from the fact that in these methods we have the last intermediate time step coefficient $c_s = 1$, such that $\mathbf{Y}'_s = \mathbf{Y}'(t^{(n+1)}) \approx \frac{d}{dt}\mathbf{x}(t^{(n+1)})$ holds. Accordingly, no additional evaluations for the time derivative vectors $\frac{d}{dt}\mathbf{x}(t^{(n+1)})$ have to be performed.

4 Hybrid extrapolation techniques

All the start vectors $\mathbf{x}_{0,i}, i = 1, ..., m$, derived from the extrapolation techniques described above for the iterative solution of (2) are computationally cheap to attain, but numerical test in Clemens et al. (2002d) indicate that just one extrapolation method used for the generation of start values may not be suitable for all kind of transient problems.

In order to solve this problem, in Clemens et al. (2002d) already a minimal residual norm selection criterion has been proposed for a set of start values $\mathbf{x}_{0,i}, i = 1, ..., m$, constructed with different extrapolation methods. In this **minimal norm hybrid extrapolation** scheme first the norms of residual vectors for the different start solutions $\mathbf{x}_{0,i}$ are evaluated with low additional computational costs. The vector $\mathbf{x}_{0,j}$ corresponding to the smallest residual norm is chosen as a start vector:

$$\begin{aligned} r_i &:= \|\mathbf{M}\mathbf{x}_{0,i} - \mathbf{b}^{(n+1)}\|_2, \quad i = 1, \ldots, m, \\ \mathbf{x}_0^{(n+1)} &:= \mathbf{x}_{0,j}, \text{ for } j: \ r_j = \min_{i=1,\ldots,m}\{r_i\}. \end{aligned} \tag{14}$$

The numerical tests show that this approach, while rather simple, is surprisingly effective for a robust reduction of the required computational work independent of the current excitation form.

A more refined technique for the construction of a start solution vector for system (2) is available with the novel **subspace projection extrapolation (SPE)** technique (SPE).

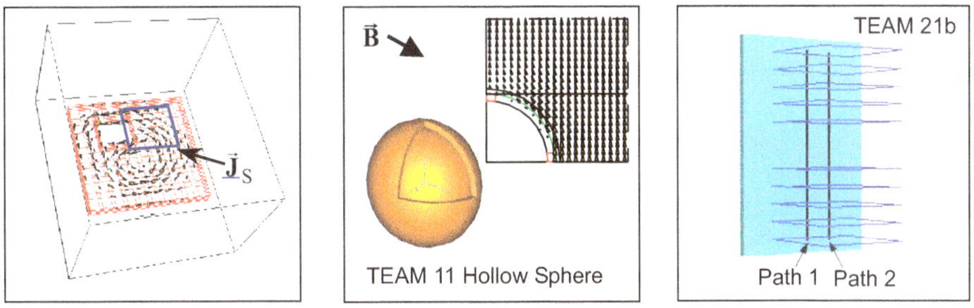

Fig. 1. Three eddy current test problems: **1)** A copper plate with a hole: Time harmonic eddy currents in a conductive plate. **2)** The TEAM 11 problem: A conductive sphere in an abruptly started homogeneous vertical magnetic field **3)** The TEAM 21b problem: Time harmonic eddy currents in a ferromagnetic nonlinear iron plate.

Table 1. Extrapolation strategies for a fixed time stepping BDF1-scheme. The numbers denote the total number of matrix-vector-multiplications during the complete time integration process.

Problem	No. 1	No. 2	No. 3
(a) Taylor 1st order $\mathbf{x}_0^{(n+1)} := \mathbf{x}^{(n)}$	3424	6663	-
(b) Taylor 2nd order $\mathbf{x}_0^{(n+1)} := \mathbf{x}^{(n)} + \Delta t \dot{\mathbf{x}}^{(n)}$	2554	16730	-
(c) Minimal norm hybrid method for (a), (b)	2714	6703	74392
(d) SPE scheme using (a), (b) ($m = 2$)	923	2889	41182

For this we require that the system matrix $\mathbf{M}$ in (2) has to be symmetric and positive (semi-)definite, which is commonly the case in discrete electro- and magneto-quasistatic field formulations. Here the initial vectors $\mathbf{x}_{0,i}$, $i = 1, ..., m$, are orthonormalized using a Modified Gram Schmitt procedure (see e.g. Meister (1999)) to yield the vectors $\mathbf{v}_{0,i}$, $i = 1, \ldots, \widetilde{m}$, which form a orthonormal basis to a vector subspace $V_{\widetilde{m}} = \text{span}\{\mathbf{v}_j\} = \text{span}\{\mathbf{x}_{0,i}\} \subseteq \mathbb{R}^N$ which is spanned already by all the different start vectors $\mathbf{x}_{0,i}$. Since the orthonormalization process will detect linear dependencies in the m original vectors $\mathbf{x}_{0,i}$, the relation $\widetilde{m} \le m$ will hold.

With the definition of the orthonormal operator $\mathbf{V} := \{\mathbf{v}_1|...|\mathbf{v}_{\widetilde{m}}\} \in \mathbb{R}^{N\times\widetilde{m}}$, we can restrict the system (2) onto $V_{\widetilde{m}}$

$$[\mathbf{V}^T\mathbf{M}\mathbf{V}]\mathbf{z} = \mathbf{V}^T\mathbf{b}^{(n+1)}. \tag{15}$$

to achieve an optimal approximative solution $\mathbf{x}_0^{(n+1)} := \mathbf{V}\mathbf{z}$ of (2) inside $V_{\widetilde{m}}$. Note, that the solution of (15) commonly involves a $\widetilde{m} \times \widetilde{m}$ matrix with $\widetilde{m} \ll 100 \ll N$. Thus a direct Gauss elimination process can be adopted and the inverse $[\mathbf{V}^T\mathbf{M}\mathbf{V}]^{-1}$ of the system matrix is available. With this the **subspace projection extrapolation** can be summarized as

$$\mathbf{x}_0^{(n+1)} := \mathbf{V}[\mathbf{V}^T\mathbf{M}\mathbf{V}]^{-1}\mathbf{V}^T\mathbf{b}^{(n+1)}. \tag{16}$$

For the resultant initial vector $\mathbf{x}_0^{(n+1)}$ the Galerkin condition $\mathbf{V}^T[\mathbf{b}^{(n+1)} - \mathbf{M}\mathbf{x}_0^{(n+1)}] = 0$ holds w.r.t. $V_{\widetilde{m}}$. Such a Galerkin condition is also essential for the construction of the iterative conjugate gradient solution process.

The subspace projection extrapolation scheme essentially yields an optimal start solution for the system (2) restricted to the linear subspace $V_{\widetilde{m}} \subset \mathbb{R}^N$ spanned by the different start vectors $\mathbf{x}_{0,i}$. In principle it may occur that there is no reduction of the iteration steps for the start values generated with this SPE method if the subsequently applied conjugate gradient process is considered as an exact method. Beneficial effects, however, will arise in approximative solution processes, where the iterations are terminated once a prescribed accuracy for the relative norms of the residual vectors $\|\mathbf{M}\mathbf{x}_k^{(n+1)} - \mathbf{b}^{(n+1)}\| / \|\mathbf{M}\mathbf{x}_0^{(n+1)} - \mathbf{b}^{(n+1)}\| \le \varepsilon_{\text{PCG}}$ is reached.

For many electro- or magneto-quasistatic problems described with (1) the matrix $\mathbf{M} = \mathbf{M}(\mathbf{x})$ will, however, depend on the solution vector. Such behavior arises with the simulation of nonlinear ferromagnetic materials in magneto-quasistatic problems or field-dependent conductivities in transient electro-quasistatic simulations. While the subspace projection extrapolation initially is derived as optimal start value generation scheme only for linear problems, in this nonlinear case it can be used to perform a few nonlinear fix-point iteration steps $k = 0, 1, ...,$

$$\mathbf{V}^T\mathbf{M}(\mathbf{x}_{0,k})\mathbf{V}\mathbf{z}_{k+1} = \mathbf{V}^T\mathbf{b}^{(n+1)}, \quad \mathbf{x}_{0,k+1} := \mathbf{V}\mathbf{z}_{k+1}, \tag{17}$$

where this iteration is started for $k = 0$ with $\mathbf{x}_{0,0} := \mathbf{x}^{(n)}$. The iteration (17) corresponds to that of a Successive-Approximation method restricted to the subspace $V_{\widetilde{m}}$.

5 Numerical results

The extrapolation strategies presented above are compared for typical transient magnetic test configurations shown in Fig. 1 simulated with a modified vector potential formulation based on the Finite Integration Technique described

Table 2. Extrapolation strategies for the adaptive SDIRK3(2)-scheme from Cameron (1999). The numbers denote the total number of matrix-vector-multiplications during the complete time integration process.

Problem	No. 1	No. 2	No. 3
(a) Taylor 1st order $\mathbf{x}_0^{(n+1)} := \mathbf{x}^{(n)}$	4636	12310	70647
(b) Taylor 2nd order $\mathbf{x}_0^{(n+1)} := \mathbf{x}^{(n)} + \Delta t \dot{\mathbf{x}}^{(n)}$	4473	10925	69946
(c) Stage-extension extrapolation	3877	7466	58799
(d) Continuous extension extrapolation	4145	10778	66660
(e) Minimal norm hybrid method for (a),(c),(d)	3777	6688	55198
(f) SPE-scheme using (a),(c),(d) ($m = 3$)	2743	6572	35743

Table 3. Extrapolation strategies for the adaptive linear–implicit scheme RODAS3(2) described in Clemens et al. (2002c). The numbers denote the total number of matrix-vector-multiplications during the complete time integration process.

Problem	No. 1	No. 2	No. 3
Zero start vector $\hat{\mathbf{v}}_0^{(n+1)} := 0$	8397	8135	87277
SPE scheme using 3 last stage solutions ($m = 3$)	4766	5537	45590

in Clemens and Weiland (1999). Problem 1 is related to the TEAM 7 problem, a copper plate with a hole featuring a ramped 50 Hz sinusoidal current excitation simulated for 80 time steps. Problem 2 is a hollow conductive, non-ferromagnetic sphere in an abrupt B-field (TEAM 11) integrated with 40 time steps. Problem 3 consists of the TEAM 21b problem, a nonlinear 50 Hz time harmonic eddy current problem, simulated over 50 time steps (see also Clemens et al. (2002d)).

In Table 1 the effect of the new SPE method is displayed for the commonly used implicit BDF1 method and in Table 2 for a time-step adaptive 4-stage SDIRK3(2)-method introduced in Cameron (1999). The solution of the linear algebraic systems is performed with a SSOR-preconditioned CG method which terminates its iterations after having reached a relative accuracy $\varepsilon_{\text{PCG}} = 10^{-5}$. The comparison of the different start strategies is given in terms of the required total number of matrix vector multiplications (MxV-operations) for the time integration process. This number corresponds well to the required computational time.

The results for both the standard fixed time step BDF1-scheme as well as the adaptive SDIRK3(2)-scheme show that the application of the SPE scheme speeds up the solution process by a factor ≥ 2 for linear problems when compared to just using the 1st order Taylor expansion, i.e., starting the PCG-iterations with the old solution $\mathbf{x}_0^{(n+1)} := \mathbf{x}^{(n)}$. The new SPE approach also outperforms all the other extrapolation methods presented above. The convergence of the nonlinear problem, in which 4 nonlinear SPE-cycles are performed before entering the nonlinear iteration scheme, is also considerably improved.

In Table 3 results for the linear–implicit time integration method RODAS3(2) described in Lang (2001) or Clemens et al. (2002c) are depicted. Its intermediate stage solutions, for which only linear systems have to be solved, no longer have interpolating character. Thus the extrapolation schemes available for SDIRK-methods are not directly applicable and zero start vectors are commonly used. The application of the new SPE scheme to its respective three last stage solutions ($m = 3$), however, allows to achieve a considerable speed-up of the PCG solver convergence.

6 Conclusion

For transient quasistatic electromagnetic field simulations with implicit time stepping schemes algorithmic improvements are presented. Different extrapolation techniques for new time step solutions are compared in combination with the preconditioned conjugate gradient algorithm. Simple extrapolation schemes are suitable for all standard implicit time integration schemes. Their ideas extend to specialized schemes for multi-stage implicit Runge-Kutta time stepping schemes. With the subspace projection extrapolation technique a new hybrid technique is proposed which allows to choose an optimal initial value independent of the problem dynamics at only moderate additional costs. Numerical tests illustrated the resulting improvements of the computational efficiency for several transient magnetodynamic problems.

References

Cameron, F.: Low-Order Runge-Kutta Methods for Differential Algebraic Equations, Ph.D. thesis, Tampere Univ. of Technology, 1999.

Cameron, F., Piché, R., and Forsman, K.: Variable step size time integration methods for transient eddy current problems, IEEE Trans. Magn., 34, 3319–3323, 1998.

Clemens, M. and Weiland, T.: Transient eddy current calculation with the FI-method, IEEE Trans. Magn., 35, 1163–1166, 1999.

Clemens, M., Gersem, H. D., Koch, W., Wilke, M., and Weiland, T.: Transient simulation of nonlinear electro-quasistatic problems using the Finite Integration Technique, in Proceedings of the 10th International IGTE Symposium on Numerical Field Calculation in Electrical Engineering, 2002, Graz, Austria, pp. 510–515, 2002a.

Clemens, M., Wilke, M., and Weiland, T.: 3D transient eddy current simulations using FI^2TD with variable time step size selection schemes, IEEE Trans. Magn., 38, 605–608, 2002b.

Clemens, M., Wilke, M., and Weiland, T.: Linear-implicit time integration schemes for error-controlled transient nonlinear magnetic field simulations, in Proc. CEFC,2002, Perugia, p. 332, to appear in IEEE Trans. Magn. (scheduled May 2003), 2002c.

Clemens, M., Wilke, M., and Weiland, T.: Extrapolation strategies in transient magnetic field simulations, in Proc. CEFC 2002, Perugia, p. 331, to appear in IEEE Trans. Magn. (scheduled May 2003), 2002d.

Hairer, E. and Wanner, G.: Solving Ordinary Differential Equations II, Stiff and Differential-Algebraic Problems, Springer, Wien, New York, 1996.

Kameari, A.: Calculation of transient 3d eddy current using edge elements, IEEE Tr. Magn., 26, 466–469, 1990.

Lang, J.: Adaptive Multilevel Solution of Nonlinear Parabolic PDE Systems: Theory, Algorithm and Application, Springer-Verlag, Berlin, Heidelberg, New York, 2001.

Meister, A.: Numerik linearer Gleichungssysteme, Vieweg-Verlag, Braunschweig, Wiesbaden, 1999.

Nicolet, A. and Delincé, F.: Implicit Runge Kutta methods for transient magnetic field computation, IEEE Trans. Magn., 32, 1405–1408, 1996.

Preis, K., Bíró, O., Supancic, P., and Ticar, I.: FEM simulation of thermistors including dielectric effects, in Proceedings of the CEFC 2002, Perugia, Italy, 16.–19. June, edited by E. Cardelli, 2002.

van der Sluis, A. and van der Vorst, H. A.: The rate of convergence of conjugate gradients, Numer. Math., 48, 543–560, 1986.

Wang, H., Taylor, S., Simkin, J., Biddlecomb, C., and Trowbridge, B.: An adaptive-step time integration method applied to transient magnetic field problems, IEEE Trans. Magn., 37, 3478–3481, 2001.

A hybrid method combining the FDTD and a time domain boundary-integral equation marching-on-in-time algorithm

A. Becker and V. Hansen

Lehrstuhl für Theoretische Elektrotechnik der Universität Wuppertal, Rainer-Gruenter-Str. 21, 42119 Wuppertal, Germany

Abstract. In this paper a hybrid method combining the FDTD/FIT with a Time Domain Boundary-Integral Marching-on-in-Time Algorithm (TD-BIM) is presented. Inhomogeneous regions are modelled with the FIT-method, an alternative formulation of the FDTD. Homogeneous regions (which is in the presented numerical example the open space) are modelled using a TD-BIM with equivalent electric and magnetic currents flowing on the boundary between the inhomogeneous and the homogeneous regions. The regions are coupled by the tangential magnetic fields just outside the inhomogeneous regions. These fields are calculated by making use of a Mixed Potential Integral Formulation for the magnetic field. The latter consists of equivalent electric and magnetic currents on the boundary plane between the homogeneous and the inhomogeneous region. The magnetic currents result directly from the electric fields of the Yee lattice. Electric currents in the same plane are calculated by making use of the TD-BIM and using the electric field of the Yee lattice as boundary condition. The presented hybrid method only needs the interpolations inherent in FIT and no additional interpolation. A numerical result is compared to a calculation that models both regions with FDTD.

1 Introduction

The FDTD-Method is a very efficient and accurate technique for solving bounded electromagnetic field problems. For the analysis e.g. of scattering or antenna problems including the open space the solution space must be kept finite by introducing an absorbing boundary condition (ABC). An often applied ABC is obtained by introducing a perfectly matched layer (PML), the effect of which is incorporated into the solution-procedure by regarding the PML-region as a part of the FDTD solution space, that means, by treating it within the FDTD scheme. As the FDTD is a "local" method the PML can be regarded as a "local" ABC, too. A "global" ABC is obtained if an integral formulation for a surface inclosing the bounded FDTD solution space is used as a starting point. As an example for such a "global" FDTD-ABC formulation (de Moerlosse et al., 1993) shall be mentioned, where the magnetic field is calculated just outside the FDTD solution space by an integral equation. As the Yee-scheme consisting of two shifted lattices does not provide the required electric and magnetic currents in the same plane, one of these must be calculated by an additional interpolation. A very interesting class of TD-integral formulations are the so called Time Domain Boundary-Integral Marching-on-in-Time Algorithms (TD-BIMs), which in the last few years draw more an more attention. By this, the two main drawbacks of TD-BIMs -stability problems and high computational cost- have been more and more alleviated (e.g. Shanker et al., 2000). Thus, it is reasonable to make use of these advantages and to develop a hybrid method that combines the FDTD/FIT and a TD-BIM, in other words, to make use of the TD-BIM as a "global" ABC for the FDTD solution space.

This paper is organized as follows:

- Section 2 gives an overview of the TD-BIM that is used here to calculate the electric current density at the boundary between the inhomogeneous and the homogeneous body.
- Section 3 shortly repeats the differences between the FDTD and the FIT. The latter is used here to model inhomogeneous bodies.
- Section 4 explains the algorithm of the hybrid method proposed here which combines both numerical techniques.
- Section 5 shows a first numerical example to test the method.
- In Sect. 6 a summary and an outlook are given.

Correspondence to: A. Becker
(abecker@uni-wuppertal.de)

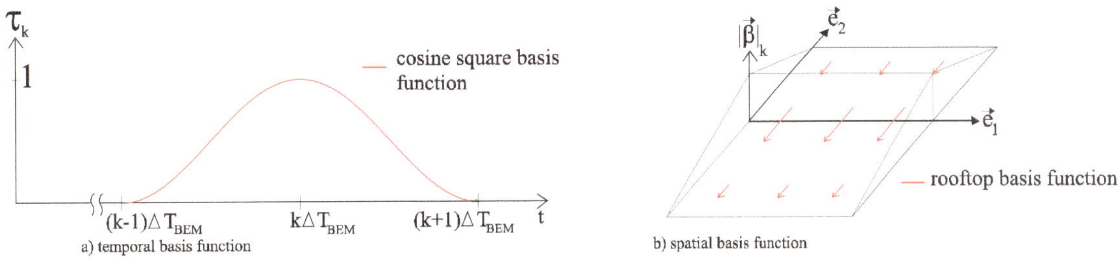

Fig. 1. Temporal and spatial basis function used for the TD-BIM.

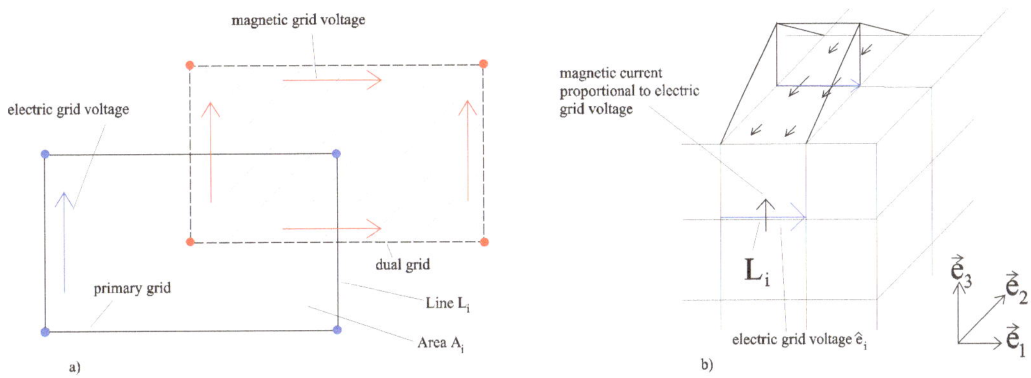

Fig. 2. Spatial allocation of the electric and the magnetic grid voltages and the straightforward allocation of the magnetic current.

2 TD-BIM

The time domain electric field integral equation (TD-EFIE) for a homogeneous infinite space is given to:

$$\boldsymbol{E}(\boldsymbol{r},t) = -\frac{\partial \boldsymbol{A}(\boldsymbol{r},t)}{\partial t} - \nabla\phi(\boldsymbol{r},t) - \frac{1}{\varepsilon}\nabla\times\boldsymbol{F}(\boldsymbol{r},t). \quad (1)$$

When applying the equivalence principle the potentials are related to the equivalent surface sources of the field by

$$\boldsymbol{A}(\boldsymbol{r},t) = \frac{\mu}{4\pi}\int_{A'}\frac{\boldsymbol{J}(\boldsymbol{r}',t-\frac{R}{c})}{R}da', \quad (2)$$

$$\phi(\boldsymbol{r},t) = \frac{1}{4\pi\varepsilon}\int_{A'}\frac{\rho(\boldsymbol{r}',t-\frac{R}{c})}{R}da', \quad (3)$$

$$\boldsymbol{F}(\boldsymbol{r},t) = \frac{\varepsilon}{4\pi}\int_{A'}\frac{\boldsymbol{M}(\boldsymbol{r}',t-\frac{R}{c})}{R}da' \quad (4)$$

with $R=|\boldsymbol{r}-\boldsymbol{r}'|$ and the region outside of A' source free. The surface A' is the surface on which the currents flow, the so called Huygens-surface. The substitution of Eq. (2), Eq. (3) (together with the continuity equation $\nabla\boldsymbol{J}+\frac{\partial}{\partial t}\rho=0$) and Eq. (4) into Eq. (1) results into a relationship between the electric field and the electric and the magnetic current densities (Rao, 1999):

$$\boldsymbol{E}(\boldsymbol{r},t) = \int_{A'}\left[-\frac{\mu}{4\pi}\frac{\partial}{\partial t}\frac{\boldsymbol{J}(\boldsymbol{r}',t-\frac{R}{c})}{R} + \nabla\frac{\int_0^{t-\frac{R}{c}}\nabla\boldsymbol{J}(\boldsymbol{r}',t)\,dt}{4\pi\varepsilon R} - \nabla\times\frac{\boldsymbol{M}(\boldsymbol{r}',t-\frac{R}{c})}{4\pi R}\right]da'. \quad (5)$$

For the numerical solution the surface A' is subdivided into small patches. The electric and magnetic current density is numerically approximated as a series of unknown coefficients multiplied with basis functions:

$$\boldsymbol{J}(\boldsymbol{r},t) = \sum_{i=0}^{N_s}\sum_{j=0}^{N_t} J_{(i,j)}\boldsymbol{\beta}_i(\boldsymbol{r})\tau_j(t),$$

$$\boldsymbol{M}(\boldsymbol{r},t) = \sum_{i=0}^{N_s}\sum_{j=0}^{N_t} M_{(i,j)}\boldsymbol{\beta}_i(\boldsymbol{r})\tau_j(t) \quad (6)$$

N_s and N_t are the numbers of space samples and the time step in which the electric field is calculated, respectively. The spatial and temporal basis functions applied here are shown in Fig. 1. The width of the temporal basis function is $2\Delta T_{BEM}$. We integrate the combination of Eqs. (5) and (6) multiplied with a test function $\boldsymbol{\beta}_k$ over the domain A_k of the test function (Sarkar et al., 2000):

$$\int_{A_k}\boldsymbol{E}(\boldsymbol{r},t)\boldsymbol{\beta}_k(\boldsymbol{r})da = \int_{A_k}\int_{A'}\sum_{i=0}^{N_s}\sum_{j=0}^{N_t}\left[-\frac{\mu}{4\pi}\frac{J_{(i,j)}\boldsymbol{\beta}_k(\boldsymbol{r})\boldsymbol{\beta}_i(\boldsymbol{r}')\tau'(t-\frac{R}{c})}{R} - \frac{\int_0^{t-\frac{R}{c}}J_{(i,j)}\gamma_k(\boldsymbol{r})\gamma_i(\boldsymbol{r}')\tau(t)\,dt}{4\pi\varepsilon R} - \nabla\times\frac{M_{(i,j)}\boldsymbol{\beta}_k(\boldsymbol{r})\boldsymbol{\beta}_i(\boldsymbol{r}')\tau(t-\frac{R}{c})}{4\pi R}\right]da'da. \quad (7)$$

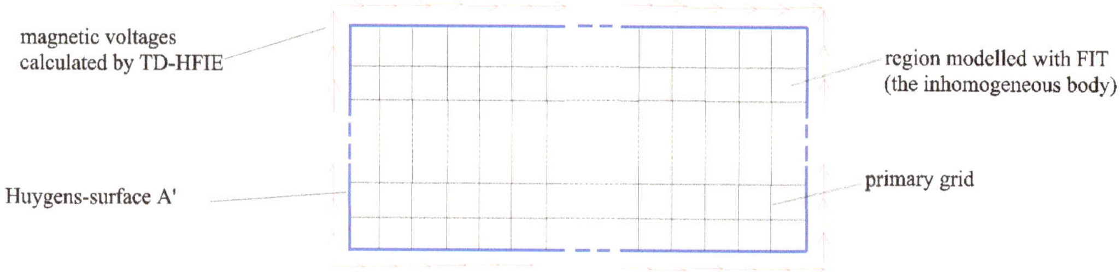

Fig. 3. The allocation of the Huygens-surface A' in the Yee lattice.

The term γ represents the surface-divergence of the basis function $\boldsymbol{\beta}$. Equation (7) connects the tangential electric field on surface A_k to the electric and the magnetic currents flowing on surface A'. The equivalence principle relates the tangential electric field on the surface A_k to the equivalent magnetic current:

$$\boldsymbol{\beta}_k \cdot (\boldsymbol{n} \times \boldsymbol{M}) = \boldsymbol{\beta}_k \cdot (\boldsymbol{n} \times (\boldsymbol{E} \times \boldsymbol{n})) \\ = \boldsymbol{\beta}_k \cdot [\boldsymbol{E}\,(\boldsymbol{n} \cdot \boldsymbol{n}) - \boldsymbol{n} \cdot (\boldsymbol{n} \cdot \boldsymbol{E})] = \boldsymbol{\beta}_k \cdot \boldsymbol{E} \tag{8}$$

Substituting Eq. (8) into Eq. (7) leads to a relationship between the electric and magnetic current density:

$$\int_{A_k} \sum_{i=1}^{N} M_{(i,N_t)} \boldsymbol{\beta}_k \cdot (\boldsymbol{n} \times \boldsymbol{\beta}_i(\boldsymbol{r}))\, da = \\ \int_{A_k} \int_{A'} \sum_{i=1}^{N} \sum_{j=1}^{N_t} \Bigg[-\frac{\mu}{4\pi} \frac{J_{(i,j)} \boldsymbol{\beta}_k(\boldsymbol{r})\, \boldsymbol{\beta}_i(\boldsymbol{r}')\, \tau'\left(t - \frac{R}{c}\right)}{R} \\ - \frac{\int_0^{t-\frac{R}{c}} J_{(i,j)} \gamma_k(\boldsymbol{r}) \gamma_i(\boldsymbol{r}')\, \tau(t)\, dt}{4\pi \varepsilon R} \\ -\nabla \times \frac{M_{(i,j)} \boldsymbol{\beta}_k(\boldsymbol{r})\, \boldsymbol{\beta}_i(\boldsymbol{r}')\, \tau\left(t - \frac{R}{c}\right)}{4\pi R} \Bigg] da'\, da. \tag{9}$$

If the *actual* magnetic current density on surface A' is known, the *actual* electric current density can be calculated from the electric current density retarded in time at least one time step and from the magnetic current density by enforcing continuity on all subdomains A_k of A'. Eq. (9) can be written as:

$$\underbrace{\begin{pmatrix} Z_{1,1} & \dots & Z_{1,N} \\ \vdots & \ddots & \vdots \\ Z_{N,1} & \dots & Z_{N,N} \end{pmatrix}}_{Z} \begin{pmatrix} I_{1,N_t} \\ \vdots \\ I_{N,N_t} \end{pmatrix} \\ = \boldsymbol{f}\left(\sum_{i=1}^{N} \sum_{j=1}^{N_t-1} J_{(i,j)}\right) + \boldsymbol{g}\left(\sum_{i=1}^{N} \sum_{j=1}^{N_t} M_{(i,j)}\right) \\ = \begin{pmatrix} \sum_{i=1}^{N} \sum_{j=1}^{N_t-1} Y_{(1,i,j)} J_{(i,j)} + \sum_{i=1}^{N} \sum_{j=1}^{N_t} X_{(1,i,j)} M_{(i,j)} \\ \vdots \\ \sum_{i=1}^{N} \sum_{j=1}^{N_t-1} Y_{(N,i,j)} J_{(i,j)} + \sum_{i=1}^{N} \sum_{j=1}^{N_t} X_{(N,i,j)} M_{(i,j)} \end{pmatrix} \tag{10}$$

The matrix $\mathbf{Z}$ describes the *actual* electric fields produced by the *actual* electric currents and thus is highly sparse. In a time-invariant system both Z and the coefficients $X_{(k,i,j)}$ and $Y_{(k,i,j)}$ have to be calculated just once, at the first time step. If the tangential electric field equals zero (that means the material is a perfect electric conductor and there is no magnetic current density) Eq. (10) will be sufficient for calculating the electric current density in a marching-on-in-time algorithm (MoT). Otherwise a second equation is needed to calculate both current densities. In our case we will use the FIT to update the magnetic current density and Eq. (10) to update the electric current density. Thus the surface A' will encase the region modelled with FIT.

3 FDTD/FIT

The TD-FIT can be considered as a special formulation of the FDTD-method (Weiland, 1996). In contrast to FDTD it is based on the Maxwell Equations in integral form. By using electric and magnetic fluxes and grid voltages as unknown coefficients and by using the Yee lattice to allocate them, these equations can be written as:

$$\mathbf{C}\hat{\boldsymbol{e}} = -\frac{\partial}{\partial t}\hat{\hat{\boldsymbol{b}}} \tag{11}$$

$$\tilde{\mathbf{C}}\hat{\boldsymbol{h}} = \frac{\partial}{\partial t}\hat{\hat{\boldsymbol{d}}} + \hat{\hat{\boldsymbol{j}}}. \tag{12}$$

$\hat{\boldsymbol{e}}$ represents the grid voltage above one Line L_i of the Yee cell. The term $\hat{\hat{\boldsymbol{d}}}$ represents the flux through one Area A_i of the Yee cell. The unknown coefficients can be considered as the analytic – and thus numerically error-free – values. The application of Eq. (12) is illustrated in Fig. 2a. A second order accurate central difference approximation for the time derivative leads to the leap frog algorithm. The definition of grid voltages as unknown coefficients is essential if using the FIT instead of FDTD for our hybrid method.

4 Hybrid method

In this section we will introduce our hybrid method. The TD-BIM acts as a "global" absorbing boundary condition enclosing the inhomogeneous bodies which are modelled by FIT.

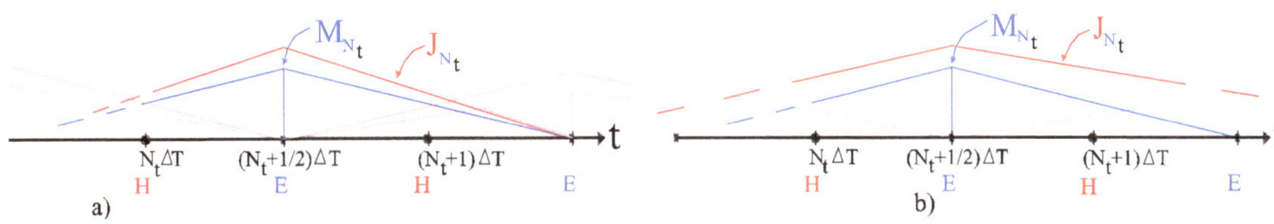

Fig. 4. The temporal allocation of the magnetic currents and possible allocations of the electric currents.

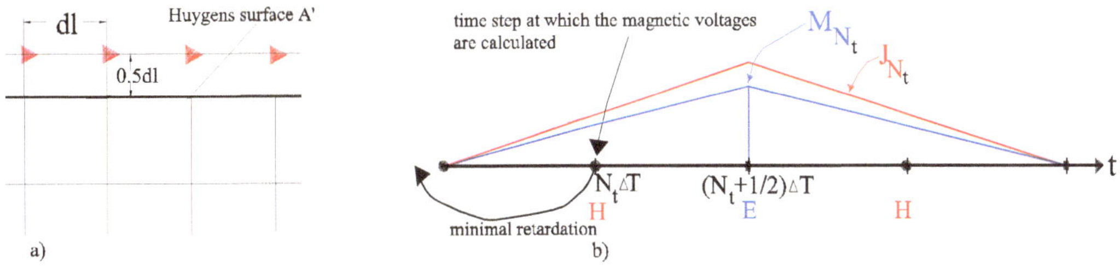

Fig. 5. The boundary of the FIT-domain and the resulting minimal retardation in the calculation of the magnetic grid voltages.

4.1 Basic features

We interconnect the inhomogeneous body and the surrounding homogeneous region by calculating tangential magnetic grid voltages just outside the inhomogeneous body using a TD magnetic field integral equation (TD-HFIE) which uses electric and magnetic currents flowing on the Huygens-surface A' whose location is shown in Fig. 3.

The magnetic current on the surface A' results directly from the electric grid voltages of the FIT, which will be shown beneath. Due to the fact that the magnetic grid voltages are spatially (and temporally) separated from the electric grid voltages the electric currents on the Huygens-surface A' can not be calculated from the magnetic grid voltages. Therefore the electric current will be calculated by using a TD-BIM -as described in Sect. 2 – which is capable of considering bodies outside the FIT volume.

To develop the algorithm of the hybrid method we start with the FIT and a Cartesian grid. The unknowns in FIT are grid voltages defined as:

$$\hat{e}_i = \int_{L_i} \boldsymbol{E} ds \;, \hat{h}_i = \int_{L_i} \boldsymbol{H} ds \tag{13}$$

As already mentioned we use the rooftop basis function for the TD-BIM because the discretization for FIT leads to a patch model of the surface A' consisting of rectangular elements (see Fig. 2) if Cartesian Coordinates are used. The unknowns in the TD-BIM are proportional to the flux orthogonal to these edges. With the location in time of the magnetic current density according to Fig. 4 this results into the following relationship between the unknown $\hat{e}_i$ of the FIT and the magnetic current of the TD-BIM (see Fig. 2b):

$$\int_{L_i} \boldsymbol{M}_{N_t} \boldsymbol{e}_3 = \int_{L_i} (\boldsymbol{e}_2 \times \boldsymbol{E}_{N_t})\, \boldsymbol{e}_3 ds =$$

$$\int_{L_i} \boldsymbol{E}_{N_t} \boldsymbol{e}_1 ds = \hat{e}_{i,N_t} \tag{14}$$

Now the magnetic current density can be calculated from the electric grid voltages by using Eq. (14). The electric currents are located at the same time step like the magnetic currents (see Fig. 4a). It is possible to locate the electric currents at the same time step like the magnetic grid voltages and even to use a time step ΔT_{BEM} for the electric currents which is larger than the time step given by FIT (see Fig. 4b). The latter decreases calculation time and improves stability because the time step given by the FIT is extremely small.

4.2 Time step $t = N_t \Delta T$

The calculation of the magnetic grid voltages interconnecting the inhomogeneous and the homogeneous region (see Fig. 3) with the TD-HFIE is equivalent to the use of the so called pulse-test-function:

$$\int_S \boldsymbol{H} ds = \int_A \boldsymbol{H} \boldsymbol{\beta}_{pulse} da \;, \text{ with } \; \boldsymbol{\beta}_{pulse} = \left. \frac{ds}{|ds|} \right|_{r \epsilon S} \tag{15}$$

The electric and magnetic current density is approximated according to Sect. 2. Thus the magnetic grid voltages can be calculated by combining Eq. (7) and the Fitzgerald-transformation/duality. The use of a pulse-test-function allows us to calculate the magnetic grid voltages without the need to calculate a term proportional to $1/r^2$. As seen in Fig. 5a the minimum distance between boundary plane and magnetic grid voltages is $0.5L_{min}$ which is half of the smallest spatial discretization. In Fig. 5a the distance between the Huygens-surface and the magnetic grid voltages is half a cell-size.

In our numerical example we will use one and a half cell-sizes between Huygens-surface and the grid voltages calculated by the TD-HFIE (see Fig. 6a). Due to the Courant condition the minimal temporal retardation can be calculated to:

$$\tau_{min} = \frac{L_{min}}{c} = \frac{\sqrt{3} c \Delta T}{c} = \sqrt{3} \Delta T > 0.5 \Delta T \tag{16}$$

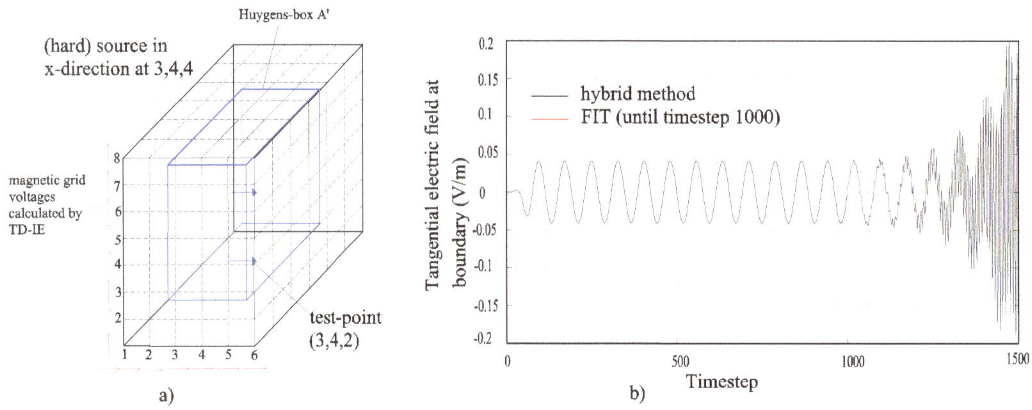

Fig. 6. Geometry and electric fields at the boundary of the FIT-volume.

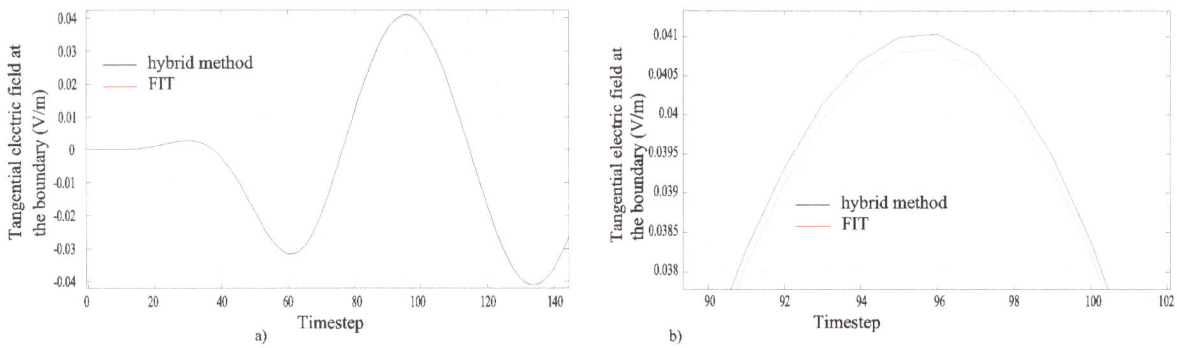

Fig. 7. The electric fields at the boundary of the FIT-volume.

As seen in Fig. 5b the *actual* magnetic grid voltages are not functions of the *actual* electric and magnetic currents. Thus the grid voltages can be calculated by using *expired* currents only.

4.3 Time step $t = (N_t + 0.5)\Delta T$

If the tangential magnetic grid voltages are known the FIT-algorithm can be completed in the interior and on the boundary as usual ($t = (N_t + 0.5)\Delta T$). Consequently, the electric grid voltages are known in the whole interior and on the boundary. With Eq. (14) the magnetic currents can be calculated in a very efficient way. Anyway, the FIT does not give a relationship between the magnetic and the electric fields in one common plane. At the next time-step ($t = (N_t + 1)\Delta T$) the magnetic as well as the electric currents at time steps $t = (N_t + 0.5)\Delta T$ are needed. The latter are calculated by applying Eq. (10) at $t = (N_t + 0.5)\Delta T$. As boundary condition we use the electric fields calculated by FIT (proportional to the magnetic flux as related by Eq. 14), i.e. the fields calculated by the TD-IE must equal the fields calculated by the FIT. Now both electric and magnetic currents are known on the surface A'.

5 Numerical results

As a first numerical evaluation of the hybrid method we simulate an electric source (sinusoidal, 500 MHz) in the interior of an air filled cube (see Fig. 6a) by impressing one electric grid voltage. This source reflects all incoming waves and so it acts like a small wire. The spatial and temporal discretization is determined by the Courant condition, assuming a 1.1 GHz source and 20 cells per wavelength. The cube is $6 \times 7 \times 8$-cells big. Between the Huygens-surface and the FIT-boundary we put one layer. As an example we calculate one electric grid voltage on the Huygens-surface. The result is compared to a calculation in which a cube with PMC-boundary condition is simulated. This simulation contains so many cells in each direction that waves at the boundary can be neglected. The results agree well despite of the extremely small distance between boundary and source. However, after approximately 12 periods the calculation with the hybrid method becomes unstable. Such instabilities are typical for TD-BIMs and the accuracy and stability strongly depends on the numerical evaluation of the matrix $\mathbf{Z}$ and the coefficients $X_{(k,i,j)}$ and $Y_{(k,i,j)}$. So it can expected be that these instabilities can be postponed in time far enough for practical purposes, e.g. by using semi-analytic integration.

6 Summary and outlook

The presented hybrid method combines the "local" FIT and a "global" TD-BIM. The TD-BIM acts like an absorbing boundary condition and allows waves to pass bidirectional. The method can be directly transformed into a frequency domain technique and needs no additional interpolation than the interpolations inherent in FIT. The accuracy and stability can be improved by using semi-analytic integration instead of the implemented numerical integration. The efficiency can be drastically increased by using techniques like the plane wave time domain algorithm (Shanker, 2000).

References

De Moerlosse, J. and De Zutter, D.: Surface Integral Representation Radiation Boundary Condition for the FDTD-method, IEEE Trans. AP., 41, 890–896, 1993.

Rao, S. M.: Time Domain Electromagnetics, Academic Press, 132, 1999.

Sarkar, T. K., Lee, W., and Rao, S. M.: Analysis of Transient Scattering From Composite Arbitrarily Shaped Complex Structures, IEEE Trans. AP, 48, 1625–1634, 2000.

Shanker, B., Ergin, A., Aygü, K., and Michielssen, E.: Analysis of Transient Electromagnetic Scattering Phenomena Using a Two-Level Plane Wave Time-Domain Algorithm, IEEE Trans. AP, 48, 510–523, 2000.

Weiland, T.: Time Domain Electromagnetic Field Computation With Finite Difference Methods, International Journal of Numerical Modelling: Electronic Networks, Devices and Fields, 9, 295–319, 1996.

18

Millimetre-wave imaging with reduced number of degrees of freedom in array signal processing: concepts for enhanced angular resolution

G. Kotyrba and H. Chaloupka

Fachbereich Elektrotechnik und Informationstechnik, Bergische Universität Wuppertal, Rainer-Gruenter-Str. 21, 42119 Wuppertal, Germany

Abstract. This paper deals with millimetre-wave imaging systems based on linear receiving arrays and is focused on the issue of angular resolution and coverage for arrays where the number M of ports is smaller than the number N of elements due to forming of subarrays. For a wide angular coverage a trade-off between a narrow beam width and a high side-lobe suppression is shown to occur which requires proper synthesis of the subarray pattern. Two concepts for an enhanced angular resolution are presented. The first uses a combination of a frequency dependent distribution network (frequency scanning) with array signal processing to enhance angular resolution by cost of a reduced range resolution. The second approach takes advantage of relative movements to enhance angular resolution via a combined angular and temporal (Doppler) processing.

1 Introduction

Active millimetre-wave imaging systems are expected to provide useful functionality in a wide range of applications, e.g. automotive cruise control (Russell,1997) and robotics if the cost for such systems can be kept at an affordable level. Depending on the special type of application the imaging radar system has to meet certain requirements with respect to the maximum range, range resolution and accuracy, the angular field of view (angular coverage), angular resolution and accuracy as well as the update rate. This paper is in particular focused on angular resolution and coverage.

In principle, the angular information can be derived from either a sequential or a simultaneous (monopulse) angular scanning mode. The first approach includes mechanically steered directional antennas as well as phased arrays and switched multi-beam antenna arrays. The latter approach comprises array signal processing (ASP) techniques based on an architecture with parallel receivers and array signal operations in the digital domain. In case of a scenario with relative movements between objects the information retrieved from sequential and simultaneous scanning may significantly differ from each other due to temporal fluctuations of the radar-cross-sections (RCS) of scattering centres.

In all approaches for estimating the angular distribution of the backscatterers from received data, the angular resolving power is related to the linear size D of the aperture in terms of the free space wavelength λ. In case of conventional beamforming the angular resolution is on the order of the 3-dB beam width. For a required resolution $\delta\Theta$ the necessary aperture size thus becomes

$$D_{conv} \geq \frac{50\lambda}{\delta\Theta/\deg}. \tag{1}$$

High-resolution approaches such as minimum variance direction estimator (MVDR) and MUSIC are capable of resolving scattering centres within the beam width. However, since for these methods the angular accuracy rapidly degrades with the number of scatterers separated by less than a beam width (Friedlander 1991, Wax 1989, Bresler 1986), only a small fraction of $N-1$ (with N as the array order) scatterers within a beam width can be resolved. Thus, the beam width influences the resolving power of high-resolution approaches as well. The second important parameter is the available number of degrees of freedom $M-1$ in beamforming which corresponds to M antenna ports possessing M mutually orthogonal radiation pattern. Furthermore, M parallel receiver branches are required for ASP systems or M electronic phase-shifters for phased array systems. For a linear array and about 180 degree angular coverage half-wave element spacing is required leading to $N = 2D/\lambda + 1$ element ports and therefore to a maximum number of $M_{max} - 1 = 2D/\lambda$ degrees of freedom.

Provided cost saving issues at the device level are successfully solved, the operational frequencies for the considered millimetre wave radar systems can be chosen to be about 75 GHz and may even move to values above 100 GHz. Consequently, in most applications the implementation of an ar-

Correspondence to: G. Kotyrba
(kotyrba@hft.uni-wuppertal.de)

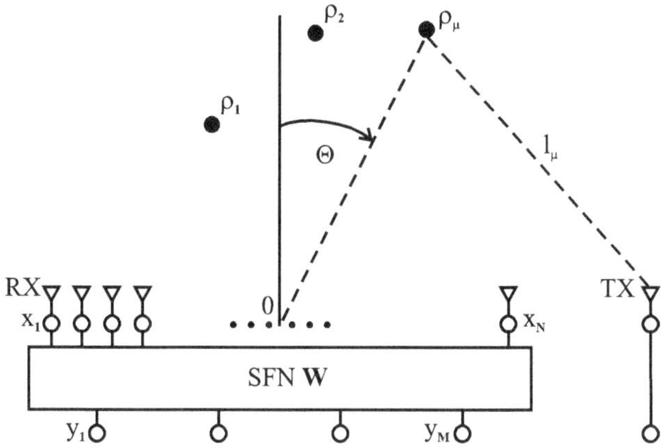

Fig. 1. Considered imaging system with transmitting antenna TX, scattering centers ρ_μ, linear receiving array with N elements and subarray forming network (SFN) with $M < N$ output ports.

ray which is according to Eq. (1) sufficiently large does not represent a serious problem anymore but the implementation of M_{max} receiver channels (or even phase shifters for phased array systems) becomes very costly. It is therefore of interest to exploit less than the available number $M_{max} - 1$ degrees of freedom for beamforming. This is accomplished by employing fixed millimetre wave beamforming networks composed of fixed phase shifters and power dividers to provide from in total N antenna elements $M < M_{max} = N$ subarrays possessing mutually orthogonal subarray radiation pattern.

This paper aims to discuss the impact of a suboptimal ($M < M_{max}$) number of degrees of freedom onto the angular resolution capabilities and presents concepts which partially allow to overcome performance degradations caused by the reduced number M. In Sect. 2 a model and the corresponding notations for the problem under consideration is introduced. Section 3 provides some general results for the system degradations and discusses criteria for an "optimum" choice of the subarrays. In Sect. 4 a novel combination of frequency scanning with array signal processing is proposed which allows to trade-off resolving in range and angle. Section 5 discusses an approach which combines array signal processing with temporal Doppler processing in order to enhance angular resolution.

2 Representation of the considered problem

The system to be discussed aims to produce a two-dimensional "mm-wave image" of a 3-dimensional scenario composed of various objects. In the used linear model the complex backscattering phenomena are replaced with a set of independent scattering centres. Scattering by a scattering centre can be characterised by a scattering coefficients $\rho_i(\boldsymbol{u}_{inc}, f)$ with $|\rho_i|^2 = \sigma_i$ representing the radar cross-section. Both quantities depend on the aspect direction $\boldsymbol{u}_{inc}$, frequency and polarisation. The scalar quantities are understood to be the values which apply to the specific polarisation of the employed TX and RX antennas. The position of scattering centre i is in two dimensions characterised by the azimuth angle Θ_i ("look direction") or the sinus of the look direction $\sin\Theta_i = u_i$ and the path length l_i or corresponding time delay $\tau_i = l_i/c$ between phase centre of transmission antenna to the scatterer and back to the centre of the receiving antenna.

The following considerations are based on the assumption that a pulse radar with centre frequency f_0, effective frequency bandwidth Δf and pulse repetition frequency $1/T_r$ is used. Furthermore, a bistatic configuration with a single transmission antenna and a linear array as receiving antenna (see, Fig. 1) is considered. Modification of the results to account for other configurations like a monostatic configuration or a multiple-input/ multiple-output (MIMO) configuration are straightforward.

With conventional (no high-resolution approach) processing two different path-lengths can be resolved if the corresponding complex-valued envelopes possess a sufficiently small overlap in time. This leads to a path length resolution of about $\Delta l \approx c/\Delta f$. If a periodically switched time gate (range gate) is allocated to each path length interval of length Δl the echos from scattering centres falling into this interval are sampled in time ("snapshots") with a time increment of T_r. The path lengths can change in time if relative movements between antenna and scattering centres occur. In order to move the corresponding echo from one range gate into the adjacent range gate the path length must change by Δl. However, path length changes on a much smaller scale lead to phase changes from snapshot to snapshot. A path length change of c/f_0 leads to a change in phase by 2π. For f_0= 75 Ghz and Δf= 1 GHz for example, a 300 mm path length change are required for a shift into another range gate but only 4 mm for a phase change of 2π.

At each instant in time ("snapshot") which corresponds to a considered path length interval (or range gate) a set of N complex amplitudes $\boldsymbol{x} = [x_1, x_2, ..., x_N]^T$, with superscript T denoting the transposition, is induced at the N antenna elements. If $M < N$ subarrays are formed via a (ideally) lossless distribution network ("subarray forming network", SFN) the N-dimensional vector $\boldsymbol{x}$ is transformed into a M-dimensional vector

$$\boldsymbol{y} = \mathbf{W} \cdot \boldsymbol{x}. \tag{2}$$

$\mathbf{W}$ exhibits $M < N$ rows and N columns and $\mathbf{W} \cdot \mathbf{W}^H = 1$ holds. In the following sections the properties of the multiport antenna consisting of the N elements without SFN (number of ports $L = N$) has to be compared with the multiport antenna build from the N-element array in combination with the SFN ($L = M < N$). For ease of consideration it is assumed that within the field of view in vertical direction the radiation pattern of the considered antennas can be assumed to be constant so that only the dependence on Θ and therefore on $u = \sin\Theta$ has to be taken into account. In the remaining part of this paper the variable u is used and for convenience referred to as "angle".

The properties of a general L-port antenna are completely described by the steering vector

$$s(u) = [C_1(u), C_2(u), ..., C_L(u)]^T \tag{3}$$

where the functions $C_l(u)$ represent the directivity in direction u associated with port l. The set of all steering vectors span a L-dimensional space $\mathbf{S}_L$. Assuming a fictious lossless distribution network connected to this L-port antenna and the weights adjusted such that the gain g (or directivity on the lossless case) is maximised towards a given look direction u_0, one finds for the weights to be given by

$$\boldsymbol{w}(u_0) = \boldsymbol{s}(u_0)/|\boldsymbol{s}(u_0)|. \tag{4}$$

Therefore the gain-function becomes

$$g(u, u_0) = \boldsymbol{w}^H(u_0) \cdot \boldsymbol{s}(u) = |\boldsymbol{s}^H(u_0) \cdot \boldsymbol{s}(u)|^2/|\boldsymbol{s}(u_0)|^2 \tag{5}$$

with the maximum gain given by

$$g_{max}(u_0) = g(u_0, u_0) = |\boldsymbol{s}(u_0)|^2 \tag{6}$$

In order to allow the properties of different multi-port antennas to be compared with each other the functions $g_{max}(u_0)$, $\Delta U(u_0)$ and $L(u_0)$ can be considered. Function $g_{max}(u_0)$ characterises the angular coverage of the L-port antenna. If a homogeneous coverage of all look angles in the interval $-u_{0,max} < u_0 < u_{0,max}$ is required the condition

$$g_{max}(u_0) \approx \text{const} \quad \text{for} \quad -u_{0,max} < u_0 < u_{0,max} \tag{7}$$

should be met. The width of the maximum of the gain-function at u_0 defines the local beam width which is introduced as

$$\Delta u(u_0) = 2\sqrt{g(u_0, u_0)/\frac{\mathrm{d}^2 g}{\mathrm{d}u^2}(u = u_0, u_0)}. \tag{8}$$

Beside the main maximum the gain-function can possess local maxima (side-lobe) with highest side-lobe at u_{sl}. This property is described by

$$L(u_0) = g(u_{sl}, u_0)/g(u_0, u_0). \tag{9}$$

Measurement of backscattered field from scattering centres in look direction u_1 to u_Q with the scattering amplitudes ρ_1 to ρ_Q are represented by the $L \times L$ covariance matrix

$$\mathbf{R} = \left\langle \sum_{v=1}^{Q} \sum_{\mu=1}^{Q} \rho_v^* \rho_\mu \boldsymbol{s}^*(u_v) \cdot \boldsymbol{s}^T(u_\mu) \right\rangle + P_{noise}\mathbf{1}. \tag{10}$$

In Eq. (10) the scattering coefficients are understood to include the path-loss in wave propagation. P_{noise} represents noise contribution. The symbol $\langle\rangle$ denotes the mean of covariance matrices which are gained from different snapshots (periodic sampling of range gate with time increment T_r). In case of a scene without relative movement the contributions from different scattering centres have to be considered as correlated . However, if scattering centres possess relative movements, their phase difference varies from snapshot to snapshot so that their contributions become de-correlated leading to

$$\mathbf{R} = \sum_{\mu=1}^{Q} \sigma_\mu \boldsymbol{s}^*(u_\mu) \cdot \boldsymbol{s}^T(u_\mu) + P_{noise}\mathbf{1}. \tag{11}$$

The results for the general L-port antenna can now be applied to the array without subarray forming network (SFN) with $L = N$ and to the array with SFN and $L = M < N$. For the linear array with N identical elements with element pattern $\mathbf{C}_{el}(\Theta, \Phi) \approx \sqrt{g_0}$ in the field of view and spacing $a = \lambda/2$ for 180 degree angular coverage the steering vector becomes

$$s_x(u) = [1, \exp(j\pi u), \exp(j2\pi u), ..., \exp(j\pi(N-1)u)]^T \exp(-j0.5(N-1)\pi u)\sqrt{g_0}. \tag{12}$$

Steering the maximum gain into the look direction u_0 results in a gain function

$$g_x(u, u_0) = \boldsymbol{s}_x^H(u) \cdot \boldsymbol{s}_x(u_0) = \frac{g_{el}}{N} \sin^2\left[\frac{N}{2}\pi(u - u_0)\right] \Big/ \sin^2\left[\frac{1}{2}\pi(u - u_0)\right] \tag{13}$$

with local (differential) beam width for $N \geq 2$ estimated via

$$\Delta u(u_0) = \frac{1}{\pi}\sqrt{\frac{24}{N^2 - 1}} \tag{14}$$

and maximum gain

$$g_{max}(u_0) = \boldsymbol{s}_x^H(u_0) \cdot \boldsymbol{s}_x(u_0) = N g_{el}. \tag{15}$$

3 Impact of reduced number of degrees of freedom on the angular resolution and coverage

At the output of the subarray forming network (SFN) a M-dimensional vector $\boldsymbol{y}$ is available, related to the N-dimensional vector $\boldsymbol{x}$ via

$$\boldsymbol{y} = \mathbf{W} \cdot \boldsymbol{x}. \tag{16}$$

Performing ASP to the M-dimensional signal vector $\boldsymbol{y}$ with the M-dimensional steering vector $\boldsymbol{s}_y(u)$ is equivalent with ASP to the N-dimensional signal vector $\boldsymbol{x}$ with the N-dimensional steering vector

$$s_x(u) = \mathbf{W}^T \cdot s_y(u). \tag{17}$$

Equation (17) defines a M-dimensional subspace $\mathbf{S}_y$ of the N-dimensional space $\mathbf{S}_x$. Consequently, $\boldsymbol{s}_x(u)$ is confined to the subspace $\mathbf{S}_y \subset \mathbf{S}_x$. Thus, at the output ports of the subarrays only a subset of gain-functions is available. It is therefore of interest to know how the restriction to this subset of gain-functions impacts on the achievable angular resolution and coverage.

Theoretically, there is an infinite number of choices for the formation of subarrays and therefore for the $M \times N$ matrix $\mathbf{W}$. Different $\mathbf{W}$s result in different steering vector functions $\boldsymbol{s}_y(u)$ and in different gain functions $g(u, u_0)$.

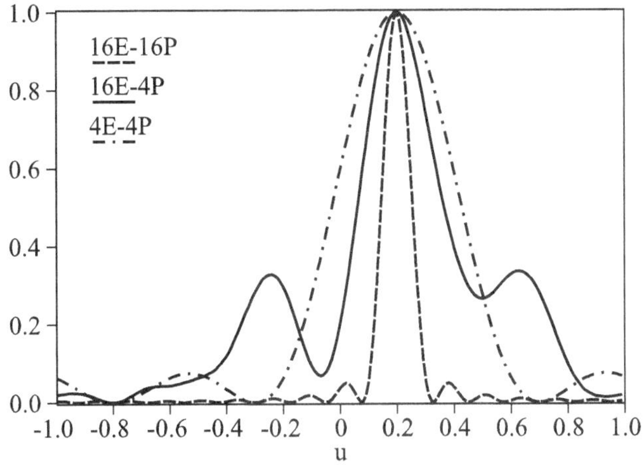

Fig. 2. Normalised gain functions for 3 different antenna configurations which have a angular coverage of $u_{0,max} \approx 1$ and an element spacing of $\lambda/2$ in common: Array with $N = M = 4$ (4E-4P), array with $N = M = 16$ (16E-16P) and array with $N = 16$ and $M = 4$ (16E-4P) with specifically chosen subarrays.

Hence, there is a need for criteria which allow to select gain-functions which "optimally" match the requirement for a given imaging problem. The realisation of these optimum gain-functions corresponds to the choice of an optimum matrix **W**. For the angular coverage requirement

$$g_{max}(u_0) \approx g_{max,0} = const \quad \text{in} \quad -u_{0,max} < u_0 < u_{0,max}$$

the general restriction

$$\int_{-1}^{1} \frac{g(u,u_0)}{g_{max,0}} du \approx 2\frac{u_{0,max}}{M} \tag{18}$$

can be derived. This means that the area under the normalised gain function for all look angle directions depends on the required angular coverage $u_{0,max}$ and the number M of subarray ports only. Furthermore, the achievable local differential beam width is limited by the aperture size (excluding superdirective antennas) and therefore by the number N of antenna elements. Eq. (14) may be used as an estimate for the lower limit for Δu:

$$\Delta u(u_0) \geq \frac{1}{\pi}\sqrt{\frac{24}{N^2-1}} \approx \frac{\lambda}{D}. \tag{19}$$

The degrees of freedom in synthesising a suitable set of gain functions $g(u, u_0)$ for a given number M of subarrays and a given angular coverage can be deduced from Eqs. (18) and (19). Figure 2 may serve to illustrate the consequences of restrictions (18) and (19). It shows the gain-function (for look-direction $u_0 = 0.2$ as an example) for 3 different array configurations which have an angular coverage of nearly $u_{0,max} \approx 1$ and a half-wave spacing between antenna elements in common. The first case is an array with $N = 4$ elements and $M = N$. The achievable gain function is governed by the area under the gain-function of about 0.5 (Eq. 18) and

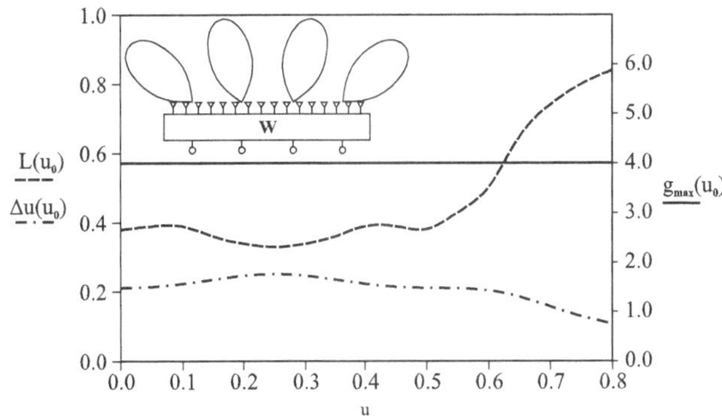

Fig. 3. Dependence of maximum gain $g_{max}(u_0)$, local beam width $\Delta u(u_0)$ and side-lobe ratio $L(u_0)$ on the steering direction u_0 for the array with $N = 16$ and $M = 4$ and specifically chosen subarrays.

the minimum local beam width of about 0.4 (see, Eq. 19). The second case is $M = N = 16$. Now the achievable local beam width becomes because of Eq. (19) lower and the area under the gain-curve by a factor of 4 lower. The third case with $N = 16$, but $M = 4$ belongs to the class of antenna configurations which are of main interest in this paper. Now, the achievable local beam width is governed by $N = 16$, but the area under the gain-curve by $M = 4$. If the same small beam width as in case 2 is chosen the larger area will lead to 3 side-lobes of equal height as the main lobe. With larger beam width the side-lobes can be made smaller. Case 3 in Fig. 3 represents an example where 4 subarrays were formed from an 16-element array. The main beam directions of these subarray are chosen such that they point into different directions. A minimisation of the variations of the maximum gain $G_{MAX}(U_0)$, the local beam width $\Delta u(u_0)$ and the side-lobe ratio $L(u_0)$ within $|u_0| < u_{0,max}$ could be a suitable criterium for the choice of the subarrays and therefore of **W**. Figure 3 illustrates the variations of these 3 quantities with u_0 in case of the array with $M = 4$ and $N = 16$ and the particular choice of the subarrays. In this case the subarrays were designed to yield constant maximum gain up to $u_0 = 0.8$.

There is a large number of different concepts for estimating the angular distribution $\sigma(u)$ of the RCS: Conventional (Bartlett estimation)

$$\sigma_1(u) = \mathbf{s}_y^H(u) \cdot \mathbf{R}_y \cdot \mathbf{s}_y(u), \tag{20}$$

Minimum variance direction estimation (MVDR) as example for a high-resolution approach

$$\sigma_2(u) = \frac{1}{\mathbf{s}_y^H(u) \cdot \mathbf{R}_y^{-1} \cdot \mathbf{s}_y(u)}. \tag{21}$$

In case of the Bartlett estimation the obtained result is significantly improved if the contributions of the different scattering centres are de-correlated, in case of the MVDR decorrelation is a necessary condition. As already explained above, pulse radar yields snapshots of received signal N-dimensional vector $\boldsymbol{x}$ at instances $t = t_0, t_0 + T_r, t_0 + 2T_r, \ldots$

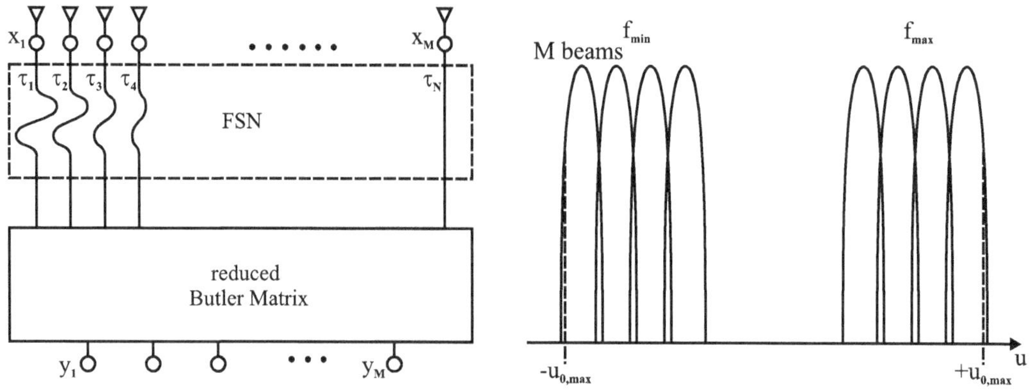

Fig. 4. Combination of frequency scanning with ASP. Left: Configuration with N-dimensional antenna array, frequency scanning network (FSN) and reduced Butler matrix with $M =< N$ output ports. Right: Illustration of frequency scanning for $M = 4$ beams.

in time. In case of relative movements, decorrelation is achieved by averaging between different snapshots. For both approaches, Eqs. (20) and (21) angular resolution improves with the local beam width of the gain-function.

4 Combination of frequency scanning and ASP

As seen from Eqs. (18) and (19), there is a trade-off for $M < N$ between the demand for a narrow beam width and a low side-lobe ratio of the gain-function on one side and a large angular coverage on the other side. Only if the angular coverage is reduced to

$$u_{0,max} \leq M/N, \tag{22}$$

the gain-function can be realised with a beam width and a side-lobe ratio in agreement with the values achieved for $M = N$. This can be accomplished by means of a reduced Butler matrix network. The "full" Butler matrix (BM) for a linear N element array is a $2N$-port comprising hybrid couplers and fixed phase-shifters with N ports connected to the N antenna elements. The N remaining ports represent the output ports of the BM. As a result of the BM-topology the N-dimensional steering vector for the output ports of the full BM becomes

$$\begin{aligned} &\mathbf{s}_{Butler,N}(u) = [C_1(u), C_2(u), ..., C_N(u)]^T \quad \text{with} \\ &C_n(u) = \sqrt{g_0}\sin[0.5\pi N(u - u_n)]/ \\ &\sqrt{N}\sin[0.5\pi(u - u_n)] \end{aligned} \tag{23}$$

with $u_n = (2n - N - 1)/N$. Hence, each output port corresponds to one of N mutually orthogonal beams and the set of N beams provide a full coverage for $1 < u_0 < 1$. For $M < N$ and coverage restricted to angles according to Eq. (22) it is sufficient to have those M of N output ports of the full Butler matrix available which correspond to M adjacent beams with coverage for $u_{0,max}$ in accordance with Eq. (23). This leads to a reduced BM with N input but only M output ports. The topology of this reduced BM is derived from the topology of the full BM by removing all hybrid couplers and fixed phase-shifters which are exclusively allocated to the $N - M$ ports which do not belong to the M selected ports.

The steering vector in case of the reduced BM follows from that of the full BM if $M - N$ components of $\mathbf{s}_{Butler}(u)$ (see, Eq. 23) are replaced with zeros. The corresponding gain-function is characterised by the same beam width and side-lobe ratio as for the full BM but the maximum gain $g_{max}(u_0)$ rapidly drops to low values for $u_0 > M/N$. A larger angular coverage can be obtained if additionally to the reduced BM a network for frequency scanning (see, Fig. 4) is introduced. This frequency scanning network (FSN) consists of parallel delay lines with different delays τ_n which are inserted between the antenna elements and the BM. Now the radiation pattern becomes frequency-dependent. For a prescribed angular coverage of $u_{0,max}$ the directional pattern of the linear array at the maximum operational frequency must be shifted by $\Delta u = 2(u_{max} - M/N)$ relative to the pattern at the minimum operational frequency. As a consequence, the received pulses from the scattering centres become a bandpass-filtered version of the transmitted pulses. The frequency bandwidth of the received pulses is by a factor of $M/Nu_{0,max}$ reduced in comparison to the bandwidth of the transmitted pulses. Therefore the range resolution is reduced by the same factor. Angular resolution is gained by cost of range resolution. Comparing the described combination of ASP with frequency scanning with conventional frequency scanning (only one transceiver channel), the range resolution is improved by a factor of M and additionally high-resolution methods can be applied. Signal processing for angular information requires ASP to be combined with digital filtering in the frequency domain. A severe practical issue associated with this concept is the realisation of relatively large delays in the FSN which tend to introduce dissipative insertion loss.

5 Combined angular and temporal (Doppler) processing

The M-dimensional vectors $\mathbf{y}(t_i)$ for the $K = \Delta t/T_r$ snapshots taken at $t_0, t_0 + T_r, t_0 + 2T_r, ..., t_0 + (K - 1)T_r$ are

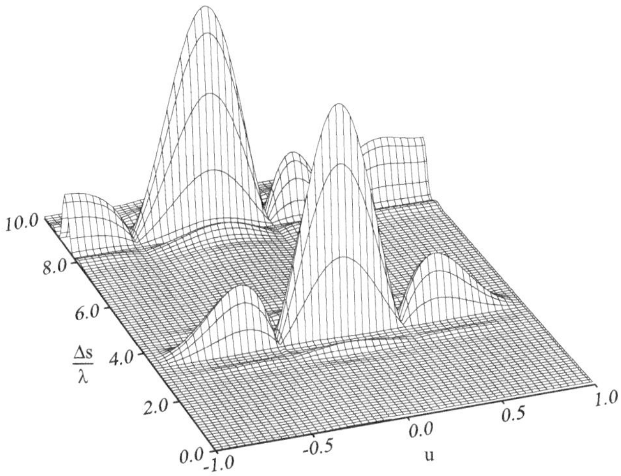

Fig. 5. Resolution of 2 scattering centers in the 2-dimensional angle-Doppler-domain.

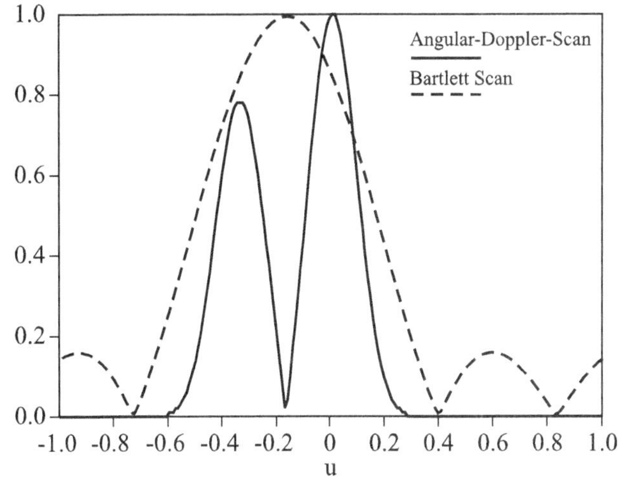

Fig. 6. Improvement of angular resolution by means of angular-Doppler-scan in comparison to conventional angular Bartlett scan.

taken as input data for a combined processing in angle and Doppler velocity (see e.g. Klemm, 2002). In the simplest processing scheme a 2-dimensional FFT is applied to the $M \times K$ complex values resulting in the 2-dimensional (angle and Doppler) distribution for the scattering centres. This is illustrated by means of an example where backscattered data from 2 scattering centres (separated in angle by $\Delta u = 0.34$) are assumed to be gained from a linear array with $M = N = 4$. Due to the low order (size) of the array the conventional Bartlett estimation (see, Eq. 20) for the angular scatterer distribution is not sufficient to resolve the 2 scattering centres. It is now assumed that within the observation time the changes in pathlengths for both scattering centres differ by about 5.5λ (e.g. by 22 mm in case of $f_0 = 75\,\text{GHz}$). Figure 5 shows the angle-Doppler distribution as result of the 2D-FFT. Instead of the Doppler velocity the relative shift $\Delta s/\lambda$ is used as a coordinate. Two maxima are clearly separated in this two-dimensional distribution. By applying a non-linear adaptive scaling to this distribution and by projecting back to the coordinate u the angular distribution as shown in Fig. 6 in comparison to the result of the Bartlett estimation is obtained. This demonstrates an improved angular resolution as result of the combined processing in angle and Doppler. With a high-resolution method, e.g. with MVDR (see, Eq. 21), the two scattering centres in this example would also have been resolved without any additional Doppler processing. However, this does not prove that the combined angle-Doppler processing is not superior to the angular processing. If a 2-dimensional high-resolution estimation scheme would be applied to the 2-dimensional angle-Doppler data, a better angular resolution than in case of the one-dimensional high-resolution estimation would be achieved.

The difference between this approach and the synthetic-aperture radar (SAR) principle can be explained as follows: In case of SAR there is some knowledge about the direction and velocity of the movement of the scenario relative to the antenna. This enables the different Doppler shifts to be transformed into angular positions and the angular information is gained from Doppler processing. The approach described in this paper does not require any knowledge about direction and velocity of the relative movements. If differential movements of the scattering centres occur an improved angular resolution is achieved but the angular information is retrieved from angular and not from Doppler processing.

References

Bresler, Y. and Macovski, A.: On the number of signals resolvable by uniform linear array, IEEE Trans. Acoust., Speech, Signal Processing, 34, 1361–1375, 1986.

Friedlander, B. and Weiss, A. J.: On the number of signals whose directions can be estimated by an array, IEEE Trans. Acoust., Speech, Signal Processing, 39, 1686–1689, 1991.

Klemm, R. : Principle of space-time adaptive processing, IEE Radar, Sonar, Navigation and Avionic Series 12, 2002.

Russell, M., Crain, A., Curran, A., Campbell, R., Drubin, C., and Miccioli, W.: Millimeter-wave radar sensor for automotive intelligent cruise control , IEE Trans. Microwave Theory Tech., 45, 2444–2453, 1997.

Wax, M. and Ziskind, I.: On unique localization of multiple sources by passive sensor arrays, IEEE Trans. Acoust., Speech, Signal Processing, 37, 996–1000, 1989.

Realistic antenna modeling for MIMO systems in microcell scenarios

C. Waldschmidt, C. Kuhnert, T. Fügen, and W. Wiesbeck

Institut für Höchstfrequenztechnik und Elektronik (IHE), Universität Karlsruhe, Germany

Abstract. This paper shows the potential of MIMO in cellular systems, where small handheld devices are used for the terminals. A complete model of a MIMO communication link is used to integrate accurate antenna modelling into MIMO system simulations. All different effects of mutual coupling between closely spaced antennas are considered. The efficiency or power budget respectively of the antenna arrays in the terminals, which are influenced by mutual coupling effects, is taken into account. Capacity simulation results based on a channel obtained from ray-tracing simulations are shown with cellular phones with up to three Inverted-F antennas.

1 Introduction

Recent studies have shown the potential of multiple transmit and multiple receive (multiple input multiple output – MIMO) antennas to reach high spectral efficiencies, see Winters (1987); Foschini and Gans (1998); Telatar (1998). MIMO systems combine classical techniques such as diversity and beamforming. Additionally they can transmit several datastreams at the same time and frequency in parallel. The capacity or spectral efficiency respectively depends on the signal to noise ratio (SNR) and the correlation properties among the channel transfer functions of different pairs of transmit and receive antennas. Uncorrelated channel transfer functions or channel coefficients in the flat fading case lead to high capacities for MIMO systems, that are based on multiplexing and use several subchannels, Chuah et al. (2002).

Many handheld devices like laptops or palmtops require small antenna spacings. This requirement is contrary to uncorrelated channel coefficients for MIMO systems based on space diversity, thus other diversity techniques have to be applied.

In almost all studies about MIMO, the SNR is assumed to be independent of the correlation properties of the channel matrix **H**, which contains the channel coefficients. Especially for small antenna spacings this assumption does not hold, as mutual coupling influences both SNR and **H**. In recent studies about mutual coupling this effect has been neglected, see Svantesson (2001), Stoytchev et al. (2001) and Waldschmidt et al. (2002). The same holds for mismatching of the antenna arrays. For MIMO systems with a constant transmit power the SNR is reduced for small antenna spacings, since the effective gain of the antennas is reduced, which decreases the capacity.

For compact antenna arrays it is necessary to analyze the SNR and the properties of **H** together. Therefore an extended channel, including transmitter (Tx), receiver (Rx), antennas and the physical channel is taken into account. The model of the whole system allows to analyze the whole radio frequency transmission chain, including mutual coupling and mismatching effects.

This paper is organized as follows. In Sect. 2 the system model, including an accurate way to model the antennas, is shown. In Sects. 3 and 4 the capacity calculation of a MIMO system and power considerations of the link are given. Sections 5 to 7 present an example of a MIMO system in a microcell scenario with realistic base station antennas and a small handheld device with up to three antennas. The correlation properties as well as the capacity and the power gain of this example are discussed in detail. Section 8 draws a conclusion.

2 System model

The complete radio frequency transmission chain consists of 5 elements: signal source (transmitter) – transmit antennas – physical channel – receive antennas – signal drain (receiver), given in Fig. 1. The number of transmit antennas is M, the one of receive antennas N. The single components are modelled as networks which are described by scattering matrices, see e.g. Pozar (1998).

Correspondence to: C. Waldschmidt
(christian.waldschmidt@ihe.uka.de)

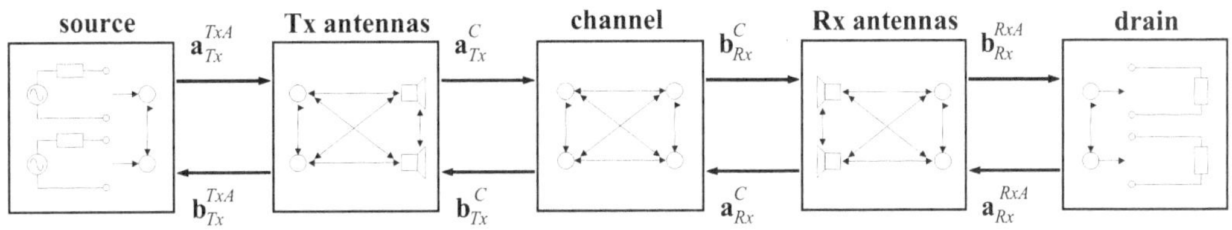

Fig. 1. System model of the complete radio frequency transmission chain. All elements are described by scattering matrices.

2.1 Signal source

The signal source is the beginning of the transmission chain and determines the power distribution among the transmit antennas. It has M ports, according to the number of transmit antennas. The output impedances of the signal source are characterized by the reflection matrix $\mathbf{r}^S_{Tx}$, where the upper index S denotes source. The outward propagating wave vector is

$$\boldsymbol{b}^S_{Tx} = \boldsymbol{b}^S_{0,S} + \mathbf{r}^S_{Tx}\boldsymbol{a}^S_{Tx}. \tag{1}$$

$\boldsymbol{b}^S_{0,S}$ is the wave vector fed into the network, which determines the power distribution among the source ports. $\boldsymbol{a}^S_{Tx}$ is the power reflected by the transmit antennas.

2.2 Transmit antennas

The antennas suffer from different effects when placed close together. They couple and interact, and cannot be considered as independent elements. The main effects, all of which have to be taken into account, are:

- The shape of the radiation pattern of the single antennas changes due to the other antennas. If e.g. omnidirectional antennas are used a pattern diversity effect occurs.
- The active gain of the single antennas has to be considered. The *active gain* is defined as the gain of a single antenna surrounded by the other terminated antennas. As the other antennas influence the pattern, the gain in terms of directivity changes, accordingly it is called the active gain. Additionally, the energy radiated from an antenna may be directly absorbed by another closely spaced antenna, see Kraus (1988), thus the active gain is reduced. With this definition the active gain is a property of each individual antenna in an array and depends on the array topology and termination of the antennas. It is not a function of the incident field.
- The signals received or transmitted by an antenna directly couple to the other antennas, only amplitude and phase shifted.

All these effects are included in the general scattering parameter description of the transmit antennas based on Wiesbeck and Heidrich (1998)

$$\begin{pmatrix} \boldsymbol{b}^{TxA}_{Tx} \\ \boldsymbol{b}^{TxA}_{F} \end{pmatrix} = \begin{pmatrix} \mathbf{S}^{TxA}_{TxTx} & \mathbf{S}^{TxA}_{TxF} \\ \mathbf{S}^{TxA}_{FTx} & \mathbf{S}^{TxA}_{FF} \end{pmatrix} \begin{pmatrix} \boldsymbol{a}^{TxA}_{Tx} \\ \boldsymbol{a}^{TxA}_{F} \end{pmatrix}, \tag{2}$$

where the upper index TxA denotes transmit antennas. Each antenna in the array is basically seen as a two port network. One port describes the excitation side of the antenna whereas the other port describes the far field properties. The submatrix $\mathbf{S}^{TxA}_{TxTx}$ describes the excitation ports of the antennas. It contains the scattering parameters corresponding to the self (diagonal elements) and mutual coupling (off-diagonal elements) impedances of the array Kraus (1988). The self impedance of an antenna is the input impedance if the antenna is remote in other words isolated from its surrounding. With other antennas in the near surrounding of an antenna the influence of these antennas on the input impedance is considered. The mutual coupling impedances describe the coupling between the antennas. The submatrices $\mathbf{S}^{TxA}_{FTx}$ and $\mathbf{S}^{TxA}_{TxF}$ describe the transmission of the signals from the excitation side of the antenna network to the far field and vice versa, thus contain information on the pattern and gain of the antennas. They are reciprocal. The submatrix $\mathbf{S}^{TxA}_{FF}$ contains the structural antenna scattering of the array with the excitation ports of the antennas terminated in matched loads. The elements of the structural antenna scattering matrix and the transmission matrices are directional, i.e. they are a function of the geometry and change for different angles of arrival and departure for different paths.

2.3 Physical channel

The channel is modelled by

$$\begin{pmatrix} \boldsymbol{b}^{C}_{Tx} \\ \boldsymbol{b}^{C}_{Rx} \end{pmatrix} = \begin{pmatrix} \mathbf{S}^{C}_{Tx} & \mathbf{S}^{C}_{TxRx} \\ \mathbf{S}^{C}_{RxTx} & \mathbf{S}^{C}_{Rx} \end{pmatrix} \begin{pmatrix} \boldsymbol{a}^{C}_{Tx} \\ \boldsymbol{a}^{C}_{Rx} \end{pmatrix}, \tag{3}$$

where the upper index C denotes channel. This scattering matrix expresses the relationship between the far field ports of the transmit and receive antenna arrays thus the coupling disappears by definition. Additionally it is assumed that there is no reflection from the far field. In other words nothing of the once radiated transmit power is received by the transmit antennas, thus the submatrices $\mathbf{S}^C_{Tx}$ and $\mathbf{S}^C_{Rx}$ equal the zero matrix.

2.4 Receive antennas and signal drain

The receive antennas are described analogue to the transmit antennas written with the upper index RxA. For the sake of completeness the mathematical description is given here

$$\begin{pmatrix} \boldsymbol{b}^{RxA}_{F} \\ \boldsymbol{b}^{RxA}_{Rx} \end{pmatrix} = \begin{pmatrix} \mathbf{S}^{RxA}_{FF} & \mathbf{S}^{RxA}_{FRx} \\ \mathbf{S}^{RxA}_{RxF} & \mathbf{S}^{RxA}_{RxRx} \end{pmatrix} \begin{pmatrix} \boldsymbol{a}^{RxA}_{F} \\ \boldsymbol{a}^{RxA}_{Rx} \end{pmatrix}. \tag{4}$$

The signal drain is similar to the signal source. It is described by the reflection matrix $\mathbf{r}^D_{Rx}$ with the dimension $N \times N$, where the upper index D denotes drain.

2.5 Merging of inner components

The five units given in the previous section are connected together now. We proceed in two steps. First, the inner three components i.e. transmit antennas, physical channel and the receive antennas are merged. The unilateral channel is introduced, which simplifies the problem. Second, the signal source and signal drain are connected to the network. After that the channel matrix $\mathbf{H}^{ext}$ of the extended channel with signal source and drain is given, which allows for capacity calculation.

The transmit antenna scattering matrix is cascade connected with the scattering matrix of the physical channel. These matrices are then connected with the scattering matrices of the receive antennas, which yields

$$\mathbf{S}^H_{TxTx} = \mathbf{S}^{TxA}_{TxTx} + \mathbf{S}^{TxA}_{TxF}\mathbf{S}^C_{TxRx} \cdot \left(\mathbf{I} - \mathbf{S}^{RxA}_{FF}\mathbf{S}^C_{RxTx}\mathbf{S}^{TxA}_{FF}\mathbf{S}^C_{TxRx}\right)^{-1} \cdot \mathbf{S}^{RxA}_{FF}\mathbf{S}^C_{RxTx}\mathbf{S}^{TxA}_{FTx} \tag{5}$$

$$\mathbf{S}^H_{TxRx} = \mathbf{S}^{TxA}_{TxF}\mathbf{S}^C_{TxRx} \cdot \left(\mathbf{S} - \mathbf{S}^{RxA}_{FF}\mathbf{S}^C_{RxTx}\mathbf{S}^{TxA}_{FF}\mathbf{S}^C_{TxRx}\right)^{-1} \cdot \mathbf{S}^{RxA}_{FRx} \tag{6}$$

$$\mathbf{S}^H_{RxTx} = \mathbf{S}^{RxA}_{RxF} \cdot \left(\mathbf{I} - \mathbf{S}^C_{RxTx}\mathbf{S}^{TxA}_{FF}\mathbf{S}^C_{TxRx}\mathbf{S}^{RxA}_{FF}\right)^{-1} \cdot \mathbf{S}^C_{RxTx}\mathbf{S}^{TxA}_{FTx} \tag{7}$$

$$\mathbf{S}^H_{RxRx} = \mathbf{S}^{RxA}_{RxRx} + \mathbf{S}^{RxA}_{RxF} \cdot \left(\mathbf{I} - \mathbf{S}^C_{RxTx}\mathbf{S}^{TxA}_{FF}\mathbf{S}^C_{TxRx}\mathbf{S}^{RxA}_{FF}\right)^{-1} \cdot \mathbf{S}^C_{RxTx}\mathbf{S}^{TxA}_{FF}\mathbf{S}^C_{TxRx}\mathbf{S}^{RxA}_{FRx}. \tag{8}$$

Equations (5) to (9) describe the extended channel without signal source and drain

$$\mathbf{S}^H = \begin{pmatrix} \mathbf{S}^H_{TxTx} & \mathbf{S}^H_{TxRx} \\ \mathbf{S}^H_{RxTx} & \mathbf{S}^H_{RxRx} \end{pmatrix}. \tag{9}$$

The back transmission of signals through the physical channel is subject to the channel attenuation, thus the power reradiated by the receiver and received by the transmitter is twice as strong attenuated as the signals at the receiver. Therefore it is justified to neglect to back transmission, see also Wallace and Jensen (2002), and to set $\mathbf{S}^C_{TxRx} = \mathbf{0}$, which simplifies Eqs. (5) to (9). We call this channel unilateral channel. The result is

$$\mathbf{S}^H = \begin{pmatrix} \mathbf{S}^{TxA}_{TxTx} & \mathbf{0} \\ \mathbf{S}^{RxA}_{RxF}\mathbf{S}^C_{RxTx}\mathbf{S}^{TxA}_{FTx} & \mathbf{S}^{RxA}_{RxRx} \end{pmatrix}. \tag{10}$$

The term $\mathbf{S}^{RxA}_{RxF}\mathbf{S}^C_{RxTx}\mathbf{S}^{TxA}_{FTx}$ in Eq. (10) describes the transmission of the signals from the input ports of the transmit antennas to the output ports of the receive antennas. Using the Heavyside transformation Pozar (1998), it can be expressed as the ratio of the voltages U at the transmit m and receive n antennas.

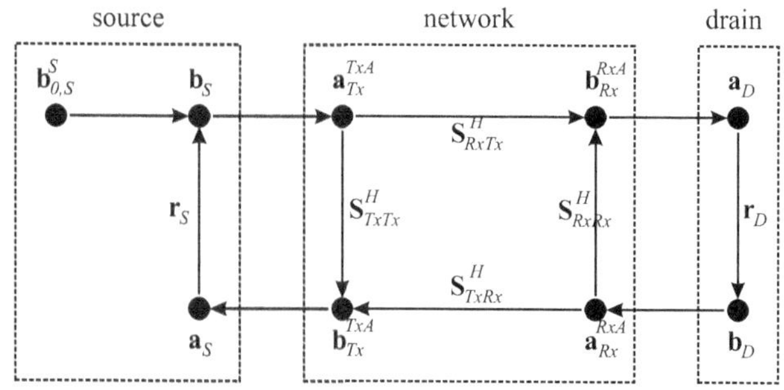

Fig. 2. Signal flow graph for the network above.

$$S^H_{RxTx,nm} = \sqrt{\frac{Z_{0,m}}{Z_{0,n}}}\frac{U_n}{U_m}\bigg|_{a_k=0} \quad \text{for all} \quad k \neq m, \tag{11}$$

where Z_0 is the characteristic impedance of the scattering parameters.

2.6 Termination with source and drain

Since the mutual coupling effects strongly depend on the termination of the transmit and receive antennas, the signal source and drain are added to the inner components of the network. Figure 2 shows the merged inner components network terminated with source and drain. With the wave vectors $\boldsymbol{a}^{TxA}_{Tx}$, $\boldsymbol{a}^{RxA}_{Rx}$, $\boldsymbol{b}^{TxA}_{Tx}$, and $\boldsymbol{b}^{RxA}_{Rx}$ at the input and output ports of the terminated network is is possible to calculate the power gain of the extended channel and the extended channel matrix $\mathbf{H}^{ext}$. Some basic scattering parameter calculations applied to the network consisting of the merged inner components (Eq. 10) and the signal source and drain yields

$$\boldsymbol{a}^{TxA}_{Tx} = \left(I - \mathbf{r}^S_{Tx}\mathbf{S}^H_{TxTx}\right)^{-1}\boldsymbol{b}^S_{0,S} \tag{12}$$

$$\boldsymbol{a}^{RxA}_{Rx} = \left(I - \mathbf{r}^D_{Rx}\mathbf{S}^H_{RxRx}\right)^{-1}\mathbf{r}^D_{Rx}\mathbf{S}^H_{RxTx} \cdot \left(I - \mathbf{r}^S_{Tx}\mathbf{S}^H_{TxTx}\right)^{-1}\boldsymbol{b}^S_{0,S} \tag{13}$$

$$\boldsymbol{b}^{TxA}_{Tx} = \left(I - \mathbf{S}^H_{TxTx}\mathbf{r}^S_{Tx}\right)^{-1}\mathbf{S}^H_{TxTx}\boldsymbol{b}^S_{0,S} \tag{14}$$

$$\boldsymbol{b}^{RxA}_{Rx} = \left(I - \mathbf{S}^H_{RxRx}\mathbf{r}^D_{Rx}\right)^{-1} \cdot \left(\mathbf{S}^H_{RxTx} + \mathbf{S}^H_{RxTx}\mathbf{r}^S_{Tx}\left(I - \mathbf{S}^H_{TxTx}\mathbf{r}^S_{Tx}\right)^{-1}\mathbf{S}^H_{TxTx}\right) \cdot \boldsymbol{b}^S_{0,S}. \tag{15}$$

The extended channel matrix $\mathbf{H}^{ext}$, which allows for the capacity calculations in the following section, expresses the ratio of the voltages at the receive antennas $\boldsymbol{U}_{Rx}$ to the voltages at the transmit antennas $\boldsymbol{U}_{Tx}$.

$$\boldsymbol{U}_{Rx} = \mathbf{H}^{ext}\boldsymbol{U}_{Tx}. \tag{16}$$

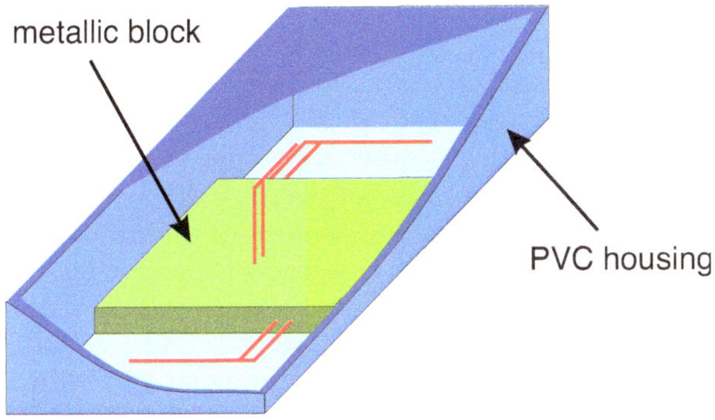

Fig. 3. Cellular phone model with three Inverted-F antennas. The size of the housing is $4.3\times10.5\times2.7\text{cm}^3$.

Using the Heavyside transformation leads to

$$Z_{0,n}^{1/2}(\boldsymbol{a}_{Rx}^{RxA}+\boldsymbol{b}_{Rx}^{RxT})=\mathbf{H}^{ext}Z_{0,m}^{1/2}(\boldsymbol{a}_{Tx}^{TxA}+\boldsymbol{b}_{Tx}^{TxA}). \quad (17)$$

To solve this equation for $\mathbf{H}^{ext}$ the voltages $\boldsymbol{U}_{Rx}$ can be expressed as a function of the voltages $\boldsymbol{U}_{Tx}$. By using Eqs. (12–16) one yields

$$\mathbf{H}^{ext}=(Z_{0,n})^{1/2}\left(\mathbf{I}+\mathbf{r}_{Rx}^{D}\right)\left(\mathbf{I}-\mathbf{S}_{RxRx}^{H}\mathbf{r}_{Rx}^{D}\right)^{-1} \cdot \mathbf{S}_{RxTx}^{H}\left(\mathbf{I}+\mathbf{S}_{TxTx}^{H}\right)^{-1}(Z_{0,m})^{1/2}. \quad (18)$$

3 Capacity calculations

The instantaneous capacity of MIMO transmission channels in the presence of spatially uncolored gaussian distributed noise can be calculated by [1]

$$C=\log_2\left(\det\left(I+\frac{SNR}{n}\mathbf{H}\mathbf{H}^{\dagger}\right)\right) \quad (19)$$

In order to assess a MIMO transmission system by the capacity, usually a fixed SNR (see e.g. Jensen, 2001) and a channel matrix, normalized with $\|\mathbf{H}\|^2=n_{Tx}n_{Rx}$ is used. $\|\mathbf{H}\|$ denotes the Frobenius Norm of $\mathbf{H}$. But if comparing different antennas in the same scenario, a fixed SNR and a normalized channel matrix any interrelation between the mean attenuation and the correlation properties of $\mathbf{H}$, that strongly influence the capacity and that are both influenced by mutual coupling as shown before, is neglected. If $\mathbf{H}$ is not normalized with the Frobenius Norm, that means the path loss and the gain of single antenna elements are included in $\mathbf{H}$, (19) can be written as

$$C=\log_2\left(\det\left(I+\frac{P_T}{\sigma^2 n}\mathbf{H}\mathbf{H}^{\dagger}\right)\right). \quad (20)$$

This formulation allows for a fair comparison of different antenna arrays, since both the correlation properties of $\mathbf{H}$ and the power budget of the link, i.e. the efficiency of the arrays, is taken into account. P_T is the total transmit power which is equally distributed among all transmit antennas, if no channel state information at the transmitter is available. If the channel is known at the receiver, the capacity can be increased by using waterfilling, see Bach-Andersen (2000). For the capacity calculations given in the following, $\mathbf{H}^{ext}$ of Eq. (18) is used in Eq. (20) and no channel state information at the transmitter is assumed.

4 Power considerations

In order to assess the efficiency in term of power of an array in a MIMO system, we use the transmission gain of the whole MIMO system model. The power gain $G_{H^{ext}}$ of the extended channel is the ratio of the real power delivered to the signal drain to the real power fed into the transmit antennas. It allows to draw conclusions on the efficiency of the whole transmission chain.

$$G_{H^{ext}}=\frac{\boldsymbol{b}_{Rx}^{RxA\dagger}\boldsymbol{b}_{Rx}^{RxA}-\boldsymbol{a}_{Rx}^{RxA\dagger}\boldsymbol{a}_{Rx}^{RxA}}{\boldsymbol{a}_{Tx}^{TxA\dagger}\boldsymbol{a}_{Tx}^{TxA}-\boldsymbol{b}_{Tx}^{TxA\dagger}\boldsymbol{b}_{Tx}^{TxA}}. \quad (21)$$

Using this definition the power gain depends on the power distribution among the elements of the excitation vector $\boldsymbol{b}_{0,S}^{S}$. Note, that the power fed into the transmit antennas is not equal to the radiated power, since there occur losses in the transmit antennas due to the losses in the single antennas and due to mutual coupling effects. Since in MIMO system simulations or measurements a large number of different channel realizations is considered the power gain of the extended channel $G_{H^{ext}}$ is a random variable.

In the following an example showing the application of the network model is given. We consider a MIMO link between a small handheld device and a base station in a microcell scenario.

5 Antennas in the handheld device

For the handheld devices half-wavelength dipole antennas and Inverted-F antennas were used. The aim of the antenna setup was to combine different diversity techniques, such as pattern, spatial and polarization diversity, to obtain uncorrelated signals at the antennas, see Waldschmidt et al. (2003). Additionally polarization diversity makes the antenna system robust against polarization mismatching. The simulation model of the handheld device consists of a metallic block, representing the battery and the display, and a PVC housing with a wall thickness of 2 mm. The size of the housing is $4.3\times10.5\times2.7\text{cm}^3$. Three different antenna configurations in the handheld device were modelled: a) two crossed dipoles, see Fig. 4, b) 3 Inverted-F antennas at a large (length=5 cm) and c) at a small (length=3 cm) metallic block, see Fig. 3. Usually Inverted-F antennas require a large ground plane, which is not given in the small handheld device. Thus the metallic block, representing the ground plane,

[1] I denotes the identity matrix $n=\min(n,m)$ $\dagger$ denotes conjugate complex transpose

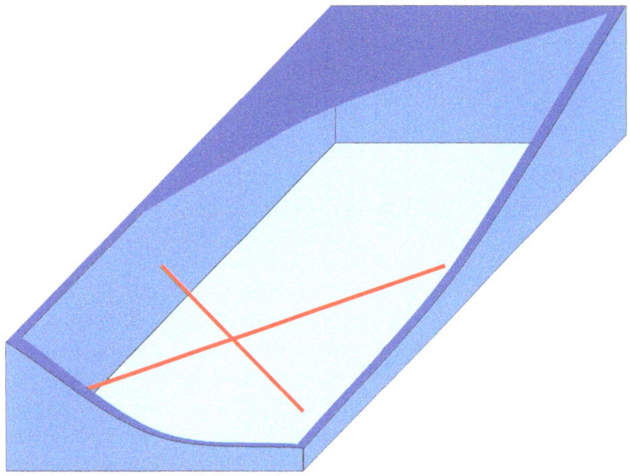

Fig. 4. Cellular phone model with 2 crossed dipole antennas.

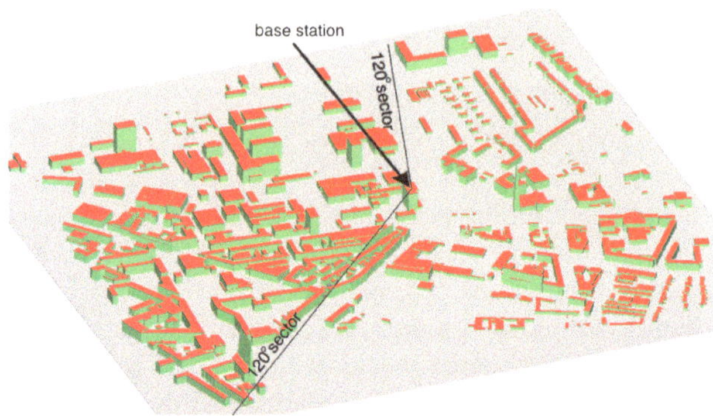

Fig. 5. Microcell scenario for the ray-tracing simulations.

acts as a part of the antenna and influences the system performance. Due to that reason, the size of the metallic block changes the pattern of the Inverted-F antennas compared to an idealized antenna and the mutual coupling impedances among the antennas. All antenna configurations were simulated with FEKO, see (www.feko.com, 2003), a standard EM-code to calculate the pattern of the coupled system and the self- and mutual coupling impedance matrix. For the system simulations, self impedance matching between the antennas and the loads (signal source and drain) was assumed. In other words, the single antennas were perfectly matched, but the mutual coupling impedances were open circuited. Perfect (in terms of power) conjugate complex matching is very unrealistic for handheld devices, since the mutual coupling impedances can hardly be matched. They change during usage of the device.

6 Microcell scenario

In the following scenario a link between a small handheld device and a base station in a microcell scenario is considered. The focus here is on the antennas in the handheld device and the base station, thus no interference or multiple users are assumed.

The base station is equipped with three antennas with spacings of 1 meter. The pattern of the commercially available base station antenna "Kathrein Antenne 735 147 (GSM 1800)" was modelled, see Baldauf et al. (2001), for the simulations. The channel data were obtained from ray-tracing simulations of a microcell scenario. The campus of the University of Karlsruhe and the surrounding buildings were used for the simulations, see Fig. 5. The base station with three Kathrein antennas radiating into the same 120° sector, is placed on one of the highest buildings in the scenario, 3 m above the roof. At 2000 randomly distributed locations on the campus the channel data are collected. The power azimuth spectrum at the base station is relatively narrow, since there are no scatterers around the antennas. At the mobile the angular spread is wide, as the mobile is surrounded by scatterers. The channel data are normalized to obtain a constant mean attenuation of the link. That means, the incoherent sum of the path weights Γ of all paths ν and polarizations of one channel realization are added and set constant.

$$const = \sum_{\nu=1}^{P} \Gamma^{\nu}_{\vartheta\vartheta} + \Gamma^{\nu}_{\vartheta\psi} + \Gamma^{\nu}_{\psi\vartheta} + \Gamma^{\nu}_{\psi\psi}, \tag{22}$$

where the subscript denotes the polarization at transmitter and receiver of the path ν. P is the total number of paths. This normalization removes the slow fading effects, but it does not influence the (spatial-) fast fading, which is essential for MIMO. Using the Frobenius Norm to normalize the channel matrices, as often done in MIMO system simulations, would change the probability distribution of the power of the fast fading signal. The normalized channel data serve as an input for the network model. The antennas are considered separately in the network, thus the power budget of the link and the efficiency of the antennas is fully taken into account.

For comparison, the same simulation was performed without complete handheld devices, but with arrays consisting of parallel $\lambda/2$-dipoles with one wavelength spacing. Thus uncorrelated channel coefficients can be expected, resulting in a optimum MIMO system.

7 Simulation results

Table 1 shows the results. In the second column the power correlation coefficient among the signals at the handheld device are given. The third column shows the 10% outage capacity in bit/s/Hz for a given constant transmit power. The transmit power was chosen to result in a 10% outage capacity for the SISO system of 3.5 bit/s/Hz. In the last column the ratio of the mean transmission gain of the MIMO system to the mean transmission gain of a SISO system with one half-wavelength dipole at the transmitter and at the receiver is given. The MIMO systems with one wavelength antenna spacings and no influence of any housing etc. perform best, as shown in the second and third row of Table 1. They can

Table 1. Simulation results.

antenna configuration	power correlation at mobile	10% outage capacity in $\frac{\text{bit}}{\text{sHz}}$	G_H/G_{SISO} at 50% in dB
2 dipoles, λ spacing	≤ 0.1	12,9	8
3 dipoles, λ spacing	≤ 0.1	14.5	11
2 crossed dipoles	≤ 0.1	11.5	6.9
3 Inv.-F, large block	≤ 0.35	13	9.6
3 Inv.-F, small block	≤ 0.2	13.5	9.6
SISO	-	3.5	0

be considered as a reference for the other antenna configurations in the handheld device for a comparison of MIMO systems with equal number of antennas. It is evident, that all small antenna arrays in the handheld device perform well in terms of correlation and power and thus lead to capable MIMO systems. Surprisingly the handheld device with the large metallic block does not outperform the one with the small block, though the antenna spacings are larger. This is due to the fact, that the pattern diversity between the antennas is stronger for the small metallic block. That clearly shows, that the whole configuration of the handheld device has to be taken into account.

8 Conclusion

The paper shows how to introduce detailed antenna modeling into MIMO system simulations, when using compact antenna arrays, that suffer from mutual coupling. Several antenna array configurations, that fit into small handheld devices are presented and discussed in detail. For the assessment of small antenna arrays it is necessary to consider both first the correlation properties of the signals and second the efficiency in terms of power of the array. It is shown, that MIMO works well with small antenna array configurations, when exploiting combinations of different diversity techniques such as polarization, pattern and spatial diversity. This framework allows for the examination of the behavior of compact arrays in MIMO system simulations.

References

Bach-Andersen, J.: Array gain and capacity for known random channels with multiple element arrays at both ends, IEEE Journal on Selected Areas in Communications, 18, 2172–2178, 2000.

Baldauf, M., Herschlein, A., and Wiesbeck, W.: Schutzabstände in der Mobilkommunikation, Frequenz, 55, 310–316, 2001.

Chuah, C., Tse, D., Kahn, J., and Valenzuela, R.: Capacity Scaling in MIMO Wireless Systems Under Correlated Fading, IEEE Transactions on Information Theory, 48, 637–650, 2002.

Foschini, G. and Gans, M.: On limits of wireless communications in a fading environment when using multiple antennas, Wireless Personal Communications, 6, 311–335, 1998.

Jensen, M.: Characteristics of measured 4x4 and 10x10 MIMO wireless channel data at 2.4 GHz, IEEE International Symposium on Antennas and Propagation, 3, 95–98, 2001.

Kraus, J.: Antennas, Electrical Engineering Series, McGraw-Hill International Editions, second edn., 1988.

Pozar, D. M.: Microwave Engineering, John Wiley & Sons, Inc., 1998.

Stoytchev, M., Safar, H., Moustakas, A. L., and Simon, S.: Compact antenna arrays for MIMO applications, IEEE International Symposium on Antennas and Propagation, 3, 708–711, 2001.

Svantesson, T.: Mutual coupling effects on the capacity of multielement antenna systems, IEEE Proceedings of Acoustics, Speech, and Signal Processing Conference, 2, 2485–2488, 2001.

Telatar, I.: Capacity of Multi-Antenna Gaussian Channels, Technical Memorandum, Bell Laboratories, Lucent Technologies, October 1995, then published in European Transactions on Telecommunications, 10, 585–595, 1998.

Waldschmidt, C., v. Hagen, J., and Wiesbeck, W.: Influence and Modelling of Mutual Coupling in MIMO and Diversity Systems, IEEE International Symposium on Antennas and Propagation AP-S, San Antonio, Texas, 3, 190–193, 2002.

Waldschmidt, C., Kuhnert, C., Schulteis, S., and Wiesbeck, W.: Compact MIMO arrays based in Polarization Diversity, IEEE internatinal Symposium on Antennas and Propagation, Columbus, Ohio, pp. 499–502, 2003.

Wallace, J. W. and Jensen, M. A.: The capacity of MIMO wireless systems with mutual coupling, IEEE Proceedings of the Vehicular Technology Conference, 2, 696–700, 2002.

Wiesbeck, W. and Heidrich, E.: Wide-band multiport antenna characteristics by polarimetric RCS measurements, IEEE Transactions on Antennas and Propagation, 46, 314–350, 1998.

Winters, J. H.: On the Capacity of Radio Communication Systems with Diversity in a Rayleigh Fading Environment, IEEE Journal on Selected Areas in Communications, 5, 871–877, 1987.

20

Speaker tracking with a microphone array using Kalman filtering

D. Bechler, M. Grimm, and K. Kroschel

Institut für Nachrichtentechnik, Universität Karlsruhe, Kaiserstr. 12, 76128 Karlsruhe, Germany

Abstract. In this publication a method for tracking a speaker with acoustical information by means of a microphone array is presented. A sound source localization algorithm based on the time delays of arrival of sound waves in microphone pairs provides initial position estimates. These significantly varying estimates are spatially filtered by an adaptive Kalman filter to obtain a smoothed trajectory of the speaker's movement. Experimental results with real data are presented for a variety of scenarios recorded in a noisy and reverberant office room environment.

1 Introduction

Integrating acoustic perception by means of a microphone array into autonomous humanoid robots is nowadays an important area of research. The aim of the robot's hearing system is not only to be able to interact with a human operator but also to create an acoustic map of the sound environment and to perform an acoustic scene analysis, i.e. the localization, separation and classification of sound sources present in the acoustic environment. Especially the task of acoustically localizing a speaker in a room environment is of great interest not only in robotics but also for teleconferencing or acoustic surveillance systems.

The technique of choice in most recent acoustic localization systems using microphone arrays is a two-step procedure. First, the time delay of arrival (TDOA) of sound signals in a pair of spatially separated microphones is estimated. In a second step the estimated TDOAs of different microphone pairs are used in combination with the microphone array geometry to localize the sound source. To avoid the computational demanding solution of a set of non-linear equations for the exact sound source position, sub-optimal closed-form localization estimators with satisfactory precision can be applied (DiBiase et al., 2001; Huang et al., 2000).

Correspondence to: D. Bechler
(bechler@int.uni-karlsruhe.de)

Due to the one-sample-precision of the TDOA estimation algorithm and due to noise and reverberation influences, the TDOA estimates and the real TDOA values are not identical which leads to relatively high variances in consecutive position estimates. In this publication a method to smooth the speaker trajectory and to assure robustness by means of an adaptive Kalman filter is evaluated for a real data single speaker scenario in a noisy and reverberant office environment.

The paper is organized as follows: In Sect. 2 the localization of sound sources based on time delay estimates is presented. Filtering the initial estimates of the sound source position by an adaptive Kalman filter is discussed in Sect. 3. Section 4 describes the experimental setup. In Sect. 5 we present and evaluate the obtained results. Finally, some conclusions are drawn and an outlook for future work is given.

2 Time- delay- based localization of sound sources

2.1 Signal model

For a given pair of spatially separated microphones M_i and M_j, the recorded sensor signals $x_i(t)$ and $x_j(t)$ for a signal $s(t)$, emanated from a remote sound source in a reverberant and noisy environment, can be modeled mathematically as

$$\begin{aligned} x_i(t) &= h_i(t) * s(t) + n_i(t) \\ x_j(t) &= h_j(t - \tau_{ij}) * s(t) + n_j(t), \end{aligned} \tag{1}$$

where τ_{ij} represents the relative signal delay of interest, $*$ signifies the convolution operator, $h_i(t)$ is the acoustic impulse response between the sound source and the ith microphone and the additive term $n_i(t)$ summarizes the channel noise in the microphone system as well as environmental noise for the ith sensor. This noise $n_i(t)$ is assumed to be uncorrelated with $s(t)$.

2.2 TDOA estimation with GCC method

The most popular approach for determining the TDOAs is called the Generalized Cross-Correlation (GCC) method Knapp and Carter (1976). The relative time delay τ_{ij} is estimated as the time lag with the global maximum peak in the GCC function $R_{ij}^{(g)}(\tau)$

$$\hat{\tau}_{ij} = \underset{\tau}{\text{argmax}} \; R_{ij}^{(g)}(\tau). \tag{2}$$

This GCC function $R_{ij}^{(g)}(\tau)$ is defined as

$$R_{ij}^{(g)}(\tau) = \int_{-\infty}^{+\infty} \psi_{ij}(\omega) X_i(\omega) X_j(\omega)^* e^{j\omega\tau} d\omega. \tag{3}$$

The weighting function $\psi_{ij}(\omega)$ intends to decrease noise and reverberation influences and tries to emphasize the GCC value at the true TDOA value τ_{ij}. For real environments the Phase Transform (PHAT) technique (Knapp and Carter, 1976) has shown the best performance. This PHAT weighting function is defined as

$$\psi_{ij}^{PHAT}(\omega) = \frac{1}{|X_i(\omega) X_j(\omega)^*|}. \tag{4}$$

2.3 Confidence criteria for TDOA estimates

To detect outliers in TDOA estimates, two confidence criteria can be used: the value of the maximum peak and the ratio between the 1st and the 2nd peak in the GCC function (Bechler and Kroschel, 2002a,b). These criteria allow a reliability scoring of individual estimates and can be used to reject erroneous measurements. The higher the value of these properties of the GCC function the higher the probability that the TDOA was estimated correctly.

2.4 Data association and clustering of TDOA estimates

With the confidence criteria described in Sect. 2.3, a tradeoff has to be made between a high number of estimates necessary for a continuous source tracking, and a high percentage of correct TDOA estimates, which is crucial for robust source localization. Additionally, the one-sample-precision of the GCC algorithm can be too imprecise, which is problematic for the localization algorithm. As a solution we propose data association and clustering techniques for the TDOA estimates. To initialize an acoustic track, a high reliability is demanded and therefore a high threshold for the value of the confidence criteria is used. Once initialized, this decision threshold can be lowered, as now it is possible to search in a region of interest around the initial TDOA estimate value. Thus, erroneous estimates outside this region of interest can be rejected. In addition, several consecutive TDOA estimates are averaged, which solves the problem of the one-sample-precision. With these data association and clustering techniques, the TDOA estimates become sufficiently accurate for the following localization algorithm to produce robust sound position estimates.

2.5 Localization algorithm

To derive from the TDOAs and the microphone array geometry the source position, the exact localization necessitates solving a set of non-linear equations, which can be computationally demanding. To accelerate the sound source position determination, the One-Step Least-Squares (OSLS) algorithm (Huang et al., 2000) is used. This closed-form location estimator approximates precisely enough accurate the exact solution to the non-linear problem.

3 Spatial filtering

For a continuous source trajectory these initial, significantly varying position estimates obtained by the localization algorithm described in Sect. 2 are spatially smoothed by a Kalman Filter (KF) as post-processing unit.

3.1 Kalman filtering

For the motion of a speaker the time-discrete state space description by means of a state and an observation equation is used. The state equation is defined as

$$\boldsymbol{x}(k) = \mathbf{A}\boldsymbol{x}(k-1) + \mathbf{B}u(k) \tag{5}$$

where $\boldsymbol{x}(k)$ is a state vector at time instant k containing the position, the velocity and the acceleration of the sound source in Cartesian coordinates. The time-invariant state transition matrix $\mathbf{A}$ as well as the variance σ_U^2 of the white noise $u(k)$ representing the system error are assumed to be known. $\mathbf{B}$ is the time-invariant system noise coupling matrix mapping the system error to the elements of the state vector $\boldsymbol{x}(k)$. In the observation equation

$$\boldsymbol{y}(k) = \mathbf{C}\boldsymbol{x}(k) + \boldsymbol{n}(k) \tag{6}$$

$\boldsymbol{y}(k)$ is the observation vector containing the source position, $\mathbf{C}$ is the time-invariant measurement matrix mapping the state of the system to the observation vector and $\boldsymbol{n}(k)$ is white noise with the covariance matrix $\mathbf{R}_{NN}(k)$ representing the measurement error. For our scenario, in which a speaker can move arbitrarily in the acoustical environment, three motion models are reasonable: a static, a constant velocity and a constant acceleration model (for the definitions of the according matrices $\mathbf{A}$, $\mathbf{B}$ and $\mathbf{C}$ cf. Grimm, 2001).

3.2 Filter algorithm

One cycle of the KF is done with the following filter algorithm. For a detailed derivation cf. Brown (1983).

In a first step, the updated estimation error covariance matrix $\mathbf{R}_{EE}(k+1|k+1)$ can be derived from the previous covariance matrix $\mathbf{R}_{EE}(k|k)$. Therefore, the *a priori* estimation error covariance matrix $\mathbf{R}_{EE}(k+1|k)$ at the $(k+1)^{th}$ iteration given k observations is calculated with

$$\mathbf{R}_{EE}(k+1|k) = \mathbf{A}\mathbf{R}_{EE}(k|k)\mathbf{A}^T + \mathbf{B}\sigma_U^2\mathbf{B}^T. \tag{7}$$

With Eq. (7), the covariance matrix $\mathbf{R}_{\nu\nu}(k+1)$ of the innovation $\nu(k+1)$ can be calculated, where $\nu(k)$ is defined as the difference between the actual observation vector $\boldsymbol{y}(k)$ and the estimated observation vector at the kth iteration given $k-1$ observations $\hat{\boldsymbol{y}}(k|k-1)$:

$$\nu(k) = \boldsymbol{y}(k) - \hat{\boldsymbol{y}}(k|k-1). \quad (8)$$

$\mathbf{R}_{\nu\nu}(k+1)$ is given by

$$\mathbf{R}_{\nu\nu}(k+1) = \mathbf{C}\mathbf{R}_{EE}(k+1|k)\mathbf{C}^T + \mathbf{R}_{NN}(k). \quad (9)$$

The *a posteriori* estimation error covariance matrix $\mathbf{R}_{EE}(k+1|k+1)$ is then found from

$$\begin{aligned}\mathbf{R}_{EE}(k+1|k+1) &= \mathbf{R}_{EE}(k+1|k)\\ &-\mathbf{W}(k+1)\mathbf{R}_{\nu\nu}(k+1)\mathbf{W}^T(k+1),\end{aligned} \quad (10)$$

where $\mathbf{W}(k+1)$ is the filter gain at instant $k+1$ defined as

$$\mathbf{W}(k+1) = \mathbf{R}_{EE}(k+1|k)\mathbf{C}^T(\mathbf{R}_{\nu\nu}(k+1))^{-1}. \quad (11)$$

Similarly, the estimate of the system state can be iteratively derived from the previous estimate. First, the *a priori* system state estimate

$$\hat{\boldsymbol{x}}(k+1|k) = \mathbf{A}\hat{\boldsymbol{x}}(k|k) \quad (12)$$

is calculated and out of it the source position is estimated with

$$\hat{\boldsymbol{y}}(k+1|k) = \mathbf{C}\hat{\boldsymbol{x}}(k+1|k). \quad (13)$$

By means of innovation and filter gain the actual state estimation results from

$$\hat{\boldsymbol{x}}(k+1|k+1) = \hat{\boldsymbol{x}}(k+1|k) + \mathbf{W}(k+1)\nu(k+1) \quad (14)$$

and the updated source position is found from

$$\hat{\boldsymbol{y}}(k+1|k+1) = \mathbf{C}\hat{\boldsymbol{x}}(k+1|k+1). \quad (15)$$

3.3 Adaptive Kalman Filter

Up to now the different motion model approaches were regarded independently. As the motion dynamics of a speaker in an office environment are variable, a single model for the KF is not suitable for all situations. Thus, a multiple model system would be advantageous. By means of the definition of a model probability described in this section, several models are used simultaneously for the determination of the final position estimate.

3.3.1 Model probability

For L models α_i with $1 \leq i \leq L$ the conditional probability $f(\alpha_i|\mathbf{y}(k))$ is the probability that the ith model based on the measurements $\mathbf{y}(1), \cdots, \mathbf{y}(k)$ is the one describing the speaker's movement correctly. With the Bayes formula we get

$$f(\alpha_i|\boldsymbol{y}(k)) = \frac{f(\boldsymbol{y}(k)|\alpha_i)\, f(\alpha_i)}{f(\boldsymbol{y}(k))}. \quad (16)$$

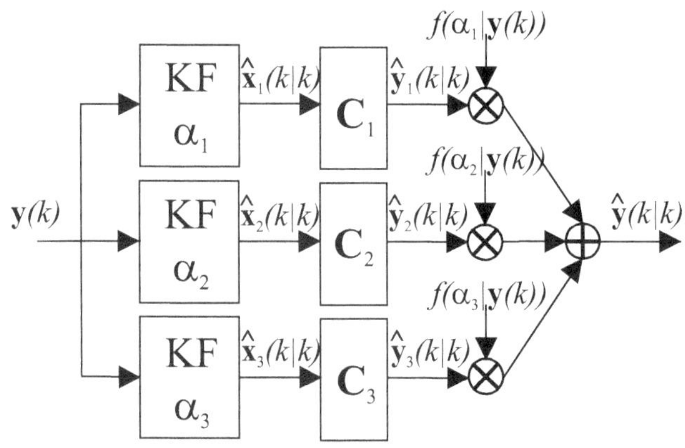

Fig. 1. Block diagram of the adaptive Kalman filter

For this conditional probability a recursive formula can be given:

$$f(\alpha_i|\boldsymbol{y}(k)) = \frac{f(\boldsymbol{y}(k)|\boldsymbol{y}(k-1),\alpha_i)\, f(\alpha_i|\boldsymbol{y}(k-1))}{\sum_{j=1}^{L} f\left(\boldsymbol{y}(k)|\boldsymbol{y}(k-1),\alpha_j\right) f\left(\alpha_j|\boldsymbol{y}(k-1)\right)}. \quad (17)$$

The terms $f(\boldsymbol{y}(k)|\boldsymbol{y}(k-1),\alpha_i)$ can be calculated in every recursive step with

$$\begin{aligned}f(\boldsymbol{y}(k)|\boldsymbol{y}(k-1),\alpha_i) &= \frac{1}{(2\pi)^{\frac{3}{2}}\sqrt{|\mathbf{R}_{\nu\nu,i}(k)|}}\\ &\cdot \exp\left\{-\tfrac{1}{2}\nu_i^T(k)\mathbf{R}_{\nu\nu,i}^{-1}(k)\nu_i(k)\right\}.\end{aligned} \quad (18)$$

3.3.2 Multiple Model Adaptive Estimator (MMAE)

The approach using several models for the final estimate simultaneously is called adaptive KF or Multiple Model Adaptive Estimator (MMAE) (Bar-Shalom, 1988). In Fig. 1 the block diagram of our adaptive KF is shown. The measurement $\boldsymbol{y}(k)$ is used in three KFs in parallel to determine one estimate of the system state $\hat{\boldsymbol{x}}_1(k|k)$, $\hat{\boldsymbol{x}}_2(k|k)$ and $\hat{\boldsymbol{x}}_3(k|k)$ for every of the three models. As mentioned, we use a static, a constant velocity and a constant acceleration model to describe the motion of a speaker. By means of the observation matrices $\mathbf{C}_1$, $\mathbf{C}_2$ and $\mathbf{C}_3$ the system state estimates are mapped to the observation estimates $\hat{\boldsymbol{y}}_1(k|k)$, $\hat{\boldsymbol{y}}_2(k|k)$ and $\hat{\boldsymbol{y}}_3(k|k)$. These position estimates are weighted with the corresponding model probabilities $f(\alpha_i|\boldsymbol{y}(k))$ according to (Eq. 17) with $1 \leq i \leq 3$ and finally summed up. Hence, the actual position estimate for the adaptive KF is determined with

$$\hat{\boldsymbol{y}}_{MMAE}(k|k) = \sum_{i=1}^{3} f(\alpha_i|\boldsymbol{y}(k)\mathbf{C}_i\hat{\boldsymbol{x}}_i(k|k). \quad (19)$$

4 Experimental setup

Real data experiments have been carried out in a (5 m × 5 m × 3 m) office room with typical environmental noise (fans, ...) and relatively strong reverberation. For

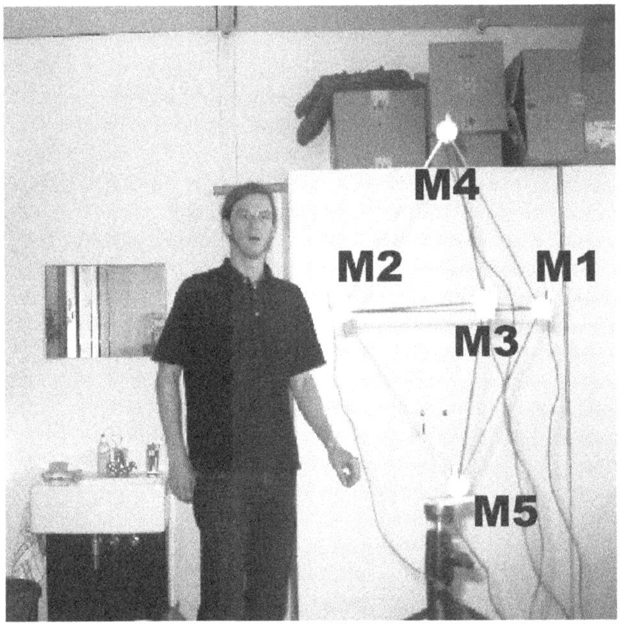

Fig. 2. Experimental setup

data recording we used a 5-microphone array in an equilateral double-tetrahedron geometry with a side length of $D = 28$ cm as shown in Fig. 2. 25 scenarios were recorded in which a speaker is uttering German sentences. In addition to stationary positions for a sitting or standing speaker, motions along given trajectories with constant velocity and transitions from stationary positions to constant walking were studied. The sampling frequency was $f_s = 16$ kHz. The recorded data were analyzed in frames of 32 ms to assure quasi-stationarity. For this data segmentation, a Hamming window with a 50% overlap was applied. For the data association and clustering techniques described in Sect. 2.4, the average of the TDOA estimates in a block out of 12 consecutive frames is calculated where blocks are overlapped by 50%. This generates a new position estimate for the sound source every 96 ms. With this averaging, robust speaker tracking is still possible since at least four estimates per second are required (Silverman and Kirtman, 1992) for a moving speaker.

5 Results

The proposed system shows robust speaker localization capabilities in a noisy and reverberant environment. The adaptive KF performs its task as post-processing unit in guaranteeing a smoothed continuous trajectory compared with the initial position estimates.

Exemplarily, Fig. 3 shows a comparison of the localization of a stationary speaker at the spherical coordinates $azimuth = 95°$, $elevation = -7.9°$ and $range = 1.5$ m before and after Kalman filtering. The temporal runs of all spherical coordinates are significantly smoothed. Note that the variance in range is reduced, but the overall range error increases by applying a KF in this case. Until 1.2 s the localization algorithm underestimates the range by about 50 cm and the KF evidently cannot remedy this systematic error.

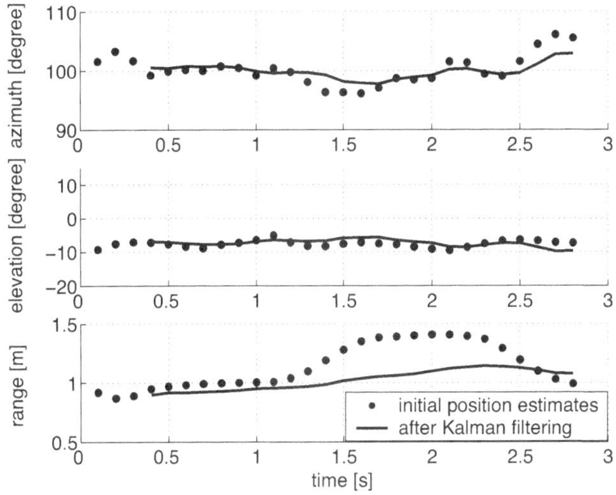

Fig. 3. Stationary speaker at $azimuth = 95°$, $elevation = -7.9°$ and $range = 1.5$ m

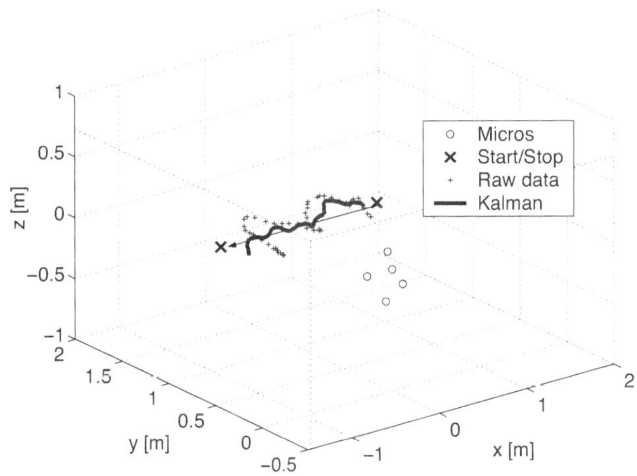

Fig. 4. Speaker moving from Cartesian coordinates $(-1m, 1m, 0.13m)$ to $(1m, 1m, 0.13m)$

An example for tracking a speaker uttering a German sentence while moving with constant velocity from (−1 m,1 m,0.13 m) to (1 m,1 m,0.13 m) is presented in Fig. 4. In this 3D-plot the true trajectory (arrow) and the positional estimates before and after applying the adaptive KF for a walking speaker are shown. Also in this example the KF functions reliably and delivers a smoothed speaker trajectory. As for this scenario there is no systematic error in the initial position estimates the overall error in all coordinates is reduced.

6 Conclusions and outlook

The good results for the determination of a smoothed continuous trajectory from the initial sound source positions justify the complexity of our adaptive Kalman filter as post-processing unit. But the reduction of the overall error of position estimates with an adaptive KF highly depends on the quality of the initial estimates used as KF input. If these estimates show low systematic errors, the KF delivers an efficient smoothing of the initial trajectory. Even if there are erroneous TDOA estimates in only one of the 4 microphone pairs, this can lead to wrong initial position estimates by the localization algorithm, which renders the operation of the filter unreliable. Therefore future work will be devoted to the improvement of the TDOA reliability by considering psychoacoustics, e.g. the precedence effect (Litovsky et al., 1999) to suppress reverberation influences. An additional advantage of the KF lies in its ability to predict source positions in case of missing current position estimates due to speech pauses or non-reliable TDOA estimates.

In this work only one speaker was present in the acoustic scene. Current investigations are concerned with the simultaneous tracking of multiple active sound sources requiring an according number of KFs working in parallel. Furthermore, it is aimed to extend this tracking system by cameras. With the fusion of acoustic and visual data strong improvements should be achieved compared with a pure acoustic tracker.

Acknowledgements. This work is part of the Sonderforschungsbereich (SFB) No. 588 *"Humanoide Roboter - Lernende und kooperierende multimodale Roboter"* at the University of Karlsruhe. The SFB is supported by the Deutsche Forschungsgemeinschaft (DFG).

References

Bechler, D. and Kroschel, K.: Confidence scoring of time difference of arrival estimation for speaker localization with microphone arrays, 13. Konferenz Elektronische Sprachsignalverarbeitung ESSV, September 2002a.

Bechler, D. and Kroschel, K.: Reliability measurement of time difference of arrival estimations for multiple sound source localization, 17th Annual Meeting of the IAR, November 2002b.

Brown, R. G.: Introduction to Random Signal Analysis and Kalman Filtering, Wiley, 1983.

Bar-Shalom Y.: Tracking and data association, Academic Press, 1988.

DiBiase, J. H., Silverman, H. F., and Brandstein, M. S.: Microphone Arrays, chapter Robust Localization in Reverberant Rooms, Springer, 2001.

Grimm, M.: Passives Audio-Tracking sich bewegender Geräuschquellen, Studienarbeit, Institut für Nachrichtentechnik, Universität Karlsruhe, 2001.

Huang, Y., Benesty, J., and Elko G. W.: Passive acoustic source localization for video camera steering, IEEE Int. Conf. Acoustics, Speech and Signal Processing (ICASSP), 909–912, June 2000.

Knapp, C. H. and Carter, G. C.: The generalized correlation method for estimation of time delay, IEEE Trans. on Acoustics, Speech and Signal Processing, 24(4):320–327, August 1976.

Litovsky, R. Y., Colburn, H. S., Yost, W. A., and Guzman, S. J.: The precedence effect, Journal of the Acoustical Society of America, 106(4):1633–1654, October 1999.

Silverman, H. F. and Kirtman, S. E.: A two-stage algorithm for determining talker location from linear microphone array data, Computer, Speech and Language, 6(2):129–152, 1992.

Broadband circular interferometric millimetre-wave ISAR for threat detection

S. Bertl, A. Dallinger, and J. Detlefsen

Technische Universität München, Lehrstuhl für Hochfrequenztechnik, Fachgebiet Hochfrequente Felder und Schaltungen, 80333 München, Germany

Abstract. To detect threats on a person's body surface the application of millimetre-waves is possible. In order to get a view of the person from all sides either the person has to be rotated standing on a turntable (Inverse Synthetic Aperture Radar, ISAR) or a sensor is moved on a circular path around the person (Synthetic Aperture Radar, SAR). The goal of the reconstruction is to obtain information about the shape of the threats and their positions in all three dimensions. At first the reconstruction is done along range and azimuth, which span the principal surface given by the sensor configuration. This paper reports on two methods to obtain the third spatial dimension. Due to the circular shape and its 3D focussing ability of the aperture this information can be obtained by the evaluation of the image focus. Secondly, if two coherent receiving channels are used, a possibility to achieve 3D spatial resolution is the processing of the interferometric phase. A comparison between these two will be presented.

1 Introduction

Broadband imaging in three dimensions is feasible using the scattered field distributions measured along curved lines. As a special case also circular apertures can provide 3D spatial resolution. A circular aperture can be realised by either using an array of sensors or by moving one sensor on a circular trajectory generating a synthetic aperture of circular shape.

In general holographic data can be focused for a known aperture and reconstruction area geometry as long as the sampling constraints are met. A direct inversion scheme of the linearised inverse scattering problem can be realised e.g. by means of correlation or backprojection algorithms. These algorithms are not optimum with respect to computation efficiency (i.e. speed) but have the advantage to work with arbitrary geometries.

In order to receive focused images of the objects placed on a person's body surface, as a first approximation a cylinder can be used as reconstruction area. This geometry is shown in Fig. 1a. One the other hand the reconstruction can also be done directly along the principal axes of the sensor, which are range and azimuth, as can be seen in Fig. 1b. It turns out that this reconstruction area is favourable when considering a three-dimensional reconstruction.

In the following after the basic considerations concerning the two-dimensional reconstructions, two approaches for an independent focussing of the third spatial dimension will be presented. The first one is the interferometry, which evaluates the phase difference of two coherent receiving channels. Secondly due to the circular shape of the aperture focussing can also be done by certain amplitude criteria. These two methods can be used separately, but they can also be joined in order to combine the specific advantages of them. For example, the limited ambiguity of the interferometric phase can be compensated by the use of an additional amplitude criterion.

All the images of reconstruction results are from measured data. The measurements were done with an FMCW-Radar with a bandwidth of 10 GHz that is described in more detail in (Detlefsen et al., 2006).

2 SAR-processing in two dimensions

Most of the applications based on the SAR-principle use a linear aperture or at least can neglect or compensate the curvature of it. As a consequence focussing reduces to a two-dimensional problem, since a rotational symmetry with respect to the aperture axis can be applied. Each object is located at its 2-parametric position (u, ρ) where u repre-

Correspondence to: S. Bertl
(sebastian.bertl@tum.de)

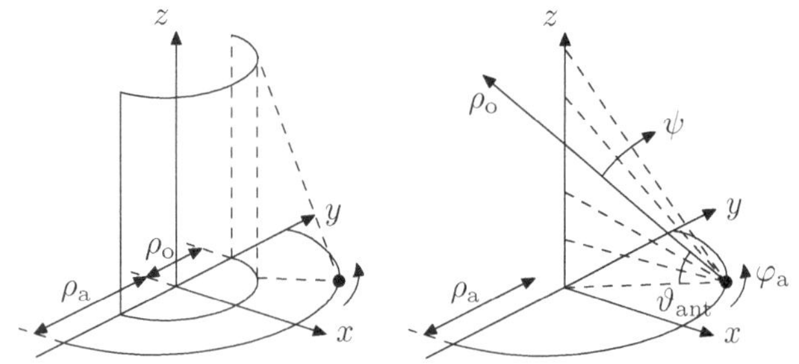

(a) Cylindrical reconstruction surface (b) Reconstruction along the principal axes

Fig. 1. SAR imaging geometry: The imaging sensor is moved on a circular path around the person. It is assumed that the person is illuminated homogeneously.

sents the azimuth variable and ρ stands for range. The back-projected signal $h(u, \rho)$ then becomes

$$h(u, \rho) = \int_{u_1}^{u_2} g(u', R)\alpha(R)\exp(j2k_c R)du' \quad (1)$$

where g stands for the calibrated data and $\alpha(R)$ is a weighting factor that compensates the propagation losses. $R{=}\sqrt{(u'{-}u)^2{+}\rho^2}$ is the range between the antenna's aperture and the object.

For circular apertures there is no longer a rotational symmetry and therefore the SAR-reconstruction becomes a three-dimensional problem. As shown in Fig. 1 the variable for the aperture angle is denoted by φ_a and the radius of the aperture is ρ_a. In cartesian coordinates the aperture can be written as

$$\mathbf{r}_a = (x, y, z)_a = (\rho_a \cos(\varphi_a), \rho_a \cos(\varphi_a), h_a) \quad (2)$$

where the z-component h_a is set to zero in the following. The index $\cdot_a$ states that the variable describes the aperture, variables with an index $\cdot_o$ belong to the object respectively the reconstruction area.

Since the focussing along the third spatial dimension should become independent from the first processing step, the two-dimensional SAR reconstruction is done along the mainaxes of the sensor, described by the along-track variable φ_o and the range variable ρ_o . The angle ψ stands for the offset from the main beam direction as illustrated in Fig. 1. Therefore the reconstruction area can be written as

$$\mathbf{r}_o = (x, y, z)_o = \begin{pmatrix} (\rho_a - \rho_o \sin(\vartheta_{ant} - \psi)) \cdot \cos\varphi_o \\ (\rho_a - \rho_o \sin(\vartheta_{ant} - \psi)) \cdot \sin\varphi_o \\ \rho_o \cos(\vartheta_{ant} - \psi) \end{pmatrix}^{\mathsf{T}} . \quad (3)$$

The distance R between the aperture and the object position then is $R{=}|\mathbf{r}_a{-}\mathbf{r}_o|{=}f(\ldots, \psi)$. The back-projected signal can be written in the same way as (1) except that the rotational symmetry no longer holds

$$h(\rho_o, \varphi_o, \psi_o) = \int_{\varphi_{a,1}}^{\varphi_{a,2}} u(r_a, \varphi_a') \cdot \alpha(R)\exp(j2k_c R)d\varphi_a' \quad (4)$$

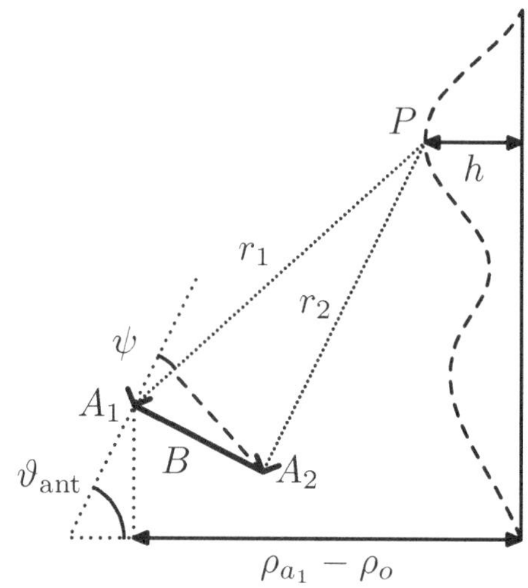

Fig. 2. Interferometric setup.

3 Options for three-dimensional focussing

After the SAR-reconstruction the position of the objects is determined along range r_o and cross-range φ_o. In the following two possibilities to determine the elevation position ψ_o will be presented.

3.1 Interferometry

For two coherent receiving channels separated by a distance B usually called baseline the phase difference between two channels holds the information about the elevation.

Due to the physical properties of the human skin (see Dallinger et al., 2004) most of the power transmitted by a sensor below the person will be reflected and not scattered back to the receiving antennas of the sensor. As a consequence only inhomogeneities will be visible. The resulting image will consist of isolated scattering centres. Phase-unwrapping algorithms can not handle these isolated scattering centres and so the intervall of a uniquely defined phase is limited to an interval of 2π and no absolute phase values with respect to a common reference phase can be determined.

From the two complex-valued SAR-images $h_i(r_o, \varphi_o, \psi_o)$, $i \in \{1, 2\}$, the interferogram $v(.)$ is formed by conjugate complex multiplication $v(.){=}h_1(.)\cdot h_2(.)^*$. It is assumed that the position offsets of the scattering centres in the two images are smaller than one resolution element and can therefore be neglected. Figure 2 shows the geometry of the interferometric sensor with two receiving antennas A_1 and A_2 separated by the distance B.

In the case of one common transmitting antenna the phase difference ϕ obtained in the interferogram is given by

$$\phi = \frac{2\pi}{\lambda}(r_2 - r_1). \quad (5)$$

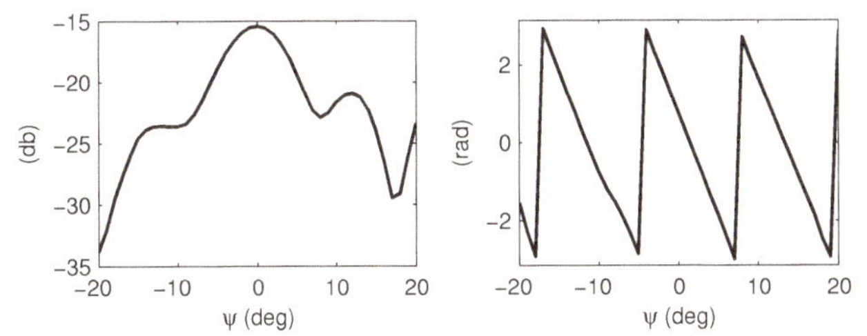

Fig. 3. Amplitude and interferometric phase along ψ at the position of a metallic sphere at a position of ψ=0° (B=1.5 cm, f=[90.5...100.5] GHz).

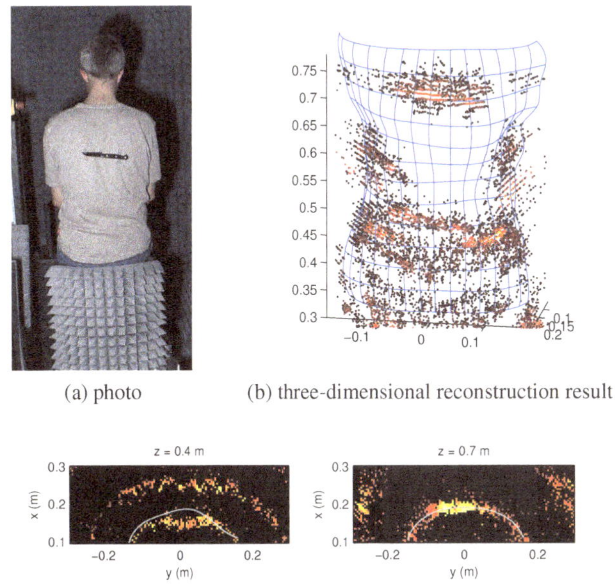

(a) photo (b) three-dimensional reconstruction result

(c) horizontal cut (d) horizontal cut

Fig. 4. Three-dimensional MMW image of a person with a horizontally orientated knife at the back.

Since the interferometric phase is limited to $[-\pi, \pi]$ the elevation angle ψ is also limited to small angles and the approximation $\sin\psi \approx \psi$ is valid. For a point scatterer located at the angle $\vartheta_{\text{ant}} - \psi_0$ the relation between the interferometric phase and the elevation angle ψ then becomes

$$\Delta\phi = \frac{2\pi}{\lambda} B \sin\Delta\psi \approx \frac{2\pi}{\lambda} B\,\Delta\psi \tag{6}$$

With a baseline B=1.5 cm and a center frequency of 95 Ghz it turns out from Eq. (6) that the unambiguousness interval $[-\pi, \pi]$ of the interferometric phase results in an unambiguousness interval of $[-6°, 6°]$ for the elevation angle ψ. Figure 3 shows that this value for ψ can also be found in the reconstruction results of measured data. Here the amplitude of one channel and the interferometric phase at the (φ_0, ρ_0)-position of a small metallic (diameter ≈3 mm) sphere along ψ is illustrated. The characteristic property of the interferometric phase at the position of a point scatterer is, that it has a zero there.

Figure 4 shows a more complex example for the interferometric reconstruction. A person is sitting on a chair on the turntable, which results in a circular aperture, as described before. The illustration of the body surface is done by spline interploation of the weighted average of the position of scattering centres in a certain intervall $\Delta\varphi_0$. Relevant reflection points are illustrated with dots with varying colour describing the intensity of scattering. Sub-figs. 4c and 4d show horizontal cuts of the reconstructed model. In addition to the contour of the body a second contour appears which is an effect of the limited unambiguity of the interferometric phase. When the approximate position of the person on the turntable is known the ambiguities can be resolved by using this information.

3.2 Options for amplitude focussing

Besides using the interferometric phase in order to focus in the third dimension, the circular shape of the aperture also provides information about the position of the scattering centres. A possibility to determine the focus of a pixel lies in the evaluation of the gradient of its amplitude. This approach for example is used for focussing in digital still cameras as described in . The difference between a camera and a SAR millimetre-wave image is that for a camera only the position of the lens has to be adjusted and in a millimetre-wave image each single pixel has to be corrected seperately. An often used gradient for the purpose of focussing is the so called Sobel-gradient. In the case of focussing a pixel with respect to ψ it becomes

$$\begin{aligned} G_\rho(\psi) &= \begin{pmatrix} -1 & -2 & -1 \\ 0 & 0 & 0 \\ 1 & 2 & 1 \end{pmatrix} \otimes h(\rho, \varphi)|_\psi \\ G_\varphi(\psi) &= \begin{pmatrix} -1 & 0 & 1 \\ -2 & 0 & 2 \\ -1 & 0 & 1 \end{pmatrix} \otimes h(\rho, \varphi)|_\psi \\ \psi_{\text{opt}} &= \max_\psi \sqrt{G_\rho(\psi)^2 + G_\varphi(\psi)^2}. \end{aligned} \tag{7}$$

Another way to use the amplitude for focussing lies in the calculation of

$$\psi_{\text{opt}}(\rho_i, \varphi_i) = \max_\psi h(\rho, \varphi, \psi)^4|_{r=r_i\pm\Delta r, \varphi=\varphi_i\pm\Delta\varphi} \tag{8}$$

in an area $r=r_i\pm\Delta r$, $\varphi=\varphi_i\pm\Delta\varphi$ around the image position (ρ_i, φ_i). At the ψ-position of a point scatterer to be focused its energy ($\sim h(\rho, \varphi, \psi)^2$) is concentrated within a minimum number of pixels. Assuming that the power transmitted from the antenna to a certain area around the scattering centre remains constant and therefore also in the reconstructed image, the term $h(\rho, \varphi, \psi)^4$ will have its extremum at the best focused position.

Figure 5a shows the measure according to Eq. (8) along ψ at the exact position of a small metallic sphere (diameter

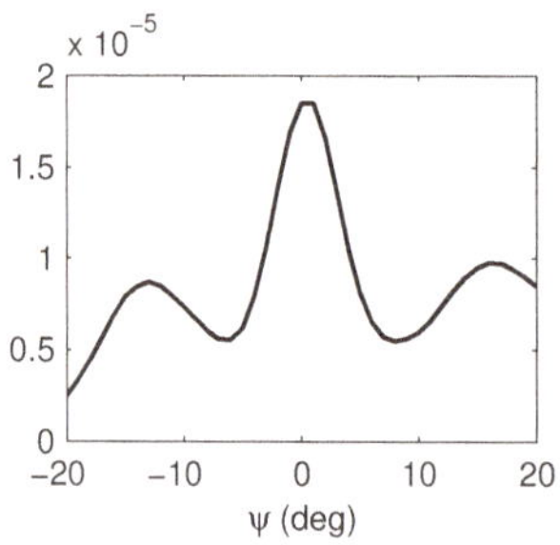

(a) measure according to eqn. (8) at the position of a sphere

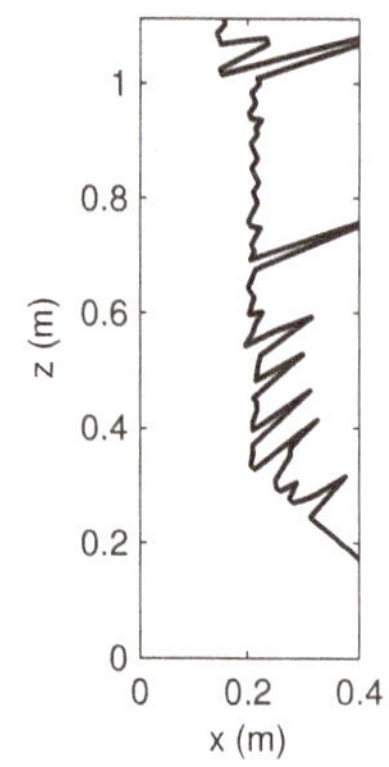

(b) $x(z = 0\,\mathrm{m})$ after projection into cartesian coordinates

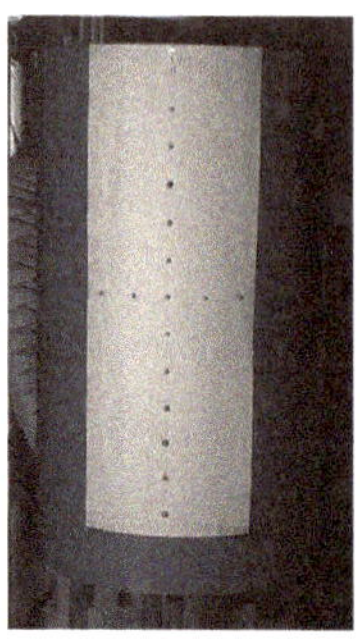

(c) photo of the measurement scene

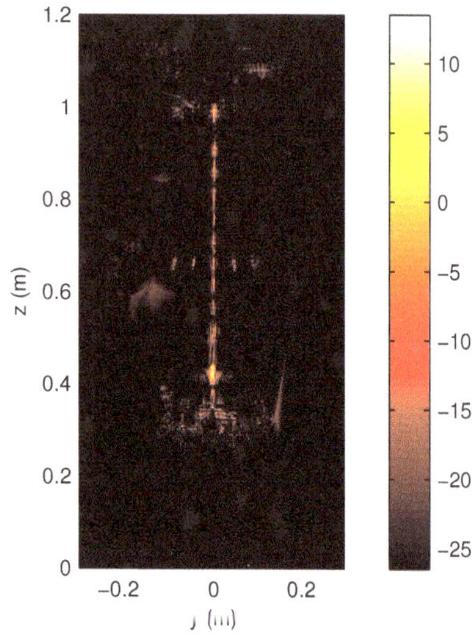

(d) projection onto yz-plane (logarithmic scaling)

Fig. 5. Example for the application of an amplitude criteria for focussing along ψ.

$\approx$3 mm) as can be seen on the photo in Fig. 5c. The whole result of the focussing process using the amplitude criterion is shown in the other sub-figures in Fig. 5.

4 Comparison of focussing methods

Table 1 gives an overview of the specific properties of the two methods of focussing, namely the interferometry and the amplitude criterion, presented in the previous section.

The obvious disadvantage of the interferometry in the case of isolated scattering centres is the fact that no global phase information can be resolved. Hence only an interval of $[-\pi, \pi]$ can be used to get an unambiguous information about the location of an object along the ψ-direction. In contrast the amplitude criteria are not limited in uniqueness and give a unique position over the whole range of ψ.

Table 1. Overview of the specific properties of the focussing methods presented in Sect.3.

	Interferometry	Amplitude criterion
Uniqueness	Uniqueness of the interferometric phase limited to $[-\pi, \pi]$ (here: $\Delta\psi \approx 12°$)	No restriction of the uniqueness
Processed area	Evaluation possible for each single pixel	Averaging within an area around a pixel
Accuracy	Accuracy can be increased by extension of the baseline (drawback: Unambiguity decreases further); Additional accuracy improvement e.g. by Multi-look SAR processing	in general limited; Further improvement leads to an increase in computational effort
	Sign of phase gives direction of necessary correction	No information about direction of correction

A big advantage of the interferometric focussing is the fact that the sign of the phase gives information about the direction of the necessary correction along ψ. Furthermore the accuracy of the interferometric phase can be increased by Multi-look processing of the SAR-data which is not explained in detail here.

Considering these facts a combination of the two focussing methods seems to make sense in the following way: The amplitude criterion is used to receive the rough position of the scattering centres along ψ. This compensates the unambiguousness of the interferometry. Then, in a second step the position along ψ is determined more exactly using the interferometric phase.

Figure 6 finally shows the example that already appeared in Fig. 4. Here the amplitude focussing and the evaluation of the interferometric phase are joined in the way proposed before. In contrast to the processing yielding the images in Fig. 4 no previous knowledge about the position of the person is necessary here. In Fig. 6 pixels with an amplitude of $0\ldots-35$ dB which mostly belong to objects like the knife or the folding of the shirt are represented with colours from yellow to red. Lower amplitudes between -35 dB$\ldots-65$ dB have grey colour. These points mostly belong to parts of the body surface.

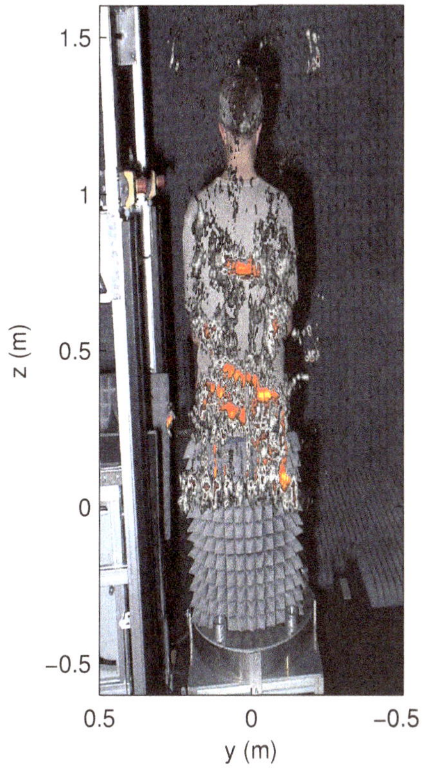

Fig. 6. Imaging results for a combined focussing using the methods two methods of Sect. 3.

5 Conclusions

Two methods for three-dimensional focussing of SAR images obtained by measurements along a circular aperture have been presented.

The use of the interferometric phase is similar to the way it is handled in the case of a linear aperture. Due to the application in imaging objects at the body surface of humans and the properties of human skin only isolated scattering centres can be measured. Hence no global phase information can be obtained.

To overcome this problem a second way of focussing has been presented which exploits the circular shape of the aperture by evaluating the amplitude of the scattering centres.

Finally it was shown that a combination of these two methods results in a possibility to determine a focussed image and the position of threats placed on person's body surface.

Further steps could be the discussion about the proper representation of the reconstructed image and the utilisation of faster reconstruction methods like $\omega - k$-algorithms which can also handle data of circular apertures as described in (Dallinger et al., 2005).

References

Dallinger, A., Schelkshorn, S., and Detlefsen, J.: Millimeter-Wave Imaging of Humans - Basic Experiments, in: IR-MMW2004/THz2004, Karlsruhe, Germany, 2004.

Dallinger, A., Schelkshorn, S., and Detlefsen, J.: Efficient omega-k-Algorithm for Circular SAR and Cylindrical Reconstruction Areas, Advances in Radio Science, 4, 85–91, 2005.

Detlefsen, J., Dallinger, A., Schelkshorn, S., and Bertl, S.: UWB Millimeter-Wave FMCW Radar using Hilbert Transform Methods, in: Proceedings of ISSSTA-06, 2006 IEEE Ninth International Symposium on Spread Spectrum Techniques and Applications, Manaus, Brazil, 2006.

Jie, H., Rongzhen, Z., and Zhiliang, H.: Modified fast climbing search auto-focus algorithm with adaptive step size searching technique for digital camera, Consumer Electronics, IEEE Transactions on, 49, 257–262, 0098–3063, 2003.

Rosen, P. A., Hensley, S., Joughin, I. R., Li, F. K., Madsen, S. N., Rodriguez, E., and Goldstein, R. M.: Synthetic aperture radar interferometry, Proceedings of the IEEE, 88, 333-382, 0018–9219, 2000.

Soumekh, M.: Synthetic Aperture Radar Signal Processing with Matlab Algorithms, John Wiley & Sons, Inc., 1999.

Ulander, L. M. H., Hellsten, H., and Stenstrom, G.: Syntheticaperture radar processing using fast factorized back-projection, Aerospace and Electronic Systems, IEEE Transactions on, 39, 760-776, 0018–9251, 2003

Extended post processing for simulation results of FEM synthesized UHF-RFID transponder antennas

R. Herschmann[1,2]**, M. Camp**[1,2]**, and H. Eul**[1]

[1]Leibniz Universität Hannover, Institute of Radiofrequency and Microwave Engineering, Appelstraße 9a, 30167 Hannover, Germany
[2]Smart Devices GmbH & Co. KG, Schönebecker Allee 2, 30823 Garbsen, Germany

Abstract. The computer aided design process of sophisticated UHF-RFID transponder antennas requires the application of reliable simulation software. This paper describes a Matlab implemented extension of the post processor capabilities of the commercially available three dimensional field simulation programme Ansoft HFSS to compute an accurate solution of the antenna's surface current distribution. The accuracy of the simulated surface currents, which are physically related to the impedance at the feeding point of the antenna, depends on the convergence of the electromagnetic fields inside the simulation volume. The introduced method estimates the overall quality of the simulation results by combining the surface currents with the electromagnetic fields extracted from the field solution of Ansoft HFSS.

1 Introduction

The development and optimization of passive transponders for application in UHF-RFID systems with wide read range is directly correlated to the supply of appropriate antenna designs. On the one hand these antennas must be frequency selective in order to suppress interferences of radio systems in adjacent wave bands. Otherwise the antennas should exhibit enough bandwidth to compensate for tolerances of chip impedance and to account for the influence on impedance from the adjacencies of the antenna. A largely stable use of a passive transponder makes sense in a moderately variable environment only in such a way. The dependence between geometry and impedance qualities of the antenna makes the synthesis of suitable antenna structures possible for the fulfilment of the qualification profile under consideration of the at most permitted geometric measurements.

Correspondence to: R. Herschmann
(herschmann@hft.uni-hannover.de)

The design process of a transponder antenna for the application in an UHF-RFID system using the Finite Element Method (FEM) allows for the consideration of extended, geometrical complex objects with arbitrary material characteristics in the near field region. The material specific qualities of these objects generally lead to an impact on the antenna impedance associated with changes of the resonant frequency, bandwidth and matching between antenna and transponder chip. Recent simulation programmes allow for an optimization of the antenna design by the variation of defined geometry parameters of simulation models parameterized correspondingly under consideration of predefined environmental conditions.

This work evaluates the quality of the simulation results of Ansoft's FEM based simulation programme High Frequency Structure Simulator (HFSS). This product is used as a design tool for the synthesis of plane transponder antennas (see Herschmann et al., 2006). Despite the various areas of application, which arise by the use of this sophisticated simulation tool, limits also exist in the calculation possibilities. For instance it is not possible to calculate the electromagnetic fields outside the simulation model in arbitrarily defined field points or field point areas. Moreover, the quality of the calculation of the surface current distribution of plane antenna structures is insufficient. Therefore a module implemented in Matlab is introduced that offers extended possibilities for the post processing of HFSS simulation results. A check of the convergence of the electromagnetic field can be carried out in the complete simulation volume by the calculation of the surface currents of the antenna structures examined here. The knowledge of the frequency dependent current distribution forms the base for calculating the electromagnetic fields in arbitrary space points.

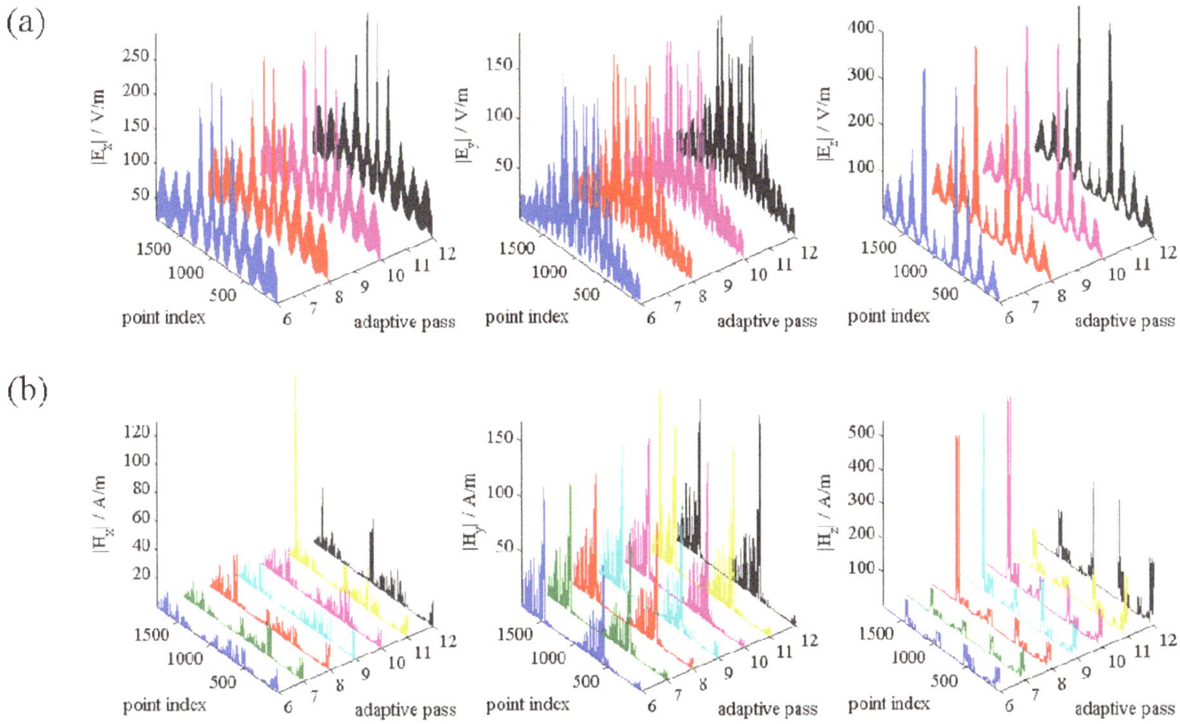

Fig. 1. Convergence of the electromagnetic field at defined observation points versus the number of adaptive passes; while the electric field shows well convergence properties in **(a)**, the magnetic field shows substantial changes in **(b)**.

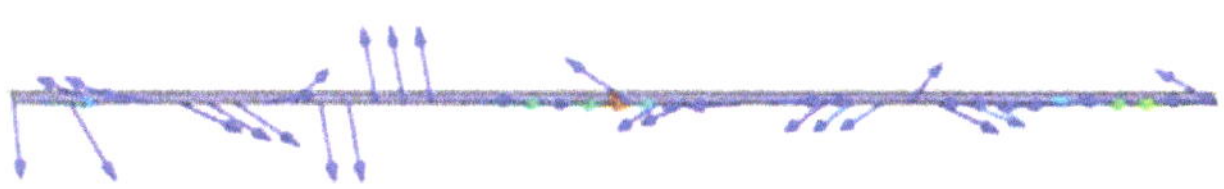

Fig. 2. Surface current distribution of a $\lambda/2$-dipole at resonant frequency extracted from Ansoft HFSS.

2 HFSS simulation results: convergence of electromagnetic fields and impedance

It is appropriate to clarify the necessity for dealing with the subject of field and impedance convergence by observing the simulation results for a simple resonant dipole antenna using HFSS. The surface current distribution of that antenna is well known to be sinusoidal along the dipole's centre line (see Stutzman and Thiele, 1998). In Fig. 1 the Cartesian components of the electromagnetic fields are represented. While the observation points of the electric field are distributed inside the simulation volume, the magnetic fields are computed along the antenna surface. The electromagnetic fields are plotted against the number of adaptive passes. In contrast to the electric fields the magnetic fields show substantial changes in field magnitudes even at a high number of adaptive passes. HFSS computes the surface current distribution by processing the magnetic fields on the antenna structure. But the results for two dimensional antennas with neglected metal thickness are not correct as depicted in Fig. 2. On the other hand convergence of the impedance at the antenna's feeding point shown in Fig. 3 is already reached within a few adaptive passes. Due to the physical relation between antenna impedance and antenna surface current distribution it is appropriate to check the field solution.

3 HFSS eXtension Module: a Matlab implemented procedure for post processing FEM based simulation results

In order to validate the simulation results of HFSS the convergence of impedance as well as the convergence of electromagnetic fields surrounding the antenna have to be considered because of the physical relation between these characteristic quantities. The electromagnetic fields inside the simulation volume are caused by the antenna's surface currents. A method implemented by use of Matlab permits the computation of a corrected distribution of the surface currents. Therefore this method permits statements about the quality of the field convergence in the complete simulation volume and puts aid at the assessment of HFSS results in the design of antenna prototypes. On the basis of a simple spiral antenna the approach is described detailed in the following sections.

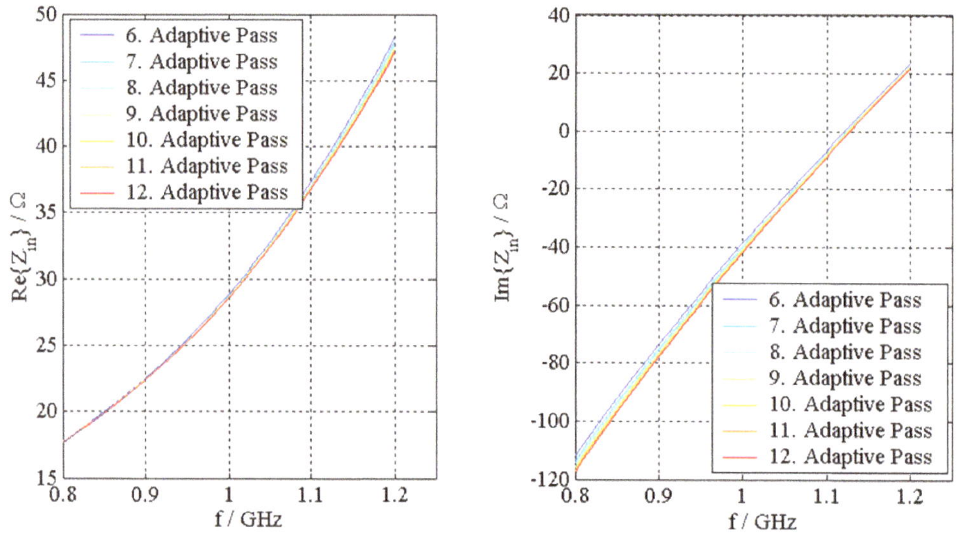

Fig. 3. Impedance convergence at the antenna's feeding point versus the number of adaptive passes.

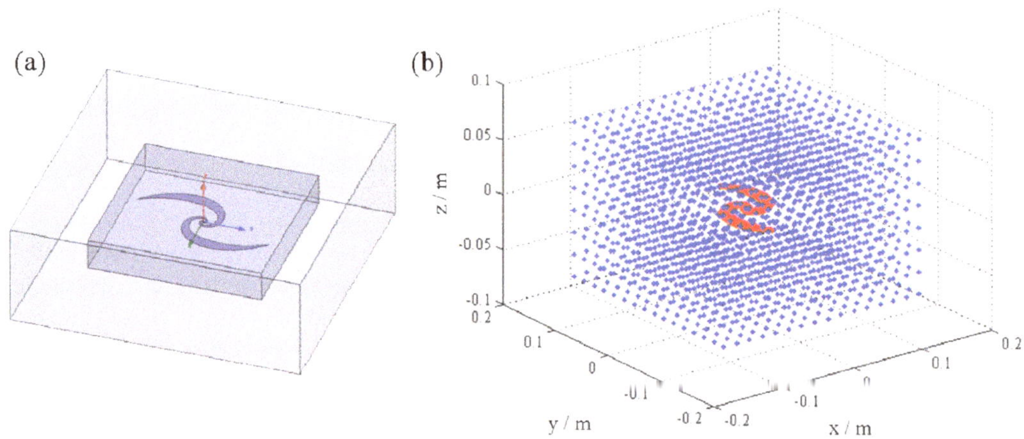

Fig. 4. Ansoft HFSS simulation model of the spiral antenna and distribution of the observation points inside the simulation volume (blue markers represent observation points for the electric field, red markers are used for the magnetic field).

3.1 The principle of the calculation method

The Matlab procedure performs a comparison between the electric fields taken directly from the solution of the finite element method implemented in HFSS and the electric fields which are derived from the distribution of the antenna's surface currents. To this end a suitable observation point area has to be defined within the boundaries which are provided by the used simulation model. It is possible to use different geometries for this observation point area, e.g. lines, planes or cuboids. The higher the dimension of this geometrical structure chosen as the observation point area the better the results regarding the corrected surface current distribution. The electric fields are used for comparison because HFSS determines the magnetic fields by a numeric calculation based on the solution of the electric fields. The electric fields are therefore computed with a higher precision. Aim of the procedure is the optimization of a scaling factor applied to calculated initial surface current distribution in order to minimize the difference between the compared electric fields. This initial current distribution of the antenna is calculated from the magnetic fields in close proximity to the antenna surface. As a working example the method is applied to the analysis of the field convergence of a two arm spiral antenna. The equally spaced observation points of the electric fields $\boldsymbol{E}=h\,(f, \boldsymbol{r}_{a2})$ are positioned between an inner and outer cuboid. The observation points of the magnetic fields $\boldsymbol{H}=g\,(f, \boldsymbol{r}_{a1})$ are distributed along the antenna surface with a defined gap between antenna plane and observation point planes. The field distribution of the observation point areas is summarised in Fig. 4. Consequently the quality of the optimized scaling factor of the surface current distribution

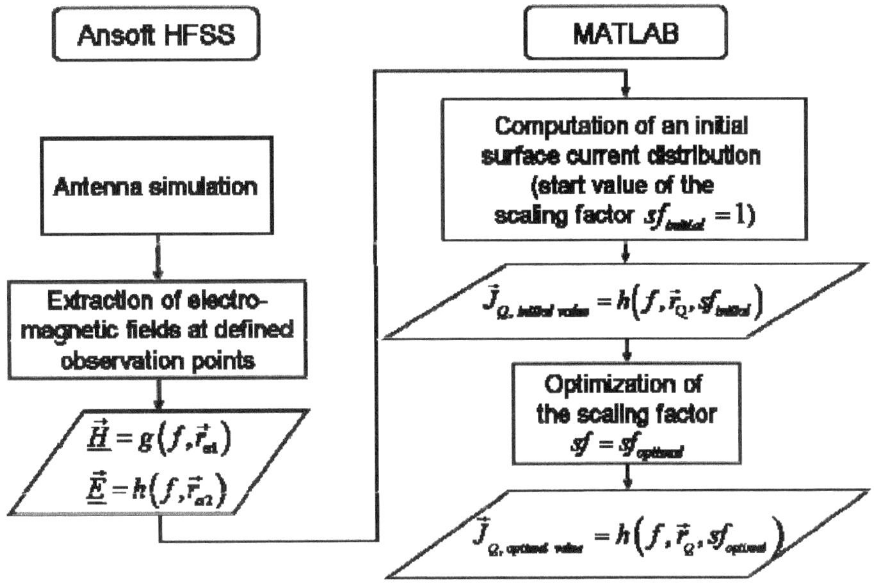

Fig. 5. Schematic representation of the surface currents calculation procedure.

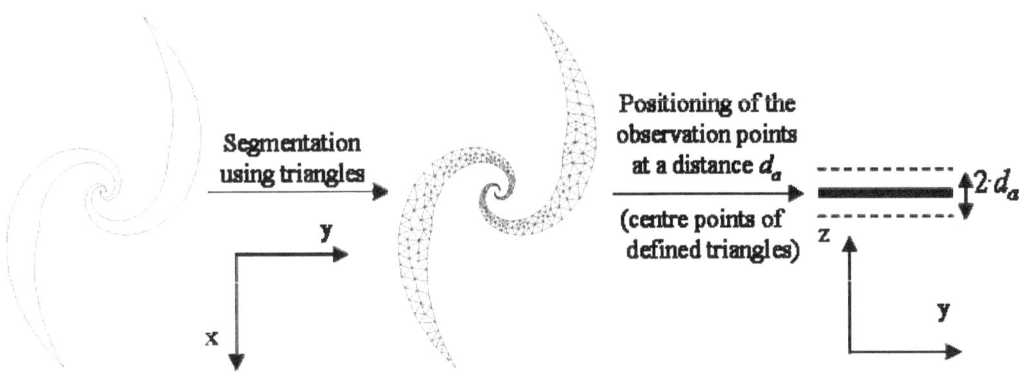

Fig. 6. Generation of observation points for the calculation of the magnetic field strength vectors.

is associated directly with the convergence qualities of the electromagnetic fields calculated by HFSS within the simulation model. Only by a sufficient number of adaptive passes a convergence of the fields can be expected besides the convergence of the scattering parameters at the antenna's port. The necessary number of iterations and therefore the quality of the results depends on the complexity of the simulation model. The flowchart in Fig. 5 represents the described procedure graphically.

3.2 Determination of the equivalent Huygens sources

The computation of the surface current distribution on a two dimensional antenna structure by assuming the antenna as a bounding plane between separated spaces. A variation of Eq. (1) leads to an initial current distribution which has to be corrected by use of the above mentioned iterative optimization process.

$$\mathrm{Rot}\boldsymbol{H}=\boldsymbol{n}_{12}\times(\boldsymbol{H}_2-\boldsymbol{H}_1)=\boldsymbol{J}_Q \quad (1)$$

Equation (1) is only valid at the boundary of adjoining space areas. In the approach discussed in this paper the observation points for extracting the magnetic fields are computed from the centre points of the triangles used for antenna segmentation and are placed at a distance d_a above and below the antenna plane as shown in Fig. 6. Therefore the initial current distribution is approximated using Eq. (2):

$$\mathrm{Rot}\tilde{\boldsymbol{H}}=\boldsymbol{n}_{12}\times\left(\boldsymbol{H}_{2,d_a}-\boldsymbol{H}_{1,d_a}\right)=\tilde{\boldsymbol{J}}_Q=\boldsymbol{J}_{Q,\text{initial value}} \quad (2)$$

To discretise the two dimensional antenna geometry a free available mesh programme introduced in Persson (2005) is applied.

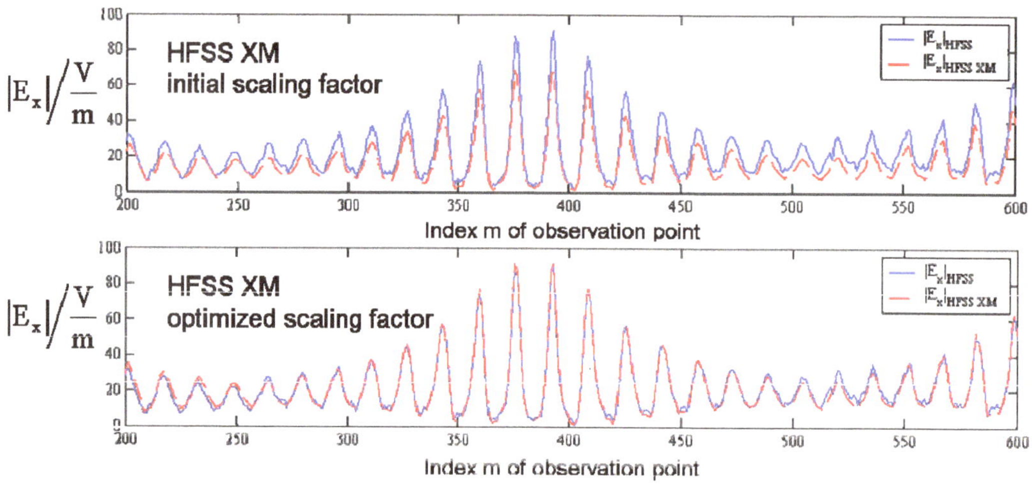

Fig. 7. Comparison of simulated (HFSS) and computed (HFSS XM) electric fields at defined observation points with regard to initial and optimized scaling factor.

3.3 Calculation of the radiation fields on basis of the equivalent Huygens sources

The optimization process for calculating the corrected values of the surface current distribution combines these currents to the electromagnetic fields at the defined observation points inside the simulation volume by applying the formalism summarised in Eq. (3).

$$\begin{aligned}
&\boldsymbol{H}(\boldsymbol{r}_A) = \frac{jk}{4\pi}\cdot(\boldsymbol{m}\times\boldsymbol{r}_A)\cdot C\cdot e^{-jkr_A}\\
&\boldsymbol{E}(\boldsymbol{r}_A) = \frac{\eta}{4\pi}\cdot\left((\boldsymbol{M}-\boldsymbol{m})\cdot\left(\frac{jk}{|\boldsymbol{r}_A|}+C\right)+2\boldsymbol{M}C\right)\cdot e^{-jkr_A}\\
&C = \frac{1}{r_A^2}\cdot\left(1+\frac{1}{jk|\boldsymbol{r}_A|}\right) \qquad \boldsymbol{M} = \frac{(\boldsymbol{r}_A\cdot\boldsymbol{m})\cdot\boldsymbol{r}_A}{|\boldsymbol{r}_A|^2}\\
&\eta = \sqrt{\frac{\mu_0}{\varepsilon_0}} = 120\pi\,\Omega
\end{aligned} \tag{3}$$

The electromagnetic fields are computed directly from an electric dipole moment $\boldsymbol{m}=\boldsymbol{J}_Q\cdot A_Q$, where A_Q represents the area of a single triangle used for segmentation of the antenna surface, without determining the magnetic vector potential and necessary application of the subsequent functions of vector calculus in order to calculate the radiation fields (see Makarov, 2002). The segmented antenna structure contains a high number of dipole moments depending on the number of elements necessary for a high quality mesh triangulation. Therefore the electromagnetic field in a specific observation point $\boldsymbol{r}_A$ inside the simulation volume is a superposition of the contributions of all N dipole moments covering the antenna surface.

$$\begin{aligned}
&\boldsymbol{H}(\boldsymbol{r}_A) = \sum_{n=1}^{N}\boldsymbol{H}_n\left(\boldsymbol{r}_A-\boldsymbol{r}_{Q,n}\right)\\
&\boldsymbol{E}(\boldsymbol{r}_A) = \sum_{n=1}^{N}\boldsymbol{E}_n\left(\boldsymbol{r}_A-\boldsymbol{r}_{Q,n}\right)
\end{aligned} \tag{4}$$

The quality of the calculated radiation fields depends on the quality of the generated mesh and the field solution inside the simulation volume of HFSS. The interrelation between the surface currents and the electromagnetic fields surrounding the antenna is the basis for the computation of an optimized scaling factor for the current distribution. The iterative process starts with an initial value of $sf_{\text{initial}}=1$. The expressions defined in Eq. (5) are used as a criterion to evaluate the difference between the electric fields based on the field solution of HFSS and the electric fields computed by applying Eqs. (3) and (4) to the surface current distribution. The optimal value of the scaling factor is found for the minimum of the residual error using a Matlab implemented optimizer.

$$\begin{aligned}
&F_{x,\text{real},m}(sf) = \left|Re\left\{E_{x,HFSS}(\boldsymbol{r}_m)\right\} - Re\left\{E_{x,HFSS\,XM}(\boldsymbol{r}_m)\right\}\right|^2\\
&F_{y,\text{real},m}(sf) = \left|Re\left\{E_{y,HFSS}(\boldsymbol{r}_m)\right\} - Re\left\{E_{y,HFSS\,XM}(\boldsymbol{r}_m)\right\}\right|^2\\
&F_{z,\text{real},m}(sf) = \left|Re\left\{E_{z,HFSS}(\boldsymbol{r}_m)\right\} - Re\left\{E_{z,HFSS\,XM}(\boldsymbol{r}_m)\right\}\right|^2\\
&F_{\text{real}}(sf) = \frac{1}{M}\cdot\sum_{m=1}^{M}\left[F_{x,\text{real},m}(sf)+F_{y,\text{real},m}(sf)+F_{z,\text{real},m}(sf)\right]\\
&\Rightarrow sf_{\text{optimal,real}} = \min\left\{F_{\text{real}}(sf) : sf\in(0,5)\right\}\\[1em]
&F_{x,\text{imag},m}(sf) = \left|Im\left\{E_{x,HFSS}(\boldsymbol{r}_m)\right\} - Im\left\{E_{x,HFSS\,XM}(\boldsymbol{r}_m)\right\}\right|^2\\
&F_{y,\text{imag},m}(sf) = \left|Im\left\{E_{y,HFSS}(\boldsymbol{r}_m)\right\} - Im\left\{E_{y,HFSS\,XM}(\boldsymbol{r}_m)\right\}\right|^2\\
&F_{z,\text{imag},m}(sf) = \left|Im\left\{E_{z,HFSS}(\boldsymbol{r}_m)\right\} - Im\left\{E_{z,HFSS\,XM}(\boldsymbol{r}_m)\right\}\right|^2\\
&F_{\text{imag}}(sf) = \frac{1}{M}\cdot\sum_{m=1}^{M}\left[F_{x,\text{imag},m}(sf)+F_{y,\text{imag},m}(sf)+F_{z,\text{imag},m}(sf)\right]\\
&\Rightarrow sf_{\text{optimal,imag}} = \min\left\{F_{\text{imag}}(sf) : sf\in(0,5)\right\}\\[1em]
&\boldsymbol{J}_{Q,\text{optimal}} = sf_{\text{optimal,real}}\cdot Re\left\{\boldsymbol{J}_{Q,\text{initial}}\right\} + j\cdot sf_{\text{optimal,imag}}\cdot Im\left\{\boldsymbol{J}_{Q,\text{initial}}\right\}
\end{aligned} \tag{5}$$

The electric fields calculated by HFSS in the defined observation point area are indicated by the index *HFSS*. The expressions indexed by *HFSS XM* represent the electric fields obtained from the surface currents correspondingly. Determining the squares of the differences in the Cartesian components at M observation points leads to a stronger weighting of high deviations compared to small errors. The separate

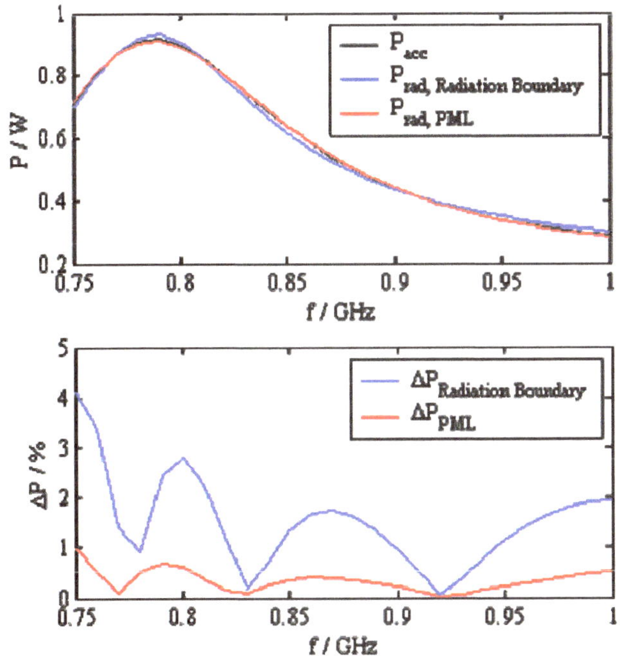

Fig. 8. Power balance of the spiral antenna; the calculations are performed using different boundary layers in the HFSS simulation.

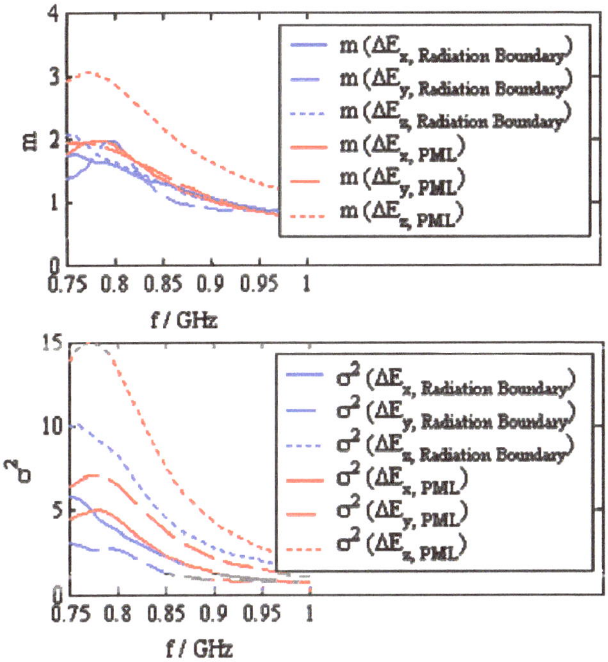

Fig. 9. Statistical evaluation of differences regarding the antenna's radiated electric fields extracted from HFSS field solution and computed from the surface current distribution; the calculations are performed using different boundary layers in the HFSS simulation.

treatment of the real and imaginary parts generally delivers different optimal scaling factors. An important reference regarding the quality of the simulator's field solution is a good agreement between $sf_{\text{optimal,real}}$ and $sf_{\text{optimal,imag}}$. There is a risk that an adaption of the electric fields is possible inside the simulation volume also at an insufficient field convergence by the surface currents of the antenna structure, if the number of degrees of freedom of the optimization process is increased. For instance it is possible to establish separate scaling factors for the Cartesian components of the electric fields. However the physical significance of the current surface distribution optimized this way is challenged. The chosen implementation of the HFSS XM method avoids this situation. An insufficient convergence of the electromagnetic fields inside the simulation volume results in remaining significant deviations between the HFSS based field components and the fields extracted from the surface current distribution.

3.4 Results and ranges of application

A comparison of the electric fields simulated by HFSS and the electric fields computed on basis of the surface current distribution by HFSS XM in the observation point area introduced in Fig. 6 is presented in Fig. 7 at a frequency near the first serial resonance. The two graphics show only an extract of all defined observation points and are restricted to the magnitude of the x-component of the complex electric field vector. Deviations and agreements of the other vector components as well as phase values range at the same order of magnitude. While distinct deviations are visible by use of the initial scaling factor, a very well agreement between the simulated and computed electric field distribution by applying the optimized scaling factor is achieved. But it is insufficient to estimate the quality of the computed surface current distribution this way only. Further criteria are necessary for an evaluation of the reliability of the computed results and the necessary validation of the field convergence of the HFSS simulation process. Therefore a power balance and a statistical evaluation with regard to the comparison of the simulated and computed electric fields are applied. To reduce the physical size of the simulation model it is possible to use perfectly matched layer instead of a radiation boundary applied conventionally. The effects on the convergence of electromagnetic fields inside the simulation volume caused by different boundary conditions is analysed as well. In Fig. 8 the calculated radiated power P_{rad} based on the surface currents is compared to the accepted power P_{acc}. The spiral antenna is simulated without any losses. Thus all accepted power should be radiated. The percentage deviation is less than ΔP=4% independent of the applied boundary condition over the frequency range of interest. The maximum differences occur in the range of the resonant frequency. Finally Fig. 9 represents an assessment of the residual errors by calculation of the mean average value m and variance σ^2 of the error vector defined in Eq. (6) regarding the differences of the simulated and computed solutions of the electric fields. The error vectors contain the absolute values of the differences of the magnitudes of the electric fields for every Cartesian component at the defined M observation points. The low mean average values as well as moderate variances for all three Cartesian components validate the sufficient number of adaptive passes and the very well converged field solution inside the simulation volume.

$$\Delta E_i(f) = \begin{pmatrix} \left|\left|E_{i,1,HFSS}(f)\right| - \left|E_{i,1,HFSS\,XM}(f)\right|\right| \\ \vdots \\ \left|\left|E_{i,m,HFSS}(f)\right| - \left|E_{i,m,HFSS\,XM}(f)\right|\right| \\ \vdots \\ \left|\left|E_{i,M,HFSS}(f)\right| - \left|E_{i,M,HFSS\,XM}(f)\right|\right| \end{pmatrix},$$
$$i = x, y, z, m = 1..M \qquad (6)$$

Based on the optimized surface current distribution it is possible to separate ohmic losses of the antenna from radiation losses and select a proper equivalent antenna circuit. Furthermore it is possible to compute the electromagnetic fields in arbitrary observation point areas and extract information on polarisation states for example. The implemented method is capable to compute the surface current distribution of geometrical complex shaped antennas for use in UHF-RFID transponder systems.

4 Conclusion

The Matlab implementation of HFSS XM makes the calculation of the surface current distribution of plane antennas as well as the computation of the electromagnetic fields and measures derived from them possible in arbitrary observation points and areas. Furthermore the convergence of the field solution computed by HFSS is checked during the determination of optimized scaling factors for the real and imaginary parts of the antenna's surface currents. Therefore it improves the capabilities implemented in HFSS already and antenna prototyping for UHF-RFID transponder becomes more sophisticated. The introduced method is capable to compute the surface current distribution of three dimensional antennas with finite metallization thickness as well as conformal antennas. Based on the surface current distribution it is possible to lay out a library containing the geometrical data of parameterized transponder antennas, the impedance at the feeding point and the related surface current distribution as a function of frequency. Therefore it is not necessary to save the memory intensive HFSS models. The quality of the computed surface current distribution depends on the convergence of the electromagnetic fields inside the simulation volume of HFSS. According to the examinations made in this paper the field convergence necessary for accurate calculation of the surface current distribution requires significant higher number of adaptive passes than a merely impedance convergence at the antenna's feeding point.

References

Herschmann, R., Camp, M., and Eul, H.: Design und Analyse elektrisch kleiner Antennen für den Einsatz in UHF RFID Transponder, Adv. Radio Sci., 4, 93–98, 2006, http://www.adv-radio-sci.net/4/93/2006/.

Makarov, S. N.: Antenna and EM Modeling with MATLAB, John Wiley & Sons Inc., 2002.

Persson, P.-O.: Mesh Generation for Implicit Geometries, in: Department of Mathematics, PhD thesis, Massachusetts Institute of Technology, 2005.

Stutzman, W. L. and Thiele, G. A.: Antenna theory and design, 2nd ed., New York, Wiley, 1998.

23

Applicability of new radio technologies for advanced UMTS networks

A. Dekorsy, M. Schacht, S. Brueck, and G. Fischer

Lucent Technologies Networks System GmbH, Thurn-und-Taxis-Strasse 10, 90411 Nuremberg, Germany

Abstract. An expanded effort is under the way to support the evolution of UMTS (Universal Mobile Telecommunication System). Apart from delivering high data rates, future UMTS releases will also require to provide high network performance in terms of system capacity, low radiated power, and high coverage. Well promising performance-enhancing technologies are smart antennas as well as multiuser detection. Although these new radio technologies have recently been subject to intense research, main UMTS network integration aspects with their specific constraints have been neglected in many cases. Especially the interaction with UMTS radio resource control being required to meet Quality of Service (QoS) constraints has to be included to assess the applicability of these technologies for UMTS.

In this paper, we study the interaction of beamforming concepts as well as multiuser detection with load and power control. We also work out UMTS specific constraints like signal-to-interference-plus-noise ratio (SINR) operating points, pilot power pollution or channel estimation, all strongly limiting network performance. Results are shown for capacity gains and power reduction for all beamforming concepts of interest as well as linear multiuser detection schemes. The results show that fix as well as user-specific beamforming significantly improves network performance gains in downlink. In uplink multiuser detection indicates fairly modest system capacity gains, while it reduces tremendously mobile station power.

1 Introduction

Future wireless mobile communication systems like the upcoming 3rd generation mobile system UMTS (Universal Mobile Telecommunication System) will be required to provide significant improvements in network performance. Rather than solely increasing the number of servable users, network improvement comprises also a reduced radiated power corresponding to an increased coverage. Thus, CDMA (code division multiple access) based future communication systems have to balance the trade-off between system capacity, coverage and emission. In the last few years several new transceiver techniques and radio resource control strategies have been developed to cope for that challenge. Some major and well promising techniques being foreseen for UMTS application are (1) smart antennas and (2) multiuser detection. Several smart antennas like fixed beamforming and user-specific beamforming are of high interest for UMTS 3GPP (2002). While straight forward solutions exist for the uplink Godara (2002); Farrokhi et al. (1998), the downlink direction is much more challenging for UMTS, mainly caused by unknown channel profiles at the base stations as well as powerful pilot transmission Farrokhi et al. (1998); Schacht et al. (2003). Both fixed and user-specific beamforming lead to a performance increase by spatially suppressing inter/intra-cell interference. A further approach to cope with high intra-cell interference, solely for the uplink, is multiuser detection Verdu (1998). In UMTS, the user are generally separated in a code division manner that offers to apply more sophisticated receivers solutions like simple single user matched-filtering (SUMF)[1]. In this paper we focus on Minimum Mean Square Error (MMSE) multiuser detection.

Recently, smart antennas as well as multiuser detection have been subject to intense research, but main network integration aspects with specific constraints have been often missed. As an example, the impact on radio resource control strategies has been rarely investigated up to now. Radio resource control is required since it is of utmost importance in UMTS to keep the air interface load under predefined thresholds to guarantee QoS requirements of already established calls Holma and Toskala (2000); Mueckenheim and Bernhard (2001). Thus, resource control strategies like power or load control has a strong impact on network performance and has to be included for performance assessments.

Correspondence to: A. Dekorsy
(dekorsy@lucent.com)

[1] Note that Rake detection represents single-user matched filtering. In this paper both terms are used in the same manner.

The main objective of this article is to discuss UMTS network enhancements by applying smart antennas in downlink and multiuser detection in uplink direction. We depict concepts that are practicable for UMTS application and we gain an insight in their performance subject to UMTS constraints. Note that we restrict our investigations on UMTS dedicated channels only, but we also cover packet data transmission over that channels as well as mixed services. We present relative gain figures in terms of system capacity gain and emission reduction. Smart antennas for downlink beamforming are shown to deliver significantly higher network performance in terms of capacity as well as power reduction. With multiuser detection the performance increase is ambivalent. We observe moderate capacity gains only, but significant power reduction or coverage gains, respectively.

The paper is structured as follows. Section 2 deals with the interaction of beamforming and linear multiuser detection with power control and it introduces UMTS load control. In Sect. 3 we illuminate the beamforming concepts and give detailed system level results. Next, in Sect. 4 we describe multiuser detection for UMTS and we specially focus on the signal-to-interference-plus-noise ratio (SINR) operating points. Detailed network simulations are shown. Finally, Sect. 6 summarizes main aspects.

2 Power- and load control

2.1 Power-control

The enhancement of UMTS networks by advanced technologies like downlink beamforming or multiuser detection in the uplink requires investigating the interaction of these more sophisticated techniques with given radio resource management algorithms. A general view on power control can be found in the fundamental work by Yates (1995). A more detailed description on combining linear multiuser detection with power control for uplink transmission is reported in Ulukus and Yates (1997). In context with downlink beamforming, Farrokhi et al. (1998) introduced an iterative algorithm that converges to the optimum solution of power control and beamforming by utilizing a virtual uplink. Thus, the more sophisticated problem of downlink beamforming can be solved by leveraging on iterative algorithms known for uplink transmission.

Both, linear multiuser detection as well as beamforming can be interpreted as linear pre-filtering of received user signals. While beamforming acts in space domain, linear multiuser detection acts in the code domain. Thus, it is possible to set up a generic view on combining either beamforming or multiuser detection with power control. In the following, we summarize main aspects for the combination of linear pre-filtering with power control being described in detail in the publications mentioned above. We restrict the description to uplink transmission, since the optimum downlink beamforming can be described by an virtual uplink Farrokhi et al. (1998).

The aim of jointly optimizing power control and linear pre-filtering the mixture of asynchronous received signals is to assign users with transmitted power levels so as to minimize the total transmitted power while ensuring a certain quality of service. We define here the quality of service (QoS) in terms of a signal-to-interference-plus-noise ratio (SINR) per user measured at the output of the linear filter. For notational convenience, we indicate the active users by $i = 1, \ldots, K$ for the remainder of the paper, e.g. the user powers are denoted by p_i with $i = 1, \ldots, K$ and vector $\boldsymbol{p} = (p_1, \ldots, p_K)^T$ denotes the vector with received powers. The transmitted power is related to the received power by the log-normal distributed shadow-fading path loss coefficient h_i, i.e. $p_i = h_i \tilde{p}_i$, with $\tilde{p}_i$ as transmitted power of the ith user. In the following, we refer our evaluation to the received power.

The SINR of user i at the output of the linear detector can be expressed by

$$SINR_i = \frac{\psi_{i,i}\, p_i}{\sum_{k \neq i}^{K} \psi_{i,k}\, p_k + \sigma_n^2}. \tag{1}$$

The element $\psi_{i,k}$ describes the cross-coupling between the i-th user and k-th interferer after linear filtering and can be collected in $K \times K$ matrix $\boldsymbol{\Psi}$. The matrix $\boldsymbol{\Psi}$ includes all system characteristics like user specific signature sequences as well as the user specific mobile radio channel profile. Note, in case of multiuser detection users are separated in a code division multiple access (CDMA) manner where the user specific signature sequences are described by the codes used in the system. UMTS applies Walsh-Hadamard codes of different length also called orthogonal variable spreading factor (OVSF) codes. With beamforming, the users get a finger print in space domain and the user specific signature is given by the array response of the beamforming array Farrokhi et al. (1998); Godara (2002). Finally, note that $\boldsymbol{\Psi}$ also depends on the linear filter applied. This filter is described by matrix $\mathbf{W}$ for the remainder of the paper, and each column of $\mathbf{W}$ represents the filter vector of one specific user. Thus, we can understand matrix $\boldsymbol{\Psi}$ as function of $\mathbf{W}$, e.g. $\boldsymbol{\Psi}$=f($\mathbf{W}$). With this relation and Eq. (1) we can formulate the optimization problem as

$$\min_{\mathbf{W}, 0<p<P_{\max}^L} \sum_i p_i \quad s.t.\ SINR_i \geq \gamma_i \tag{2}$$

with γ_i as required and $SINR_i$ as measured signal-to-interference-plus-noise ratio. Each power is upper limited by the maximum allowable link power $P_{\max}^L$. The SINR requirements in Eq. (2) can be reformulated by an inequality of interference constraints Yates (1995):

$$p_i \geq \mathbf{I}_i(\mathbf{p}, \mathbf{W}) \tag{3}$$

and

$$\mathbf{I}_i(\mathbf{p}, \mathbf{W}) = \frac{\gamma_i}{\Psi_{ii}} \left(\sum_{k \neq i}^{K} \Psi_{ik}\, p_k + \sigma_n^2 \right) \tag{4}$$

describes the effective interference seen by the ith user. It is important to note that the problem statement in Eq. (2) is equivalent to the following one, where the optimization with respect to the linear filter is inserted in the constraint set Ulukus and Yates (1997). With Eqs. (2) and (3) the optimization problem can be rewritten in

$$\min_{\mathbf{p}} \sum_i p_i \quad s.t. \quad p_i \geq \min_{\mathbf{W}} \mathbf{I}_i(\mathbf{p}, \mathbf{W}), \quad i = 1, \ldots, K \tag{5}$$

In Eq. (5) the optimization problem is separated in an outer optimization that is defined over the power vector only, while the inner optimization assumes a fixed power vector and is defined over the filter coefficients. Since the powers are fixed over the inner minimization of the interference functions, this minimization can be re-formulated with Eqs. (1) and (4) as follows:

$$\min_{\mathbf{W}} \mathbf{I}_i(\mathbf{p}, \mathbf{W}) = \min_{\mathbf{W}} \frac{\mathbf{I}_i(\mathbf{p}, \mathbf{W}) p_i}{p_i} = \frac{p_i \cdot \gamma_i}{\max_{\mathbf{W}} SINR_i}. \tag{6}$$

Interpretation of Eq. (6) yields that for fixed power vectors the minimization of effective interference with respect to the filter matrix in the power constraints is equivalent to maximizing the SINR given by Eq. (1). Hence, the MMSE filter is the optimum linear filter solution of the optimization problem Eq. (2) Farrokhi et al. (1998); Yener et al. (2001); Visotsky and Madhow (1999).

In Foschini and Mijanic (1993) it has been recognized for single antenna reception and simple matched-filtering in the code domain ($\mathbf{W} = \mathbf{I}$) that the optimal solution can also be approached iteratively:

$$p_i(l) = \frac{\gamma_i}{SINR_i(l-1)} \cdot p_i(l-1) \tag{7}$$

for $i = 1, \ldots, K$. Parameter l denotes the l-th iteration step.

Based on the the concept of *standard interference function* introduced in Yates (1995) it has been shown in Ulukus and Yates (1997) that the optimal solution to the power control problem can still be found by an iteration when the matrix filter $\mathbf{W}$ depends on $\mathbf{p}$ as it is the case for MMSE. In this case the optimal solution is given by a two-stage iterative algorithm, where within each power update Eq. (7) the MMSE filter has to be re-calculated. Again, the above statement is valid for MMSE multiuser detection as well as MMSE based beamforming Farrokhi et al. (1998); Ulukus and Yates (1997); Yener et al. (2001); Dekorsy and Brueck (2003).

2.2 Load control

Beside link specific power restrictions in Eq. (2) it is of utmost importance in UMTS to keep the air interface load under predefined thresholds, especially in order to guarantee the needed QoS requirements of already established calls. Load control for UMTS dedicated channels composes admission and congestion control. While the first one controls the set up of new calls the latter one is responsible for already established calls. In UMTS the cell specific load can be monitored by measuring the totally radiated or transmitted power of a sector utilizing the power rise in downlink and noise rise in uplink Holma and Toskala (2000); Mueckenheim and Bernhard (2001). The power rise PR is defined as a sector's total transmit power P^{tot} over the power P_{pilot} dedicated to the primary common pilot channel (P-CPICH):

$$PR := \frac{P^{tot}}{P_{Pilot}}, \tag{8}$$

where the total power P^{tot} is composed of the sum of all user powers p_i and the P-CPICH power[2]. Note that the P-CPICH consumes 10% of the maximum transmitted sector power. If Secondary Common Pilot Channels (S-CPICH) exist as used for the fixed beam approach Schacht et al. (2003), they also contribute to P^{tot}.

For the uplink, the noise rise is defined as:

$$NR := \frac{I_0}{I_{th}}. \tag{9}$$

Here I_0 denotes the totally received wide band power at the antenna connector including user signal, interference as well as noise. I_{th} is the thermal noise power in the 3.84 MHz frequency band. A new call will be established if both the NR as well PR remains lower than a maximum allowable threshold. This decision is due to admission control. Moreover, congestion control drops an active subscriber if the NR or PR exceeds a given congestion threshold. A typical value for admission control is 6 dB while for congestion control a typical value is 10 dB Mueckenheim and Bernhard (2001).

3 UMTS downlink beamforming

3.1 Beamforming concepts

The concepts to be investigated are 3-sectorization as reference, fixed beam forming (FxB), and adaptive (user-specific) beamforming (ABF). For the remainder of the paper we assume a uniform linear array (ULA) at the center of each cell composed of four antenna elements with half-wavelength spacing. Since UMTS is designed with low data rate signalling feedback channels to obtain high bandwidth efficiency Holma and Toskala (2000), neither the instantaneous channel coefficients nor array response vectors are known at the base station for downlink beamforming. In this paper we assume that the beamforming weights per user are computed in advance by exploiting the long-term spatial characteristics of the uplink channel Godara (2002); Czylwik and Dekorsy (2001). This approach is especially valid when considering practicable beamforming solutions suitable for UMTS applications. However, it requires the same data mix in uplink and downlink. With this assumption of pre-computed, and thus, fixed beamforming weights in the downlink, we only have to control the power.

[2] Further downlink channels are not considered herein.

Table 1. Traffic mix.

12.2 kbit/s	64.4 kbit/s	384 kbit/s
50%	40%	10%

Table 2. SINR requirements.

Data rate	Downlink SINR requ.	BLER in %
12.2 kbit/s	7.4 dB	0.7
64 kbit/s	4.2 dB	5
384 kbit/s	2.7 dB	5

3.1.1 Adaptive beamforming

The adaptive beamforming strategy investigated is to maximize the received uplink signal power while maintaining the total interference consisting of intra-, inter-cell interference, and thermal noise. The optimization yields a generalized eigenvalue problem where the optimum beamforming solution $\mathbf{w}_k^{opt}$ per user k is the generalized eigenvector corresponding to the largest eigenvalue Godara (2002). Note, UMTS suffers strong pilot transmission in the downlink due to powerful P-CPICH transmission. We consider the strong pilot transmission by diagonal loading concepts. Thus, we simply add P-CPICH signal power (weighted by the uplink channel pathloss) on the diagonal elements of the given uplink interference covariance matrix. This diagonal loading weaks nulling of interferers and transfers the beamforming strategy to maximization of the signal-to-noise ratio (SNR). Results not shown herein indicate maximization of SNR to be the preferable adaptive beamforming solution.

A problem ABF has to cope with is the degradation of the channel estimation which can no longer be based on the P-CPICH due to user specific antenna patterns, but has to use the pilot bit sequence on the dedicated physical control channel (DPCCH). Since this sequence is not continuously and sent out with less power than the P-CPICH we assume a worst case degradation of 3 dB 3GPP (2002).

3.1.2 Fixed beamforming

For fixed beamforming the ULA is used to preform a number of fixed beams which cover the area of a standard 120°-sector 3GPP (2002). Assignment of users to beams is based on the maximum received uplink power per beam. In order to allow proper channel estimation and beam identification, an additional S-CPICH is established per beam. Their powers are adjusted to provide the same SINR as on the P-CPICH.

To accomplish downlink beamforming the characteristics of the individual antenna elements have to be considered Czylwik and Dekorsy (2001). Here, we apply antenna diagrams with squared cosine shapes Czylwik and Dekorsy (2001); Schacht et al. (2002). If not declared differently, a 3 dB angular aperture of 120° is used.

Table 3. SINR gains of beamforming concepts.

	Standard Sectorization	FxB	ABF
SINR in dB	0	4	3

Table 4. Parameters.

Path loss	COST-HATA
Shadow fading stand. dev.	8dB
Max. base station power	43 dBm
Cell radius	1000m
Size active set	4
Hand-Over Add-margin	4 dB
Hand-Over Drop-margin	6 dB
PC step size	1 dB
PC error, stand. dev.	1.5 dB

3.2 System level investigations

Within this section we work out the performance improvement by smart antennas when combined with UMTS specific radio resource management (RRM) algorithms as power control, handover, and load control Holma and Toskala (2000). It should be stressed that all RRM algorithms investigated are practicable for UMTS. Thus, the main objective here is to work out the performance of beamforming methods applicable for UMTS under the constraint of RRM algorithms and their corresponding parameter settings. We also consider heterogeneous traffic with different data rates listed in Table 1.

The investigation approach was as follows. In a first step, we worked out required SINR values for different UMTS services by detailed physical layer simulations. Note that for UMTS the QoS is expressed by a block error rate (BLER) that corresponds to an SINR on the Dedicated Physical Data Channel (DPDCH). Table 2 lists these required SINRs and their corresponding BLERs.

In order to include spatially formed interference (inter- as well as intra-cell interference) intensive system level simulations were carried out for the beamforming concepts of interest. Table 3 summarizes the gain in terms of an averaged SINR per user after RAKE reception for mixed traffic scenarios. Since fixed beamforming significantly decreases with larger angular spread Schacht et al. (2002) we focus here on an angular spread of 1°. Finally, with the results summarized in Table 3 dynamic system level simulations were carried out using a system level simulator including all RRM algorithms mentioned above. The main simulation parameters are listed in Table 4. Note that a UMTS applicable power control algorithm with 1 dB stepsize and log-normal distributed power control error with 1.5 dB standard deviation is used.

Aside from the reduction of the totally emitted power, another important parameter to be investigated is the call-error-rate (CER). The CER is the sum of blocking as well as drop-

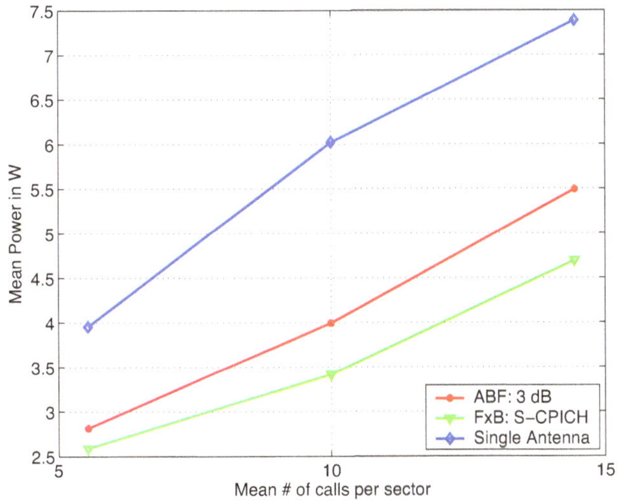

Fig. 1. Mean power vs. load per sector, ULA, $M = 4$ elements.

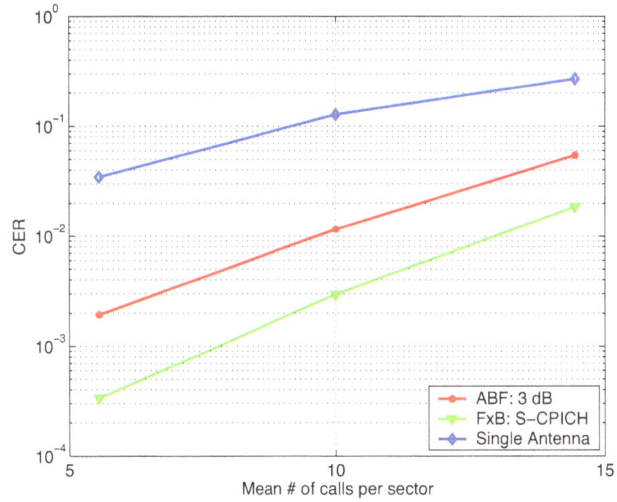

Fig. 2. CER vs. load per sector, ULA, $M = 4$ elements.

ping rate of the network. The blocking rate itself is defined by the ratio of the number of blocked calls over the total number of calls. Same holds for the dropping rate, respectively. A call is blocked if at least one of the following events occur: SINR of the P-CPICH is lower than the required one or if the actual power rise is higher than the maximum allowable limit (admission control). We assumed a power rise of 6 dB for blocking. A call drops if at least one of the following events occur: The transmitted power of the user reaches its maximum and the DPDCH (Dedicated Physical Data Channel) can not meet the SINR requirement, or the power rise exceeds its upper limit. Here, the latter aspect defines the congestion control Mueckenheim and Bernhard (2001) and the upper limit is set to 10 dB. To assess the reduction of radiated power, we used the mean power of a sector. Fig. 1 displays the mean power of a sector versus the network load for the different antenna concepts. It is clearly indicated that the beamforming concepts reduce the emitted power by the base stations compared to the reference of 3 sectorization. For adaptive beamforming we find a power reduction of about 2 dB and for fixed beam switching 2.3 dB. The reason for fixed beam to be superior can be explained by (1) an angular spread of 1° that is much more narrow than the beamwidth of a fixed beam, and (2) ABF suffers a high SINR loss of 3 dB due to weak channel estimation on the DPCCH. The results clearly show the potential of beamforming concepts to reduce the emission of UMTS base stations. In addition to emission reduction, beamforming concepts can tremendously improve system capacity. Figure 2 shows the corresponding call-error-rate for the different antenna concepts versus the load per sector. For UMTS services a maximum CER of 2% is often demanded by service providers. The results in Fig. 2 clearly demonstrate that a network with ordinary sectorization is not capable of meeting that restriction for the traffic mix assumed and 10 calls per sector in average. In contrast beamforming concepts can handle the traffic. For user specific beamforming we observed a CER of 1%, while for fixed beam switching the CER is of app. 0.3%. In summary, applying beamforming concepts for UMTS downlink transmission leads to a reduction of an averaged totally transmitted power per base station of approx. 2–2.5 dB with an additional reduced call-error-rate.

4 Multiuser detection for UMTS

Besides beamforming multiuser detection represents another approach to cope with high co-channel interference of CDMA based systems. In this paper, we consider MMSE multiuser block detection for asynchronous mixed service UMTS networks. The basic modulation scheme of the UMTS uplink is BPSK whereby the user data are mapped onto antipodal in-phase bits, while control (pilot, power control) information are mapped onto antipodal quadrature-phase bits Holma and Toskala (2000). The control and data bit sequences are spread by mutually orthogonal Walsh codes (Orthogonal Variable Spreading Factor (OVSF) codes). The resulting spread time-slot signal is scrambled by a mobile-specific scrambling code. UMTS offers to apply long as well as short scrambling codes. Short scrambling codes with length of 256 chips (one control symbol) allow significant reduction in complexity and are assumed in this paper. Multiuser detection in asynchronous transmission can only be performed optimally by considering a signal block of appropriate length Verdu (1998), here denoted by L. Depending on its data rate user k transmits $N(k)$ symbols within this period of length L. As an example, a voice user of rate 12.2 kbit/s applies Walsh codes of length 64 resulting in $N(k) = 40$ symbols per block of length $L = 2560$ chips. Moreover, in this paper inter-cell interference is assumed to be proportional to the power in the cell of interest and is modelled by the inter-cell interference factor f Holma and Toskala (2000). This model ensures coupling between sectors of a fully deployed sectorized network.

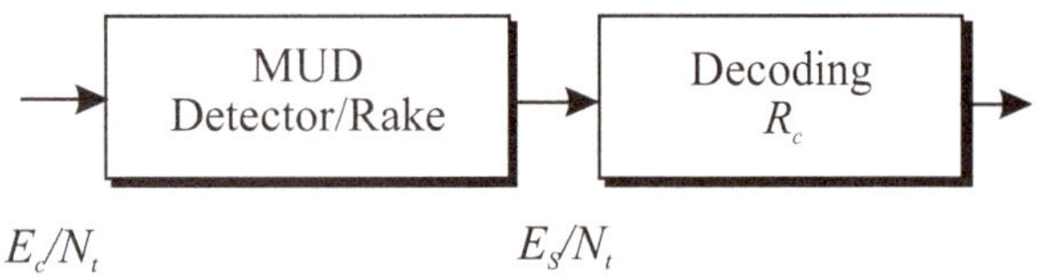

Fig. 3. Basic signal processing diagram of DPDCH.

5 SINR operating points in UMTS

This section gives some essential comments on the SINR operating points of multiuser detectors. The link quality in UMTS is defined by the BLER that has to be ensured for the service offered. This BLER corresponds to an averaged SINR. This averaged SINR requirement is measured on the DPCCH, the inner loop (fast) power control has to adjust the transmitted powers in order to ensure the link quality. Note that directly associated with the SINR on the DPCCH is an SINR on the DPDCH. It is common to express the SINR requirement on the DPDCH in terms of energy per information bit versus overall interference, often denoted by E_b/N_t Holma and Toskala (2000). Be aware that this E_b/N_t requirement strongly depends on the environment (PedA-, VehA-channel, etc.), on receiver performance as channel estimation algorithms as well as the service offered. Figure 3 shows the basic signal processing diagram of the DPDCH and the corresponding SINR requirements. At the input of the decoder the SINR can be expressed in terms of energy per symbol E_s while before MUD detection we have the energy per chip E_c.

These different SINRs are given in linear domain by:

$$\frac{E_s}{N_t} = \frac{E_b}{N_t} R_c, \quad \text{with } R_c < 1, \tag{10}$$

where R_c is the channel coding rate. Moreover, for simple Rake detection we have $E_c/N_t = E_s/N_t \, 1/SF$ with SF as spreading factor of the service. Parameter N_t takes into account the overall interference seen by the user signal before decoding. If multiuser detection is applied the MUD detector operates on symbol energy basis as indicated by Fig. 3. Hence, the operating point for the MUD detector is given by the E_s/N_t value that has to be ensured by the inner power control. Table 5 lists some typical E_s/N_t values for different uplink UMTS services in single rate systems. Under inclusion of these UMTS constraints we obviously obtain very low E_s/N_t values, and, hence, very low operating points of the multiuser detectors. Note that these SINR operating points represent the former introduced SINR requirement denoted by $\gamma = E_s/N_t$. In addition, Table 5 lists required SINR values E_c/N_t before despreading as well. It is has been shown above that the joint optimization problem Eq. (2) ends up in a MMSE filter solution. But the problem arising with linear MUD is that at low SNR operating points simple RAKE detection (single user matched filtering) converges to the optimum MMSE solution. This is because of the trade-off between cancellation of intra-cell interference and the drawback of noise enhancement that is inherent in linear filtering in general. This property of linear MUD was intensively worked out in the past and is well known in literature, e.g. Verdu (1998).

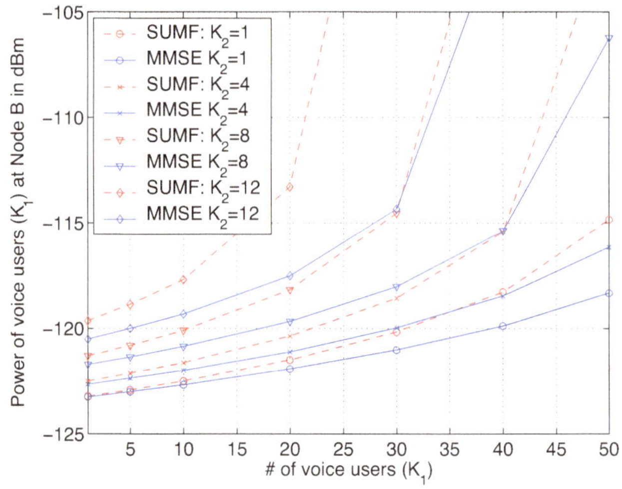

Fig. 4. Power of one voice user, $f = 0.83$, and $L = 512$.

5.1 System-level investigations

In the following we show results of linear MMSE block detection in asynchronous mixed-service UMTS networks. For asynchronous transmission we have to consider a signal block of appropriate length that is here chosen to be $L = 512$. Within this paper, the service mix composes voice with rate 12.2 kbit/s (service 1) and data transmission with rate 64 kbit/s (service 2), refer to Table 5. The relative user delays are up to $\tau_{\max} = 30$ chips to take into account asynchronous transmission. We focus on 6-sector networks with cell radius of 1000 m, and the fractional cell loading factor is chosen to be $f = 0.83$ Holma and Toskala (2000). Thus, the amount of inter-cell interference a user has to deal with is of 83% of the intra-cell interference the user exposes to. Finally, K_1 and K_2 denotes the load of the voice and data service, respectively, which will both be varied in our investigations in the following. Finally SUMF performance is used as reference throughout the investigations.

Figure 4 shows the required power at the base station receiver of one voice user for different service mixes. In general, we observe that the higher the load is the more power can be saved by MMSE compared to SUMF. Especially an increase of data users results in higher MMSE performance gains. As an example, for K_1=20 and K_2=1 MMSE reduces the power in the order of 1 dB only, while for same number of voice but higher number K_2=12 of data users MMSE results in 3–4 dB less power. Note that voice users have less impact on the MMSE performance gain due their significant lower interference contribution. Although the spreading factor of the data service is four times the voice spreading factor, the SINR requirement of the data service is more relaxed

Table 5. Required SINR values for different UMTS services.

Data Rate [kbit/s]	BLER [%]	SF	R_c	E_b/N_t [dB]	$\gamma = E_s/N_t$ [dB]	E_c/N_t [dB]
12.2	0.7	64	$\frac{1}{4.9}$	4.5	−2.4	−20.46
64	5	16	$\frac{1}{3.75}$	2	−3.7	−15.7

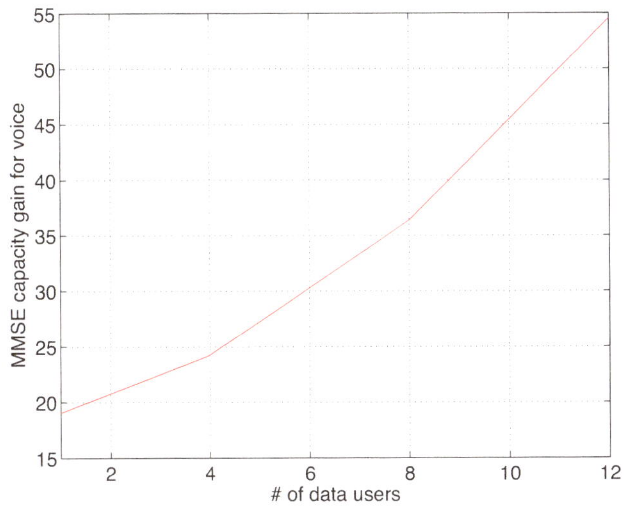

Fig. 5. MMSE capacity gain for voice, $f = 0.83$ and $L = 512$.

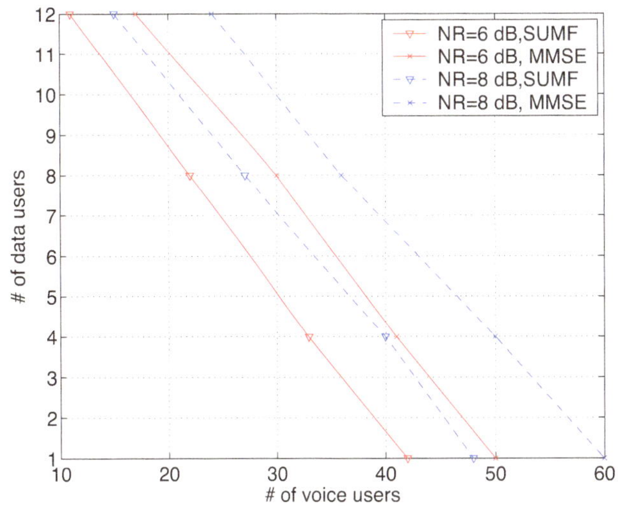

Fig. 6. Maximum service mix load for different noise rise levels.

yielding to approximately 3–4 dB more power consumption of the data service to meet its requirement, refer to Table 5. Compared to SUMF the MMSE results in almost identical power reduction for both services which is in the range of 2–3 dB measured for service loads on a noise rise level of 6 dB, refer to Fig. 6. A noise rise level of 6 dB is typical for quoting system capacity.

Also for this noise rise, Fig. 5 depicts the MMSE capacity gains for voice versus data load. The results indicate the higher the data service load the higher the gain for MMSE in terms of voice capacity. But, due to the very low SNR operating points we obtain fairly modest MMSE gains only. As an example, for $K_2 = 1$ (almost single rate), we observe a capacity increase of voice users in the order of 18% only, while for $K_2 = 8$ we obtain an MMSE gain of 36%. Even for very high data loads the capacity stays below 60% (note that for single rate data system, i.e. no voice users, the maximum number of servable data users for SUMF at NR=6 dB is 14 users, refer to Dekorsy and Brueck (2003).

Finally, Fig. 6 shows the maximum service loads that can be served by the two detectors for noise rise levels 6 dB and 8 dB. The graphs can be interpreted as "rate-regions" for the given service mix taken of data (64 kbps) and voice (12 kbps) subject to noise rise threshold 6 dB or 8 dB, respectively. Figure 9 clearly indicates how MMSE extends the "rate region".

6 Conclusion

This paper investigated the applicability of fixed and adaptive beamforming concepts as well as multiuser detection for mixed service UMTS networks. Special interest was given to the interaction of these promising techniques with UMTS power- and load control strategies. For beamforming the focus was on downlink, while multiuser detection was applied in the uplink direction. With the application of smart antennas (ULA with 4 elements) we observed a tremendous reduction of the call-error-rate. On top of higher quality the radiated base station power can be reduced by at least 2 dB. In the uplink, multiuser detection performance enhancements are severely limited by very low SINR operating points. With this restriction, MMSE error block detection shows fairly modest capacity gains. Nevertheless, MMSE results in fair to high power reduction at the mobile station in the range of 2–3 dB.

References

Czylwik, A. and Dekorsy, A.: System Level Simulations for Downlink Beamforming with Different Array Topologies, GLOBECOM, San Antonio, USA, 3222–3226, 2001.

Dekorsy, A. and Brueck, S.: On System Capacity and Coverage Improvements by Linear MUD for UMTS, VTCS, Jeju, South Korea, 2003.

Farrokhi, R. F., Liu, K. J. R., and Tassiulas, L.: Transmit Beamforming and Power Control for Cellular Wireless Systems, JSAC, 16, 8, 1437–1449, 1998.

Foschini, G. J. and Miljanic, Z.: A simple distributed autonomous power control algorithm and its convergence, ITVT, 42, 641–646, 1993.

3GPP, Technical Specification Group Radio Access Network: Proposal for user specific beamforming for UTRA FDD, Document for Discussion and Decision, TSGR4-17, ETSI, 2001.

Godara, L. C.: Handbook of Antennas in Wireless Communications, CRC Press, 0-8493-0124-6, 2002.

Holma, H. and Toskala, A.: WCDMA for UMTS, John Wiley & Sons, LTD., Chichester, England, 2000.

Mueckenheim, J. and Bernhard, U.: A framework for load control in 3rd generation CDMA networks, GLOBECOM, San Antonio, USA, 2001.

Schacht, M., Dekorsy, A., and Jung, P.: System Capacity from UMTS Smart Antenna Concepts, VTCF, Orlando, USA, 2003.

Schacht, M., Dekorsy, A., and Jung, P.: Downlink Beamforming Concepts in UTRA FDD, Kleinheubacher Tagung, Kleinheubach, Germany, 2002.

Ulukus, S. and Yates, R. D.: Adaptive Power Control and MMSE Interference Suppression, Wireless Networks, Special Issue on Multiuser-Detection in Wireless Communications, 1997.

Verdu, S.: Multiuser Detection, Cambridge University Press, UK, 1998.

Visotsky, E. and Madhow, U.: Optimum Beamforming Using Transmit Antenna Arrays, VTC, 851–856, 1, 1999.

Yates, R. D.: A Framework for Uplink Power Control in Cellular Radio Systems, JSAC, 13, 7, 1341–1347, 1995.

Yener, A. and Yates, R. D., and Ulukus, S.: Interference Management for CDMA Systems Through Power Control, Multiuser Detection, and Beamforming, ITC, 49, 7, 1227–1239, 2001.

24

Synchronization in MIMO OFDM systems

L. Häring and A. Czylwik

Department of Communication Systems, University of Duisburg-Essen, Bismarckstr. 81, 47057 Duisburg, Germany

Abstract. In this paper, an overview of carrier frequency offset (CFO) estimation algorithms for Orthogonal Frequency Division Multiplexing (OFDM) systems is presented. It is well-known that multicarrier systems suffer from their high sensitivity to mismatches of transmitter and receiver oscillator frequencies. The performance degrades since the CFO destroys the orthogonality of the subcarriers. Hence, extensive research has been done on the estimation and correction of the CFO in Single-Input Single-Output (SISO) systems. Mainly, the proposed algorithms can be categorized into data-aided and blind techniques. Several estimation techniques have been extended to the Single-Input Multiple-Output (SIMO) case where multiple receive antennas can be utilized to gain diversity. However, less attention has been paid on synchronization in the attractive Multiple-Input Multiple-Output (MIMO) case which is topic of tremendous interest in current research. The present paper concentrates on aspects of this new scenario. Starting with algorithms for SISO and SIMO, this contribution reviews briefly proposed carrier frequency synchronization techniques which could be implemented in forthcoming MIMO systems.

1 Introduction

In recent years, OFDM has become increasingly popular for future high data-rate wireless communication systems because of its advantages in a multipath environment. It has been standardized for Digital Audio Broadcasting (DAB), Terrestrial Digital Video Broadcasting (DVB-T) as well as for broadband wireless local area networks (WLAN) i.e. Hiperlan/2 and IEEE 802.11a/g. However, one of the main drawbacks is its high sensitivity to frequency offsets since the orthogonality between subcarriers is destroyed thus leading to a considerable system performance degradation (Pollet et al., 1995). Hence, the estimation and correction of frequency offsets have been subject of intensive research in the last decade. But it has not been investigated extensively in systems with multiple antennas at transmitter and receiver site. Recently, such MIMO systems have become attractive since the capacity was proven (under ideal conditions) to scale with the minimum number of transmit and receive antennas (Foschini and Gans, 1998). Therefore, the contribution introduces thoroughly a MIMO system model that considers CFOs. Furthermore, it gives an overview of important existing estimation algorithms for SISO systems and extensions to multiple antenna systems.

The remainder of the paper is organized as follows. Section 2 presents the OFDM system model for the SISO and MIMO case as well as impacts of frequency offsets on the system performance. In Sect. 3, important classes of SISO algorithms are reviewed. Extensions of these algorithms to multiple antenna systems are discussed in Sect. 4. To illustrate some important aspects, Sect. 5 presents a brief simulation example. Finally, in Sect. 6 conclusions are drawn.

2 OFDM signal model

We start with the OFDM model formulation for the SISO case. Subsequently, we extend the model to a MIMO system (see Fig. 1).

2.1 System model in SISO

In OFDM, information data are transmitted blockwise. A sequence of complex data symbols is split into blocks and fed to different subcarriers. For the k-th block, an IDFT operation on the symbols of all carriers is carried out, which can be expressed as

$$x_i(k) = \frac{1}{\sqrt{N}} \sum_{n=0}^{N_u-1} s_n(k)\, e^{j\frac{2\pi}{N}ni}\,, \quad 0 \le i \le N-1\,, \qquad (1)$$

Correspondence to: L. Häring
(haering@sent5.uni-duisburg.de)

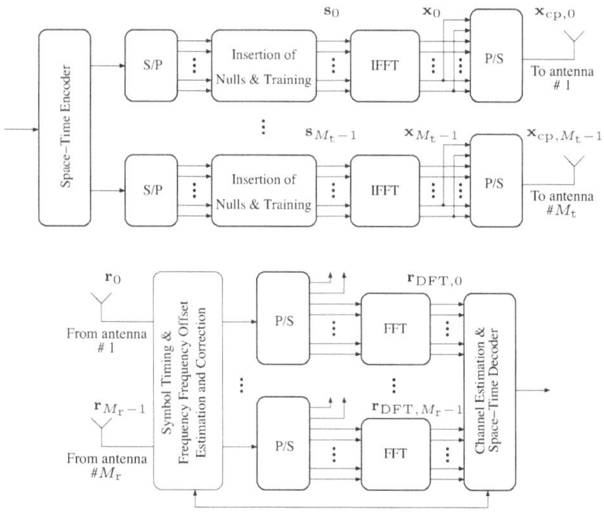

Fig. 1. Block diagram of transmitter and receiver in a MIMO OFDM system.

where N, N_u ($\leq N$) and $s_n(k)$ denote the IDFT-size, the number of used subcarriers and data symbols of the n-th subcarrier, respectively. If $N_u < N$, the residual $N - N_u$ subcarriers referred to as *null* or *virtual* subcarriers are filled with zeros. With the defined vectors

$$\mathbf{s}(k) = \left[s_0(k), s_1(k), \ldots, s_{N_u-1}(k)\right]^T \tag{2}$$

$$\mathbf{x}(k) = \left[x_0(k), x_1(k), \ldots, x_{N_u-1}(k), \ldots, x_{N-1}(k)\right]^T$$

and the IDFT-matrix

$$\mathbf{W}_{N_u} = \frac{1}{\sqrt{N}} \begin{pmatrix} 1 & 1 & \ldots & 1 \\ 1 & e^{j\frac{2\pi}{N}} & \ldots & e^{j\frac{2\pi}{N}(N_u-1)} \\ \vdots & \vdots & \ddots & \vdots \\ 1 & e^{j\frac{2\pi}{N}(N-1)} & \ldots & e^{j\frac{2\pi}{N}(N-1)(N_u-1)} \end{pmatrix},$$

the k-th block can be written in a more compact way:

$$\mathbf{x}(k) = \mathbf{W}_{N_u}\mathbf{s}(k) \ . \tag{3}$$

Subsequently, a cyclic prefix of length G is preceded, which is just a duplication of the last G samples:

$$\mathbf{x}_{cp}(k) = [\underbrace{x_{N-G}(k), \ldots, x_{N-1}(k)}_{\text{cyclic prefix}}, x_0(k), \ldots, x_{N-1}(k)]^T \ .$$

In order to avoid interblock interference (IBI), G must be chosen larger than the length L of the channel impulse response. After passing through a pulse-shaping filter with the impulse response $g_t(t)$ (e.g. root-raised cosine), the transmitted time-continuous waveform yields

$$x_t(t) = \sum_{k=-\infty}^{\infty} \sum_{i=-G}^{N-1} x_{cp,i+G}(k) g_t\big(t - kN_bT - iT\big), \tag{4}$$

where T is the sampling period and $N_b = N + G$ denotes the length of one OFDM block.

At the receiver, mismatches between transmitter and receiver oscillators as well as Doppler effects result in a frequency offset f_o. Usually, the most dominant frequency offset arises from oscillator imperfections.

Assuming a frequency-selective channel with the time-invariant impulse response $h_c(t)$ and a receiver filter with $g_r(t)$ which is matched to the transmit filter, we can write for the *effective* channel impulse response $h(t) = g_t(t) * h_c(t) * g_r(t)$. The received time-continuous waveform containing a frequency offset yields then

$$r(t) = e^{j(2\pi f_o t + \phi)} \sum_{k=-\infty}^{\infty} \sum_{i=-G}^{N-1} x_{cp,i+G}(k) h\big(t - kN_bT - iT\big),$$

where noise contributions are omitted and ϕ is an arbitrary phase. Without loss of generality, we set $\phi = 0$.

Sampled at time instants $t = k'N_bT + i'T$[1] and using the normalized offset $\varepsilon = f_o/\Delta f$ related to the carrier subspacing $\Delta f = 1/(NT)$, one obtains with Eq. (1)

$$r_{i'}(k') = e^{j\frac{2\pi}{N}\varepsilon(k'N_b+i')} \frac{1}{\sqrt{N}} \sum_{k=-\infty}^{\infty} \sum_{i=-G}^{N-1} \sum_{n=0}^{N_u-1} s_n(k) \cdot e^{j\frac{2\pi}{N}ni} h\big((k'-k)N_bT + (i'-i)T\big) \ . \tag{5}$$

Furthermore, we assume that the duration of the combined channel impulse response is limited such that $h(t) = 0 \ \forall t \backslash [0, LT]$, where $L < G$. Hence, the only nonzero contribution in Eq. (5) comes from $k = k'$:

$$r_{i'}(k') = e^{j\frac{2\pi}{N}\varepsilon(k'N_b+i')} \frac{1}{\sqrt{N}} \sum_{n=0}^{N_u-1} s_n(k') \sum_{i=i'}^{i'-L} h\left(i' - i\right) e^{j\frac{2\pi}{N}ni} .$$

The sampling period T was omitted for notational convenience. Using the combined channel transfer function $H_n = \sum_{i=0}^{L} h(i)e^{-j\frac{2\pi}{N}ni}$, the k'-th received OFDM-block for $i' \in [0, \ldots, N-1]$ after removal of the cyclic prefix is given by

$$r_{i'}(k') = e^{j\frac{2\pi}{N}\varepsilon(k'N_b+i')} \frac{1}{\sqrt{N}} \sum_{n=0}^{N_u-1} s_n(k') \cdot \underbrace{\sum_{i'-i=0}^{L} h\big(i'-i\big)e^{-j\frac{2\pi}{N}n(i'-i)}}_{H_n} e^{j\frac{2\pi}{N}ni'}$$

$$= e^{j\frac{2\pi}{N}\varepsilon(k'N_b+i')} \cdot \mathrm{IDFT}_{i'}\left\{s_n(k')H_n\right\} \ . \tag{6}$$

Alternatively, in matrix formulation if k' is substituted by k and a block fading channel is assumed, we obtain

$$\mathbf{r}(k) = e^{j\frac{2\pi}{N}\varepsilon k N_b} \mathbf{C}_\varepsilon \mathbf{W}_{N_u} \mathbf{H}(k) \mathbf{s}(k) \tag{7}$$

[1] For reasons of simplicity and space limitation, in this paper, perfect timing synchronization is assumed. Timing synchronization in OFDM systems involves finding the optimal position of the DFT-window for demodulation. Due to margins between the length of guard interval and channel impulse response in practical systems, requirements with respect to timing synchronization are relaxed.

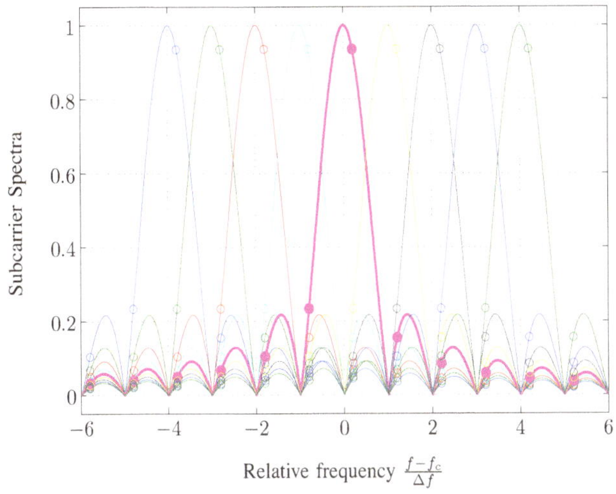

Fig. 2. Effects of a CFO ($\varepsilon = 0.2$).

with $\mathbf{C}_\varepsilon = \operatorname{diag}\left(1, e^{j\frac{2\pi}{N}\varepsilon}, \ldots, e^{j\frac{2\pi}{N}\varepsilon(N-1)}\right)$ denoting the diagonal $N \times N$ carrier frequency offset matrix and $\mathbf{H}(k) = \operatorname{diag}(H_0(k), \ldots, H_{N_u-1}(k))$ representing the diagonal $N_u \times N_u$ channel matrix. A CFO turns out to rotate the phase of the received samples linearly with time. To demodulate, we left-multiply $\mathbf{r}(k)$ with the DFT matrix $\mathbf{W}_{N_u}^H$:

$$\mathbf{r}_{\mathrm{DFT}}(k) = e^{j\frac{2\pi}{N}\varepsilon k N_b}\mathbf{W}_{N_u}^H \mathbf{C}_\varepsilon \mathbf{W}_{N_u}\mathbf{H}(k)\mathbf{s}(k)\ . \tag{8}$$

If the matrix $\mathbf{W}_{N_u}^H \mathbf{C}_\varepsilon \mathbf{W}_{N_u}$ is non-diagonal, the orthogonality between subcarriers is lost, thus resulting in inter-carrier interference (ICI). To analyze this further and using some straightforward derivation, we can find

$$\mathbf{r}_{\mathrm{DFT}}(k) = e^{j\frac{2\pi}{N}\varepsilon k N_b}\mathbf{B}\mathbf{H}(k)\mathbf{s}(k) \tag{9}$$

with

$$\mathbf{B} = \begin{pmatrix} b_0 & \ldots & b_{N_u-1} \\ \vdots & \ddots & \vdots \\ b_{-N_u+1} & \ldots & b_0 \end{pmatrix}, \tag{10}$$

where

$$b_n = \frac{\sin(\pi(\varepsilon+n))}{N\sin\left(\frac{\pi}{N}(\varepsilon+n)\right)}\, e^{j\pi\frac{N-1}{N}(\varepsilon+n)}\ . \tag{11}$$

Equations (9)–(11) indicate two effects of CFOs: The amplitudes of the *desired* subcarriers multiplied by b_0 are reduced and cross terms (for $n \neq 0$) are introduced thus confirming the above mentioned statement. Figure 2 illustrates both issues.

In his work (Moose, 1994), MOOSE investigated the impact of a CFO analytically. He calculated a lower bound of the signal-to-interference-and-noise power ratio (SINR) visualized in Fig. 3 for different signal-to-noise power ratios (SNR). It can be clearly seen, that especially at high SNR regions, the performance degradation caused by synchronization errors becomes considerable.

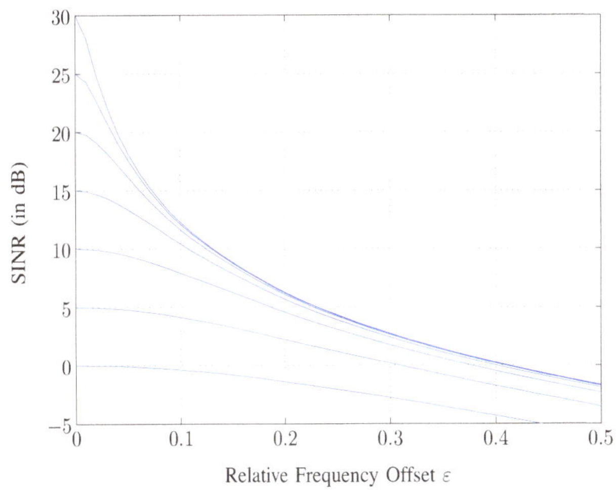

Fig. 3. Lower Bound of SINR (Moose, 1994).

2.2 Extension to MIMO

In the following, it is considered that M_t transmit and M_r receive branches share the same oscillator, respectively. With this practical assumption, still only one CFO emerges. Hence, the k-th signal block at the m_r-th receive antenna after cyclic prefix removal can be written as

$$\mathbf{r}_{m_r}(k) = e^{j\frac{2\pi}{N}\varepsilon k N_b}\mathbf{C}_\varepsilon \mathbf{W}_{N_u}\sum_{m_t=0}^{M_t-1}\mathbf{H}_{m_r,m_t}(k)\mathbf{s}_{m_t}(k)\ ,$$

where $\mathbf{s}_{m_t}(k)$ denotes the block vector transmitted from the m_t-th transmit antenna. $\mathbf{H}_{m_r,m_t}(k)$ is a diagonal $N_u \times N_u$ matrix describing the frequency channel characteristic from the m_t-th transmit antenna to the m_r-th receive antenna. Stacking blocks from all antennas together,

$$\tilde{\mathbf{s}}(k) = \operatorname{vec}\left(\mathbf{s}_0(k), \ldots, \mathbf{s}_{M_t-1}(k)\right) \tag{12}$$

$$\tilde{\mathbf{r}}(k) = \operatorname{vec}\left(\mathbf{r}_0(k), \ldots, \mathbf{r}_{M_r-1}(k)\right)\ , \tag{13}$$

we can find the following compact notation:

$$\tilde{\mathbf{r}}(k) = e^{j\frac{2\pi}{N}\varepsilon k N_b}\tilde{\mathbf{C}}_\varepsilon \tilde{\mathbf{W}}_{N_u}\underbrace{\operatorname{diag}\left(\mathbf{H}_0(k), \ldots, \mathbf{H}_{M_r-1}(k)\right)}_{\tilde{\mathbf{H}}(k)}\check{\mathbf{s}}(k)$$

with $M_r M_t N_u \times 1$ vector $\check{\mathbf{s}}(k) = \operatorname{vec}\left(\tilde{\mathbf{s}}(k), \ldots, \tilde{\mathbf{s}}(k)\right)$ and

$$\tilde{\mathbf{C}}_\varepsilon = \mathbf{I}_{M_r} \otimes \mathbf{C}_\varepsilon \tag{14}$$

$$\tilde{\mathbf{W}}_{N_u} = \mathbf{I}_{M_r} \otimes \mathbf{W}_{N_u} \tag{15}$$

$$\mathbf{H}_{m_r}(k) = \left[\mathbf{H}_{m_r,0}(k), \ldots, \mathbf{H}_{m_r,M_t-1}(k)\right]\ . \tag{16}$$

Here, $\otimes$ denotes the Kronecker product and $\mathbf{I}_K$ the $K \times K$ identity matrix. Taking the DFT by left-multiplication $\tilde{\mathbf{r}}(k)$ with $\tilde{\mathbf{W}}_{N_u}^H = \mathbf{I}_{M_r} \otimes \mathbf{W}_{N_u}^H$ yields

$$\begin{aligned}\tilde{\mathbf{r}}_{\mathrm{DFT}}(k) &= e^{j\frac{2\pi}{N}\varepsilon k N_b}\tilde{\mathbf{W}}_{N_u}^H \tilde{\mathbf{C}}_\varepsilon \tilde{\mathbf{W}}_{N_u}\tilde{\mathbf{H}}(k)\check{\mathbf{s}}(k) \\ &= e^{j\frac{2\pi}{N}\varepsilon k N_b}(\mathbf{I}_{M_r} \otimes \mathbf{B})\tilde{\mathbf{H}}(k)\check{\mathbf{s}}(k)\ .\end{aligned} \tag{17}$$

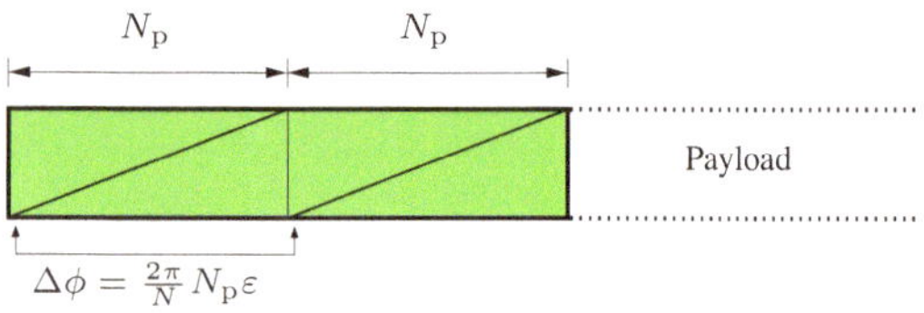

Fig. 4. Preamble structure consisting of two identical halves (Moose, 1994).

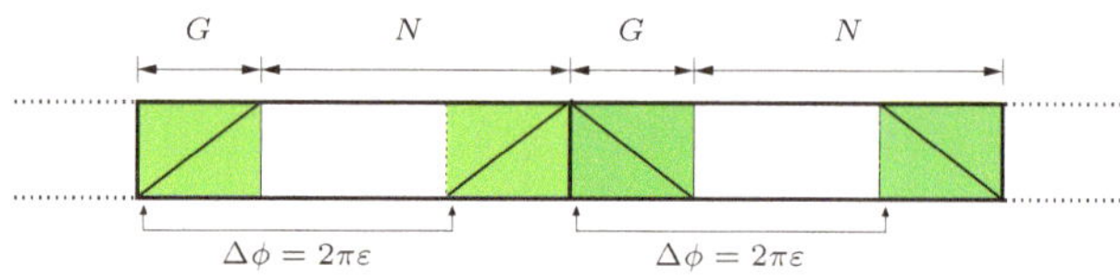

Fig. 5. OFDM block structure with cyclic prefix.

It can be seen that the influence of a CFO is the same for all receive antennas. Similarly to Eq. (8), to restore the orthogonality between subcarriers, the unknown parameter ε in $\mathbf{C}_\varepsilon$ needs to be estimated, preferably before carrying out the DFT. The estimate $\hat{\varepsilon}$ can be used to derotate the received samples before demodulation.

Algorithms that find an estimate $\hat{\varepsilon}$ can be mainly classified in *data-aided* and *blind* methods (dependent on the information that is available). Although also possible in the frequency-domain (after DFT processing) (e.g. Santella, 2000, and references therein), this paper only deals with methods working on time-domain samples.

3 Frequency synchronization in SISO systems

3.1 Data-aided approaches

In many data-aided synchronization approaches, algorithms are based upon the observation (e.g. in Eq. 6), that CFOs only cause linear (related to time) phase shifts of the received symbols. Hence, frequency offsets can be estimated by "measuring" the phase rotation (unless the phase rotation does not exceed an angle of 2π). To follow this approach, MOOSE proposed in Moose (1994) two identical training symbols as a preamble (see Fig. 4).

From one symbol to the other, the phase changes by $\Delta\phi = \frac{2\pi}{N} N_p \varepsilon$:

$$r_{i+N_p} = r_i e^{j\frac{2\pi}{N} N_p \varepsilon} + n_i \,, \quad i = 0, \ldots, N_p - 1 \,, \tag{18}$$

where N_p denotes the length of one training symbol. In the original work, MOOSE proposed $N_p = N$ und used samples in the frequency domain. For n_i representing additive white Gaussian noise (AWGN), he calculated the Maximum-Likelihood (ML) estimator:

$$\hat{\varepsilon} = \frac{1}{2\pi} \frac{N}{N_p} \angle R = \frac{1}{2\pi} \frac{N}{N_p} \angle \left(\sum_{i=0}^{N_p - 1} r_i^* r_{i+N_p} \right) \,, \tag{19}$$

where $\angle R$ denotes the angle of the complex correlation sum.

The fact that some timing synchronization techniques rely upon computing the correlation sum in Eq. (19), argues for the practical usefulness of this algorithm. Furthermore, although mathematically shown only for an AWGN channel, the estimator in Eq. (19) turns out to perform well even in frequency-selective fading channels since the periodicity of the training structure is still preserved (if a cyclic prefix is preceded).

However, one major drawback of this method is its short estimation range of $|\varepsilon| < \frac{N}{2N_p}$ caused by phase ambiguities. Shortening the length of the training symbols increases the acquisition range, but lowers the estimation performance as well. Several researchers have investigated possibilities to extend the range by modifying the preamble – however, mainly by cost of some increase in computational load. Moreover, two-step procedures with two different training sequences consisting of a course and fine estimation step have also been proposed as well as symbols with more than two identical halves. The interested reader is referred to e.g. Schmidl and Cox (1997), Morelli and Mengali (1999), Li et al. (2001).

3.2 Blind approaches

In van de Beek et al. (1997), VAN DE BEEK *et al.* proposed to exploit the cyclic prefix which is essential anyway to prevent IBI and furthermore helps to simplify the channel equalization. Originally, the authors found an optimal joint estimator for both timing and frequency synchronization. With the correlation between received symbols of the cyclic prefix and of its duplicate for the k-th block (signal power σ_s^2)

$$\mathrm{E}\left\{ r_i^*(k) r_{i+N}(k) \right\} = \sigma_s^2 e^{j2\pi\varepsilon} \,, \quad i = 0, \ldots, G-1 \,, \tag{20}$$

the ML estimate of the CFO in an AWGN channel yields

$$\hat{\varepsilon} = \frac{1}{2\pi} \angle R(k) = \frac{1}{2\pi} \angle \left(\sum_{i=0}^{G-1} r_i^*(k) r_{i+N}(k) \right) \,. \tag{21}$$

In order to enhance the estimation performance, averaging over subsequent OFDM blocks is beneficial. Besides the fixed small estimation range of half of the carrier spacing, the estimator in Eq. (21) – in contrast to the data-aided approach – suffers from considerable performance degradation in frequency-selective fading channels. At high SNR, it exhibits a large error floor. Recently proposed methods try to adapt VAN DE BEEK's approach to multipath channels. The idea is to use statistical properties of the received signal to weight contributions in the correlation sum dependent on how heavily they are corrupted by IBI.

Another important class of blind estimation methods utilizes redundancy introduced by *virtual* subcarriers. These carriers can be regarded as spectral guard regions that are often used in practical OFDM systems to relax requirements of the analog radio frequency (RF) front-end. Additionally, a major advantage of these algorithms is the larger acquisition range.

The basic idea was found by LIU and TURELI (Liu and Tureli, 1998) first: Let us assume $N - N_u$ virtual carriers, thus the "full" IDFT-matrix becomes $\mathbf{W} = [\mathbf{W}_{N_u}, \mathbf{w}_{N_u+1}, \ldots, \mathbf{w}_N]$.

Due to the orthogonality property of column vectors of $\mathbf{W}$, we obtain in absence of CFO and noise ($n = 1, \ldots, N - N_u$)

$$\mathbf{w}^H_{N_u+n}\mathbf{r}(k) = \mathbf{w}^H_{N_u+n}\mathbf{W}_{N_u}\mathbf{H}(k)\mathbf{s}(k) = 0 \; . \quad (22)$$

This basic observation leads to the following problem formulation to estimate ε:

$$J(e) = \sum_{k=0}^{K-1}\sum_{n=1}^{N-N_u}\left|\mathbf{w}^H_{N_u+n}\mathbf{C}_e^{-1}\mathbf{r}(k)\right|^2$$
$$\hat{\varepsilon} = \arg\min_e J(e) \; . \quad (23)$$

The estimator in Eq. (23), similar to the (in array processing) well-known MUSIC algorithm, is equivalent to the non-linear least squares (NLS) solution. Under some uncritical assumptions, the cost function in Eq. (23) is shown to yield a unique minimum. However, MA *et al.* have proven in Ma et al. (2001) the loss of identifiability in case of "channel nulls". To guarantee channel-independent identiafibility, hopping virtual subcarriers (positions of these carriers change from block to block) have been suggested.

Yet another blind synchronization algorithms proposed by BÖLCSKEI that also has got a large acquisition range is based upon the second-order statistics of OFDM signals. BÖLCSKEI has shown in Bölcskei (2001) that cyclostationarity is introduced by a cyclic prefix, pulse shaping or by the use of different transmit powers on the subcarriers such that the second-order statistics contain information on the synchronization parameters.

The interested reader is referred to Bölcskei (2001) and Gini and Giannakis (1998).

4 Frequency synchronization in SIMO and MIMO systems

As stated before, if transmitter and receiver branches share a common oscillator, respectively (which was considered throughout this paper), evidently still only one frequency offset in Eq. (14) must be estimated. Thus in principle, all the described algorithms can be also used for such a MIMO scenario. However, *spatial diversity* can be exploited in order to enhance the estimator performance.

4.1 Data-aided approaches

Unfortunately, for the synchronization unit, channel state information is not available which complicates any diversity approach. This general problem was investigated by CZYLWIK in Czylwik (1999) for SIMO systems; the synchronization algorithm used in his paper was based upon MOOSE's approach. He found out that the optimum of any linear combination of uncorrelated received signals leads to a *selection combing* (SC) concept, where only the strongest branch signal passes. However, if a non-linear operation is performed first, also information from other antennas can be advantageously utilized by a method similar to *maximum ratio combining* (MRC), in which branch weights are set in order to maximize the SNR.

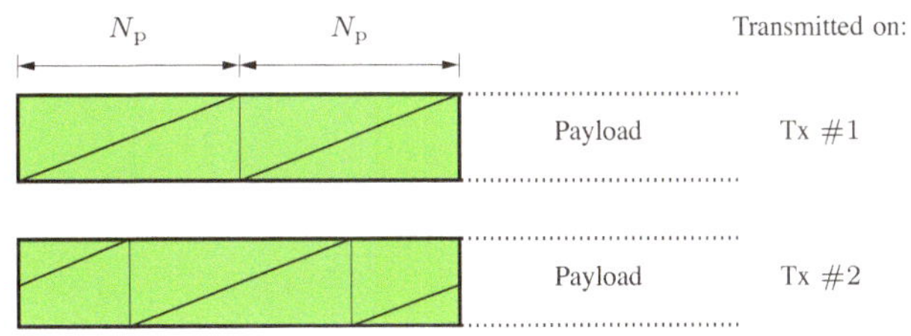

Fig. 6. Modified preamble suitable for MIMO transmission (Schenk and van Zelst, 2003).

A synchronization unit for MIMO systems based upon MOOSE's and VAN DE BEEK's approach was suggested by MODY *et al.* (Mody and Stüber, 2001). But diversity concepts have not been sufficiently explored in this proposal. The main novelty was the modification of training sequences. Generally, the same training sequences as for synchronization are also used for channel estimation. Hence, additional requirements have to be met. The authors suggested the usage of chirp like orthogonal sequences proposed by Suehiro (Suehiro and Hatori, 1998).

Recently in 2003, SCHENK *et al.* (Schenk and van Zelst, 2003) analyzed a MIMO extension of Moose (1994) also from an analytical point of view. In this approach, they used constant-envelope orthogonal codes with good periodic correlation properties, such as FRANK-ZADOFF codes (Frank and Zadoff, 1962). Training sequences on different antennas are cyclically shifted as illustrated in Fig. 6 for $M_t = 2$.

Instead of averaging estimates on different antenna signals which is equivalent to *equal gain combining* (EGC), the estimator is calculated

$$\hat{\varepsilon} = \frac{1}{2\pi}\frac{N}{N_p}\angle\tilde{R} = \frac{1}{2\pi}\frac{N}{N_p}\angle\left(\sum_{m_r=0}^{M_r-1} R_{m_r}\right) , \quad (24)$$

where R_{m_r} denotes the complex correlation sum of each antenna branch. R_{m_r} is composed in the same way as in Eq. (19). In a frequency-selective fading channel, the estimator yields a MRC-like performance since antenna signals with a higher SNR contribute more to the estimate.

For an AWGN channel, the Cramer-Rao bound (CRB) is Schenk and van Zelst (2003)

$$\operatorname{var}\hat{\varepsilon} \geq \frac{N}{(2\pi)^2 \cdot N_p^3 \cdot M_r \cdot \mathrm{SNR}} \; . \quad (25)$$

Clearly, the CRB depends on the number of receive antennas rather than on the number of transmit antennas. We will see later, however, that in a fading channel also transmit diversity can enhance the estimator performance.

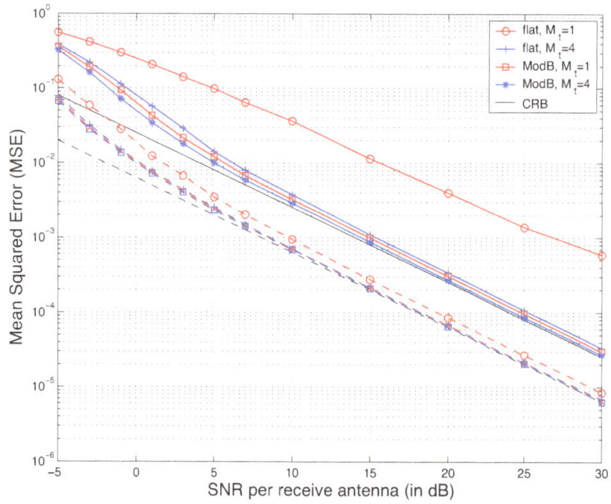

Fig. 7. CRB and simulated MSE of the CFO estimation for different combinations of M_t and $M_r = 1$ (solid, —) and $M_r = 4$ (dashed, – –) averaged over 20000 realizations.

4.2 Blind approaches

Again motivated by Czylwik (1999), HONAN *et al.* also extended the blind estimation algorithm based on virtual subcarriers to the SIMO case (e.g. in Honan et al., 2003). Path gains $|\gamma_{m_r}|$ are introduced in the cost function:

$$J(e) = \sum_{k=0}^{K-1} \sum_{m_r=0}^{M_r-1} \sum_{n=1}^{N-N_u} |\gamma_{m_r}|^2 \left| \mathbf{w}_{N_u+n}^H \mathbf{C}_e^{-1} \mathbf{r}_{m_r}(k) \right|^2$$

$$\hat{\varepsilon} = \arg\min_e J(e) \,. \qquad (26)$$

The authors proved that the weights $|\gamma_{m_r}|$ facilitate estimator MRC diversity gain when selected proportional to the branch SNR. They validated also in an experimental setup the superior performance of this approach compared to EGC especially at low SNRs. At higher SNR, EGC and MRC coincide.

The cost function in Eq. (26) can also be used in MIMO scenarios. In Honan and Tureli (2003) it is shown that the CRB in AWGN channels does not change with the number of transmit antennas. However, the frequency offset estimator benefits from transmit diversity which leads to an essential performance improvement at lower SNR. In addition, the MIMO diversity gain helps the estimator to be less sensitive in case of channel nulls.

5 Simulation example

Due to space limitation, only a short simulation example of the MIMO estimation algorithm proposed by SCHENK is following to demonstrate the main aspects.

To have a fair comparison, the total power radiated from M_t transmit antennas in MIMO is restricted to σ_s^2, which means that the transmitted power per antenna is reduced by the factor M_t. Simulations have been carried out for an AWGN and frequency-selective fading channel. The multipath channel is implemented as a tapped-delay line with Rayleigh fading coefficients and a power delay profile given by typical indoor models used for WLANs at 5 GHz (Medbo and Schramm, 1998). Impinging signals at different antennas are assumed to be uncorrelated. We choose the DFT-size $N = 64$, the sampling period $T = 50$ ns (e.g. in IEEE 802.11a) and periodicity length $N_p = 16$. As a cyclic prefix, one period of 16 samples is preceded. The relative CFO error in all simulations is $\varepsilon = 0.2$. To study the influence of multiple antennas, the mean squared error (MSE) of $\hat{\varepsilon}$ versus SNR per receive antenna for a different number of M_t and M_r is simulated.

In Fig. 7, the simulation results for both a flat and frequency-selective (Model B for typical large open space and office environment (Medbo and Schramm, 1998), average rms delay spread of 100 ns) fading channel are depicted. For reference, also the CRB for AWGN is displayed.

As expected, we can observe a performance degradation in fading channels compared to the CRB (as well as for a simulated AWGN channel which is not displayed here), especially at low SNRs. (At high SNRs, apart from the SISO flat fading case, the MSE approximates the CRB, which reveals a ML estimator.) The degradation decreases with increasing delay spread due to the gain of *frequency* diversity. It is interesting to note that the estimator does not suffer from multipath since the periodicity of training symbols is preserved.

However, performance improvement is also achieved by *space* diversity. The effect of *transmit* diversity becomes apparent for $M_t = 4$ versus $M_t = 1$ in a flat fading channel using only a single receive antenna. Nevertheless, *receive* diversity plays the dominant role since for $M_r > 1$ the estimator is less sensitive to fading and furthermore lowers the bound of the MSE. This simulation example shows that frequency offset estimation in MIMO systems is more robust under different fading conditions, which can be explained by the space diversity introduced by multiple antennas.

6 Conclusion

A MIMO OFDM system model including a carrier frequency offset has been developed and the effects of such offsets on the system performance in OFDM systems have been illustrated. Basic SISO data-aided and blind estimation algorithms using time-domain samples have been reviewed. Moreover, some extensions to a SIMO and MIMO scenario that efficiently exploit spatial diversity have been presented. Finally, a simulation example has been shown to demonstrate the main aspects.

References

Bölcskei, H.: Blind estimation of symbol timing and carrier frequency offset in wireless OFDM systems, IEEE Trans. Comm., 49, 6, 988–999, 2001.

Czylwik, A.: Synchronization for systems with antenna diversity, Proc. of IEEE Veh. Technol. Conf., 2, 728–732, 1999.

Foschini, G. and Gans, M.: On limits of wireless communications in a fading environment when using multiple antennas, Wireless Personal Communications, 6, 311–335, 1998.

Frank, R. and Zadoff, S.: Phase Shift Pulse Codes with good periodic correlation properties, IRE Trans. on Inform. Theory, IT-8, 381–382, 1962.

Gini, F. and Giannakis, G. B.: Frequency offset and symbol timing recovery in Flat-Fading channels: A cyclostationary approach, IEEE Trans. Comm., 46, 400–411, 1998.

Honan, P. J. and Tureli, U.: Blind Carrier Offset Estimation for MIMO OFDM systems: Cramer Rao Bound and Nonlinear Least Squares, in Proc. of Conf. on Inform. Sciences and Systems, 2003.

Honan, P. J., Ambati, R., and Tureli, U.: Performance Analysis of Diversity Combining Method for OFDM Blind Carrier Synchronization, Proc. of IEEE Veh. Technol. Conf., 4, 2672–2676, 2003.

Li, J., Liu, G. Q., and Giannakis, G. B.: Carrier frequency offset estimation for OFDM-based WLANs, IEEE Signal Proc. Letters, 8, 80–82, 2001.

Liu, H. and Tureli, U.: A high-efficiency carrier estimator for OFDM commucations, IEEE Commun. Letters, 2, 104–106, 1998.

Ma, X. L., Tepedelenlioglu, C., Giannakis, G. B., and Barbarossa, S.: Non-Data-Aided Carrier offset estimators for OFDM with null subcarriers: Identifiability, algorithms, and performance, IEEE J. Select. Areas Commun., 19, 2504–2515, 2001.

Medbo, J. and Schramm, P.: Channel Models for HIPERLAN/2 in Different Indoor Scenarios, ETSI/BRAN document no. 3ERI085B, 1998.

Mody, A. N. and Stüber, G. L.: Synchronization for MIMO systems, Proc. of IEEE Globecom Conf., 1, 509–513, Texas, USA, 2001.

Moose, P. M.: A technique for Orthogonal Frequency Division Multiplex frequency offset correction, IEEE Trans. Comm., 42, 1994.

Morelli, M. and Mengali, U.: An improved frequency offset estimator for OFDM applications, IEEE Comm. Letters, 3, 75–77, 1999.

Pollet, T., van Bladel, M., and Moeneclaey, M.: BER sensitivity of OFDM systems to carrier frequency offset and Wiener phase noise, IEEE Trans. Comm., 43, 191–193, 1995.

Santella, G.: A Frequency and Symbol Synchronization System for OFDM Signals: Architecture and Simulation Results, IEEE Trans. Veh. Technol., 49, 254–275, 2000.

Schenk, T. and van Zelst, A.: Frequency Sychronization for MIMO OFDM wireless LAN systems, in Proc. of IEEE Veh. Technol. Conf., 509–513, Florida, USA, 2003.

Schmidl, T. and Cox, D.: Robust frequency and timing synchronization for OFDM, IEEE Trans. Comm., 45, 1613–1621, 1997.

Suehiro, N. and Hatori, M.: Modulatable Orthogonal Sequences and their Application to SSMA Systems, Trans. on Inform. Theory, 34, 93–100, 1998.

van de Beek, J., Sandell, M., and Börjesson, P.: ML estimation of time and frequency offset in OFDM systems, IEEE Trans. Signal Proc., 45, 1800–1805, 1997.

25

Sievenpiper HIS and its influence on antenna correlation

S. K. Hampel, O. Schmitz, and I. Rolfes

Leibniz Universität Hannover, Institut für Hochfrequenztechnik und Funksysteme, Appelstr. 9A, 30167 Hannover, Germany

Abstract. This paper deals with the influence of artificial magnetic conductors (AMC), so-called Sievenpiper High Impedance Surfaces (HIS), on the MIMO and Diversity performance of a planar linear-polarized 2×2 dipole array in the ISM-band at 2.45 GHz. The characteristic performance criteria such as envelope correlation coefficient, spectral efficiency, Mean Effective Gain (MEG) and Diversity gain of a coupled 2×2 dipole array are investigated. By means of full-wave electromagnetic analysis as well as Monte-Carlo simulations applying statistical channel models the characteristic antenna pattern just as the MIMO and Diversity analysis is performed, respectively. The obtained results show that the application of Sievenpiper High Impedance Surfaces to planar antenna arrays enables good MIMO and Diversity performance compared to ideal configurations in free-space while offering the design of low profile antennas with simultaneously enhanced characteristics.

1 Introduction

With nowadays wireless communication techniques and the upcoming new standards like IEEE 802.11n (MIMO WLAN), UWB or LTE it is obvious that a huge demand for higher data rates still exists. Concerning Edholm's Law the border of data transmission with 1 Gbit/s will be reached in the near future. On the one hand modulation schemes and coding algorithms in the classical way cannot achieve much more spectral efficiency. On the other hand bandwidth as well as transmit power are regulated. Consequently new concepts in terms of MIMO and Diversity have to be investigated in order to reach the goal of high data rates and reliable transmission links.

Correspondence to: S. K. Hampel
(hampel@hft.uni-hannover.de)

Furthermore, Moores Law indicates higher integration levels with enhanced transistor technology. This in turn paves the way for increasing functionality with constant chip area consumption or decreasing die sizes while keeping functionality constant.

These two aspects for performance enhancement of wireless communication systems call for appropriate low profile antenna designs with MIMO and Diversity capability. Therefore, this paper deals with the impact of so-called Sievenpiper HIS (see Sievenpiper, 1999) on antenna correlation, which is a critical factor when considering the MIMO and Diversity performance of multi-element-antenna setups (MEA). The Sievenpiper HIS themselves exhibit the possibility of designing low profile antennas with high efficiency while using constructive interference mechanisms based on the characteristic behavior of so-called artificial magnetic conductors (AMC). Furthermore the shielding effect of this spatial filter prevents electromagnetic interactions with biological systems (user) or circuitry in the handheld. This eases the perturbation occuring in the communication terminal while conserving the radiation efficiency of the antenna as well as shielding the environment against unwanted radiation.

Therefore the main goal is to investigate the influence of these HIS on the correlation properties affected by the antenna coupling as well as the consequences considering the spectral efficiency of a 2×2 MIMO system and its Diversity gain and Mean Effective Gain (MEG) mainly controlled by the gain pattern of the antenna setup.

The organization of the paper is as follows. In Sect. 2 a brief description of the underlying theory concerning Sievenpiper HIS is presented. Sections 3 and 4 introduce the Diversity and MIMO analysis with its characteristics in general terms. Initial and detailed results regarding the antenna characterization as well as Diversity and MIMO performance based on stochastic envelope processes for the spatial profile of the incident and radiated waves at the receive and the

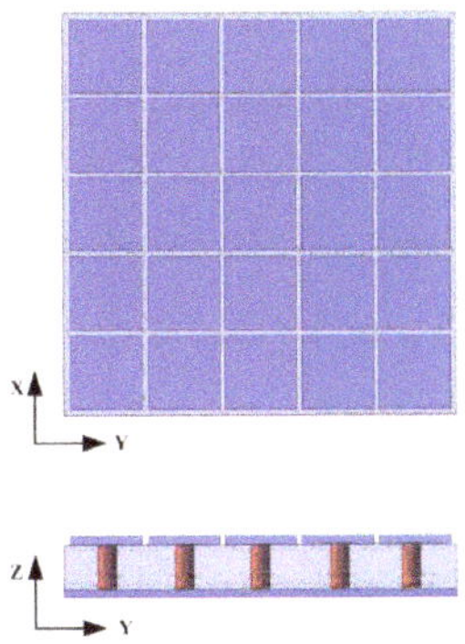

Fig. 1. Top and side view of a Sievenpiper HIS with square patches placed in the xy-plane.

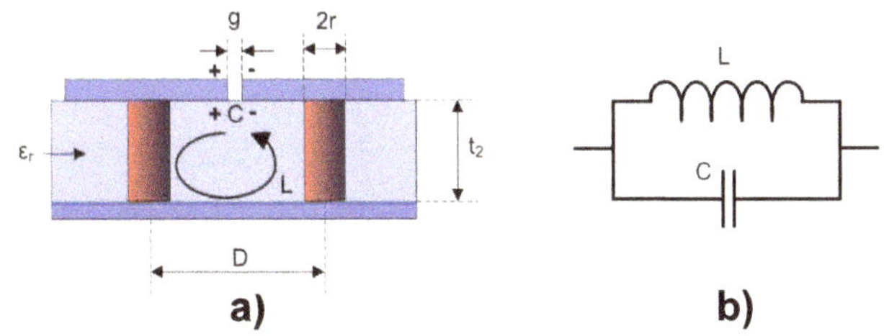

Fig. 2. **(a)** Side view of two unit cells. **(b)** Equivalent parallel resonant circuit.

transmitter are given in Sect. 5.

2 Basics of Sievenpiper HIS

The Sievenpiper HIS can be regarded as composite material consisting of a frequency selective surface (FSS) on a conducter backed dielectric substrate with periodically embedded cylindrical vias as shown in Fig. 1.

Its underlying theory is in general based on the effective medium theory (see Merill, 1999) and moreover on the transmission line theory. Assuming that the physical sizes of all geometry parameters are much smaller than the effective wavelength corresponding to the frequency of operation, i.e. $t_2, r, D, g \ll \lambda_{\mathrm{eff}}$ (see Fig. 2a), a description of the electromagnetic properties of the Sievenpiper HIS by the use of the above mentioned transmission line theory is applicable. Therefore, only a fraction of the whole structure, a so-called unit cell (see Fig. 2a), is sufficient to describe the behavior of the complete periodic two dimensional filter with the periodicity D. The metallic FSS on top of the grounded dielectric host substrate with permittivity ϵ_{r} forms a parallel resonant circuit as depicted in Fig. 2b. Here the capacitive part of the surface impedance is provided by the FSS grid whereas the inductive surface impedance is a property of the substrate (see Clavijo, 2003). The effective capacitance C_{FSS} and effective inductance L_{Sub} forming the parallel resonant circuit can be expressed as follows:

$$C_{\mathrm{FSS}} = \epsilon_{\mathrm{eff}}\epsilon_0 \frac{2D}{\pi} \ln\left(\frac{2D}{\pi g}\right) \qquad (1)$$
$$L_{\mathrm{Sub}} = \mu_0 \mu_r t_2$$

For simplicity in a first approximation polarization as well as angular dependency are dropped. From Eq. (1) it can be obtained that the reactive characteristics of the FSS layer and the host substrate are mainly related to the gap-width g, the periodicity D, the effective dielectric constant ϵ_{eff} and the substrate height t_2, respectively. At resonant frequency of the resulting parallel resonant circuit, $\omega_0 = 1/\sqrt{L_{\mathrm{Sub}} C_{\mathrm{FSS}}}$, the extremely high surface impedance of the structure leads to a vanishing magnetic field (AMC). This yields to a reflection coefficient with a magnitude of $\Gamma=1$ and a phase of $\phi_\Gamma=0°$. The operational bandwidth of the in-phase reflection is determined by the corner frequencies with $\phi_\Gamma=\pm\pi/2$. These borders arise from the assumption of a radiator placed infinitely close above the surface of the HIS. This will result in more constructive than destructive interference of a directly radiated and reflected wave in the upper hemisphere within these bounds.

The surface wave suppression phenomenom of the Sievenpiper HIS leading to electromagnetic band-gap structures (EBG) will be omitted here. For the interested reader we refer to Clavijo (2003) and Goussetis (2006) for more information about the EBG behavior of the Sievenpiper HIS.

3 Diversity analysis

The intention of applying Diversity with multiple receive and/or transmit antennas in wireless communication systems is to overcome the drawbacks like fading and depolarisation comprising the signal transmitted over the wireless channel (see Jakes, 1974). Therefore different Diversity techniques like spatial-, pattern- or polarization-Diversity can be implemented in a transceiver to reduce these signal impairments with the benefit of either increasing the transmission link reliability or the effective data rate both by increasing the available SNR at the receiver. An important characteristic influencing the Diversity performance is the complex correlation coefficient of incoming signals resulting from signal correlation within the channel and the antenna correlation between each element of a MEA system. The correlation influences the achievable Diversity gain, which is a measure for the reduction of the channel imposed data link impairments. The complex correlation coefficient between antenna element i and j of a receiver is given in Eq. (2). Here R and σ represent the covariance term and the standard deviation, respectively.

$$\rho_{i,j} = \frac{R_{i,j}}{\sqrt{\sigma_i^2 \sigma_j^2}}, \qquad (2)$$

The definition of the covariance $R_{i,j}$ in Eq. (2) is given by Fujimoto (2001) and shown in Eq. (3).In this expression

$C_{\vartheta_i,\varphi_i}(\vartheta,\varphi)$ denotes the far field pattern of the specific antenna i and j, respectively, decomposed into its two polarization components in elevation ϑ and azimuth φ corresponding to a spherical coordinate system. Moreover * indicates the conjugate complex, $\boldsymbol{k}$ is the wavevector and r_{ij} is the distance between the indicated antenna elements. XPR is called cross polarization ratio and is a channel property which indicates the conversion of a linear-polarized radiated wave in its corresponding orthogonal component. The two dimensional probability distribution $p_{\vartheta,\varphi(\vartheta,\varphi)}$ is corresponding to the anlge of arrival (AOA) and angle of departure (AOD) behaviour of the incoming and radiated waves, respectively, in a statistical channel model scenario. Here AOA and AOD are given by a Laplacian distribution in azimuth and a Gaussian distribution in elevation (see Waldschmidt, 2004).

$$\begin{aligned} \mathrm{R_{i,j}} = \int_0^{2\pi}\int_0^{\pi} & \Big[C_{\vartheta_i}(\vartheta,\varphi)\cdot C^*_{\vartheta_j}(\vartheta,\varphi) + \cdots \\ & \cdots + \mathrm{XPR}\cdot C_{\varphi_i}(\vartheta,\varphi)\cdot C^*_{\varphi_j}(\vartheta,\varphi)\Big] \times \cdots \\ & \cdots \times p_{\vartheta,\varphi}(\vartheta,\varphi)\cdot e^{\mathrm{jkr_{ij}}}\sin(\vartheta)d\vartheta d\varphi\,. \end{aligned} \tag{3}$$

In case of a fading environment with Rayleigh distributed signal envelope and equally distributed signal phase the so-called envelope correlation $\rho_{\mathrm{e_{i,j}}}$ can be approximated by

$$\rho_{e_{i,j}} \approx \left|\rho_{i,j}\right|^2. \tag{4}$$

Furthermore the characteristic known as MEG, which denotes the ratio of the mean received power to the mean incident power of an antenna, is the second figure of merrit while evaluating Diversity performance of MEA systems. The MEG can be expressed as (see Fujimoto, 2001)

$$\begin{aligned} \mathrm{MEG}_i = \int_0^{2\pi}\int_0^{\pi} & \left[\frac{\mathrm{XPR}}{1+\mathrm{XPR}}G_{\vartheta,i}(\vartheta,\varphi) + \cdots \right. \\ & \left. \frac{1}{1+\mathrm{XPR}}G_{\varphi,i}(\vartheta,\varphi)\right] p_{\vartheta,\varphi}(\vartheta,\varphi)\sin(\vartheta)d\vartheta d\varphi, \end{aligned} \tag{5}$$

with $G_{\vartheta,\varphi_i(\vartheta,\varphi)}$ the absolute antenna gain pattern in azimuth and elevation of element i.

4 MIMO analysis

Similar to Diversity techniques MIMO systems try to take advantage of the multipath propagation properties of wireless communication channels. Here parallel transmission of data streams over M transmit antennas should lead to a boost in data rate while exploiting the spatio-temporal properties of the channel by an MEA arrangement. The use of N receive antennas leads to a $M\times N$ channel matrix $\mathbf{H}$ with complex coefficients h_{mn} (Rayleigh distributed amplitude and equally distributed phase), mapping the M transmitted signals of the transmit vector $\boldsymbol{x}$ on N received signals of the receive vector $\boldsymbol{y}$, as stated in Eq. (6). The reduction of the $M\cdot N$ channel impulse responses to simple complex coefficients is valid as long as the channel is assumed to be narrowband (see Tse, 2004).

$$\begin{pmatrix} y_1 \\ y_2 \\ \vdots \\ y_m \end{pmatrix} = \begin{pmatrix} h_{11} & h_{12} & \dots & h_{1n} \\ h_{21} & h_{22} & \dots & h_{2n} \\ \vdots & \vdots & \ddots & \vdots \\ h_{m1} & h_{m2} & \cdots & h_{mn} \end{pmatrix} \cdot \begin{pmatrix} x_1 \\ x_2 \\ \vdots \\ x_n \end{pmatrix} \tag{6}$$

Expanding the well known Shannon formula for channel capacity C of a SISO system to MIMO channel capacity C_{MIMO} yields the following equation

$$\begin{aligned} C_{\mathrm{MIMO}} &= \log_2 det\left(\bar{\mathbf{I}} - \frac{P_T}{M\sigma_n^2}HH^H\right) \\ C_{\mathrm{MIMO}} &= \sum_i \log_2\left(1 + \frac{P_T}{M\sigma_n^2}\lambda_i\right) \end{aligned} \tag{7}$$

Here $\mathbf{I}$ represents the identity matrix, P_T is the total transmit power equally allocated to the M transmit antennas, σ_n^2 is the noise power of the assumed white gaussian noise affecting the received signal, H^H represents the hermitian of the channel matrix $\mathbf{H}$ and λ_i is the i-th eigenvalue of the underlying eigenvalue problem. The maximum number of useable subchannels is limited to the maximum number of eigenvalues ($\min(M,N)$). The rank of the matrix and the strength of the eigenvalues λ_i as well as the number of useable subchannels depends on the signal correlation introduced in the former section. The higher the correlation the lower the distict subchannel capacity which results in a waste of transmit power allocated to this subchannel. Likewise to the afore presented Diversity analysis the success and benefit of MIMO transmission is bound to the strength of signal correlation, which means: low correlation leads to a maximum in performance, here MIMO channel capacity.

5 Simulation results

In this section a coupled MEA setup with HIS will be evaluated in comparison to its equivalent coupled free-space model (FS) with respect to its Diversity and MIMO performance. Based on full-wave electromagnetic simulation and Monte-Carlo analysis, the specific antenna characteristics, like input-matching and gain pattern as well as the antenna correlation, the Diversity gain, the MEG and the channel capacity will be presented, respectively.

5.1 Array setup and input matching

A picture of the investigated setup is shown in Fig. 3. A coupled 2×2 planar dipole array with element spacing d is placed Δh=4 mm above a customized HIS. The dipoles and HIS frequency of operation is defined to be f_0=2.45 GHz in FS. Therefore the distance between dipoles and HIS is approximately $\Delta h=\lambda_0/30$ with λ_0 being the corresponding wavelength to f_0. The input return loss and the resulting

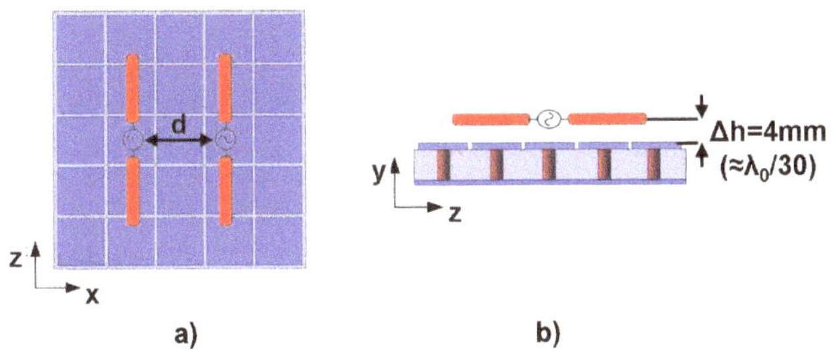

Fig. 3. **(a)** Top and **(b)** side view of the array configuration with HIS.

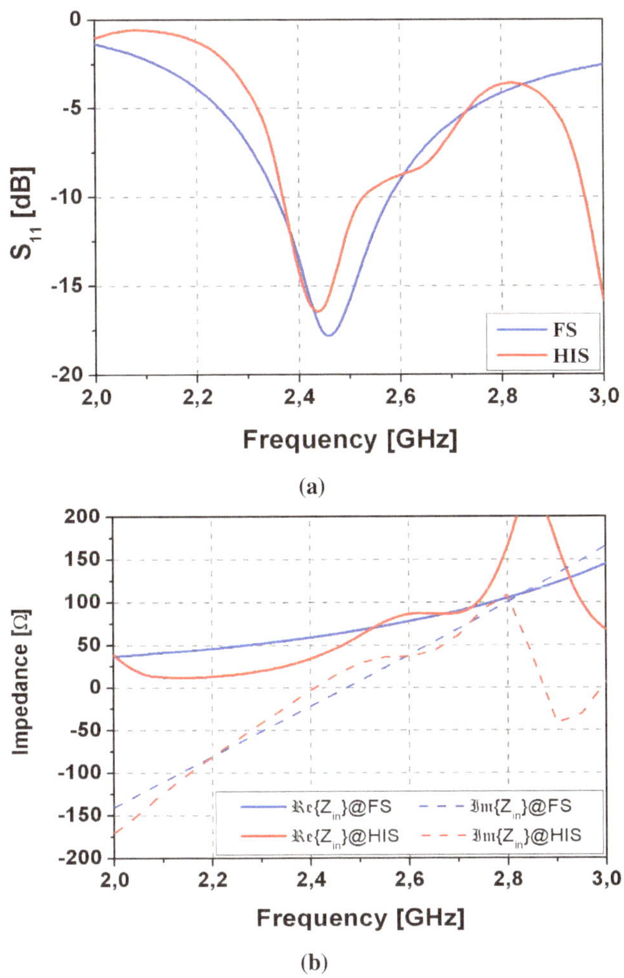

Fig. 4. Input matching **(a)** and resulting input impedance **(b)** over frequency of one dipole above HIS (red) and in FS (blue).

input impedance of one dipole is depicted in Fig. 4 and compared to an equivalent free-space dipole resonating at 2.45 GHz to underline the functionality of the HIS. For the HIS setup a good input match with a reflection coefficient S_{11} less then -15 dB and a purely real input impedance of $Re\{Z_{in}\}\approx 40\ \Omega$ can be observed.

Table 1. Statistical parameter of applied channel scenarios.

	Statistical parameter			
	m_ϑ	σ_ϑ	m_φ	σ_φ
CS1	90°	10°	60°	20°
CS2	90°	10°	90°	20°

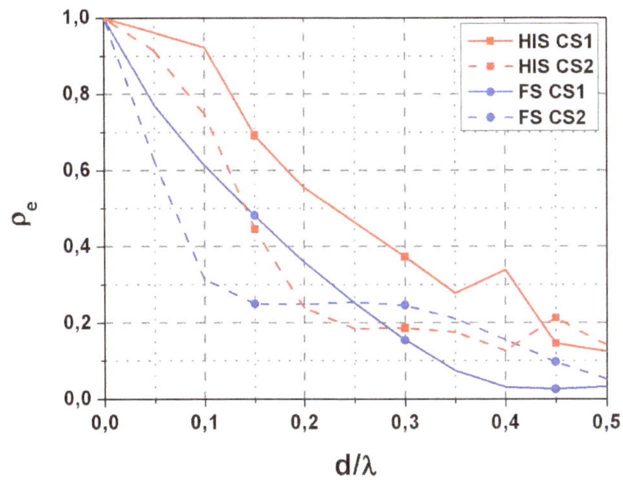

Fig. 5. Envelope correlation ρ_e over interelement spacing d of dipole array above HIS (red) and in FS (blue) for CS1 (solid) and CS2 (dashed).

5.2 Correlation properties and MIMO performance

Since an adequate functionality of the dipole in conjunction with the HIS is proven, the antenna correlation of both setups will be investigated while varying the interelement spacing d between both antennas from $d=0$ to $d=\lambda/2$. Furthermore two different channel scenarios (CS) will be used during the Monte Carlo simulation with the statistical parameters in Table 1 while the XPR was kept constant with 0 dB. Figure 5 shows the resulting curves for the envelope correlation over d in both CS for both arrangements while the SNR in Eq. (7) was set to 20 dB. It can be seen that, with increasing interelement spacing, the correlation relaxes. Furthermore the correlation for CS2 in main beam direction reveals a decrease in correlation as well (for HIS and FS). This behaviour is well known in literature and was expected due to the decorrelating influence of the exponential term in Eq. (3). Moreover, the setup with HIS offers higher correlation than the FS array except for the case of $d/\lambda\geq=0.25$ in CS2. The higher correlation of the HIS setup in contrast to the FS arrangement can be explained while considering the radiation pattern for a fixed spacing (here $d/\lambda=0.25$) shown in Fig. 6. As can be seen by the red lined pattern the constructive interference due to the reflection of the radiated wave on the HIS surface leads to a loss in pattern Diversity with respect to the FS pattern. Here

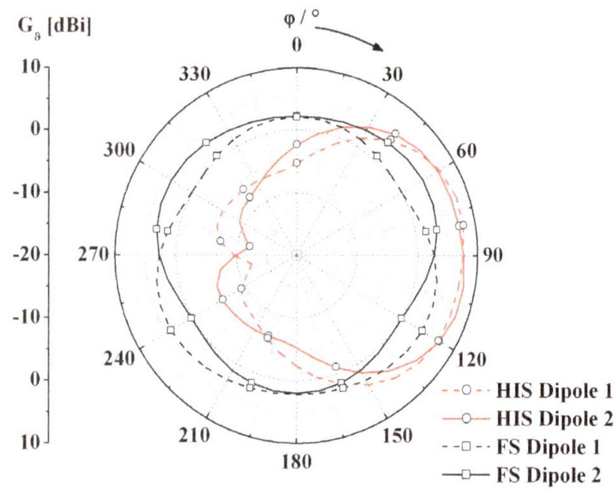

Fig. 6. Gain pattern for interelement spacing $d=\lambda/4$ for each dipole of both dipole array arrangements: HIS (red) and FS (black).

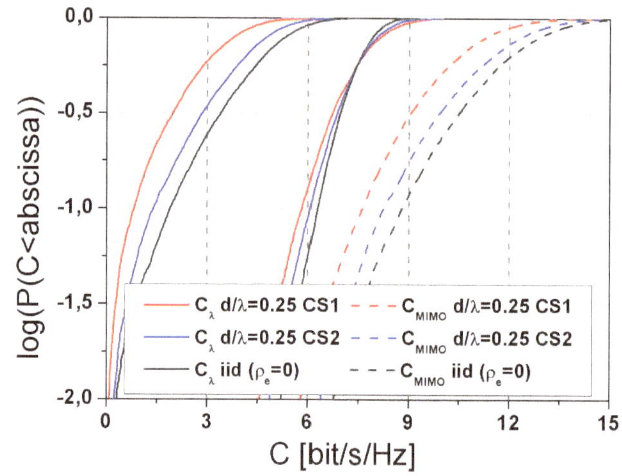

Fig. 7. CDF of subchannel capacity (C_λ) and total capacity (C_{MIMO}) for the dipole array with HIS for $d=\lambda/4$ in CS1 (red) and CS2 (blue) compared with uncorrelated Rayleigh process (iid, grey).

Table 2. Envelope correlation and 10%-outage capacity for the dipole array with HIS for $d=\lambda/4$ in CS1&CS2 compared to an uncorrelated Rayleigh process iid.

	ρ_e	$C_{10\%\mathrm{out}}$ [bit/s/Hz]	% of $C_{10\%out,max}$
CS1	0.46	7.61	86.5
CS2	0.18	8.24	98.4
iid	0	8.8	100

Table 3. MEG and MEG-ratio for both dipoles in both arrangements for $d=\lambda/4$ and CS1.

	MEG_1 [dBi]	MEG_2 [dBi]	R_{MEG} [dB]
HIS	3.35	3.55	0.2
FS	0.46	−2.4	2.86

the difference in the shape of the two radiating antennas is more distinct which yields to less correlation. Furthermore, the differences in pattern shape over the azimuth angle φ in the case of the FS array results in less sensitivity to a change in mean of AOA.

The estimated MIMO channel capacity in terms of the cumulative distributed function (CDF) and the so-called 10%-outage capacity $C_{10\%\mathrm{out}}$ for $d/\lambda=0.25$ in both CS is given in Fig. 7 and Table 2, respectively. Here only the dipoles with HIS filter are considered due to the fact that the results for the FS array offer approximately the same correlation in both CS as the HIS arrangement for CS2. Furthermore, the Rayleigh-curve for identically independently distributed (iid) channel coefficients without any correlation is given, which can be regarded as the maximum achievable capacity for a 2×2 array.

The CDF can be interpreted as probability of the capacity to be higher than the corresponding abscissa value. The analysis of these results reveals the expected worse performance of the HIS setup due to higher correlation. Nevertheless, these results are not as bad as it seems, taking into account that only normalized patterns ($C_{\vartheta,\varphi,i(\vartheta,\varphi)}$) have been used for the estimation of the correlation coefficient and the enhanced antenna gain and the associated benefits were dropped during this analysis. This aspect will be investigated in the next subsection dealing with the achieveable Diversity performance. Due to this lack of gain performance, values of $C_{10\%\mathrm{out}}>86\%$ of $C_{10\%\mathrm{out,max}}$ can be considered as adequate result.

5.3 Diversity performance

For the purpose of comparison and conclusiveness the following results dealing with the Diversity performance of the dipole array were carried out for the same antenna spacing as for the channel capacity in the preceding subsection, $d/\lambda=0.25$. Furthermore, the analysis is limited to CS1 where the poor MIMO performance was realized, while here Diversity offers a promising alternative for performance enhancement of a supposed data transmission link. This is straightforward because of the afore mentioned discrepancy of dropping the absolute gain value while processing the MIMO capacity. The results for the distinct MEG of each antenna as well as the MEG-ratio, based on the gain pattern depicted in Fig. 6 with a maximum gain in main beam direction of the distinct arrangements of $G_{i,\mathrm{HIS}}=8$ dB and $G_{i,\mathrm{FS}}=4$ dB for HIS and FS, respectively, are summarized in Table 3.

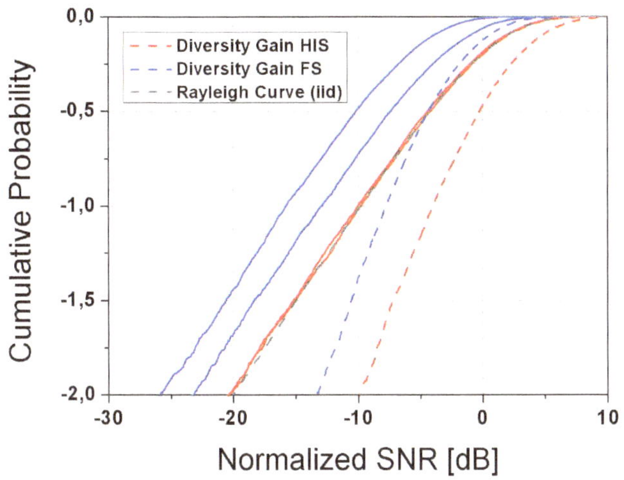

Fig. 8. CDF of available SNR for both dipole arrays, HIS (red) and FS (blue), for $d=\lambda/4$ in CS1 compared to an uncorrelated Rayleigh process (iid, grey).

Due to the higher gain of the MEA setup in conjunction with the HIS in the proposed primary radiation hemisphere ($0°\leq\varphi\leq180°$) compared to the FS array, the increase of the MEG of about 3 dB and 6 dB is obvious keeping in mind the depency of the MEG on the antenna gain (see Eq. 5).

A variation of the XPR is not appropriate within this investigation, due to the fact, that the power imbalance is only interesting in case of polarization Diversity, which is not applicable when using two parallel oriented linearly polarized antennas with vanishing cross-polarized component. In Fig. 8 the CDF of the normalized SNR is depicted applying maximum ratio combining as Diversity technique. It can be seen that both arrangements achieve a good performance concerning the Diversity gain with G_{HIS}=5.88 dB and G_{HIS}=7.86 dB for the HIS and FS configuration, respectively, in case of a given outage probability level of 1%. More important is the additional shift in available SNR of approximately 4 dB when applying the HIS. This can be ascribed to the increased MEG. This consequently leads to the conclusion that the dipole array with HIS performs better compared to the reference FS setup in case of spatial directivity of the wireless channel even though the envelope correlation is higher.

6 Conclusions

In this article the influence of Sievenpiper HIS on the correlation properties of a 2×2 array consisting of planar dipoles was investigated. The obtained results concerning envelope correlation over interelement spacing, MIMO channel capacity, MEG and Diversity gain have been compared to a reference array in FS. Here two different spatial channel scenarios of incident waves, CS1 and CS2, were regarded to evaluate the array performance. Although the correlation properties of the dipoles in conjunction with the HIS were not as good as the correlation of the FS setup, the MIMO capacity reveals good results anyhow. Furthermore the Diverstity performance of the HIS setup, given by the MEG and the Diversity gain, outperforms the FS array due to a huge increase in antenna gain based on constructive interference of radiated and reflected waves. Therefore when considering MIMO and Diversity systems the use of conventional and well known planar antenna concepts in conjunction with Sievenpiper HIS is applicable. In addition to the enhanced antenna performance the HIS offers the possibility of designing planar low profile antennas with small geometrical dimensions for integration in wireless transmission terminals instead of conventional planar conductor backed or external vertical monopole antennas.

References

Clavijo, S.: Design Methodology for Sievenpiper High-Impedance Surfaces: An Artificial Magnetic Conductor for Positive Gain Electrically Small Antennas, IEEE Trans. Antennas and Propagation, 51, 2678–2690, 2003.

Fujimoto, K.: Mobile Antenna Systems Handbook, Artech House, Boston, 2001.

Goussetis, G.: Tailoring the AMC and EBG characteristics of periodic metallic arrays printed on grounded dielectric substrate, IEEE Trans. Antennas and Propagation, 54, 82–89, 2006.

Jakes, W. C.: Microwave Mobile Communications, IEEE Press. Inc., New York, 1974.

Merill, W. M.: Effective Medium Theories for Artificial Materials Composed of Multiple Sizes of Spherical Inclusions in a Host Continuum, IEEE Trans. Antennas and Propagation, 47, 142–148, 1999.

Sievenpiper, D. F.: High-Impedance Electromagnetic Surfaces, Dissertation, University of California, Los Angeles, 1999.

Tse, D.: Fundamentals of Wireless Communications, University of California, Berkeley, 2004.

Waldschmidt, C.: Compact Wide-Band Multimode Antennas for MIMO and Diversity, IEEE Trans. Antennas and Propagation, 52, 1963–1969, 2004.

Multidimensional measurement by using 3-D PMD sensors

T. Ringbeck, T. Möller, and B. Hagebeuker

PMDTechnologies GmbH, Am Eichenhang 50, 57076 Siegen, Germany

Abstract. Optical Time-of-Flight measurement gives the possibility to enhance 2-D sensors by adding a third dimension using the PMD principle. Various applications in the automotive (e.g. pedestrian safety), industrial, robotics and multimedia fields require robust three-dimensional data (Schwarte et al., 2000). These applications, however, all have different requirements in terms of resolution, speed, distance and target characteristics. PMDTechnologies has developed 3-D sensors based on standard CMOS processes that can provide an optimized solution for a wide field of applications combined with high integration and cost-effective production. These sensors are realized in various layout formats from single pixel solutions for basic applications to low, middle and high resolution matrices for applications requiring more detailed data. Pixel pitches ranging from 10 micrometer up to a 300 micrometer or larger can be realized and give the opportunity to optimize the sensor chip depending on the application.

One aspect of all optical sensors based on a time-of-flight principle is the necessity of handling background illumination. This can be achieved by various techniques, such as optical filters and active circuits on chip. The sensors' usage of the in-pixel so-called SBI-circuitry (suppression of background illumination) makes it even possible to overcome the effects of bright ambient light. This paper focuses on this technical requirement. In Sect. 2 we will roughly describe the basic operation principle of PMD sensors. The technical challenges related to the system characteristics of an active optical ranging technique are described in Sect. 3, technical solutions and measurement results are then presented in Sect. 4. We finish this work with an overview of actual PMD sensors and their key parameters (Sect. 5) and some concluding remarks in Sect. 6.

Correspondence to: B. Hagebeuker
(b.hagebeuker@PMDTec.com)

1 Basic principle

1.1 Time-Of-Flight basic principle

As the Time-Of-Flight principle has already been described in detail in many technical publications (Kraft et al., 2004; Ringbeck et al., 2003; Buxbaum et al., 2001), only the basic principles will be discussed here.

Figure 1 shows the basic Time-Of-Flight principle. In its most simple form, a light pulse is transmitted by a sender unit and the target distance is measured by determining the turn-around time the pulse needs to travel from the sender to the target and back to the receiver. With knowledge of the speed of light the distance can then easily be calculated. However, the receiver needs to measure with picosecond-accuracy the delay between start and stop, if millimeter-precision is required. To realize such system solutions with discrete components, as is done in today's TOF rangefinders, each component in the signal chain must have a very high system bandwidth.

In contrast Figs. 1 and 2 show the basic working principle of a PMD based range imaging camera. Rather than using a single laser beam (which would have to be scanned over the scene to obtain 3-D) the entire scene is illuminated with modulated light. The advantage of PMD devices is that we can observe this illuminated scene with an intelligent pixel array, where each pixel can individually measure the turnaround time of the modulated light. Typically this is done by using continuous modulation and measuring the phase delay in each pixel. In addition to this robust method of obtaining 3-D without scanning, the realization of the phase measurement in a quasi optical domain offers huge advantages compared to the above mentioned discrete solutions. This is one reason, why we do not require an optical reference channel for most applications.

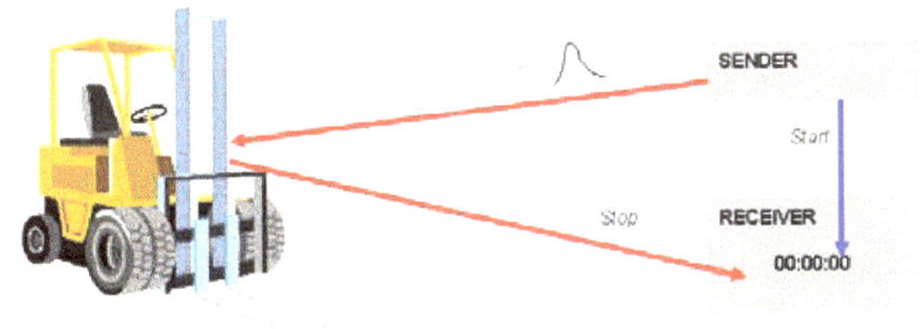

Fig. 1. Time-Of-Flight measurement principle with pulsed light.

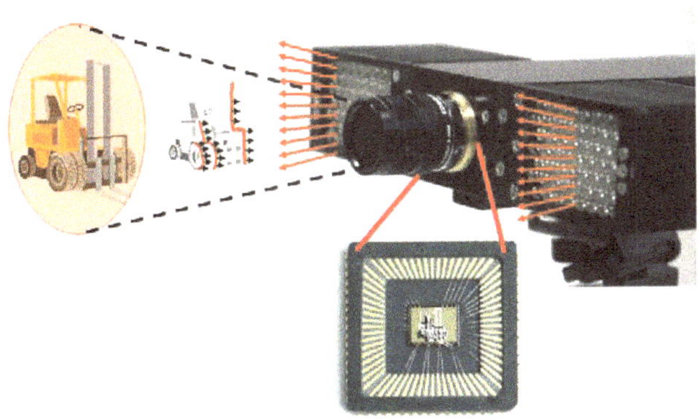

Fig. 2. Imaging Time-Of-Flight, measurement based on PMD.

1.2 Operation principle of PMD

This section describes the principle of a simplified PMD sensor realized in CMOS technology. Because the complete mixing process of the electric and optical signal takes place within each pixel we call the PMD elements "smart pixels". Figure 3a shows an illustration of a single pixel PMD sensor element. It is a five-terminal device with an optical input window, i.e. two transparent modulation electrodes in the middle of the illustration. These light sensitive Photogates are isolated from the substrate by a thin oxide layer. The gates are conductive and transparent for the received light. On the left and the right there are readout diodes which are connected to the pixel readout circuitry. In a PMD pixel the movement of generated charge carriers can be controlled by the reference signal applied to the modulation gates. This way one can influence a charge transport to the left or to the right side. The potential distribution in the surface region is influenced by these push-pull voltages leading to a "dynamic seesaw" for the generated charge carriers (Kraft et al., 2004).

If the incident light is constant and the modulation is a rectangular signal with a duty cycle of 50% the generated charge carriers within a modulation period move to the left and to the right equally. At the end of such a modulation process the output voltages at the readout nodes are the same as those shown in Fig. 3b.

If, however, the incident light is modulated (e.g. also as a rectangular signal), and there is no phase (phase = 0°) delay between modulation of light and detector, then all charge carriers will be moved to one of both readout diodes (Fig. 3c). For other phase delays, the modulation of the light and its phase delay compared to the electrical reference signal results in a difference between the two output voltages. This difference of both output nodes is directly dependent on the phase delay between light and pixel modulation. This data can be used to calculate the distance from a light-reflecting object to the sensor as described in a later section.

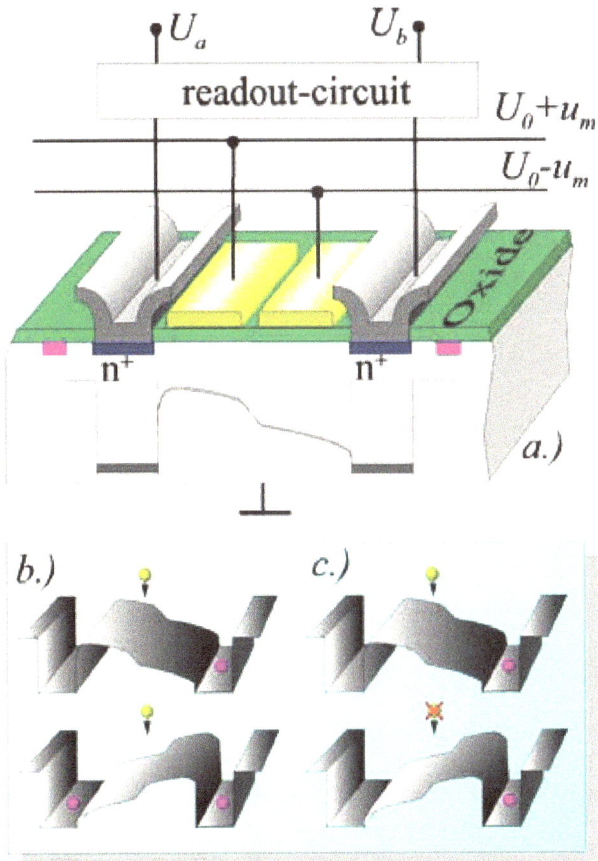

Fig. 3. Operation principle of PG-PMD.

Figure 4 gives an impression of how both readout nodes will behave for different target distances.

1.3 Analysis of the ACF (autocorrelation function)

To calculate the distance between target and camera, the autocorrelation function of electrical and optical signal is analyzed by a phase-shift algorithm (Fig. 5). Using four samples A_1, A_2, A_3 and A_4 – each shifted by 90 degrees – the phase, which is proportional to the distance, can be calculated using the following equation.

$$\varphi = \arctan\left(\frac{A_1 - A_3}{A_2 - A_4}\right) \tag{1}$$

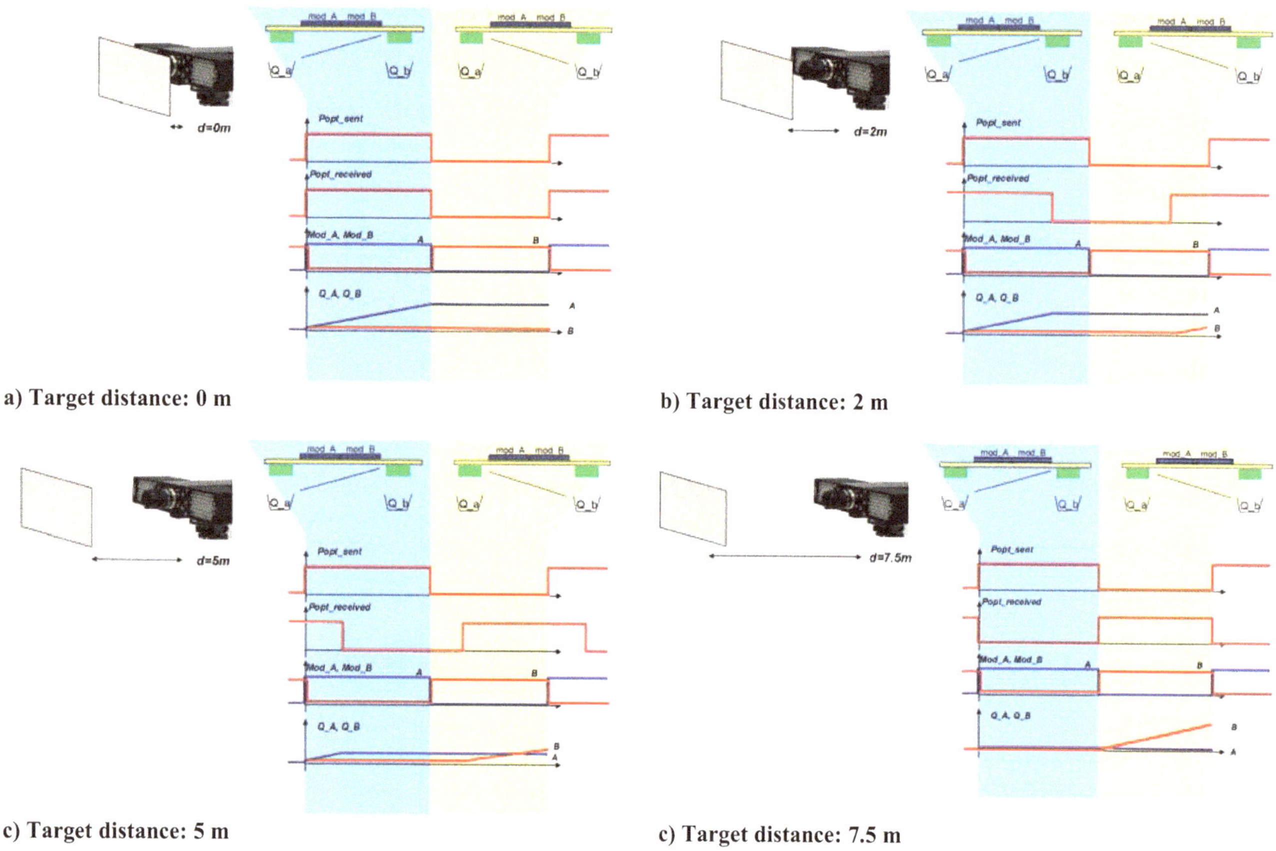

a) Target distance: 0 m

b) Target distance: 2 m

c) Target distance: 5 m

c) Target distance: 7.5 m

Fig. 4. PMD operation with different target distances.

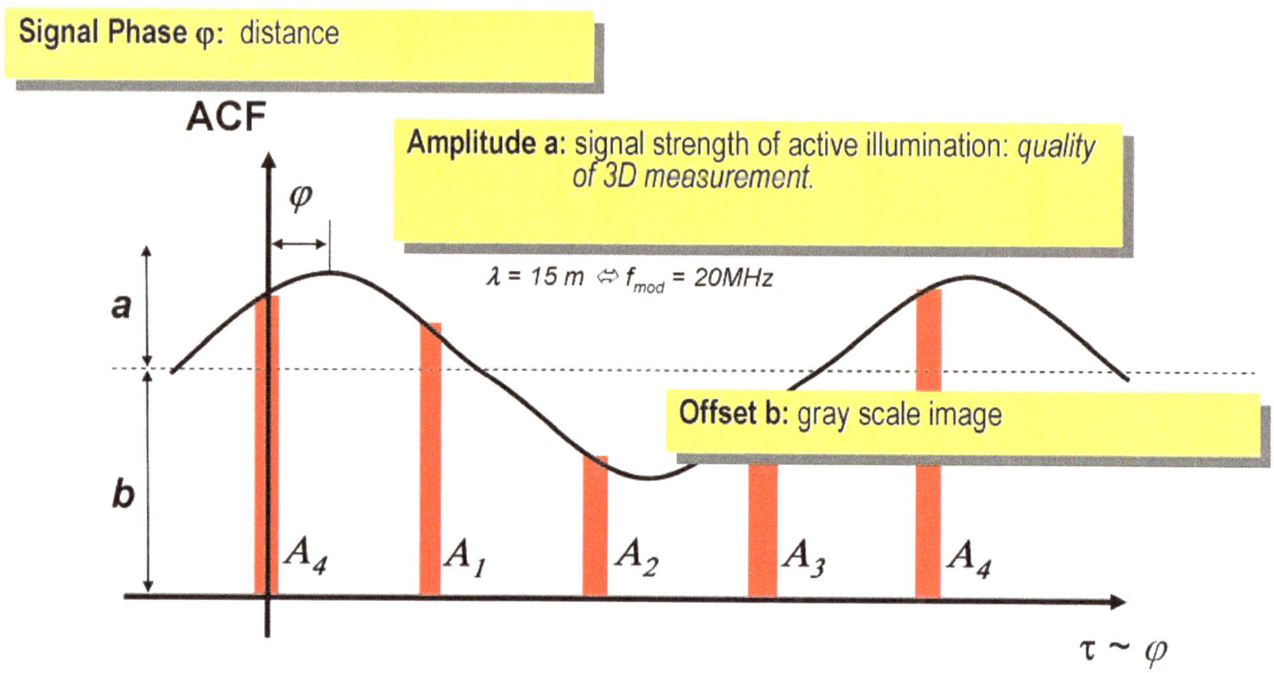

Fig. 5. ACF (autocorrelation function), Signal Phase, Amplitude and Offset.

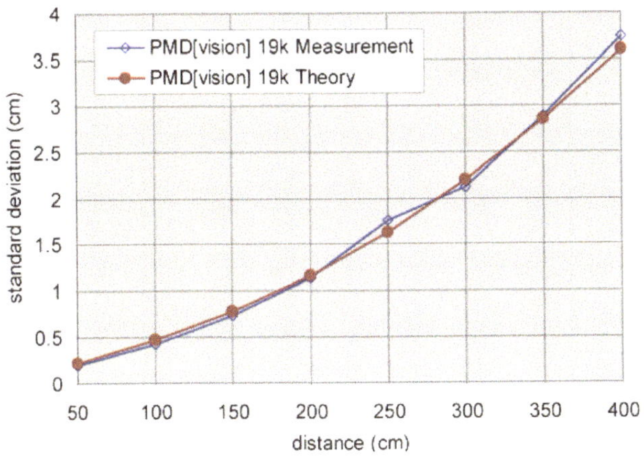

Fig. 6. Calculated standard deviation versus measurement results.

In addition to the phase-shift of the signal, two other values can be extracted. At first, the signal strength of the received signal (amplitude):

$$a = \frac{\sqrt{(A_1 - A_3)^2 + (A_2 - A_4)^2}}{2} \quad (2)$$

And the second value, the offset b of the samples, represents the gray-scale value of each pixel.

$$b = \frac{\sum_1 + \sum_2 + \sum_3 + \sum_4}{4} \quad (3)$$

The distance d to the target is given by Eq. (4). At a modulation frequency of f_{mod}=20 MHz, for example, the wavelength λ_{mod} is 15 m. The maximum distance for the target is $d_{max}=\lambda_{mod}/2$ because the active illumination has to cover the distance twice: from the sender to the target and back to the sensor chip.

$$d = \frac{c \cdot \varphi}{4\pi \cdot f_{mod}} \quad (4)$$

2 Challenges

2.1 Technical requirements: how much light for accurate measurement?

It is obvious that for an active optical measurement system the reliability is directly influenced by the amount of (active) light which arrives at the sensor. In Lange (2000) and Buxbaum (2002) the dependency of statistical measurement uncertainty as a function of optical signal strength and system noise is derived. The measurement uncertainty dR of a PMD time of flight system can be calculated as:

$$dR = \frac{1}{\sqrt{N_{phase}}} \cdot \frac{1}{k_{tot} \cdot \frac{S}{N}} \cdot \frac{\lambda_{mod}}{\sqrt{8} \cdot \pi} \quad (5)$$

Where k_{tot} is the modulation contrast, S is the number of (active) signal electrons, N is the equivalent number of noise electrons (including shot noise of active light, background light and dark current as well as thermal noise, reset noise, system noise and ADC noise), N_{phase} is the number of measurements and λ_{mod} is the wavelength of the modulation signal in meters. An alternative notation of this product of k'_{tot} and S' is fully equivalent, where S' represents the complete number of integrated electrons, including background light and dark current and k'_{tot} represents the overall system contrast, also including dark current and background light offset. Please note that for both cases the product of k_{tot} and S represents the "active signal share" after integration (demodulation amplitude).

The helpful thing is that for each measurement, all parameters of the above equation are known. That means, that each single measurement more or less directly offers the information of its own reliability, without requiring the use of statistics. Subsequently, thresholds can be set to ignore measurement results with great uncertainty.

One can also easily see from the equation above, what modifications of the system and sensor parameters can be made to improve the measurement accuracy. The following section describes the influence of the parameters mentioned above.

High PMD contrast and high modulation contrast of the light sources improves the distance accuracy. Optimizing the contrast doesn't influence the signal-to-noise ratio but rather decreases the inaccuracy by a factor of $1/k_{tot}$. The maximum value of k_{PMD} and k_{light} is 1.

$$k_{tot} = k_{PMD} \cdot k_{light} \quad (6)$$

It is always desirable to lower the total noise without affecting the signal. This can be done by reducing the dark current noise. Therefore special opto-CMOS processes with low dark current should be used. Especially in automotive applications where the temperature can be up to 125°C, dark-current-optimized CMOS processes are essential for satisfactory performance because dark current doubles with every 8°C. Improvements are also achieved by low amplifier- and reset-noise on-chip and low system-noise, as well as decreasing background illumination noise. However, since background illumination is an environmental parameter and not a system parameter, it is difficult to decrease the background illumination in an observed scene. The number of photons reaching the sensor can be reduced though. This is done by proper band pass filtering of the incoming light.

Increasing the signal by using stronger light sources increases the accuracy linearly if the photon shot noise of the uncorrelated light or other noise sources are dominant rather than the photon shot noise of the active illumination. Otherwise, if the photon shot noise N_{active} of the active illumination is the most dominant part of the total noise N, the distance accuracy increases by the square root of the signal.

$$\frac{S}{N} \propto \left\{ \begin{matrix} \sqrt{S}, & \text{if } N_{active} > N \\ S, & \text{else} \end{matrix} \right\} \quad (7)$$

LED		
LED	0.3	nm/K
Temp. Automotive min	-40	°C
Temp. Automotive max	120	°C
Temp. Automotive range	160	K
	48	nm
LED	50	nm
Filter	98	nm
Viewing angle	20	deg (diagonal)
Range	40	m
Image Format	1:1	
edge of image (object space)	10.0	m
area of image (object space)	99	m^2
#LEDS	100	LEDs
optical power per LED	0.04	W/LED
total optical power of LEDs (CW)	4	W
power density in object space (CW)	0.040	W/m^2
power density in object space (CW, filtered)	0.039	W/m^2
Burst-Operation	5	Factor
	20	W
power desnisty in object space (Burst)	0.20	W/m^2
power desnisty in object space (Burst, filtered)	0.19	W/m^2

SUN		
726	W/m^2	power density of sun
83	W/m^2	power density of sun (filtered)

Fig. 7. Calculation sunlight vs. active light.

It must be kept in mind, that using more optical power comes at a price, and the overall costs of a TOF camera system can significantly rise by adding more LED or lasers. To avoid this, burst operation can be implemented. In burst operation, the light sources have the same average power and therefore the same number of light emitting elements, but higher peak power than in continuous operation. The shorter exposure time leads to the same number of signal electrons but less noise electrons due to less accumulated background light electrons. Thus, burst operation is effective under strong background light conditions, such as outdoor applications with sunlight. Another advantage of burst operation is the ability of detecting fast moving objects.

Large lens apertures lead to more photo generated electrons, both information holding electrons from the light source and unwanted ones from background light. In most cases the accuracy can be improved linearly with aperture enlargement since the number of photo generated electrons is a square function of the aperture. In very special cases, when the noise floor due to dark current, quantization etc is dominant, the improvement of accuracy increases with the square of the aperture.

Sensitivity of the PMD sensor can be enhanced by high fill factor and micro lenses. Good reflectivity and close targets will increase the signal, too. However, this is strongly dependent on the application and therefore not an essential part of design.

Another powerful method to improve the measurement accuracy is to use higher modulation frequencies (lower modulation wavelength λ_{mod}). For example doubling the frequency doubles the measurement accuracy. Physical limitations are the bandwidth of the light source and the bandwidth of the sensor. Furthermore the contrast may decrease at high frequencies, which will lower the effectiveness of this method.

Figure 6 shows the comparison between the calculated accuracy and the measured one. For the calculation, the parameters described above were identified for a 160×120 pixel camera system. The measurement was done under well defined environment conditions. A good match of both curves can be observed. Hence, the performance of new applications can be easily estimated.

We can summarize that high modulation contrast of PMD and light source, optimized optical filters, high modulation frequency, good fill factor, large pixels and low noise floor are important design goals for PMD sensors and systems.

2.2 Influence of sunlight

For most outdoor applications sunlight is much brighter than the active illumination. This has two negative impacts on the TOF-system:

- Shot noise is increased (results in distance noise)
- A huge amount of the dynamics of the sensor is occupied by the sunlight (the sensor may saturate)

Figure 7 gives an overview of the calculation of the optical power of 100 LEDs in comparison to the sun for a typical application scenario. In this example the ratio between sun and active light is 18 000=85 db at a distance of 40 m (20° viewing angle). In the following section we will show how this ratio can be improved by intelligent system design. Firstly, the optical input signal is spectrally filtered. Although LEDs in the near infrared have a typical spectral width of 50 nm, thermal drift of LEDs and automotive temperature range requirements must also be taken into account. These facts allow only a medium spectral filter width of approximately 100 nm to be used. In this case, the power density of the sun on the sensor is lowered and the ratio sun/active light goes down to

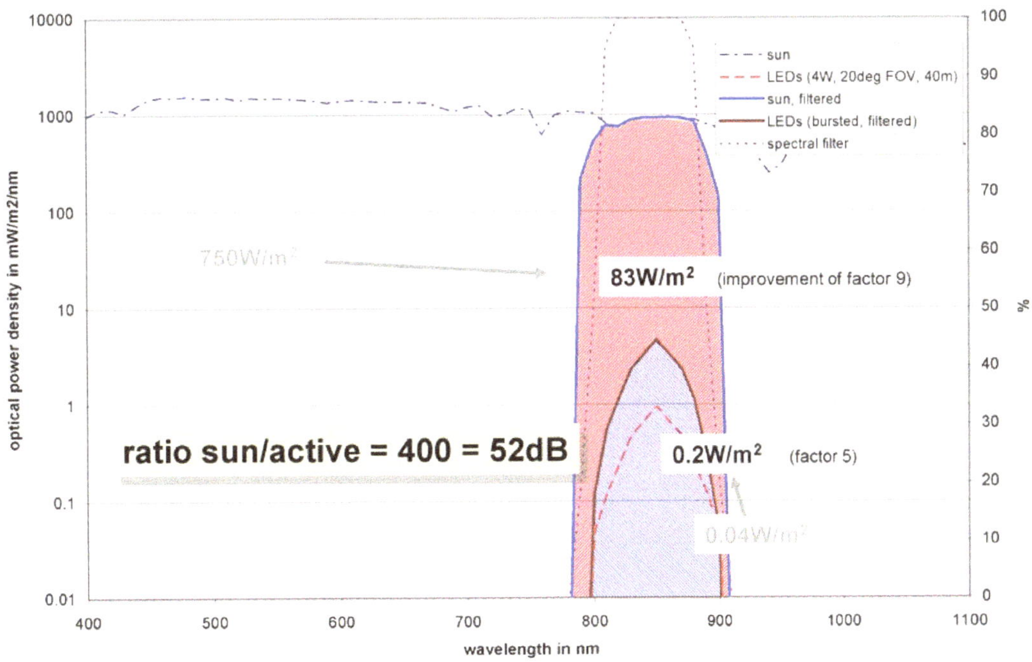

Fig. 8. Optical spectrum of sunlight with spectral filtering and active light in burst-mode.

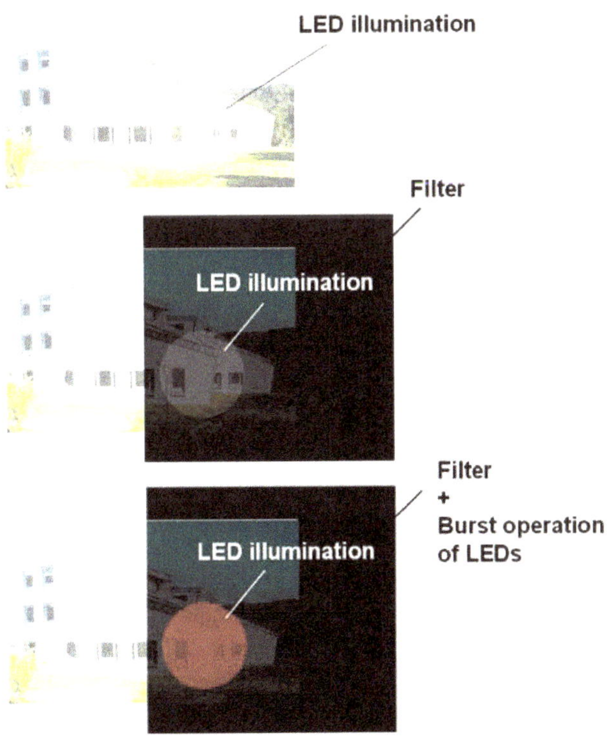

Fig. 9. Usage of optical filters and burst mode.

2000=66 db. Further improvements can be achieved by using a burst mode for the LEDs. A burst operation of factor 5 results in a sunlight/active-light ratio of 400=52 db.

Figure 8 shows the result of both enhancements in a diagram and Fig. 9 shows the effects on an example picture. Spectral filtering and burst operation of the light source are effective methods to improve the "optical" SNR. The reasonable sunlight rejection of these introduced methods is about −33 db. But the remaining ratio of sunlight/active signal is still poor: Even the filtered sun is still a factor of 400 brighter than the pulsed LEDs. Additional sunlight rejection methods have to be found.

The key challenges for optical TOF measurement are sensors with a high bandwidth and high sensitivity and a system with high dynamics to achieve sunlight robustness.

3 Technical solutions

3.1 Definition of contrast

Considering the output channels of the PMD sensor, two values are of importance: the difference of the readout channels delivering the phase information, and the sum of both channels which is proportional to the incident light and the overall photo generated charge carriers. As mentioned above, an important figure of merit for distance measurement is the ratio of the measured signal difference compared to the signal sum, the PMD contrast k_{PMD} .

$$k_{\mathrm{PMD}} = \frac{|Q_A - Q_B|}{Q_A + Q_B} = \frac{|\Delta V_A - \Delta V_B|}{\Delta V_A + \Delta V_B} \tag{8}$$

The contrast has its maximum value when using phase shifts of 0° and 180° in terms of electrical and optical signal.

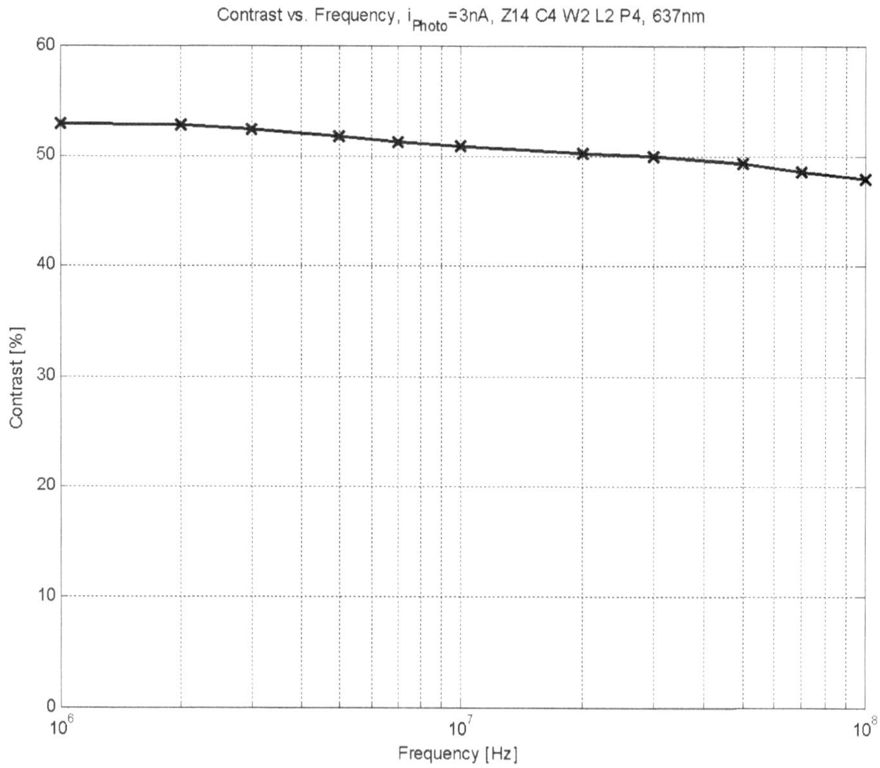

Fig. 10. Exemplary contrast behaviour showing high contrasts combined with high bandwidth.

Depending on modulation waveform, an ideal device can have a maximum contrast of 100%, that is, all photo generated charge carriers contribute to the phase measurement since they are properly distributed to the corresponding readout channels (perfect device and system). Therefore, the contrast reflects the separation ability of the device in terms of charge carriers generated by correlated light. By definition the contrast is significantly affected by the signal sum, therefore determination of the device contrast is only valid without any uncorrelated illumination, and so the measurement must be carried out with a modulated light signal only.

The contrast of a PMD device is mainly dependent on pixel architecture and geometry. The main focus during pixel design is always the active region of the device which is defined by the light sensitive area and the light sensitive modulation gates. Of course, other requirements must also be met, such as overall pixel fill factor and photoresponsivity. Our development has led to PMD pixels with scalable pixel pitch without a loss of performance. The active pixel size can be modified based on application requirements from $10\,\mu m \times 10\,\mu m$ up to $300\,\mu m \times 300\,\mu m$ or even larger. This feature is based upon a special kind of pixel layout – PMD pixels with large dimensions can therefore offer the same performance as small designs. The pixels are optimized for high bandwidth (beyond 100 MHz) to increase pixel performance and accuracy. Figure 10 shows exemplary contrast behaviour of a typical 2-gate PMD pixel. The contrast of 3-gate pixels is typically higher (60%). It can be seen that a high contrast can be attained over several decades of modulation frequency which is mandatory for accurate distance measurements (Eq. 5).

3.2 SBI – Suppression of background illumination

The mentioned optical filtering and bursting of light sources might not be sufficient to reduce disturbing uncorrelated sunlight. The remaining background illumination can still saturate the sensor for typical integration times which can lead to a decrease of accuracy or pixel failures. In addition, there are other causes for uncorrelated disturbances. The unavoidable dark current of a semiconductor device especially, can have a considerable influence particularly at higher temperatures. For example, an increase in temperature of 80 degrees leads to a dark current which can be 1000 times higher and thus would significantly decrease measurement accuracy. Obviously solutions must be found, in order to reduce the disturbing influence of any uncorrelated signal.

The possibility to combine the sensor with on-chip application-specific circuitry is fundamental for an enhanced smart optical sensor. To improve the dynamic range of the PMD device, additional circuitry for the instantaneous suppression of background illumination (SBI) has been introduced. Using such a special readout technique allows precise

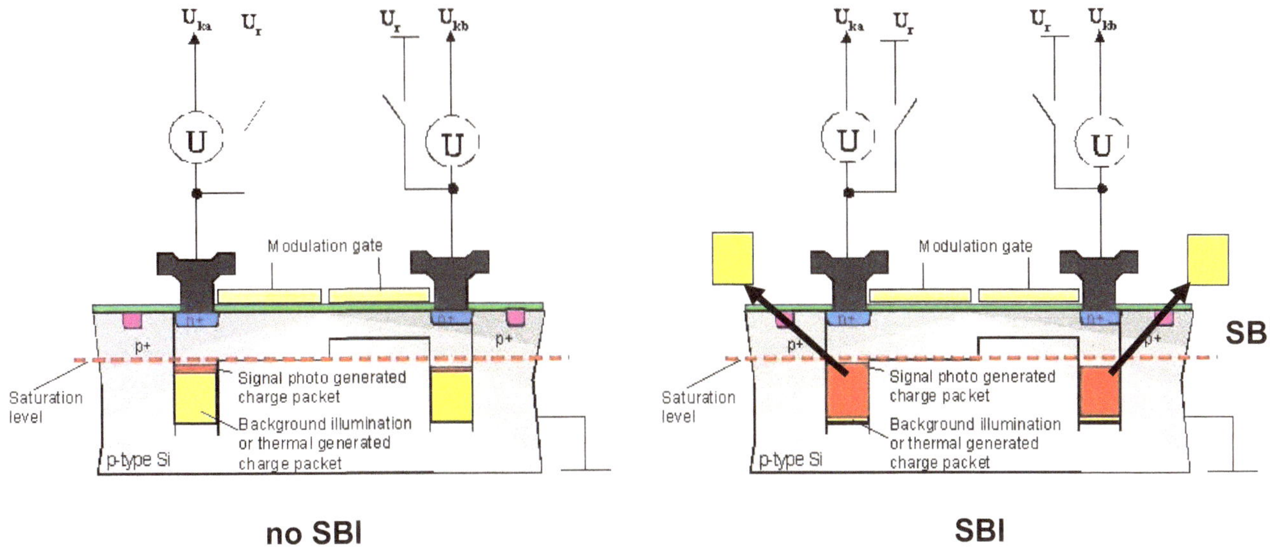

Fig. 11. Working Principle of PMDTec's patented SBI circuitry.

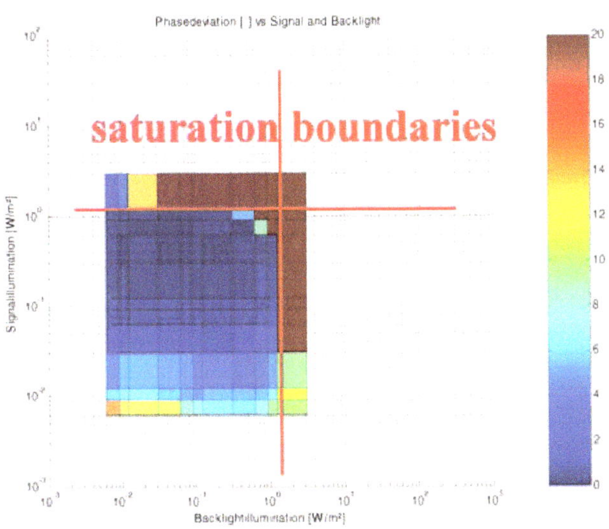

Fig. 12. Dynamic Measurements (SP, no SBI).

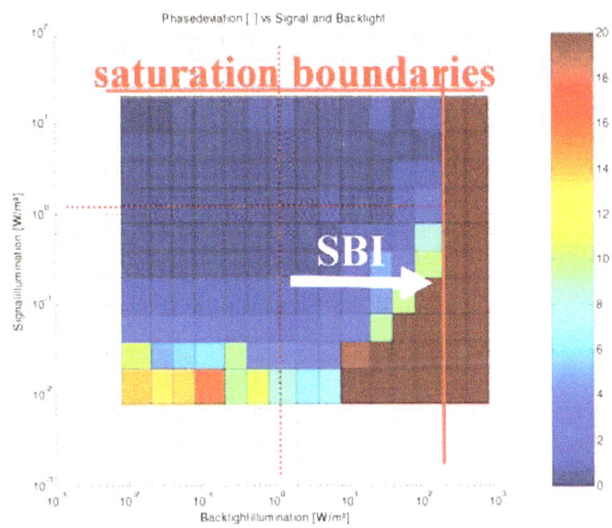

Fig. 13. Dynamic Measurements (SP, SBI).

3-D measurements even in environments with huge uncorrelated signals at high temperatures. Therefore this principle is a key feature for a number of range imaging applications where the existence of disturbing uncorrelated signals is unavoidable. The working principle of PMDs patented SBI circuitry is illustrated in the following figure:

To eliminate uncorrelated signals which are distributed on both readout sides equally, additional compensation sources have been added. These sources inject additional charges in both readout terminals in order to instantaneously compensate for the saturation effects of uncorrelated signals during the integration process. This compensation means that the dynamic range of the following readout stage can be completely used by the correlated signal information only. In other words, the readout circuitry of the pixel ideally sees only the correlated signal and no unwanted background DC component. To offer a wide range of applications and flexibility this concept has been implemented in most of our PG-PMD arrays (Fig. 16). An active SBI circuitry will reduce the background DC component, and so the scenes ambient light greyscale information will be lost. Alternatively a similar greyscale image can also be extracted out of the acquired amplitude data (active light greyscale image), even with active SBI. Also the SBI feature can be electronically deactivated or adjusted so that it is always possible to acquire both high-quality 3-D measurement and 2-D greyscale information, depending on the application needs.

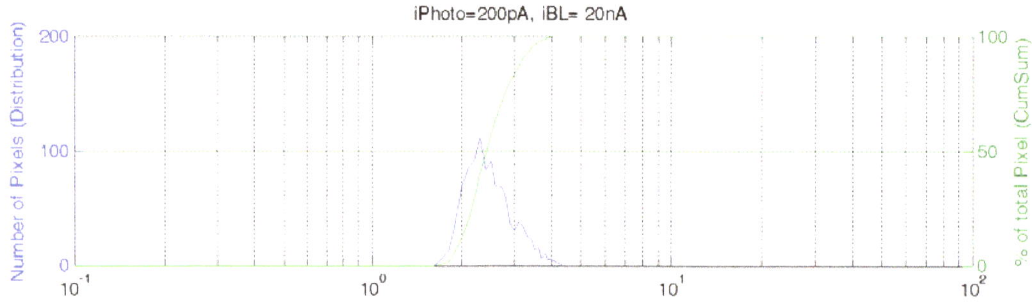

Fig. 14. Statistic analysis (64×16 Matrix, SBI).

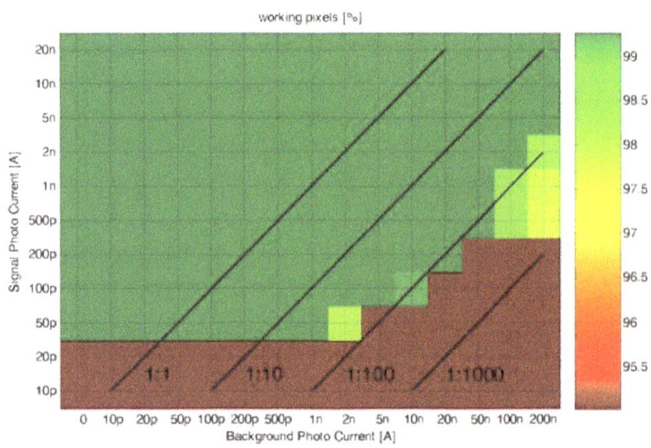

Fig. 15. Dynamic range (64×16 Matrix, SBI).

PhotonICs® PMD 768-S
3D Video Sensor Array with Active SBI
- Array format 24(H) x 32(V)
- Optical format 1/2'', 16:9 image ratio
- Effective image area 6.75mm (H) x 3.73mm (V)
- Package Standard CLCC68
- Pixel size 210.8μm (H) x 155.3μm (V)
- Fill factor (FF) 42%
- 78dB dynamic range for active illumination
- 50dB dynamic range increase by SBI circuitry

PhotonICs® PMD 1k-S 2
3D Video Sensor Array with Active SBI
- Array format 64(H) x 16(V)
- Optical format 2/3'', 3:1 image ratio
- Effective image area 9.94mm (H) x 3.37mm (V)
- Package Standard CLCC68
- Pixel size 155.3μm (H) x 210.8μm (V)
- Fill factor (FF) 42%
- 78dB dynamic range for active illumination
- 50dB dynamic range increase by SBI circuitry

PhotonICs® PMD 3k-S
3D Video Sensor Array with Active SBI
- Array format 64(H) x 48(V)
- Optical format 1/2'', 4:3 image ratio
- Effective image area 6.4mm (H) x 4.8mm (V)
- Package Standard CLCC68
- Pixel size 100μm (H) x 100μm (V)
- Fill factor (FF) 32%
- 74dB dynamic range for active illumination
- 50dB dynamic range increase by SBI circuitry

PhotonICs® PMD 19k
High Resolution 3D Video Sensor Array
- Array format 160(H) x 120(V)
- Optical format 1/2'', 4:3 image ratio
- Effective image area 6.4mm (H) x 4.8mm (V)
- Package Standard CLCC68
- Pixel size 40μm (H) x 40μm (V)
- Fill factor (FF) 31%
- 65dB dynamic range for active illumination

Fig. 16. PMD Sensor-Chips based on Photogate Technology.

3.3 Measurements

To measure the functionality of PMD sensors with SBI circuitry, a phase measurement procedure must be used. This is due to the fact that the total number of collected and stored electrons is modified by the usage of an active SBI, and so the standard definition of contrast (Eq. 8) can no longer be used as characteristic device performance parameter. To examine the achieved dynamic range (with regard to tolerable signal and background intensity levels), a real (Time-Of-Flight) phase measurement using a 4-phase algorithm is performed under controlled conditions (Eq. 9).

$$\begin{aligned} \varphi_{\mathrm{TOF}} &= \arctan\left(-\frac{\sum_i u_{\mathrm{Int}}(\varphi_i)\cdot\sin(\varphi_i)}{\sum_i u_{\mathrm{Int}}(\varphi_i)\cdot\cos(\varphi_i)}\right) \\ \varphi_i &= 0,\, \tfrac{1}{2}\pi,\, \pi,\, \tfrac{3}{2}\pi \end{aligned} \tag{9}$$

Using an automated measuring station, the intensities of signal and background light required for measurement can be adjusted very reproducibly in our optical laboratory. To verify reliable phase measurement using subsequent statistic analysis, a high number of measurements are made for several combinations of signal and background intensity.

Figure 12 shows the resulting standard deviation as a function of signal and background intensity for a single PMD pixel with disabled SBI. The overlaid saturation boundaries indicate the achieved dynamic range, limited by saturation of integration capacitors. In contrast, Fig. 13 shows the increased dynamic range with enabled SBI circuitry. The dynamic range of a typical pixel can be increased by a factor of 300 – that is, 50 dB. Since quality of SBI performance may vary from pixel to pixel due to technical fabrication tolerances, each single pixel may deliver a slightly different dynamic range. Please note, that there are always pixels in the array with far higher SBI performance (factor 1000 and even

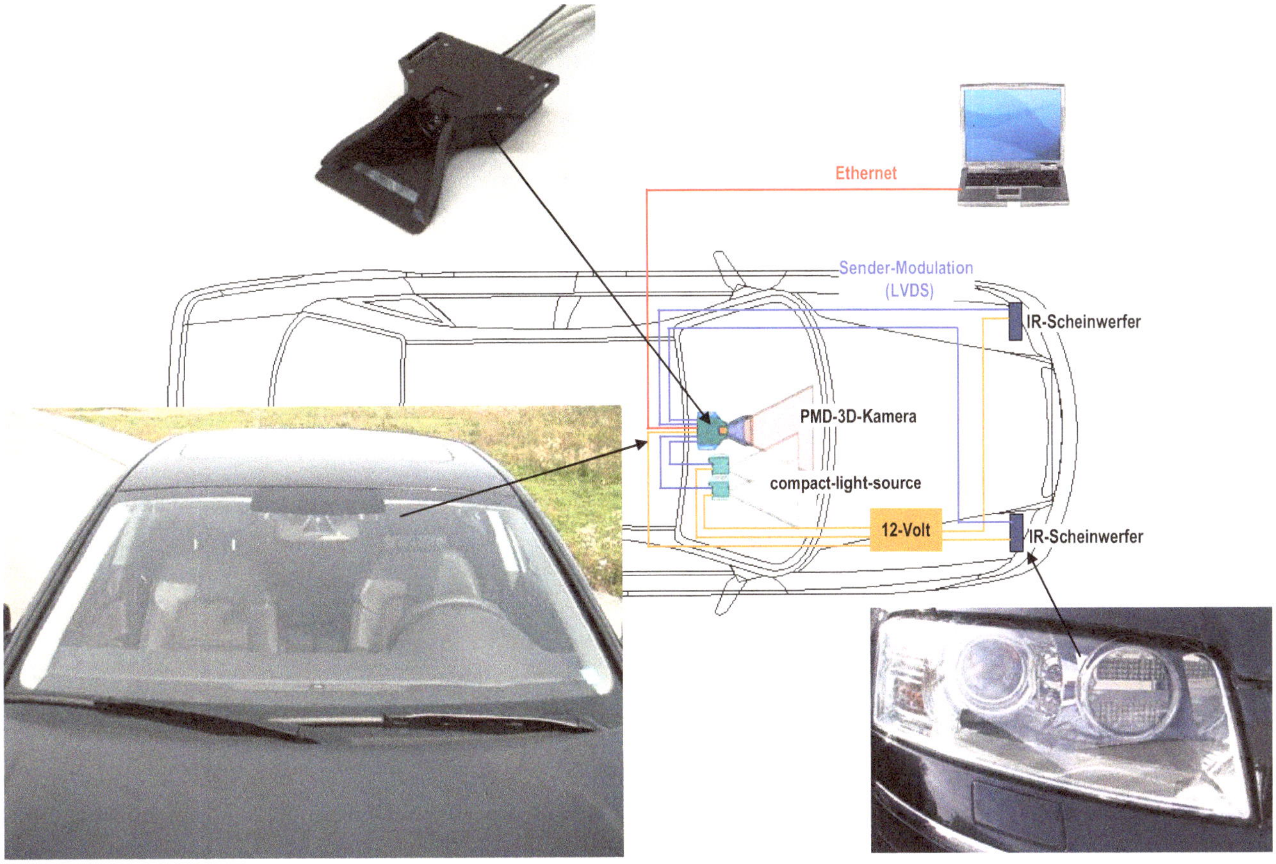

Fig. 17. Test car from 2005 with an A-Muster PMD camera system for pedestrian safety.

more). Therefore, in order to yield a more complete result, measurement should be extended from single pixel characterisation to parallel characterisation of all pixels of a PMD array with SBI circuitry.

The diagram in Fig. 14 shows the distribution of phase value standard deviation for a complete PMD array (sensor PhotonICs 1kS2) for fixed signal and background illumination levels. Statistical evaluation indicates that, 100% of the pixels achieve a phase accuracy of less than 4 degrees for a signal photo current of 200 pA, a background photo current of 20 nA and an integration time of 1 ms.

Figure 15 illustrates the SBI array performance for all tested intensities, for a given modulation frequency of 10 MHz, an integration time of 1 ms and a standard deviation of 5°, which corresponds to a distance resolution of 20 cm. Combinations of intensities that result in at least 99.3% working pixels (5σ-range) are marked in green. That means for 99.3% of all pixels, the standard deviation is less than 5°. Combinations that result in less than 95% working pixels (3σ-range) are coloured red. To facilitate measurement results, ledger lines of constant background-to-signal ratio are overlaid. As can be seen, a background-to-signal ratio up to a factor of 400, that is 52 db, is achieved for more than 95% of all pixels. Please consider, that the measurement certainty can be further increased by statistics and intelligent filter algorithms. These measurement results are based on an integration time of 250 μs.

4 Technical implementation

4.1 Realized PMD Sensors

For indoor and outdoor applications with different requirements, PMDTechnologies has developed sensors with specific features, like resolution, optical format and SBI. A short overview of four sensors with optical and mechanical dimension is given in the following figure. The sensors can be ordered for evaluation purposes.

All of these PMD Sensors are implemented in camera prototypes which shows the basic function principle or furthermore they are implemented in camera products like the PhotonICs® PMD 3k-S. In the next chapter, a camera system for pedestrian safety is introduced, which based on the PMD Sensor PhotonICs® PMD 1k-S.

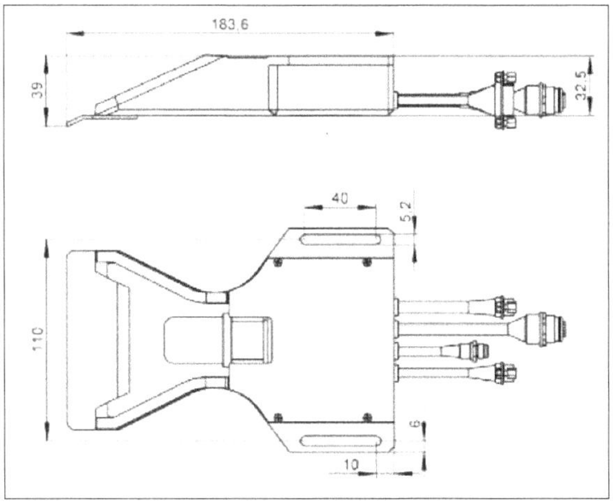

Measurement Range	10m / 35m
Field of view	52° vertical 18° horizontal
DataUpdate Rate	10,4ms (96Hz)
Distance Resolution per Pixel	±10cm
Operation Current of camera module	Max. 600mA
Temperature Range	-10° to 85°C (-40°to 105° in 2007)
Operating Voltage	11 bis 18 DC

Fig. 18. Technical Specification of A-Muster.

4.2 Pedestrian safety camera system

In Fig. 17, a camera system for pedestrian safety is depicted. This A-Muster camera is designed to demonstrate solutions for several front view application like Stop&Go, PreCrash and Pedestrian Saftey therefore additional IR-illumination sources are implemented inside the head lights to increase transmitted power and mearsurement range respectively. Other design spaces for illumination are determined and tested until beginning of 2007.

An actual sensor is available with the specification depicted in Fig. 18. The measurement range depends on illumination source. A range of 10 m is realized by an Illumination Source with 1 W optical power next to the camera system. A measurement range of above 35 m could be realized with illumination source in front of the car (i.g. inside the head lights) with and optical power more than 8 W.

5 Conclusion

Time-Of-Flight systems based on the PMD-principle give the possibility of fast 3-D measurement with customizable resolutions depending on the application. High bandwidth and high sensitivity of the PMD pixels are the fundamentals for enhanced predictable measurement accuracy. Many applications (like outdoor) require robust operation even under very bright ambient light conditions. Spectral filtering combined with an active illumination in burst-mode improve the signal to noise ratio and are effective methods to reduce the impact of ambient light. However, for typical outdoor applications, additional sunlight rejection methods are necessary. PMDTechnologies has realized an outstanding ambient light rejection method, the active SBI circuitry realized as compact in-pixel electronics. This SBI circuitry removes all uncorrelated signals instantaneously during the integration process. It is approved in the 6th sensor generation and works very reliable. The combination of all the presented techniques provides robust 3-D measurement under full sunlight conditions.

References

Buxbaum, B.: Optische Laufzeitentfernungsmessung und CDMA auf Basis der PMD-Technologie mittels phasenvariabler PN-Modulation, Ph.D. dissertation, University of Siegen, Germany, 2002.

Buxbaum, B. and Hagebeuker, B.: Dreidimensionale Umfeld-Erfassung, Elektronik Automotive, Nr. 5, pp. 77, Sep., 2005.

Buxbaum, B., Schwarte, R., Ringbeck, T., Grothof, M., and Luan, X.: MSM-PMD as correlation receiver in a new 3D-ranging system, SPIE – Remote Sensing, Laser Radar Techniques: Ranging and Atmospheric Lidar, Toulouse, 2001.

Kraft, H., Frey, J., Moeller, T., Albrecht, M., Grothof, M., Schink, B., and Hess, H.: Universität Siegen, B. Buxbaum, PMDTechnologies GmbH Siegen, 3D-Camera of High 3D-Frame Rate, Depth-Resolution and Background Light Elimination Based on Improved PMD (Photonic Mixer Device)-Technologies, OPTO 2004, AMA Fachverband, Nürnberg, 2004.

Lange, R.: 3D Time-of-flight distance measurement with custom solid-state image sensors in CMOS/CCD technology, Ph.D. dissertation, University of Siegen, Germany, 2000.

Lange, R., Seitz, P., Schwarte, R., et al.: Time-of-flight range imaging with a custom solid state image sensor, Proceedings of the SPIE, vol. 3823, pp. 180–191, Munich, 1999.

Ringbeck, T., Albrecht, M., Frey, J., Grothof, M., Heß, H., Kraft, H., Möller, T., Mosen, J., and Schink, B.: Time-of-Flight 3D-camera for autonomous navigation and industrial automation, Nürnberg, Sensor, 2003.

Schwarte, R., Xu Z., Heinol H., et al.: A new active 3D-Vision system based on rf-modulation interferometry of incoherent light,

SPIE – Intelligent Robots and Computer Vision XIV, vol. 2588, p. 126–134, Philadelphia, 1995.

Schwarte, R., Buxbaum, B., Heinol, H., Xu, Z., Schulte, J., Riedel, H., Steiner, R., Scherer, M., Schneider, B., and Ringbeck, T.: New Powerful Sensory Tool in Automotive Safety Systems Based on PMD-Technology, AMAA – Advanced Microsystems for Automotive Applications 2000, Berlin, 2000.

Spirig, T., Marley, M., Seitz, P., et al.: The multitap lock-in CCD with offset subtraction, IEEE Transactions on electron devices, vol. 44, no. 10, pp. 1643–1647, 1997.

Over-the-air demonstration of spatial multiplexing at high data rates using real-time base-band processing

V. Jungnickel, T. Haustein, A. Forck, U. Krueger, V. Pohl, and C. von Helmolt

Fraunhofer Institute for Telecommunications, Heinrich-Hertz Institut, Einsteinufer 37, D-10587 Berlin, Germany

Abstract. Over-the-air transmission experiments with a real-time MIMO test-bed are reported. We describe in principle a hardware architecture for spatial multiplexing at high data rates, discuss in detail the implementation on a hybrid FPGA/DSP platform and show measured bit error rates from indoor transmission experiments. Per-antenna rate control and joint transmission are enabled as well using an ideal feed-back link. A functional test of these new techniques is described while detailed transmission experiments are still ongoing.

1 Introduction

Some years ago, basic requirements were outlined for a new air interface which may become effective after the year 2010. A striking challenge is the need to deliver 10 times the data rate at similar costs [1].

There are spectrum, hardware and infrastructure costs in cellular radio systems. It is another issue to reduce the infrastructure costs, but there may be a trade-off between spectrum and hardware costs using multiple antennas both at the base station and at the mobile terminal forming a multiple-input multiple-output (MIMO) system [2, 3]. Various measurements have revealed the potential of MIMO to increase the spectrum efficiency at the same transmitter (Tx) power (see e.g. [4, 5, 6]). To bring these new MIMO techniques into application, the hardware effort must be evaluated, based on implementation trials.

Here we report on a real-time MIMO test-bed where the data rate is limited only by the coherence bandwidth of the indoor propagation channel and not by the digital signal processing as in an earlier implementation [7]. We report on over-the-air transmission experiments in an indoor environment and show some new features as per-antenna rate control and joint transmission.

The prototype described here has recently been shown at the *IEEE Globecom* conference in San Francisco in December 2003.

Correspondence to: V. Jungnickel
(jungnickel@hhi.de)

2 The real-time MIMO testbed

The real-time MIMO test-bed was developed in the German *HyEff* project. The intention is to show the feasibility of MIMO in real-time at first in a single carrier link, based on the well-known flat fading algorithms, and to speed-up the signal processing substantially so that the system becomes limited by the coherence bandwidth of typical indoor channels. Various hardware architectures were evaluated therefore and a promising approach is implemented and fully operational now.

In particular, we use field-programmable gate arrays (FPGAs) instead of digital signal processors (DSPs) as in [7], for the real-time processing of the base-band signals. The reason therefore is the need to process multiple data streams in a single unit. The limited number of in- and output ports of current DSPs may not allow multiple high data rate streams in parallel. For the most simple MIMO algorithms, as the minimum mean-square error (MMSE) detector, all required components for the real-time processing of the received base-band signals can be integrated on current FPGAs. However, each component must carefully be programmed in VHDL, to allow a proper timing control. [(Calculation of beam steering weights, power allocation and bit loading vector). The typical MIMO algorithms are implemented on a standard DSP.] It is also used to control the system.

The complete system is shown in principle in Fig. 1. It comprises of a transmitter (Tx) unit with 4 antennas and a receiver (Rx) unit with 5 antennas. Digital processing of the discrete-time base-band signals is done using an FPGA both at the Tx and at the Rx, and at the Rx it is complemented by a DSP. The link is currently unidirectional and so a feed-back link from the Rx to the Tx is realized by cable. Since the current implementation focuses on the real-time processing, the frame and symbol synchronization are realized by cables as well.

2.1 Transmitter

The home-made 5.2 GHz Txs perform direct up-conversion using four IQ mixers each followed by +20 dBm power amplifiers (see Fig. 2).

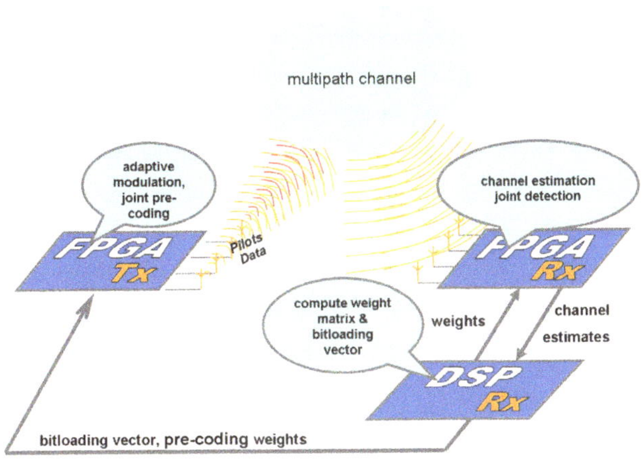

Fig. 1. Principle of the real-time MIMO testbed.

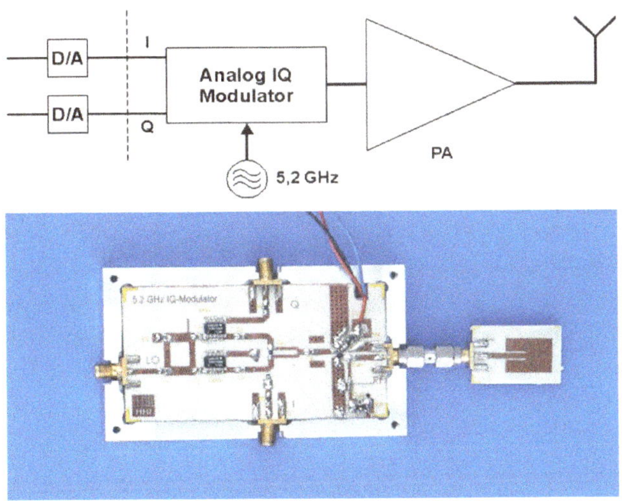

Fig. 2. One of the four 5.2 GHz transmitters.

The pilot and data signal generation, the modulation as well as the joint pre-processing using a matrix-vector multiplication unit are all realized in a XILINX XC2V8000 (see Fig. 3). The 12-bit output signals were DA converted and used to modulate the carrier (see Fig. 3).

Each of the 8 (4 times I and Q each) periodically generated signals consists of a pilot and a data block. A different 127-bit Gold sequence is used as a pilot in each signal to identify at the Rx each I and Q branch from each Tx antenna. Relatively long pilots are used to get sufficiently precise channel estimates for the linear joint transmission experiments. The requirements therefore are investigated in [8]. Each pilot is followed by a different pseudo-random data block with 1024 symbols on each stream. The joint symbol clock for all data streams is variable and so the transmission can be adapted to the coherence bandwidth in the channel. The modulation is set independently on each I and Q branch for each antenna with up to 16 PAM levels. The modulation is individually controlled using a binary vector provided externally by the DSP via cable. Note that a branch may be switched off if desired. Also the weights for the joint transmission can be controlled by the DSP via the feed-back link.

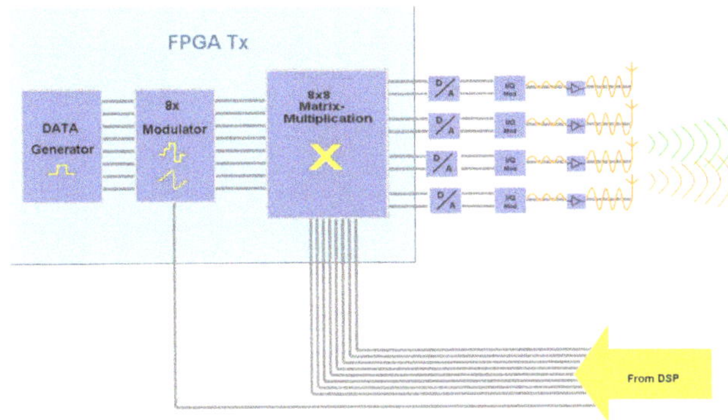

Fig. 3. Transmitter setup.

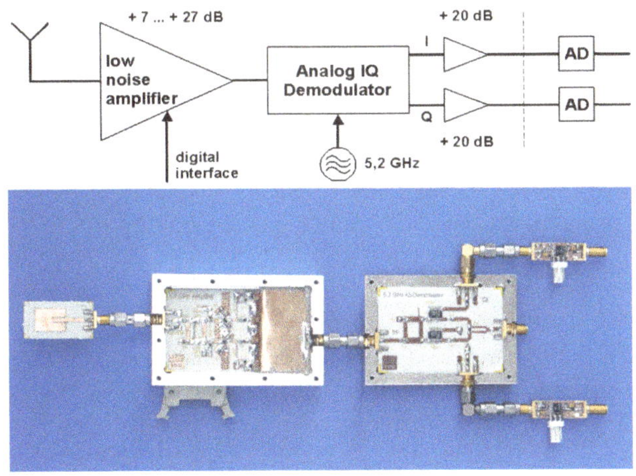

Fig. 4. One of the five 5.2 GHz receivers.

2.2 Receiver

The received signals from 5 antennas are amplified in a digitally controlled low-noise amplifier and directly down-converted in a home-made analogue IQ demodulator (see Fig. 4). The base-band signals are again amplified and digitized with 12-bit at a 25 MHz sampling rate.[1]

An overview of the Rx setup is shown in Fig. 5. It comprises of the channel estimation circuit, the channel tracking by the DSP, the joint detector, the multiple demodulators and the unit for the bit error rate (BER) measurements.

2.3 Channel estimation

80 correlation circuits (CC) using the known pilot sequences are implemented in the Rx FPGA (XILINX XC2V6000).

[1]The analogue RF design creates an IQ imbalance comparable to commercial products but taken into account in the entire MIMO signal processing concept. In principle, we treat the complex-valued MIMO base-band system with 4 Txs and 5 Rxs as a real-valued system having 8 Txs and 10 Rxs.

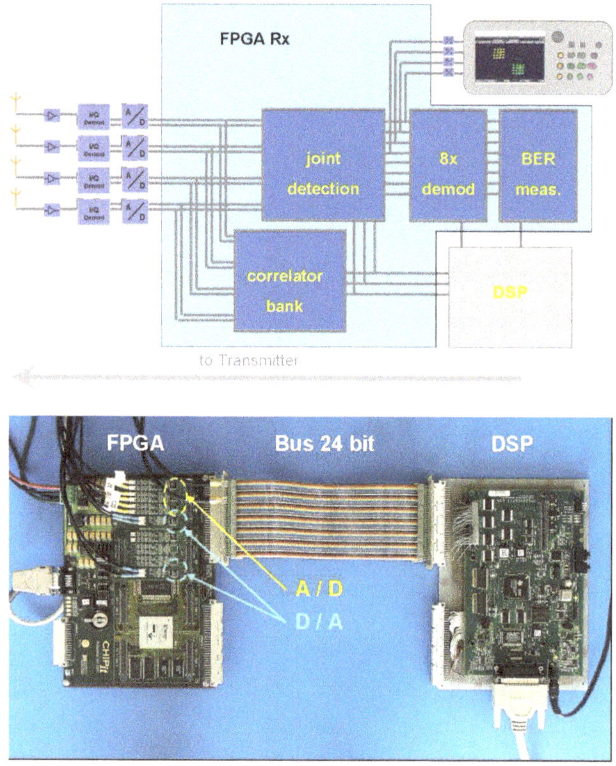

Fig. 5. Top: Receiver setup. Bottom: A/D converters are on top of the FPGA (left). A DSP (right) is connected via a parallel bus.

Fig. 6. Over-the-air transmission experiments shown at the Globecom 2003 conference. The Tx unit is located in the centre with the antennas in the top left corner. The receive antennas are top right and the Rx unit is left from the scope tower.

Since binary sequences are used, the CCs need no multiplication. The next bit in the sequence may eventually change the sign of the signal to be accumulated and then the CC switches from addition and subtraction. One such CC requires 23 out of 33.792 slices in the FPGA and so the entire channel estimation needs less than 6% of the total resources. The noise variance for the MMSE detector is estimated in an off-line measurement.

The new channel estimates are immediately available after the last bit in the sequence and stored in registers. These registers are read-out by the DSP (Texas Instruments 6713). The read/write operations of the DSP are fully asynchronous to the frame structure, enabled by back-up register pages. For a few ns after a new estimate is finished, the estimation results are transferred to these registers and access from the DSP is not permitted.

2.4 Channel tracking

The DSP is then used to calculate the coefficients of a linear MMSE filter which are then sent back to the weight registers for the joint detection in the FPGA. To achieve a higher mobility, it is critical in general to adjust the coefficients in the MMSE detector sufficiently fast. The most time consuming part of the weight calculation is a matrix inversion initially implemented in the DSP using C based on Greville's method [9]. One such inversion takes some ms depending on the antenna configuration. For multy carrier systems, the total time must be in the order of a ms carrier, as well. A detailed analysis revealed that the same task can in principle be sped up by a factor of 50 or so by better using the special matrix-vector skills of the DSP. In this way, the value for one inversion could reach some 10 μs with the same DSP and so the channel tracking rate would become sufficiently fast for indoor and pedestrian applications. Another way might be the interpolation of the inverse matrices between adjacent carriers. But our simulations indicate that the numbers of inversions can noticeably be reduced so only in the case of rather large coherence bandwidths.

2.5 Joint detection

A matrix-vector multiplication unit is implemented and used as a MMSE filter in the Rx FPGA to separate the spatially multiplexed data streams. The unit consumes 80 out of the 144 dedicated multipliers and operates synchronous to the sampling clock (25 Mio. matrix-vector multiplications per second).

2.6 Demodulation

The separated streams are individually demodulated using individual hard decisions in each I and Q branch according to the modulation format set by the DSP at the Tx, and the recovered data are compared inside the FPGA with the transmitted data. There are counters for the bit- and frame error rate in each stream which can be read out by the DSP and displayed on a screen or stored to the hard disk. The entire set-up as presented at the Globecom 2003 is shown in Fig. 6.

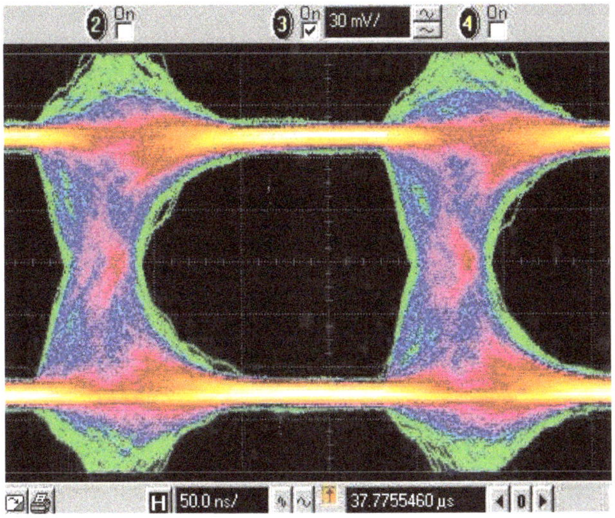

Fig. 7. Eye pattern after the separation of streams in a single I branch with 3 Tx and 4 Rx at 5 million symbol vectors per second (QPSK modulation on all streams, 30 Mbit/s).

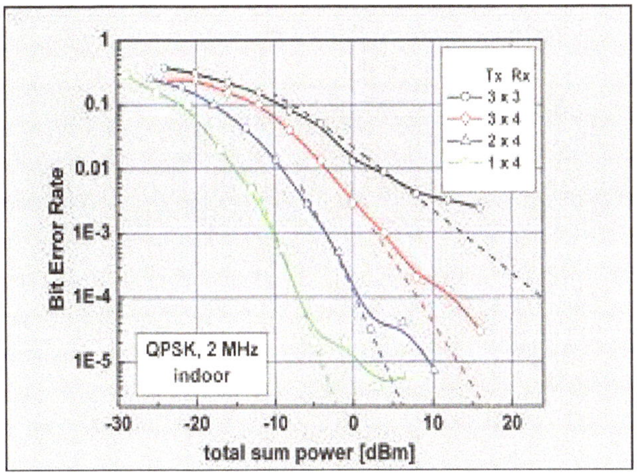

Fig. 8. Fixed rate experimental results of the mean uncoded bit error rates in an indoor scenario. The increase of the diversity is clearly evident when more receivers than transmitters are used.

3 Functional test

The eye pattern in a single I branch after the transmission with 3 Tx and 4 Rx over a 3 m distance in an indoor environment and the subsequent MIMO detection is shown in Fig. 7 at a modulation rate of 5 Mio. symbol vectors per second. The temporal dispersion in the multi-path indoor channel sets an upper limit to the maximal clock rate. It leads to the exponential fall after the start of the bit as observed in Fig. 7. Note that there is additional jitter of about 40 ns which corresponds to the 25 MHz sampling rate.

So our set-up is limited by the coherence bandwidth of the channel and not by the signal processing. Further increase of the symbol vector rate is possible in principle by increasing the sampling rate and the clock of the joint dection unit, but it requires the use of multi-carrier techniques as OFDM.

More efficient iterative MIMO detection schemes as V-BLAST were not yet implemented. Unlike in DSPs, a dedicated processing and memory architecture is required for real-time operation. Even if this might be possible in principle, the effort to implement these algorithms on the register transfer level is not negligible.

4 Transmission experiments

4.1 Fixed rate results

For the transmission experiments, up to 4 Tx and 5 Rx antennas were used in an indoor scenario. In order to realize a sufficient channel statistics, the transmitter is placed on a wagon and moved along a 5 m line through the room, which measures about 90 full carrier wavelength at 5.2 GHz. To measure the average bit error rate (BER), the bit errors are accumulated over the whole run. But at low speed it is possible to investigate the instantaneous BER as well. In the results shown in Fig. 8, the transmitter power is held fixed in each run, and a separate run was made for each power value. The Rx antennas were always more than 2 m away from the Tx and even though the line-of-sight (LOS) was free at most locations, the variation of the mean Rx power is within a few dB, due to the rich multi-path propagation. It leads to a more or less homogeneous power distribution in the entire room [6]. So this variation is left included in the results.

When changing the number of Tx and Rx antennas, the influence of the diversity is clearly obvious from the results in Fig. 8. It is observed that the diversity order increases as N_{Rx}-N_{Tx}+1 as predicted for the MMSE detector at large SNR. So the transmission becomes the more reliable the larger the difference between the numbers of receivers N_{Rx} and transmitters N_{Tx} is. The dashed lines correspond to the theory curves and they agree fairly well with the measured data.

Note that there is always an error floor at high Tx powers due to the frequency-selective nature of the channel (It disappears when the symbol rate is reduced). The position of these floors agrees very well with the theoretical analysis in [10]. In particular, the floor reduces when the diversity order increases. These floors indicate again that now we need multi-carrier techniques to further increase the symbol vector rate.

4.2 Variable rate results

Per-antenna rate control (PARC, [11]) is proposed for MIMO links to approach the channel capacity. With optimal but very complex schemes, as maximum likelihood detection, the PARC is just required to achieve this goal. But with the reduced complexity schemes, the PARC may be required also to save a reliable communication over the MIMO channel. Note that the MIMO channel matrix may eventually become close to singular. With reduced complexity schemes, a sudden increase of the noise after the MIMO detector can then be observed increasing the bit error rate. At first one

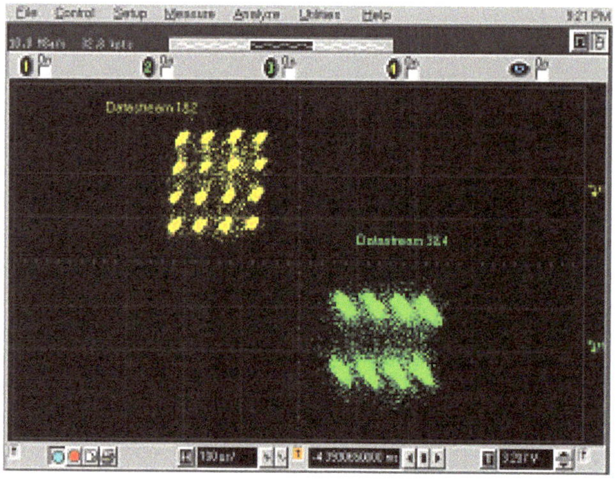

Fig. 9. Variable rate results with 4 Tx and 5 Rx antennas. A snapshot from the modulation of the third and fourth antenna signals is shown after the MMSE filter. While the former uses the same modulation in the I and Q branches, the signal on the fourth antenna is more favourably transmitted when 4 and 2 levels are used in the I and Q branches, respectively.

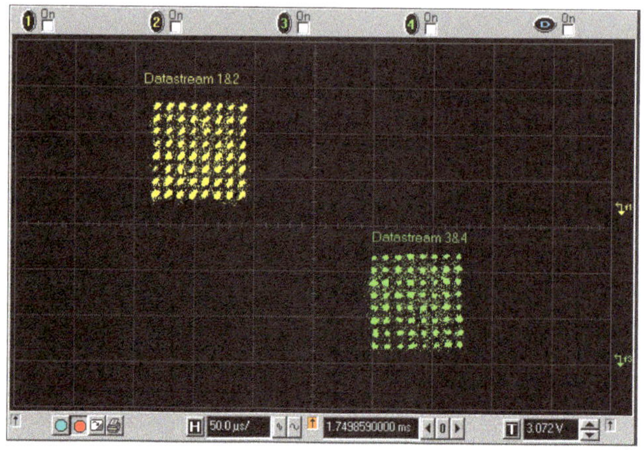

Fig. 10. Two user's signals after jointly pre-processing them based on adaptive channel inversion and transmitting the resulting signals with 4 antennas at the base station. These signals shown were immediately tapped after the AD conversion at the receivers indicating no noticeable cross-talk.

might attribute these events to some instability in the signal processing. But as soon as the PARC is switched on, this apparent instability is no longer noticeable, and the system is operational in a more stable mode.

The PARC is implemented as follows in our test-bed. As mentioned above, there are IQ-imbalances due to the RF design and so the noise in the I and Q branches corresponding to a given Tx antenna may be different after the separation of the data streams with the real-valued MMSE detector[2]. So it is natural to use multi-level modulation in each I and Q branch and steer the number of bits per branch individually. This allows a smooth link adaptation in steps of 1 bps/Hz. The demodulator design is more simple due to the individual 1-D decisions, compared to the 2-D decisions normally required, at least for 8-PSK, 32- and 128-QAM.

The PARC implementation is based on the signal to noise and interference ration (SINR) in each branch after the joint MMSE filter. The SINR is a side result from calculating the filter coefficients in the DSP. In order to guarantee a certain quality of service in each branch, i.e. a certain BER, the effective SNR is compared to a list of pre-defined values required for that BER and then the modulation is set so that the currently available SNR satisfies the requirement. If data streams are not used, the weights are again obtained using the reduced channel matrix.

A typical IQ diagram of two separated antenna signals after the MMSE filter is shown in Fig. 9. The mixed-mode modulation formats observed in the oscilloscope screen-shot are used by the scheduling algorithm on occasion. Note that the noise is anisotropic in the IQ diagram of the particular antenna, where the scheduling algorithm has decided to use a mixed modulation.

In the indoor set-up with 4 Tx and 5 Rx antennas and at high SNR, the scheduling algorithm occasionally decides in favour of 256-QAM constellation in favourable channels. But the BER in these streams is too large due to the nonlinearity of the IQ mixers at the Tx. After a careful adjustment of the local oscillator power and the modulation amplitudes, error-free detection of 64-QAM module signals is possible. So with 4 Tx antennas the maximal spectral efficiency is 24 bps/Hz. Depending on both the channel condition and the SNR, typical values range between 10 and 20 bps/Hz corresponding to data rates between 50 and 100 Mbit/s at 5 Mio. symbol vectors per second.

Note that the adaptation tends to switch off one or more antenna signals at low SNR, while the remaining streams may be transmitted at higher-level modulation formats.

Altogether we have observed that the PARC makes the MIMO communication with simple detection schemes more reliable and it should be used when the channel changes slowly enough to provide feed-back, as in indoor or pedestrian applications. With respect to the signal processing effort we feel that introducing the individual rate control for the parallel streams at the transmitter may be more efficient than using a more complex detection scheme at the receiver. Such pre-adaptation at the transmitter may be a distinct feature in future wireless systems since it reduces the signal processing effort significantly while it maintains a high quality of service.

4.3 Joint transmission using adaptive channel inversion

In the down-link of mobile communication systems, it is desirable to jointly pre-process the data for multiple users and transmit them at the same time at the same frequency using

[2]This is observed also in the estimated real-valued channel matrices. Without IQ imbalance, there are always pais of singular values which are degenerate. Including the imbalance, they are no longer degenerate.

multiple antennas at the base station. Of course, such a joint transmission (JT) needs channel side information at the transmitter, which can be realized via a feed-back link or by exploiting the channel reciprocity available in TDD systems up to moderate mobility. The most simple way to realize JT is to multiply the transmitted data streams with the pseudo-inverse of the channel matrix and in this way the striking new feature of JT can best be explained.

In particular, the JT separates the data streams in advance so that each user receives only the desired signal, while the signals of all other users are forced to zero. But this forcing of zeros is costly in terms of the Tx power. When the channel matrices become close to singular on occasion, the Tx power tends to infinity. These outage events are less likely when more base station antennas than users are in the system. But a scheduling approach to achieve this at least temporarily may be useful as well. If the channel becomes singular, there are rather frequently just one or two users for which the zero-forcing criterion substantially increases the transmitter power. By interrupting the link to those users and transmitting with the whole power to the other users for a while, on average a higher throughput can be achieved for all users. This is called adaptive channel inversion (ACI) and investigated [12, 13]. In particular it was found that the channel capacity can so be achieved, at least for small SNR, while for high SNR there is some penalty.

Using the perfect feed-back link in our test-bed, we have realized the JT based on ACI. A screenshot in Fig. 10 shows the base-band signals from two Rx antennas immediately after the AD conversion. Obviously there is no more cross-talk among the user's signals noticeable, as desired. At least when the users move slowly, the 64-QAM signals can be detected without joint signal processing at the two receivers. Consequently, the two streams can be assigned to two separate users.

5 Conclusions

In conclusion, we have described a software-reconfigurable real-time test-bed for MIMO transmission at high data rates in which the speed of the signal processing is limited only by the coherence bandwidth in the channel. Using the well-known flat-fading linear MMSE algorithm with 4 transmitters and 5 receivers, over-the-air data transmission was demonstrated in the up-link. We have shown that per-antenna rate control may significantly improve the stability of the MIMO radio link. It avoids already in advance the frequent outage events occurring after reduced-complexity MIMO detection schemes. For the down-link, joint transmission experiments using adaptive channel inversion have been described as well, still based on a perfect feed-back link. Future work is directed to the demonstration of JT in time-division duplex mode based on the channel reciprocity and to the implementation of MIMO-OFDM to further increase the symbol vector rate.

Acknowledgements. This work was supported by the German Ministry of Research and Education (BMBF) in the project *HyEff*.

References

[1] available: www.wireless-world-research.org/general_info/Bookofvisions/Bov.html.

[2] Foschini, G. J. and Gans, M. J.: On Limits of Wireless Communications in a Fading Environment When Using Multiple Antennas, Wireless Personal Communications, 6, 3, 311, 1998.

[3] Golden, G. D., Foschini, G. J., Valenzuela, R. A., and Wolniansky, P. W.: Detection Algorithm and Initial Laboratory Results using the V-BLAST Space-Time Communication Architecture, Electronics Letters, 35, 1, 14–15, 7 January, 1999.

[4] Martin, C. C., Winters, J. H., and Sollenberger, N. R.: Multiple-Input Multiple-Output (MIMO) Radio Channel Measurements, Proc. VTC Fall 2000, 2, 774–779, 2000.

[5] Chizhik, D., Ling, J., Wolniansky, P. W., Valenzuela, R. A., Costa, N., Huber, K.: Multiple Input Multiple Output Measurements and Modeling in Manhattan, IEEE JSAC, 21, 3, 2003.

[6] Jungnickel, V., Pohl, V., Nguyen, H., Krüger, U., Haustein, T., and von Helmolt, C.: High Capacity Antennas for MIMO Radio Systems, Proc. WPMC, Honolulu, Hawaii, on CD-ROM, 2002.

[7] see [3].

[8] Haustein, T., Pohl, V., Jungnickel, V., Jorswieck, E., and von Helmolt, C.: Performance of MIMO Systems with Channel Inversion, IEEE VTC-Spring 2002, Birmingham, Alabama, USA, May, 2002.

[9] Gantmacher, F. R.: Matrizentheorie, Berlin Heidelberg New York Tokyo, Springer, 48, 1986.

[10] Pohl, V., Nguyen, P. H., Jungnickel, V., von Helmolt, C.: Limits of the Achievable Symbol Rate in Flat Fading MIMO Systems, PIMRC, Beijing, China, 7–10 September, 2003.

[11] Chung, S. T., Lozano, A., and Huang, H. C.: Approaching Eigenmode BLAST Channel Capacity Using V-BLAST with Rate and Power Feedback, Proc. VTC Fall 2001 Atlantic City, NJ, USA, 2, 915–919, 2001.

[12] Jungnickel, V., Haustein, T., Jorswieck, E., and von Helmolt, C.: On Linear Pre-Processing in Multi-Antenna Systems, Proc. IEEE Globecom 2002, Taipeh, Taiwan, 2002.

[13] Jungnickel, V., Haustein, T., Pohl, V., and von Helmolt, C.: Link Adaptation in a Multi-Antenna System, IEEE VTC Spring 2003, Jeju, Korea.

28

79 GHz UWB automotive short range radar – Spectrum allocation and technology trends

H.-L. Bloecher[1], A. Sailer[2], G. Rollmann[3], and J. Dickmann[1]

[1]Daimler AG, Group Research and Advanced Engineering, 89081 Ulm, Germany
[2]University of Ulm, Institute of Microwave Techniques, 89081 Ulm, Germany
[3]GR-Consulting/SARA, 71732 Tamm, Germany

Abstract. Automotive UWB (Ultra-Wideband) short range radar (SSR) is on the market as a key technology for novel comfort and safety systems. SiGe based 79 GHz UWB SRR will be a definite candidate for the long term substitution of the 24 GHz UWB SRR. This paper will give an overview of the finished BMBF joint project KOKON and the recently started successing project RoCC, which concentrate on the development of this technology and sensor demonstrators. In both projects, the responsibilities of Daimler AG deal with application based sensor specification, test and evaluation of realized sensor demonstrators. Recent UWB SRR frequency regulation approaches and activitites will be introduced. Furthermore, some first results of Daimler activities within RoCC will be presented, dealing with the packaging and operation of these sensors within the complex car environment.

1 Introduction

Automotive radar facilitates various functions which increase the drivers safety and convenience. Exact measurement of distance and relative speed of objects in front, beside, or behind the car allows the realization of systems which improve the drivers ability to perceive objects during bad optical visibility or objects hidden in the blind spot during parking or changing lanes. Radar technology has proved its ability for automotive applications for several years. When compared to its optical counterpart video (with image processing) or lidar, the advantages of radar are obvious:

- direct distance and speed measurement
- robust against weather influences and pollution
- unaffected by light
- measurement of stationary and moving objects on and in the vicinity of the road
- invisible integration behind electromagnetically transparent materials (e.g. bumpers).

Meanwhile, many car manufacturers use 77 GHz radar for autonomous cruise control (ACC) or recently even for precrash or collision mitigation. It was first introduced to the market in the Mercedes-Benz S-class under the name "Distronic". In addition, UWB short range radar operating at 24 GHz has been developed and introduced (Mercedes-Benz S-class, Distronic Plus, 2005) and is a key enabling technology for actual and novel driver assistance and safety systems. Figure 1 gives an overview on possible (UWB) SRR based vehicle applications.

The need for a UWB bandwith is given by the application. E.g. the prediction of a possible collision needs a reliable object tracking capability. Thus, SRR has to have sufficiently high range resolution in the centimeter range to detect smaller objects and vulnerable road users such as motor cyclists and children.

2 UWB SRR frequency allocation in europe

In 2004 and 2005, the EC and the ECC (Electronic Communications Committee of the CEPT) adopted decisions that regulate the temporary introduction of vehicular UWB SRR using 24 GHz spectrum in Europe until 1 July 2013 and the unlimited allocation at 79 GHz (see Fig. 2). This regulation was the basis for the worldwide first market introduction of UWB SRR in the Mercedes-Benz S-class in 2005.

The EC added a fundamental review of its regulation, to be carried out by 31 December 2009, at the latest, to verify the continuing relevance of the initial assumptions concerning the operation of vehicular SRR and development progress

Correspondence to: H.-L. Bloecher
(hans-ludwig.bloecher@daimler.com)

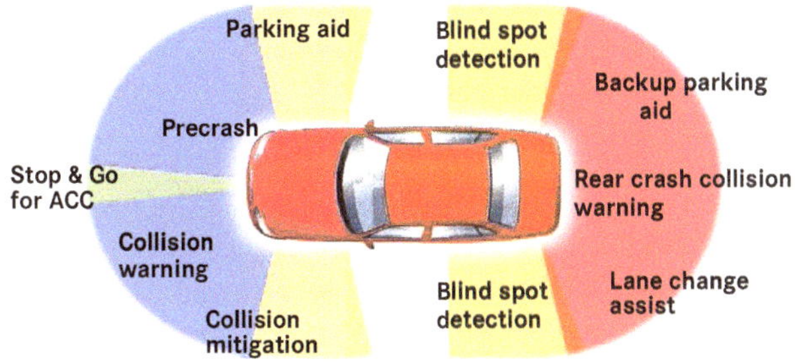

Fig. 1. SRR based assistance and safety applications.

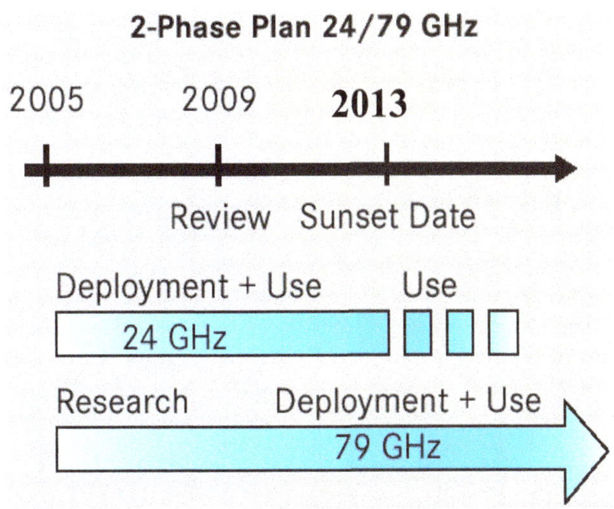

Fig. 2. European 2-phase-solution for 24 GHz UWB SRR frequency allocation.

in the 79 GHz range technology. Automotive manufacturers base product lines on long lead times and assured access to technology. Meanwhile, there seem to be indications that 79 GHz UWB SRR products are probably not mature enough in time for car lines being launched in the years 2010/2011 and produced beyond 2013. Thus, the EC recently mandated the CEPT to review the present 24 GHz UWB SRR regulation and to consider flexible regulatory approaches to avoid a technology gap until availability of 79 GHz UWB SRR sensors.

3 BMBF joint project KOKON

In comparison to 24 GHz, the frequency range around 77 GHz to 81 GHz offers different advantages, e.g.:

- one common technology platform for SRR and LRR available (frequency range 76–81 GHz)
- decreased dimensions and weight
- increased doppler sensitivity

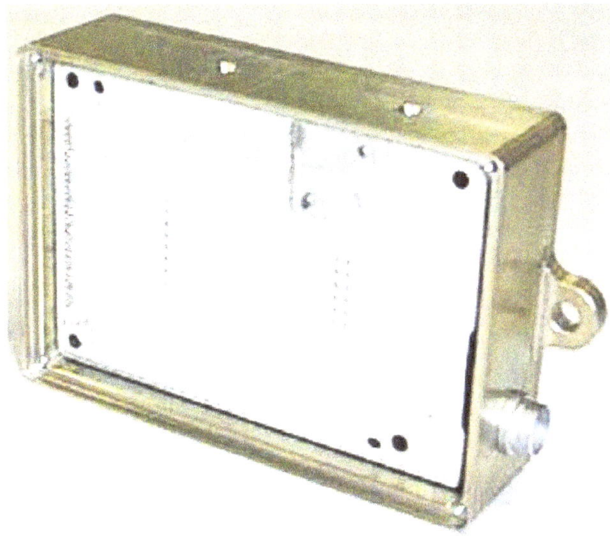

Fig. 3. 79 GHz UWB SRR demonstrator based on SiGe technology (source: Continental AG).

- higher angular resolution with moderate antenna aperture dimensions possible.

Due to this, and as a consequence of the EU regulation, the BMBF joint project KOKON (2004–2007) was launched to develop the basis technologies of 79 GHz UWB SRR and of 77 GHz advanced LRR. SiGe technology was chosen to achieve the required cost reduction and to establish the basis for highly integrated and compact radar front-ends. KOKON was a very successful project. The feasibility of 79 GHz UWB SiGe based technology and components was clearly demonstrated. Continental realized UWB SRR demonstrators (Fig. 3) and Bosch decided to use the SiGe technology in the new LRR3 ACC-radar (Fig. 4).

4 KOKON successing project RoCC

Within the recently launched BMBF joint project RoCC ("Radar-on-Chip for Cars", 2008–2011), the SiGe technology developed in KOKON shall be refined for production and application in sensor products. The RoCC consortium partners are BMW, Daimler, Bosch, Continental, and Infineon (Fig. 5). Main goals of RoCC are:

- affordable vehicle and road safety for everyone
- strengthening of technological leadership in automotive high frequency products and processes
- application based sensor specifications, test and evaluation of realized 79 GHz SiGe UWB SRR sensors
- investigation and optimization of conditions of covered sensor packaging and sensor operation in the vehicular environment

Fig. 4. Bosch 76.5 GHz ACC demonstrator realized in KOKON: RF board of the LRR2 sensor using SiGe-MMICs instead of split block components (source: Robert Bosch GmbH).

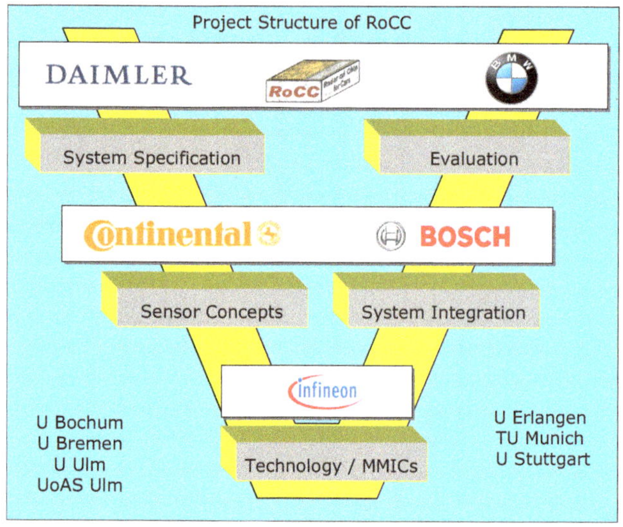

Fig. 5. Organizational structure of BMBF joint project RoCC, project partners and work packages.

- further enhancement of sensor performance and reliability (e.g. noise reduction, angular resolution)
- development of a cost-competitive radar sensor technology in the 76 to 81 GHz range with special emphasis to 79 GHz SRR and evaluation beyond 100 GHz
- foundation of a sound technology base for migration from 24 GHz SRR to 79 GHz SRR
- development of highly integrated universally applicable radar MMIC demonstrators (RoCC: "Radar-on-Chip for Cars")
- improvement of energy efficiency of SiGe-MMICs
- demonstration of radar sensors with superior signal to noise ratio (S/N)
- adaptive (smart) sensors from short range to long range (multi-mode, multi-range)
- pave way to SMD-Package for SiGe-MMICs in the frequency range of 76 to 81 GHz
- benchmarking versus 24 GHz solutions
- demonstration of feasibility of 500 GHz SiGe technology for automotive radar applications
- extended self-test, diagnosis and calibration features.

As an OEM, Daimler is active in the fields of sensor specification, test and evaluation. Further responsibilities include frequency regulation, standardization and the evaluation of frequencies above 100 GHz for future SRR/LRR applications.

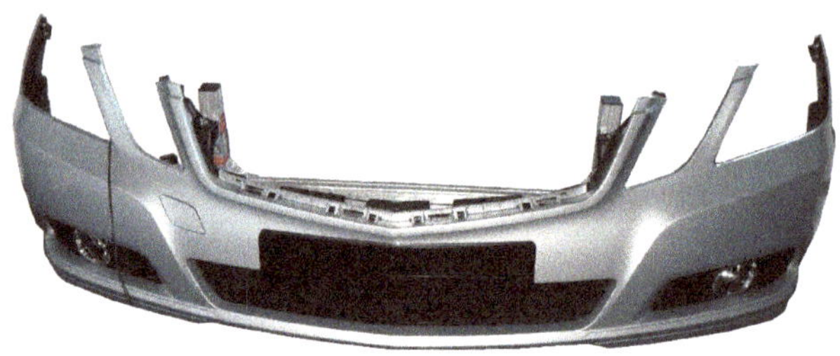

Fig. 6. Front bumper of current Mercedes S-class.

5 Current activities of Daimler AG in RoCC

A. Sensor integration and packaging aspects

SRR sensors are typically mounted invisibly behind painted bumpers or other layered components. Any cover of an antenna of a radar sensor has to be carefully designed in order to avoid performance degradation due to transmission loss, reflections, and edge effects. Even more than for lower operating frequencies, this is a crucial issue concerning sensor systems working in the frequency range from 76 GHz to 81 GHz. Bumpers and other kinds of components mounted on a vehicles front- or rear-end have to be considered as radome structures (Figs. 6 and 7). For the desired frequency range, their thickness is commonly in the order of not more than a couple of wavelengths. The OEMs among the project partners, Daimler and BMW, have to consider a wide range of sensor packaging aspects in RoCC. Special attention has to be turned on the impact of the necessary frequency bandwidth around 79 GHz:

- electromagnetic characteristics of bulk material and painting (permittivity and loss tangent)

Fig. 7. Transmission measurement of LRR radome structure with vectorial, polarimetric, quasioptical setup.

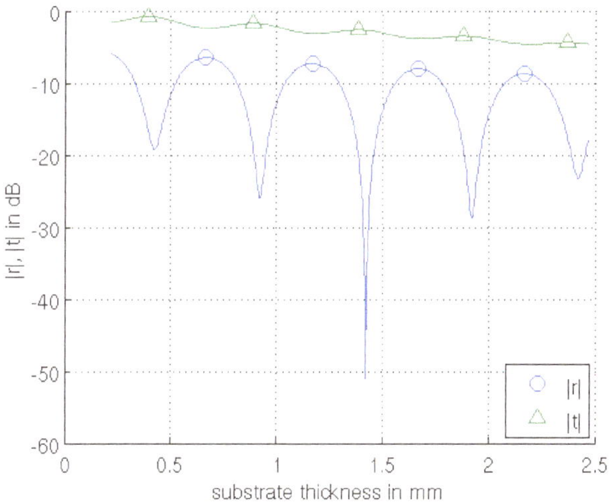

Fig. 8. Simulated transmission and reflection coeffixient for variation of substrate thickness in a multi-layer radome ($\varepsilon_{r,\text{substrate}}$=2.4, $\varepsilon_{r,\text{paint}}$=80 and f=79 GHz).

- manufacturing tolerances
- multiple paintings, especially repair paintings
- covering of radomes with water, snow, ice, dust or salt etc.

Within KOKON, it was shown that even metallic paintings with very large permittivity will not influence the radar operation inordinately, if certain design rules are applied. The main focus of the former investigations aimed towards optimizing metallic paint in terms of reduced loss and permittivity. Still, it is necessary to optimize the transmittive multi-layered radome structure for very low reflection. Varying the substrate thickness, it is possible to find a setup with

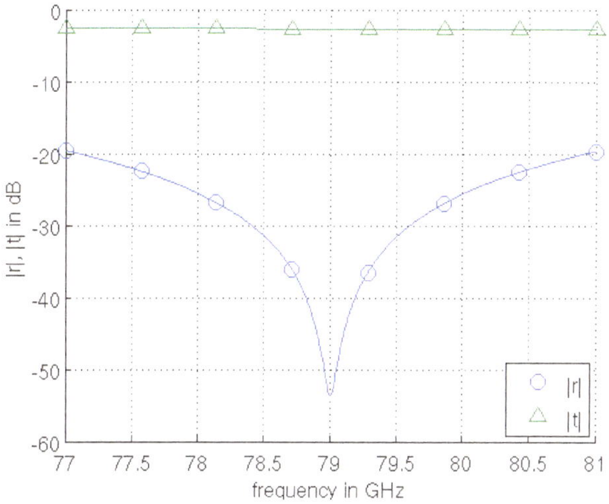

Fig. 9. Simulated transmission and reflection coefficient of the multi-layered radome shown in Fig. 8 with optimized substrate thickness d=3.44 mm.

a reflection coefficient of less than −20 dB in a very broad frequency band even for paint with permittivity as high as ε_r>80 (Figs. 8 and 9).

B. Investigation of interdependence of radar and object tracking in single- and multi-sensor systems

The request for driver assistance and active safety applications is increasing significantly. The functions shown in Fig. 1 require varying sensor output performance. Various suppliers and research groups offer and report sensors based on various principles and with very individual performance.

Either single sensors or sensor networks can accomplish environment perception around the car. One important parameter with severe influence on the functionality is the operating frequency bandwidth. The competing systems are narrow band and UWB radar sensors. Because of the bandwidth that is used, both systems will show different behavior concerning range resolution. Still, from the point of view of a safety application, this is not the only important factor. The sensor has always to be considered in its interaction with an object tracker. As a starting point, the focus lay on the application of SRR systems in pre-crash applications (Fig. 10). An ideal trajectory of a point target was corrupted with noise to simulate the inaccuracy of a real sensor (two upper curves in Fig. 11). Then, the surroundings of the sensor were discretized, and the detected obstacle was matched to a range cell (third curve in Fig. 11). This was done for two different antenna concepts. In the case of a beamforming antenna, one can consider angular and range cells, whereas in the case of a monopulse system, only range cells can be assigned. The cell that has been assigned is the input data of a tracker, a Kalman filter or extended Kalman filter, respectively. Varying parameters such as update rate, field of view, maximum

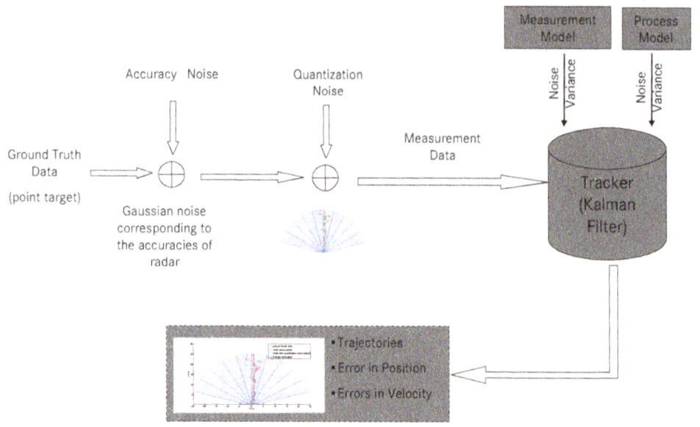

Fig. 10. Radar and tracking model for the investigation of sensor concepts in single- and multi-target scenarios (switched beam radar modeled using antenna coverage diagram and addition of Gaussian noise).

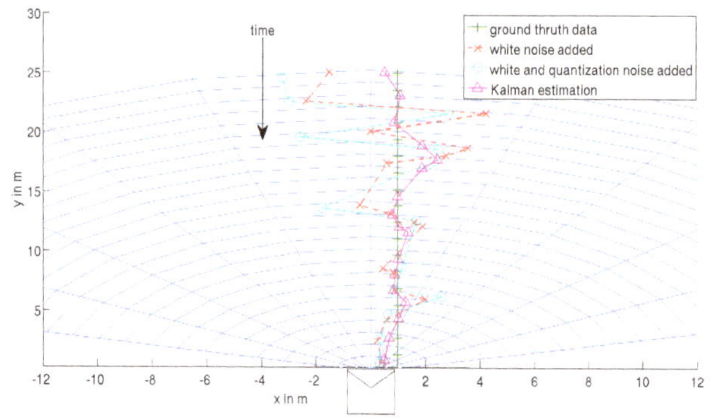

Fig. 11. Investigation of the influence of tracking on target detection in range and angle (edges of the vehicle at x=+/−80 cm). Point target position is corrupted (white noise), assigned to an angular and range cell (quantization noise) and estimated with a Kalman filter.

range, sensor resolution and accuracy allows to verify proposed sensor concepts and to create a knowledge base for the RoCC work packages sensor specification and sensor evaluation. Further work will be done by implementing typical multi-target scenarios for both beamforming and monopulse systems.

6 Conclusions

Automotive short range radar is an important sensor technology for present and future automotive active safety and comfort functions. The UWB approach provides a real-time high range resolution, which is of particular importance for the time critical safety functions, e.g. pre-crash. The European frequency regulation for UWB automotive SRR requires the shift from 24 GHz to the 79 GHz band in 2013. Beneath this, offers the application of the same the 79 GHz frequency range offers application of the same technology platform for LRR and UWB SRR. Furthermore, frequency dependent parameters as angular and velocity resolution, are improved significantly. SiGe technology has been chosen to realize low-cost sensors. Within the BMBF public funded joint project KOKON, the feasibility of SiGe based SRR/LRR has been shown successfully. The successing project RoCC has been started recently to commercialize the technology developed within KOKON and to support the availability of cost efficient and reliable 79 GHz UWB SRR products after 2013.

Acknowledgements. The authors wish to thank the German Federal Ministry of Education and Research (BMBF) for the funding of the KOKON and RoCC joint projects.

References

The ECC Decisions ECC/DEC/(04)10 (24 GHz SRR) and ECC/DEC/(04)03 (79 GHz SRR) are available in the Internet: ERO-Homepage (http://www.erodocdb.dk/doks/doccategoryECC.aspx?doccatid=1), 2004.

The EC Decisions 2005-50-EC (24 GHz SRR) and 2004-545-EC (79 GHz SRR) are available in the Internet: EC-Homepage: (http://www.europa.eu.int/eur-lex), 2005.

Proceedings of the BMBF joint project KOKON final meeting (Workshop WSW5 of the EuMW 2007), Munich, October, 2007.

Final reports of the KOKON project partners, German National Library of Science and Technology (online ressource); http://tiborder.gbv.de/psi/DB=2.63/LNG=DU/SID=1ea59db2-3b/CMD?ACT=SRCHA&IKT=1016&SRT=YOP&TRM=kokon, 2008.

Evaluation of wave propagation effects for detection of vulnerable road users

A. Fackelmeier, C. Morhart, and E. M. Biebl

Fachgebiet Höchstfrequenztechnik, Technische Universität München, Munich, Germany

Abstract. The protection of vulnerable road users (VRU) is a key issue of future driver assistance systems. An important point is to determine whether a VRU is visually hidden to the driver. This additional information is essential for the driver to prevent an accident or to reduce its severity. We show that by using a transponder attached at the VRU, wavelength dependent diffraction effects can be utilized to distinguish between visible and hidden VRUs. Several diffraction measurements were carried out taking place in open space areas with simple big scattering objects and in different road traffic scenarios demonstrating a reliability of nearly 100%. If future driver assistance systems for detecting VRUs are complemented with automotive electronics which enable distinction between visible and hidden VRUs, accidents can be avoided.

1 Introduction

The protection of vulnerable road users (VRU) is a key issue of future driver assistance systems (Rasshofer, 2007). Currently, several different systems are under investigation, based on passive sensing techniques (radar, lidar, camera) as well as on cooperative sensor systems. In a cooperative system the VRU is required to wear a transponder in order to identify themselves as a VRU as well as to send a beacon signal for localization. In contrast to passive sensing, the beacon signal can also be detected if the person wearing the transponder is visually hidden, for instance by a parked car or a roadside tree. Of course, information about this situation, i.e. whether the transponder is visually hidden or not, will be of great value for alerting the driver of the car. If the person wearing the transponder is visually hidden, the situation is much more hazardous.

Our method to identify visually hidden persons is based on dual-frequency discrimination. A transponder (according to the terms of radar technology we call it "active target") attached at the VRU emits two signals with different frequencies which are detected by the receiver in the car. If the VRU is invisible for the driver, the signals are diffracted around the scattering object and get an additional attenuation. Signals with different wavelengths experience different attenuation at the same observation angle in the shadow zone due to the wavelength dependent diffraction. The difference in signal power at the two frequencies compared to the difference at line of sight (LOS) received at the driver side is an indicator whether the emitter is hidden or not. There are many parameters to be investigated affecting the measurement result such as orientation of the VRU to the scattering object and to the receiver, height of the antennas, multipath propagation due to buildings or objects nearby and the size and shape of the scattering object which hides the VRU. Our objective was to determine whether this diffraction effect is measurable in real road traffic scenarios, how reliable a detection of hidden VRU is, and how the parameters mentioned above influence the measurement results.

2 Wave propagation effects

For theoretical analysis, the dimension of this scattering problem compared to the wavelength is too big for applying EM field simulation methods. The Geometrical Theory of Diffraction (GTD), first introduced by Keller in 1962 (Keller, 1962; Borovikov, 1992), is an extension to the geometrical optics which adds diffracted rays E_d to the geometrical optics. Dimensional considerations show that the edge diffraction is the strongest besides corner and surface diffraction (Keller, 1962) and therefore, the latter are neglected. Edge

Correspondence to: A. Fackelmeier (fackelmeier@tum.de)

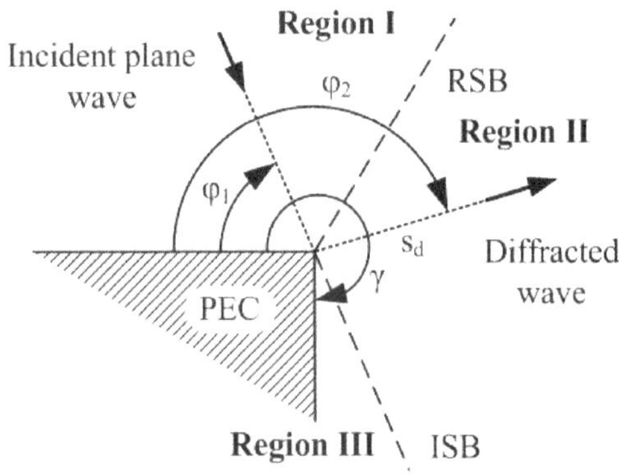

Fig. 1. Edge diffraction (ISB: Incident Shadow Boundary, RSB: Reflection Shadow Boundary).

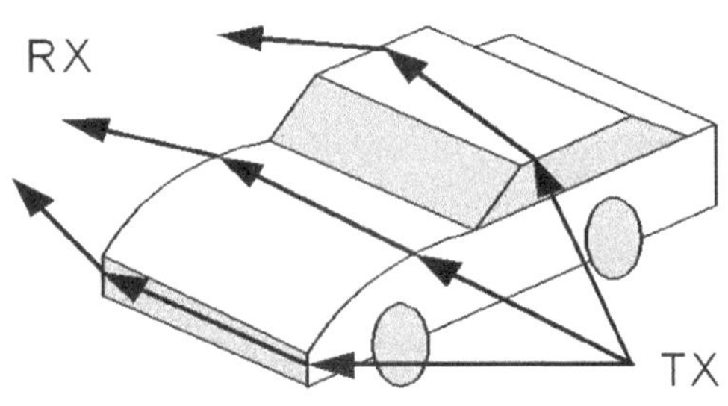

Fig. 2. Double edge diffraction at a car.

diffraction occurs with an edge radius small compared to λ. The incident wave is diffracted at the edge and spreads out as a cylindrical wave (Fig. 1).

With the diffraction coefficient R, the signal power difference in the shadow region can be calculated for different frequencies. E_i is the field incident at the point of diffraction and $A(s_d)$ $(=1/\sqrt{s}$ for edge diffraction) is the spreading factor with the distance s_d between the scattering point and the observation point P were the diffracted field strength E_d is present.

The diffraction coefficient R has to be calculated for each frequency with the angles indicated in Fig. 1. Between f_u=433 MHz and f_o=2.45 GHz the ratio in signal power of the diffracted waves is about 7.5 dB. If there is a second edge, double edge diffraction occurs and the power difference amounts to about 15 dB. Double diffraction occurs quite often in road traffic situations with cars as scattering objects (Fig. 2).

$$E_d(P) = R \cdot A(s_d) \cdot E_i \cdot \exp(-jk_0 s_d) \tag{1}$$

$$\left.\begin{matrix} R_\perp \\ R_\parallel \end{matrix}\right\} = \frac{\exp(-j\frac{\pi}{4})\sin(\frac{\pi^2}{\gamma})}{\frac{\gamma}{\pi}\sqrt{2k_0\pi}} \cdot \left[\frac{1}{\cos(\frac{\pi^2}{\gamma})-\cos(\frac{(\varphi_1-\varphi_2)\pi}{\gamma})} \pm \frac{1}{\cos(\frac{\pi^2}{\gamma})+\cos(\frac{(\varphi_1-\varphi_2)\pi}{\gamma})}\right] \tag{2}$$

Multipath propagation can cause an attenuation of more than 15 dB which is very problematic in this application as it has

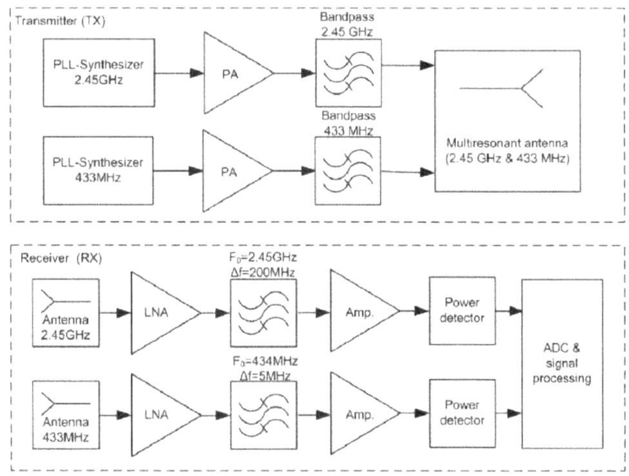

Fig. 3. Emitter (above) and receiver (below).

to be distinguished between attenuation due to scattering objects in the signal path and attenuation due to multipath propagation. In road traffic situations, multipath propagation can occur through every object near the road such as buildings, vehicles or other obstacles nearby.

3 The RF-power measurement system

A RF-power measuring system was developed and built up for the ISM-frequency bands at 2.45 GHz and 433 MHz with commercial IC devices to show that the target detection is realizable also without high-precision measurement equipment (Fig. 3). For each emitter (TX), a Phased Locked Loop (PLL) frequency stabilisation is used to avoid a frequency drift which can result in an output power variation and therefore decrease the measurement accuracy. At the receiver (RX) a high dynamic range is necessary to cover different distances of RX and TX and the additional attenuation due to the scattering objects in between. A low pass filter is implemented to flatten the interfering fast fading peaks in the signal waveform.

The transmitter antennas have to be quite omni-directional enabling detection of the target at every orientation. Other parameters like polarisation also affect the measurement results.

The small distance of the transmitting antennas to the object leads to distorted measurement results due to displacement between the antennas. This problem does not occur at the receiver which is located in a huge distance to the object. Simply adding the 2.45 GHz signals to the 433 MHz monopole is disadvantageous because measurements have shown that the side lobes arising at the 2.45 GHz antenna pattern in special cases are reflected, for instance at the headlining inside the car. To overcome this problem, a multi-resonant antenna was developed which provides a nearly coinciding phase centre (Fig. 4). For the presented

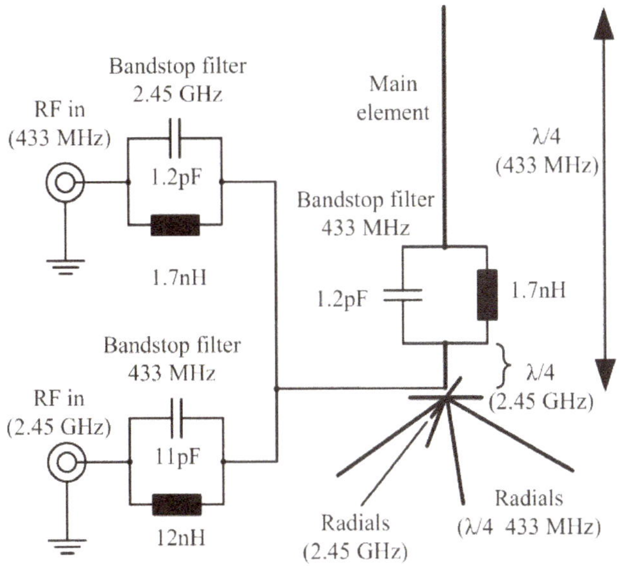

Fig. 4. Multi-resonant antenna for 2.45 GHz and 433 MHz.

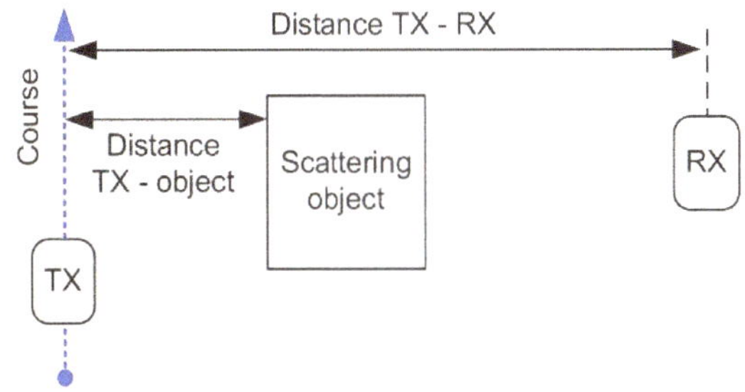

Fig. 5. Double edge diffraction at a car.

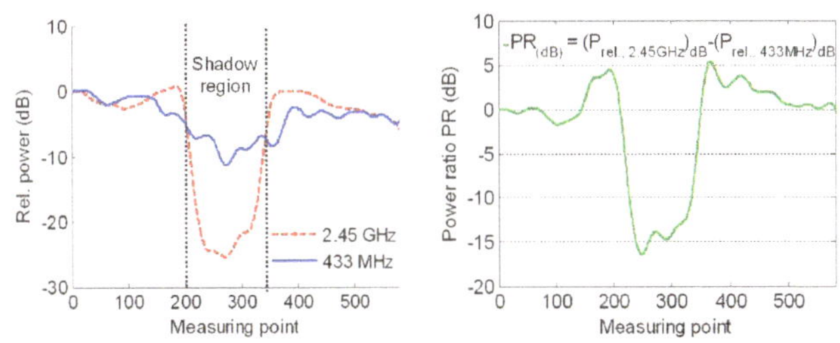

Fig. 6. Detection of a target hidden by a single object left: normalized received power at the two frequencies, below: power ratio for the detection (distance TX–RX=36 m, antenna-high$_{TX}$=1.2 m, antenna-high$_{RX}$=0.8 m).

measurements the wave polarization is always vertical. This antenna is derived from the λ/4 ground-plane antenna with 4 radials, each measuring 1/4 wavelength. With the angle between the radials and the main element the impedance can be adjusted. The main element consists of a λ/4 long silvered copper wire of 2 mm diameter (in terms of 433 MHz). On the lower end there is a band pass mounted that blocks the 2.45 GHz signal. The position of this filter is about λ/4 over the feeding point (in terms of 2.45 GHz). To allow separate matching for both frequencies to 50Ω, short and long radials are attached. The long radials provide 50Ω at an angle of about 135° for 433 MHz. With the short ones the impedance of the 2.45 GHz antenna part is adjusted. The short radials have almost no influence to the impedance of the longer wavelength. With this separate radial configuration both impedances can be tuned to 50Ω separately.

The signals have to be combined without mutual impedance mismatch. This is achieved by band-pass filters in every signal path isolating one frequency against the other. After connecting the band pass to the antenna the impedance has slightly misaligned which has to be reversed by adjusting the radials again.

The phase centre is not exactly at the same point for both frequencies in vertical direction but it is much better than with two separate monopole antennas.

4 Detection in case of single objects in the signal path

To verify radiation effects, several measurements with simple objects were carried out in open space areas (Fig. 5). The transmitter (TX) emits continuous wave signals at both frequencies that travel to the receiver (RX) either directly or disturbed by an scattering object. A relative movement of the target, receiver or object causes a position-dependent signal power at the receiver antennas. This signal power is measured separately for both frequencies at RX. The scattering object used is a metal sculpture with sharp edges and dimensions of 2 m in each direction. Figure 6 shows the diffraction measurement with a distance of 35 m between the receiver and the target and 1 m distance between target and object. After measurement point (MP) 200, the target person gets invisible for the receiver, and there are only waves present at RX which are double diffracted at the edges of the scattering object. The 433 MHz signal experiences only a small attenuation due to the object compared to the 2.45 GHz one. The lower graph indicates the ratio in signal power of both frequencies. As the theory predicts, this value is about 15 dB if the target is hidden relating to 0 dB for line of sight (LOS). If the threshold is placed for instance at −5 dB, it is easy to figure out in which interval the object is hidden by the target.

At certain distances, strong peaks in signal power arise in the shadow region resulting from interference between waves diffracted around different sides of the scattering object which can be verified through comparing propagation paths. At scattering objects with big radii, creeping waves have been detected. This has been verified through calculation of ray propagation paths leading to points with destructive interference which agree with measured peaks in the diffracted wave signal (McNamara et al., 1989).

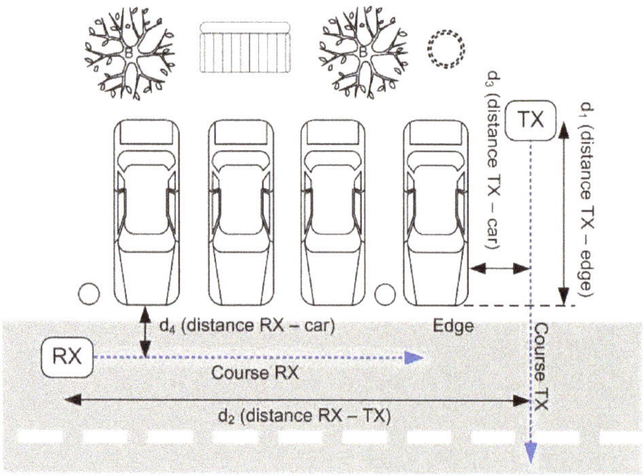

Fig. 7. Detection Measurement configuration for traffic situations.

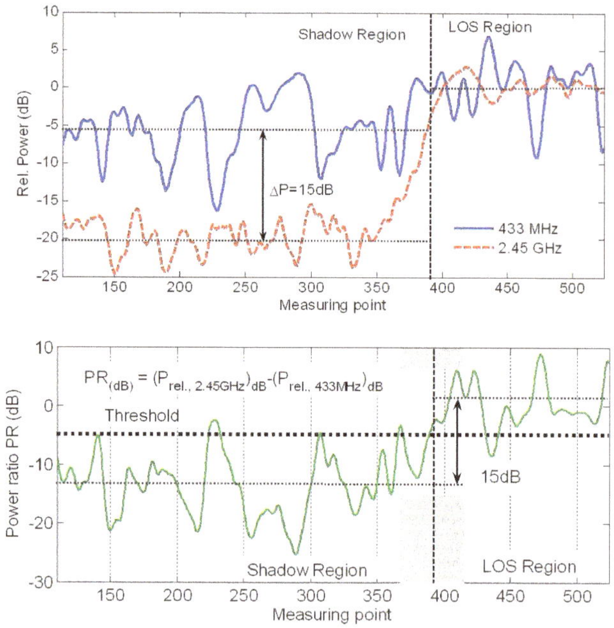

Fig. 8. Detection measurement of a moving target; above: normalized received power at the two frequencies, below: power ratio for the detection (d_2=30 m, d_3=1 m, d_4=1 m, antenna-height$_{TX}$=0.8 m, antenna-height$_{RX}$=0.6 m).

5 Detection of hidden targets in road traffic scenarios

Several different road traffic scenarios have been analyzed to verify that the results are repeatable. Figure 7 shows a typical situation – parked cars at the roadside. In the first measurement, the emitter (TX) moves normal to the road in a certain distance to the parking car. The receiver (RX) is placed at a fixed position at the road. With this configuration, the scenario of a VRU moving from invisible positions to visibility is analyzed at a fixed position of the car. The measurement starts at $d1$=10 m corresponding to the measuring point (MP) zero in Fig. 8, where the target is hidden by the cars. After MP 390, the target person becomes visible for the receiver, and TX and RX are in LOS. By setting the threshold again at −5 dB, the target person can be identified as hidden (before MP 390) or as visible (afterwards). In this case, the detection has a reliability of about 95% for the target positions regarded. By averaging the signal over a certain time period, the error probability can be minimized.

Disturbing ripples of up to 20 dB arise due to constructive and destructive interference between direct and reflected waves in LOS case between diffracted and reflected waves or also between several diffracted waves in the shadow region. By averaging the signal power over a certain period of time, the power ratio between the two frequencies for hidden targets amounts always to about 15 dB relating to 0 dB for LOS. This result was confirmed in different measurements with moving targets and coincides with the theoretical calculation of Sect. 2.

The following measurement takes place in the same scenario indicated in Fig. 7 but with a moving receiver (RX) and an emitter at fixed positions. The receiver moves parallel to the street as can be seen from Fig. 7. During this measurement, the emitter remains at a fixed position. Two measurement results with different target – edge distances (d_1) have been put together in Fig. 9: $d_{1,LOS}$=0 m (target just visible) and $d_{1,HIDDEN}$=2 m (target hidden). With this configuration the scenario of a moving car is analyzed in two cases – one with a hidden VRU and one with a VRU just visible. The movement starts at a target distance of d_2=35 m (MP 0) and stops at d_2=2 m (MP 400). At the last few meters there is a big misalignment of the antennas because TX is displaced from the RX moving path by 3 m (target hidden) or 1 m (target in LOS) respectively. So these values after MP 350 are not used for calculation of the mean value. In contrast to the prior measurements which show relative power levels this time the absolute signal power as a function of distance is shown in the upper graph of Fig. 9. The difference in the mean values of the power ratio in LOS and hidden case range with 18 dB in the same size as in the previous measurements. The signal difference is at all distances big enough for detection of hidden targets. By setting the threshold again to −5 dB, the detection has a reliability of more than 95%.

The height of the antenna over the ground has a big impact on the measurement results. For instance if both antennas are at a height where LOS can be achieved through the windows of the cars, the signal difference is less because the signals experience double slit diffraction. For 2.45 GHz, the slit (car window) is big enough that at RX a signal power level is measured which is even higher as for LOS because of Fresnel diffraction.

6 Conclusion

In this paper, the detection of hidden persons by using a transponder attached at the person is demonstrated. It has been shown that wavelength dependent diffraction effects can be utilised for the detection whether an emitter is hidden. In order to show that this effect is measurable without measurement equipment, a RF-measuring system was developed and built up for 2.45 GHz and for 433 MHz. In open space areas with the used system configuration, detection of targets hidden by single big objects can be achieved up to 60 m distance between TX and RX. Here the power difference amounts to about 15 dB between hidden emitters relating to 0 dB for LOS confirming the theoretical calculation of Sect. 2.

Several target detection measurements in different road traffic scenarios have been carried out. The strong ripple in the power measurement arises due to multipath propagation. By averaging the signal over a certain period of time, the power ratio between the two frequencies for hidden targets amounts to about 12–18 dB relating to 0 dB for LOS. The detection whether a target is hidden behind a car is possible at any orientation of the car but depends from the height of the emitter. Relatively small objects like persons or street lamp poles in the signal path do not lead to detection which may be desirable. The results were confirmed in many different measurements with moving targets and also with moving receivers.

Multipath propagation causes strong interferences. Especially in situations with buildings or other objects at the roadside, a detection with-non averaged measurement values is in some cases impossible because of the strong ripple in signal power. A solution to overcome this problem is using diversity reception. In this case, for the reception of a continuous wave signal only, this can be accomplished by switching two or more antennas consecutively to one LNA and always taking the highest signal value.

An important aspect is to set the threshold at a level with the best detection reliability between LOS and hidden target. A variable threshold depending on other parameters such as distance between RX and TX which has to be measured simultaneously may be advantageous.

If the body of the target person turns into the signal path, both frequencies are attenuated up to 30 dB. This attenuation does not arise at the same turning angle for both frequencies, so an additional power difference of more than 15 dB must be taken into account, too. For this reason, the error probability in detection for statistical target orientation increases significantly. Further investigations have to be made to determine the impact of the target person's orientation at different road traffic situations.

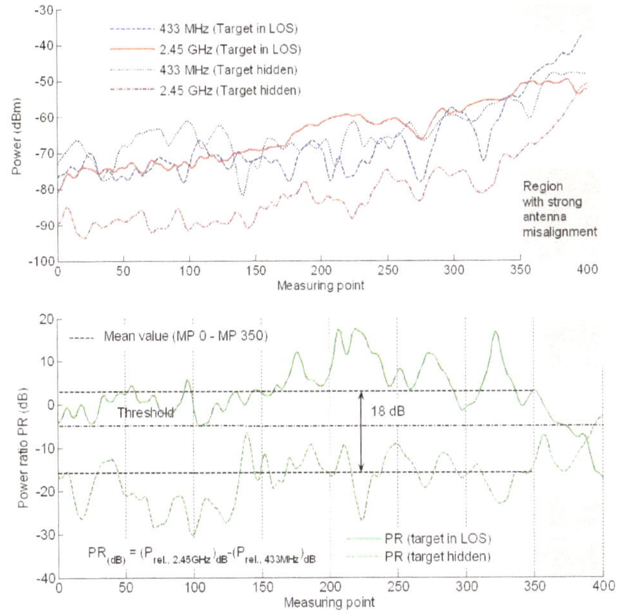

Fig. 9. Detection of a moving target, above: received power at the two frequencies, below: power ratio (d_3=1 m and d_4=1 m, $d_{1,\mathrm{LOS}}$=0 m, $d_{1,\mathrm{HIDDEN}}$=2 m antenna-height$_{\mathrm{TX}}$=0.8 m, antenna-height$_{\mathrm{RX}}$=0.6 m).

References

Borovikov, V. A. and Kinber, B. Y.: Geometrical theory of diffraction, Bath: Bookcraft, 1992.

Keller, J. B.: Geometrical Theory of Diffraction, J. Opt. Soc. Am., 5(2), 116–132, 1962.

McNamara, D. A., Pistorius, C. W. I., and Malherbe, J. A. G.: The Uniform Geometrical theory of diffraction, Boston: Artech House, 1989.

Rasshofer, R. H., Schwarz, D., Biebl, E., Morhart, C., Scherf, O., Zecha, S., Grünert, R., and Frühauf, H.: Pedestrian Protection System using Cooperative Sensor Technology, Proc. Advanced Microsystems for Automotive Applications AMMA 2007, 135–145, Berlin, Germany, 2007.

Efficient evaluation of antenna fields by a time-domain multipole analysis

J. Adam and L. Klinkenbusch

Institute for Electrical and Information Engineering, Christian-Albrechts-Universität zu Kiel, Germany

Abstract. The contribution describes a systematic method to efficiently determine frequency-domain electromagnetic antenna fields and characteristics for a broad spectrum via a single time-domain (e.g., Finite-Difference Time-Domain, FDTD) calculation. From a time-domain simulation of an antenna driven by a wide-band signal, a single modified Fourier transformation yields the frequency-domain multipole amplitudes. The corresponding multipole expansions are valid for the entire spectrum of the input pulse and at any point outside a minimum sphere enclosing the antenna. This allows a computationally cheap and elegant post-processing of arbitrary antenna characteristics. As an example of use the method is applied to determine high-resolution three-dimensional radiation patterns of an antipodal *Vivaldi* antenna.

1 Introduction

Multipole techniques have often been employed in the context of antenna fields. For example, they have been used to derive general properties of linear antennas such as the relations between the field modal amplitudes and the elements of the antenna's equivalent circuit (Chu, 1948), and to derive expressions for the antenna Q and the bandwidth (Fante, 1969; Yaghjian and Best, 2005). Moreover, multipole expansions in spherical, in cylindrical, and in Cartesian coordinates are the basis to calculate the probe-corrected far-field from measured or estimated values of the antenna near field (Hansen, 1988). The present method provides a receipt to obtain the frequency-domain spherical-multipole amplitudes from given time-domain near-field data. According to the surface equivalence theorem (Balanis, 1989) the electric far-field is completely determined by the tangential components of the field data (and resulting surface currents) on a closed surface, surrounding all scattering objects. Here, a discrete version of this theorem, namely the spherical-multipole interface (Klinkenbusch, 1995), is applied: In parallel to the time-stepping process of a time-domain near-field solver the tangential near-field on an arbitrarily chosen *Huygens* surface (that is, an imaginary closed surface enclosing the antenna structure) is replaced by a suitably chosen number of equivalent electric and magnetic elementary dipoles. The dipoles' amplitudes are temporally linearly interpolated. The corresponding time-domain spherical-multipole amplitudes are then deduced by means of the inverse Fourier transform of the bilinear form of the free-space dyadic Green's function in spherical coordinates, valid at far-field observation points.

Conventional time-domain near-field solvers (like FDTD) have intrinsic near-to-far-field routines that are based on the method of retarded potentials (Luebbers et al., 1991) and therefore make use of the closed form of this Green's function. As a consequence, a new integration over the near-field values is necessary for each observation point, and/or suitable interpolation routines are necessary (Boag and Shlivinski , 2007), which might be ineffective for large numbers of desired far-field points. In contrast, the time-domain multipole amplitudes have to be calculated only once, and the field values in arbitrary observation points are obtained by a simple superposition. Moreover, in the case of a wide-band transient pulse as the antenna signal, a numerically performed modified Fourier transform of the time-domain multipole amplitudes directly yields the frequency-domain multipole amplitudes of the antenna field for the entire frequency spectrum of this transient pulse. These frequency-domain multipole amplitudes can be used to analyze the antenna field (and other quantities of interest) at any point outside a minimum sphere enclosing all radiating elements (not just in the far-field) and at all frequencies of interest.

For the description of the present method, we first give a brief introduction to spherical-multipole expansions and a short description of the spherical-multipole interface. After that, we show how to efficiently implement the necessary convolution integrals into the time-stepping process of the

Correspondence to: J. Adam
(ja@tf.uni-kiel.de)

near-field solver. Finally we deduce the frequency-domain multipole amplitudes directly from their time-domain counterparts. To validate the present method and to demonstrate its capability of treating radiation problems of practical interest, the implementation of the method into a standard FDTD solver is applied to determine high resolution three dimensional radiation patterns of an ultra wide-band balanced antipodal *Vivaldi* antenna.

2 Frequency- and time-domain spherical multipole field representations

In any homogenous domain the frequency-domain electromagnetic field at circle frequency ω can be represented in spherical coordinates (r, ϑ, φ) by the spherical-multipole expansion

$$\begin{aligned}\boldsymbol{E}(\boldsymbol{r},\omega)=&\\ =\sum_{n,m}\Big\{&A_{n,m}(\omega)\boldsymbol{N}_{n,m}(\boldsymbol{r})+\frac{Z}{j}B_{n,m}(\omega)\boldsymbol{M}_{n,m}(\boldsymbol{r})\Big\},\end{aligned} \tag{1}$$

$$\begin{aligned}\boldsymbol{H}(\boldsymbol{r},\omega)=&\\ =-\sum_{n,m}\Big\{&\frac{j}{Z}A_{n,m}(\omega)\boldsymbol{M}_{n,m}(\boldsymbol{r})+B_{n,m}(\omega)\boldsymbol{N}_{n,m}(\boldsymbol{r})\Big\}.\end{aligned} \tag{2}$$

Here, $Z=\sqrt{\mu/\epsilon}$ denotes the intrinsic impedance of the surrounding media, and we have used the abbreviation

$$\sum_{n,m}=\sum_{n=1}^{\infty}\sum_{m=-n}^{n}.$$

The coefficients of these expansions are referred to as the (frequency-domain) electric and magnetic multipole amplitudes $A_{n,m}(\omega)$ and $B_{n,m}(\omega)$, respectively. The vector spherical-multipole functions are deduced from particular solutions of the Helmholtz equation in spherical coordinates as

$$\begin{aligned}\boldsymbol{M}_{n,m}(\boldsymbol{r}) &= z_n(\kappa r)\boldsymbol{m}_{n,m}(\vartheta,\varphi),\\ \boldsymbol{N}_{n,m}(\boldsymbol{r}) &= -\frac{z_n(\kappa r)}{\kappa r}n(n+1)Y_{n,m}(\vartheta,\varphi)\hat{\boldsymbol{r}}\\ &\quad -\frac{1}{\kappa r}\frac{d}{dr}\left[r z_n(\kappa r)\right]\boldsymbol{n}_{n,m}(\vartheta,\varphi).\end{aligned}$$

Here, the angular-dependent factors consist of the vector functions

$$\begin{aligned}\boldsymbol{m}_{n,m}(\vartheta,\varphi) &= -\frac{1}{\sin(\vartheta)}\frac{\partial Y_{n,m}(\vartheta,\varphi)}{\partial\varphi}\hat{\boldsymbol{\vartheta}}+\frac{\partial Y_{n,m}(\vartheta,\varphi)}{\partial\vartheta}\hat{\boldsymbol{\varphi}},\\ \boldsymbol{n}_{n,m}(\vartheta,\varphi) &= \frac{\partial Y_{n,m}(\vartheta,\varphi)}{\partial\vartheta}\hat{\boldsymbol{\vartheta}}+\frac{1}{\sin(\vartheta)}\frac{\partial Y_{n,m}(\vartheta,\varphi)}{\partial\varphi}\hat{\boldsymbol{\varphi}}\end{aligned}$$

and are based on the surface spherical-harmonics

$$Y_{n,m}(\vartheta,\varphi)=\sqrt{\frac{2n+1}{4\pi}\frac{(n-m)!}{(n+m)!}}P_n^m(\cos\vartheta)e^{jm\varphi}.$$

They particularly form a complete set of orthogonal vector functions on the $\mathbb{R}^3$-sphere. Here, P_n^m denotes an associated Legendre function of the first kind, see (Abramowitz and Stegun, 1972) for details. Moreover, $z_n(\kappa r)$ represents a spherical Bessel function, where $\kappa=\omega\sqrt{\mu\varepsilon}$ denotes the wavenumber. In particular, spherical Bessel function of the first kind $z_n=j_n$ are chosen to match regularity at the origin, whereas spherical Hankel functions of the 2nd kind $z_n=h_n^{(2)}$ are chosen to coincide with Sommerfeld's radiation condition.

With a suitable approximation of the mentioned spherical Hankel function, the electric far-field can be written as

$$\begin{aligned}\boldsymbol{E}_\infty(\boldsymbol{r},\omega)=-\frac{e^{-j\kappa r}}{\kappa r}\sum_{n,m}j^n\Big\{&A_{n,m}(\omega)\,\boldsymbol{n}_{n,m}(\vartheta,\varphi)\\ &-Z_0\,B_{n,m}(\omega)\,\boldsymbol{m}_{n,m}(\vartheta,\varphi)\Big\}.\end{aligned}$$

This field is easily Fourier transformed to the time-domain electric far-field representation (Oetting and Klinkenbusch, 2005):

$$\begin{aligned}\boldsymbol{e}_\infty(\boldsymbol{r},t)=-\frac{1}{r}\sum_{n,m}\Big\{&a_{n,m}(t-\tfrac{r}{v_c})\,\boldsymbol{n}_{n,m}(\vartheta,\varphi)\\ &-Z_0\,b_{n,m}(t-\tfrac{r}{v_c})\,\boldsymbol{m}_{n,m}(\vartheta,\varphi)\Big\},\end{aligned}$$

where $v_c=1/\sqrt{\mu_0\epsilon_0}$ denotes the velocity of light. The time-domain multipole amplitudes $a_{n,m}$ and $b_{n,m}$ are formally related to their frequency-domain counterparts $A_{n,m}$ and $B_{n,m}$ by the (modified) Fourier transforms

$$a_{n,m}(t)=j^n\frac{v_c}{2\pi}\int_{-\infty}^{\infty}\frac{A_{n,m}(\omega)}{\omega}e^{+j\omega t}d\omega,$$

$$b_{n,m}(t)=j^n\frac{v_c}{2\pi}\int_{-\infty}^{\infty}\frac{B_{n,m}(\omega)}{\omega}e^{+j\omega t}d\omega.$$

3 Time-domain near-to-far-field transformation by a spherical-multipole interface

In order to determine the time-domain multipole amplitudes $a_{n,m}(t)$ and $b_{n,m}(t)$, the spherical-multipole interface (Klinkenbusch, 1995) is employed: The field components on a surface S, surrounding all radiating sources (*Huygens-surface*), are replaced by a finite number of appropriately chosen electric and magnetic elementary dipoles. Assuming N_{el} electric dipoles at locations $\boldsymbol{r}_{el}^{[i]}, i\in\{1,\ldots,N_{el}\}$ and N_{mag} magnetic dipoles at $\boldsymbol{r}_{\mathrm{mag}}^{[i]}, i\in\{1,\ldots,N_{\mathrm{mag}}\}$, the corresponding current moments $\boldsymbol{c}_{el}^{[i]}$ and $\boldsymbol{c}_{\mathrm{mag}}^{[i]}$ are obtained by the near-field data as

$$\begin{aligned}\boldsymbol{c}_{el}(\boldsymbol{r}^{[i]}) &= \hat{\boldsymbol{n}}_S^{[i]}\times\boldsymbol{H}(\boldsymbol{r}^{[i]})\Delta_f^{[i]},\\ \boldsymbol{c}_{\mathrm{mag}}(\boldsymbol{r}^{[i]}) &= -\hat{\boldsymbol{n}}_S^{[i]}\times\boldsymbol{E}(\boldsymbol{r}^{[i]})\Delta_f^{[i]}.\end{aligned}$$

$\Delta_f^{[i]}$ denotes the corresponding surface area of the particular surface element, while $\hat{\boldsymbol{n}}_S^{[i]}$ represents the outwardly oriented unit vector perpendicular to that surface element. Now the time-domain multipole amplitudes are easily found by a summation of all of the dipoles fields via suitable convolution integrals containing the current moments $\boldsymbol{c}_{el}^{[i]}$ and $\boldsymbol{c}_{\text{mag}}^{[i]}$ (Oetting and Klinkenbusch, 2005; Klinkenbusch and Oetting, 2007). In a discrete form (at time-steps $k\Delta_t$ (k=0, 1, 2, ...)) and applying a temporal linear interpolation of the discrete current moments $\boldsymbol{c}_{el}^{k\,[i]}:=\boldsymbol{c}_{el}^{[i]}(k\Delta_t)$ and $\boldsymbol{c}_{\text{mag}}^{k\,[i]}:=\boldsymbol{c}_{\text{mag}}^{[i]}(k\Delta_t)$, we arrive at the following numerically implementable representations:

$$a_{n,m}(k\Delta_t)=\frac{(-1)^{n+m}}{2n(n+1)}\times \left\{ Z\sum_{i=1}^{N_{el}}\sum_{l=-w}^{w-1}\left[\psi_n^{l,[i]}\left(\boldsymbol{c}_{el}^{k-l\,[i]}+l\boldsymbol{\Delta}_{ce}^{k-l}\right)\cdot\boldsymbol{\zeta}_{n,m}^{[i]} - \frac{\chi_n^{l,[i]}}{\Delta_t}\boldsymbol{\Delta}_{ce}^{k-l}\cdot\boldsymbol{\xi}_{n,m}^{[i]}\right] + \sum_{i=1}^{N_{\text{mag}}}\sum_{l=-w}^{w-1}\frac{\psi_n^{l,[i]}}{\Delta_t}\boldsymbol{\Delta}_{cm}^{k-l}\cdot\boldsymbol{\pi}_{n,m}^{[i]}\right\}, \tag{3}$$

and

$$b_{n,m}(k\Delta_t)=\frac{(-1)^{n+m}}{2n(n+1)}\times \left\{ \frac{1}{Z}\sum_{i=1}^{N_{\text{mag}}}\sum_{l=-w}^{w-1}\left[\psi_n^{l,[i]}\left(\boldsymbol{c}_{\text{mag}}^{k-l\,[i]}+l\boldsymbol{\Delta}_{cm}^{k-l}\right)\cdot\boldsymbol{\zeta}_{n,m}^{[i]} - \frac{\chi_n^{l,[i]}}{\Delta_t}\boldsymbol{\Delta}_{cm}^{k-l}\cdot\boldsymbol{\xi}_{n,m}^{[i]}\right] - \sum_{i=1}^{N_{el}}\sum_{l=-w}^{w-1}\frac{\psi_n^{l,[i]}}{\Delta_t}\boldsymbol{\Delta}_{ce}^{k-l}\cdot\boldsymbol{\pi}_{n,m}^{[i]}\right\}. \tag{4}$$

Here we have used the abbreviations

$$\boldsymbol{\Delta}_{ce}^{l} = \left(\boldsymbol{c}_{el}^{l,[i]} - \boldsymbol{c}_{el}^{l-1,[i]}\right)$$

and $$\boldsymbol{\Delta}_{cm}^{l} = \left(\boldsymbol{c}_{\text{mag}}^{l,[i]} - \boldsymbol{c}_{\text{mag}}^{l-1,[i]}\right)$$

for all $l\in\mathbb{Z}$, where the three vector functions $\boldsymbol{\zeta}_{n,m}^{[i]}, \boldsymbol{\xi}_{n,m}^{[i]}$ and $\boldsymbol{\pi}_{n,m}^{[i]}$ are defined by

$$\boldsymbol{\zeta}_{n,m}^{[i]} = \frac{1}{r^{[i]}}\left(n(n+1)Y_{n,-m}^{[i]}\hat{r}+\boldsymbol{n}_{n,-m}^{[i]}\right),$$

$$\boldsymbol{\xi}_{n,m}^{[i]} = \frac{1}{v_c}\left(n(n+1)Y_{n,-m}^{[i]}\hat{r}+2\boldsymbol{n}_{n,-m}^{[i]}\right),$$

$$\boldsymbol{\pi}_{n,m}^{[i]} = \frac{1}{v_c}\boldsymbol{m}_{n,-m}^{[i]}.$$

Note that $Y_{n,-m}^{[i]}=Y_{n,-m}(\vartheta^{[i]},\varphi^{[i]})$, $\boldsymbol{m}_{n,-m}^{[i]}=\boldsymbol{m}_{n,-m}(\vartheta^{[i]},\varphi^{[i]})$ and $\boldsymbol{n}_{n,-m}^{[i]}=\boldsymbol{n}_{n,-m}(\vartheta^{[i]},\varphi^{[i]})$. The two scalar factors $\chi_n^{l,[i]}$ and $\psi_n^{l,[i]}$ represent integrals within the interval $[l\frac{\Delta_t v_c}{r^{[i]}}, (l+1)\frac{\Delta_t v_c}{r^{[i]}}]$, $(l\in\mathbb{Z})$, which involve Legendre polynomials and which can be evaluated completely analytically. The first factor reads

$$\psi_n^{l,[i]}=\frac{\tau_{l+1}^{[i]}P_n'(t_{l+1}^{[i]})\text{rect}(t_{l+1}^{[i]})-\tau_l^{[i]}P_n'(t_l^{[i]})\text{rect}(t_l^{[i]})}{n(n+1)},$$

where P_n denotes a Legendre polynomial of the first kind and order n, $t_l^{[i]}=l\frac{\Delta_t v_c}{r^{[i]}}$, and $\tau_l^{[i]}=\left({t_l^{[i]}}^2-1\right)$. The rectangle function is defined in the usual way as

$$\text{rect}(x)=\begin{cases}1: & |x|<1/2\\ 1/2: & |x|=1/2\\ 0: & |x|>1/2.\end{cases}$$

In the case $n>1$ the second factor reads

$$\chi_n^{l,[i]}=\frac{\lambda_n^{l+1,[i]}\text{rect}(t_{l+1}^{[i]})-\lambda_n^{l,[i]}\text{rect}(t_l^{[i]})}{(n-1)(n+2)},$$

with

$$\lambda_n^{l,[i]}=\tau_l^{[i]}\left(t_l^{[i]}P_n'(t_l^{[i]})-P_n(t_l^{[i]})\right).$$

In the case $n=1$ we finally derive

$$\chi_1^{l,[i]}=\frac{1}{3}\left[{t_{l+1}^{[i]}}^3\text{rect}(t_{l+1}^{[i]})-{t_l^{[i]}}^3\text{rect}(t_l^{[i]}) + \left(\text{rect}(t_{l+1}^{[i]})-\text{rect}(t_l^{[i]})\right)^2\right].$$

In order to efficiently fit the calculation of this expressions into the time-stepping of the time-domain near-field solver, the Legendre polynomials can be calculated by a recently proposed time-recursive scheme (Adam and Klinkenbusch, 2006):

$$\begin{pmatrix}P_n(t_{l+1}^{[i]})\\ \tau_{l+1}^{[i]}P_n'(t_{l+1}^{[i]})\end{pmatrix}=\mathbf{M}_l^{[i]}\begin{pmatrix}P_n(t_l^{[i]})\\ \tau_l^{[i]}P_n'(t_l^{[i]})\end{pmatrix},$$

where the update operator $\mathbf{M}_l^{[i]}$ is given by

$$\mathbf{M}_l^{[i]} = \frac{4\tau_{l+1}^{[i]}}{4\tau_{l+1}^{[i]}-{\Delta^{[i]}}^2 n(n+1)}\times\begin{pmatrix}\frac{4\tau_{l+1}^{[i]}+{\Delta^{[i]}}^2 n(n+1)}{4\tau_{l+1}^{[i]}} & \frac{\Delta^{[i]}\tau_{l+1}^{[i]}+\Delta^{[i]}\tau_l^{[i]}}{2\tau_l^{[i]}\tau_{l+1}^{[i]}}\\ \Delta^{[i]}n(n+1) & \frac{4\tau_l^{[i]}+{\Delta^{[i]}}^2 n(n+1)}{4\tau_l^{[i]}}\end{pmatrix},$$

with temporal increment $\Delta^{[i]}=\frac{\Delta_t v_c}{r^{[i]}}$.

4 Evaluation of antenna far- and near-fields

Consider an arbitrary linear antenna which radiates into the free unbounded space. The related electromagnetic problem is solved by a suitable near-field solver (as FDTD) using a

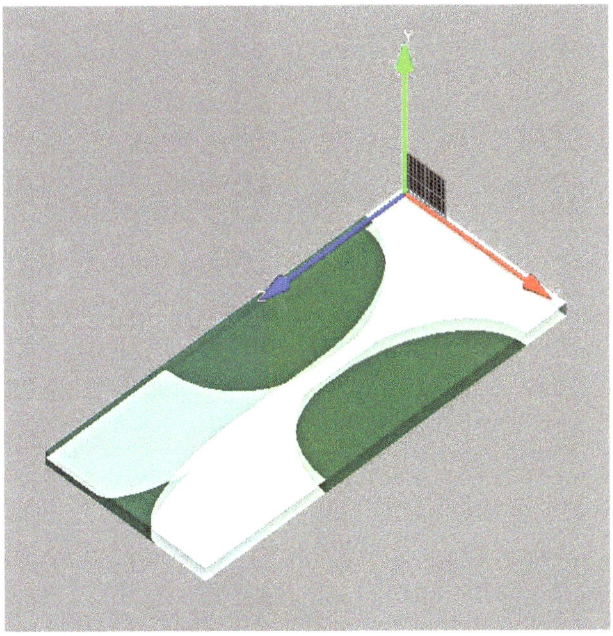

Fig. 1. CAD drawing of the antipodal Vivaldi antenna.

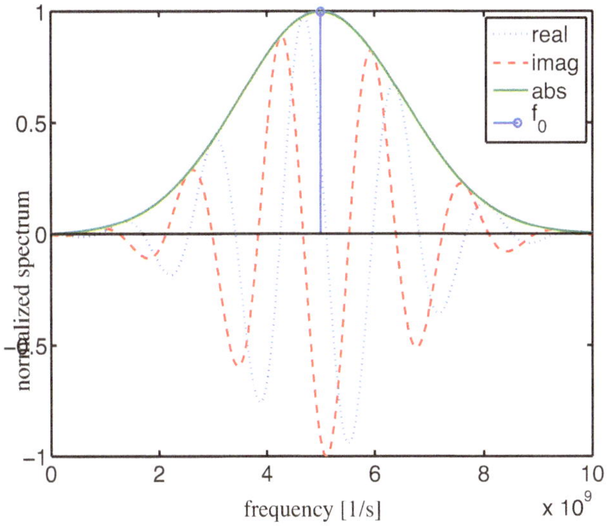

Fig. 2. Spectrum of the transient feed pulse, defined by Δ_t=2.9345 ps, f_c=5 Ghz, and β=200 time-steps.

wide-band pulse as the antenna input signal. The corresponding time-domain far-field multipole amplitudes are then evaluated as proposed in Sect. 3. The corresponding frequency-domain multipole amplitudes $A_{n,m}$ and $B_{n,m}$ are formally related to the time-domain ones by the modified Fourier transforms

$$A_{n,m}(\omega) = j^{-n}\frac{\omega}{v_c}\int_{-\infty}^{\infty} a_{n,m}(t)e^{-j\omega t}dt, \tag{5}$$

$$B_{n,m}(\omega) = j^{-n}\frac{\omega}{v_c}\int_{-\infty}^{\infty} b_{n,m}(t)e^{-j\omega t}dt. \tag{6}$$

Thus, we obtain the frequency-domain multipole amplitudes for the entire range of the spectrum of the input pulse by a (generally numerically performed but cheap) Fourier transform. The frequency-domain electromagnetic field at any desired frequency and any desired point outside the minimum sphere is then analytically obtained by means of simply evaluating (1) and (2).

As a consequence, these representations can be used to conveniently and analytically post-process the antenna near- and far-field and to obtain any related parameters.

5 Numerical Results: Vivaldi Antenna Fields by means of an FDTD solver

The proposed multipole-based near-to-far-field algorithm has been integrated into a (for research purposes provided) source-code of XFdtd, a commercially available 3-D FDTD code (Remcom Inc., 1998) (scattered field formulation, PML). In this case, the near-field data is given in discrete form (Taflove and Hagness, 2005; Kunz and Luebbers, 1993). The discrete convolution sums (3) and (4) have been implemented in the FDTD time-stepping process. Hence, the multipole amplitudes $a_{n,m}$ and $b_{n,m}$ are calculated as "on the fly" with the FDTD nearfield solver. During the post-processing stage, the frequency-domain multipole amplitudes have been computed by the means of equations (5) and (6) through a numerical (discrete) Fourier transformation.

The implementation has been applied to determine the radiation pattern of an ultra wide-band balanced antipodal Vivaldi antenna [1]. The antenna is comprised of three perfectly electrically conducting (PEC) layers where the outer ground plane layers are flared in the opposite direction of the Vivaldi. The substrate of the antenna is $49\times90\times3.15$ mm^3 and is constructed of a Duroid material with a relative permittivity of 2.32; see Fig. 1 for a CAD drawing of the structure. In order to obtain a wide-band and well defined spectrum, the antenna feed is driven by a Gaussian modulated sine pulse

$$s(t) = E_0 e^{-\alpha(t-\beta\Delta_t)^2}\sin(2\pi f_c(t-\beta\Delta_t)),$$

with center frequency f_c=5 GHz. Δ_t=2.9345 ps denotes the temporal increment (in seconds) given by the FDTD solver, the pulse width β (in time-steps) is defined by $\beta=\frac{4}{\sqrt{\alpha}\Delta_t}=200$, and the loss factor reads $\alpha=\left(\frac{4}{\beta\Delta_t}\right)^2$. The corresponding frequency spectrum is shown in Fig. 2. The feed is applied in line with the center conductor at the base

[1] http://www.remcom.com/xfdtd/examples/vivaldi-antenna-geometry.html

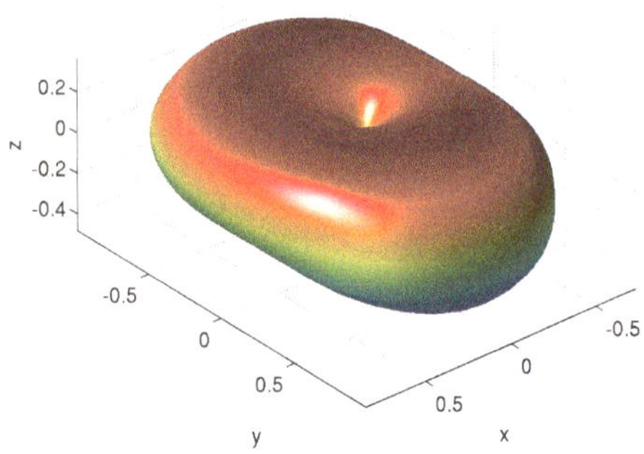

Fig. 3. Three dimensional radiated power pattern (normalized) at 1 GHz.

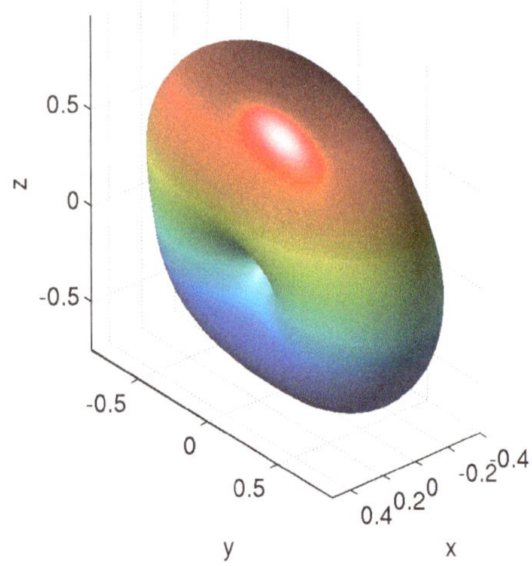

Fig. 4. Three dimensional radiated power pattern (normalized) at 2 GHz.

of the strip line. The source resistance at the feed is set to 50 Ohms to best match the input impedance of the antenna. The XFdtd antenna is simulated using $1.5\times1.575\times1.5\ \mathrm{mm}^3$ cells ($\hat{\mathbf{y}}$ dimension modified to match the 3.15 mm thickness) in a space of $68\times43\times101$ cells (plus eight cells of PML in each direction). The obtained frequency-domain multipole amplitudes are employed to determine high-resolution three-dimensional power radiation patterns of the provided antenna within the entire (discrete) spectrum of the input pulse, which is easily seen to be

$$P(\omega,\vartheta,\varphi)=\frac{4\pi}{\kappa_0^2 E_0^2}\left|\sum_{n,m} j^n\big[-\,A_{n,m}(\omega)\ \boldsymbol{n}_{n,m}(\vartheta,\varphi)\right.$$
$$\left. -Z_0\ B_{n,m}(\omega)\ \boldsymbol{m}_{n,m}\vartheta,\varphi))\big]\right|^2.$$

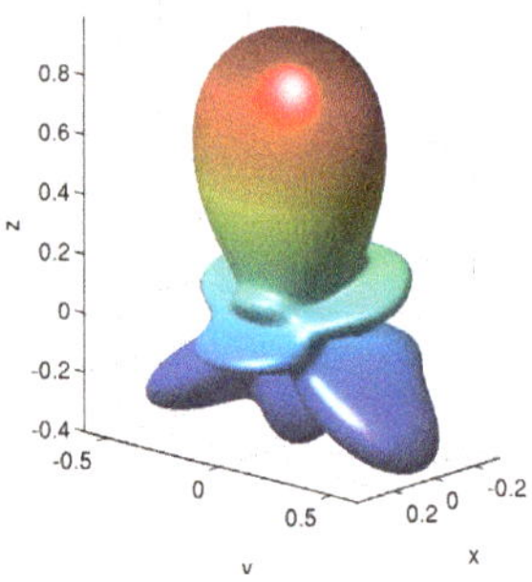

Fig. 5. Three dimensional radiated power pattern (normalized) at 5 GHz.

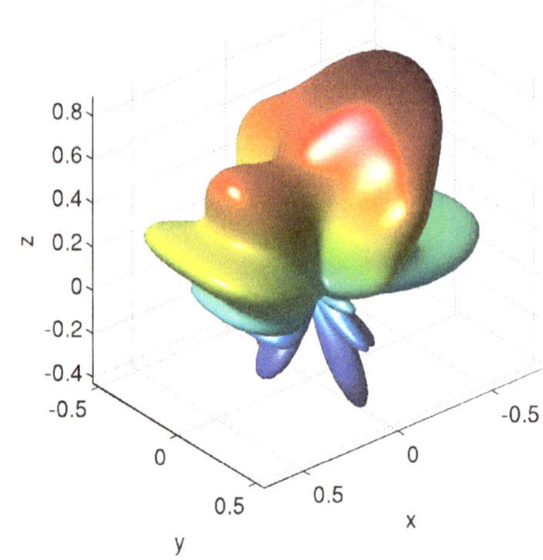

Fig. 6. Three dimensional radiated power pattern (normalized) at 9 GHz.

The low computational cost of this post-processing even admits to produce radiation-pattern-films within a few minutes. Figs. 3–6 show the radiated power patterns for an input frequency of 1 GHz, 2 GHz, 5 GHz and 9 GHz, respectively.

6 Conclusions

We have presented an efficient method to calculate frequency-domain near-and far-fields of antennas for a wide frequency spectrum by a single time-domain run of an arbitrary near-field solver. We have demonstrated the applicability of the present method by the implementation into a standard FDTD code including an analysis of an ultra wide-band balanced antipodal Vivaldi antenna.

Acknowledgements. This work has been supported by the Deutsche Forschungsgemeinschaft.

References

Abramowitz, M. and Stegun, I. E.: Handbook of Mathematical Functions, New York: Dover Pub., 10 edn., 1972.

Adam, J. and Klinkenbusch, L.: Argument-Recursive Computation of Legendre Polynomials with Applications in Computational Electromagnetics, in: Proc. of the IEEE Antennas and Propagation Society International Symposium with USNC/URSI National Radio Science and AMEREM Meetings, Albuquerque, New Mexico, 2006.

Balanis, C. A.: Advanced Engineering Electromagnetics, Wiley, New York, 1989.

Chu, L.: Physical Limitations of Omni-Directional Antennas, J. App. Physics, 19, 1163–1175, 1948.

Fante, R.: Quality Factor of General Ideal Antennas, IEEE Trans. Antennas Propagat., 17, 151–155, 1969.

Hansen, J.: Spherical Near-Field Antenna Measurements, London: Peter Peregrinus Ltd., 1988.

Remcom Inc.: XFDTD (R) Vers. 4.06, state College, PA 16805 USA, 1998.

Klinkenbusch, L.: A Spherical Multipole Interface for Numerical Methods in Electromagnetic Field Theory, in: Proceedings of the Latsis Symposium on Computational Electromagnetics, 242–247, Zürich, Switzerland, 1995.

Klinkenbusch, L. and Oetting, C.-C.: Correction to: Near-to-Far-Field Transformation by a Time-Domain Spherical-Multipole Analysis, IEEE Trans. on Antennas and Propagation, 56, 3367, 2007.

Kunz, K. and Luebbers, R.: The Finite Difference Time Domain Method for Electromagnetics, Boca Raton: CRC Press, 1993.

Luebbers, R., Kunz, K., Schneider, M., and Hundsberger, F.: A Finite-Difference Time-Domain Near Zone to Far Zone Transformation, IEEE Trans. on Antennas and Propagation, 39, 429–433, 1991.

Oetting, C.-C. and Klinkenbusch, L.: Near-to-Far-Field Transformation by a Time-Domain Spherical-Multipole Analysis, IEEE Trans. on Antennas and Propagation, 53, 2054–2063, 2005.

Boag, A. and Shlivinski, A.: Fast Evaluation of the Radiation Patterns of True Time Delay Arrays With Beam Steering , IEEE Transactions on Antennas and Propagation, 55, 3421–3432, 2007.

Taflove, A. and Hagness, S. C.: Computational Electromagnetics: The Finite-Difference Time-Domain Method, Artech House, 3 edn., 2005.

Yaghjian, A. and Best, S.: Impedance, Bandwidth, and Q of Antennas, IEEE Trans. Antennas Propagat., 53, 1298–1324, 2005.

Zhang, S. and Jin, J.: Computation of Special Functions, New York: Wiley, 1996.

Cooperative multi-user detection and ranging based on pseudo-random codes

C. Morhart and E. M. Biebl

Fachgebiet Höchstfrequenztechnik, Technische Universität München, Munich, Germany

Abstract. We present an improved approach for a Round Trip Time of Flight distance measurement system. The system is intended for the usage in a cooperative localisation system for automotive applications. Therefore, it is designed to address a large number of communication partners per measurement cycle. By using coded signals in a time divison multiple access order, we can detect a large number of pedestrian sensors with just one car sensor. We achieve this by using very short transmit bursts in combination with a real time correlation algorithm. Futhermore, the correlation approach offers real time data, concerning the time of arrival, that can serve as a trigger impulse for other comunication systems. The distance accuracy of the correlation result was further increased by adding a fourier interpolation filter. The system performance was checked with a prototype at 2.4 GHz. We reached a distance measurement accuracy of 12 cm at a range up to 450 m.

1 Introduction

There is a growing market for advanced driver assistance in modern automotive technology. Systems like Lane Change Support or Active Cruise Control are already fully integrated in today's car production. They allow the driver to extend his view or to adapt the car speed on the current traffic situation. These systems use sensors with radar, lidar or vision-based technology which offer a good positioning information in line of sight scenarios. For security applications with the goal to protect vulnerable road users the above named technologies provide insufficient results. It is very hard for them to detect objects when there is no line of sight with partially or fully hidden pedestrians.

To solve this problem an approach with cooperative sensor technology was presented (Rasshofer et al., 2007). The pedestrian is equipped with an active sensor which responds to the transmitted signal of the car. This ensures that communication occurs even in situations when the sensor is visually hidden and also disables failure detection of non-pedestrian objects. The disadvantage of this approach is the fact that every person needs to wear an active sensor. In a future application an attachment in a cell phone or in a wrist watch would be thinkable.

The position of the pedestrian is then determined by the distance and angle information relative to the car. In our earlier work (Morhart and Biebl, 2008) we presented a distance measurement sensor to fullfill this task. This sensor was capable of detecting a huge number of targets in every measurement cycle. This is possible by using very short transmit signals in a time slot method. The range of this system was 250 m at an accuracy of 1.2 m. This precision is not sufficient for successful tracking of a pedestrian in a city environment. Therefore we present an improved approach with a new transmission scheme in combination with an interpolation technique. A prototype of this system was able to achieve a 10 times smaller accuracy of 12 cm at an overall detection range of 450 m.

Correspondence to: C. Morhart
(morhart@tum.de)

2 General principle

2.1 Round Trip Time of Flight

Autonomous distance measurement between two objects mostly uses the Round-Trip Time-of-Flight (RTOF) principle. For our application, it corresponds to the time for the signal from the car to the pedestrian and vice versa. Additionally, we add a fixed waiting time T_w on the side of the pedestrian to eliminate the influence of reflections and to distinguish the answers from different pedestrian sensors. The distance Δs can then be computed by the totally elapsed time T_p:

$$\Delta s = \frac{T_p - T_w}{2} c_0. \tag{1}$$

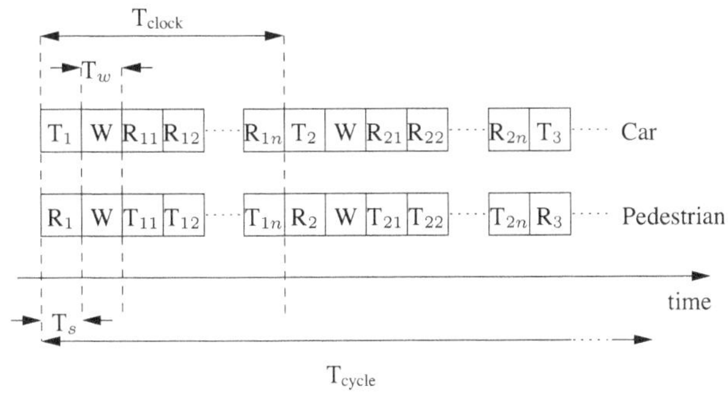

Fig. 1. Time slot order in one transmission cycle.

Δs is the distance, T_p the elapsed time, T_w the waiting time and c_0 the speed of light.

In a real application one has to take care to limit the maximum passed time as it can only be determined by a limited accuracy due to clock errors. For example with a crystal oszillator with a clock error of 10 parts per million (ppm) one gets a maximum time deviation of 1 ns at a waiting time of 100 μs. This clock error leads to a maximum distance uncertainty $\Delta\varepsilon$ of 15 cm:

$$\Delta\varepsilon = \frac{c_0}{2} T_p \cdot 10\,\text{ppm} \tag{2}$$

That means if you use that crystal and you want a precision in the range of 15 cm per measurement cycle, the passed time should not exceed 100 μs. One can improve this error by averaging over several measurement cycles, but this would decrease the system speed and also the number of adressable pedestrian sensors.

For our distance measurement system we followed a different approach. We used very short transmit pulses in combination with fast processing times. This made it possible that the maximum passed time is still in a region where the crystal inaccuracy may be neglected.

2.2 Time division multiple access

One of the main requirements for a cooperative localization system in an automotive environment is a multi-user ability. It should be possible to address with each measurement cycle a huge number of pedestrians. One solution to achieve this is to order car and pedestrian sensor systems in time slots (Fig. 1).

The car system initiates a measurement cycle by sending an pn-coded data burst with length T_s at the time slot T_1. At the same time all pedestrian systems are in receiving mode and listen in their slot R_1. Following, a waiting time T_w is added on both sides to exclude reflections on scattering objects. In the next time slots T_{11} to T_{1n} consecutively all pedestrian systems answer to the car systems in a fixed multiple of the waiting time T_w. In that way the car system can subtract each individual waiting time as a multiple of $n \cdot T_w$.

To avoid the influence of clock errors the car system has to refresh the data burst after a time T_{clock}. This time has to be chosen in accordance with the crystal precision and the maximum needed accuracy. By refreshing the data burst one may avoid that a longer waiting time causes a loss of noticeable distance precision.

2.3 Correlation

The core element of the proposed distance measurement system is the correlation algorithm. It is a time domain measurement principle which could be easy implemented on most RF modulation and demodulation hardware, e.g. WLAN transceivers. It needs a more complex digital hardware to compute the correlation algorithm, but this can be achieved with modern microprocessors.

In detail, the correlation is computed in digital form as the sum of:

$$\varphi[n] = \sum_N c[N]s[n+N]. \tag{3}$$

φ is the correlation result, $c[n]$ the correlation code, and $s[n]$ the received input signal. The equation above can be carried out as a matrix vector product or implemented in real time system as a FIR-Filter.

Another advantage of a correlation is the masking of the measurement signal. That's useful in two ways: On the one side the measurement signal can be coded so that only system sensors can read out the signal and furthermore define the distance. On the other side signals that don't belong to the measurement system can be suppressed by the correlation algorithm as it is unlikely that they ressemble the distance code.

Another advantage of this method is the insensibility to ADC jitter. The sampled input data $s[n]$ can be described by

$$s[n] = s(nT + \Delta T[n]). \tag{4}$$

T is the cycle duration of the ADC Clock and $\Delta T[n]$ is the variation of this time by clock jitter. By assuming a normally distributed clock jitter, the correlation is the average over the jitter error. This leads to an improvement in the distance uncertainty.

Figure 2 shows this result. We simulated the influence of the codelength on the standard deviation of the distance error by assumption of a jitter of $\sigma_{\Delta T}$=0.1T. The result shows that even with a correlation of small codelength the distance error can be drastically reduced.

2.4 Transmission scheme

To measure the Round-Trip Time it is very important for our system to maintain the right order of transmission. To get the distance the signal has to follow the sequence: car$\rightarrow$pedestrian$\rightarrow$car system. That means that pedestrian

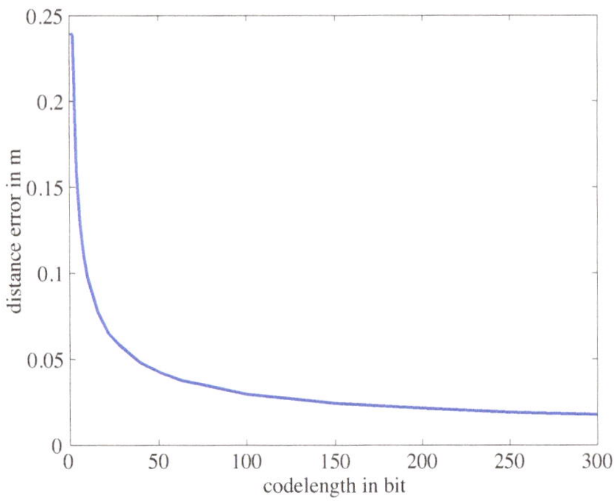

Fig. 2. Max. distance error caused by jitter in relation to the codelength.

sensors should only react on car signals and car sensors should only listen to pedestrian signals.

Figure 3 shows our solution approach. The car system initiates a measurement cycle by sending a data burst containing the correlation code plus additional transmit data. To distinguish between different car systems this code can be partly or totally varied. The pedestrian system knows that code and gets the time of arrival by correlating the incoming signal. To avoid discretisation errors and to get a simpler detection hardware, the pedestrian sensor does not generate a new correlation code but sends the inverted car code back. The advantage of this is that the sensor does not have to define the exact time-of-arrival. It has only to delay and invert the received signal. In contrast to the first advance (Morhart and Biebl, 2008) this method does not generate sampling errors on the pedestrian side and leads to higher system accuracy.

2.5 Interpolation

With the correlation result we get a high quality sensor response which is almost free of jitter and has a big difference between main and side lobes. Unfortunately, the pure correlation result has a relatively coarse distance resolution as it is limited by the sampling rate. To increase this resolution we used a interpolation approach.

Interpolation can be carried out as a polynomial interpolation or as a fourier interpolation (Kronmüller, 1991). A polynomial term can be easily processed but leads to big interpolation errors as the algorithm performs best with flat functions. A better solution is to use a fourier interpolation. Assuming that the sampling theorem is fulfilled every signal form can be reconstructed. The interpolation can be carried out with a lowpass filter either as a multiplication in frequency or as a convolution in time domain. We used a time domain approach with a low latency FIR-Filter to get an as fast as possible implementation.

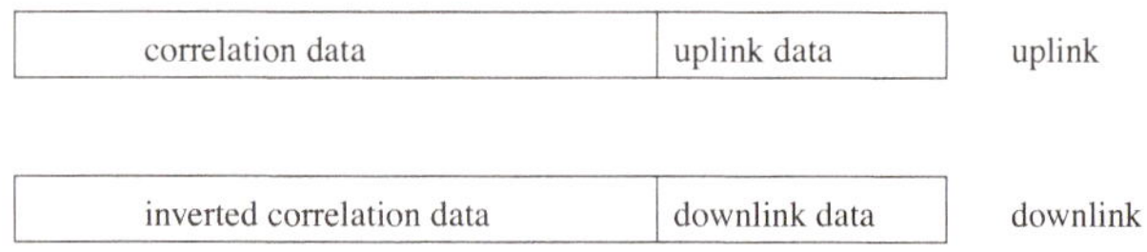

Fig. 3. Transmission scheme in up- and downlink.

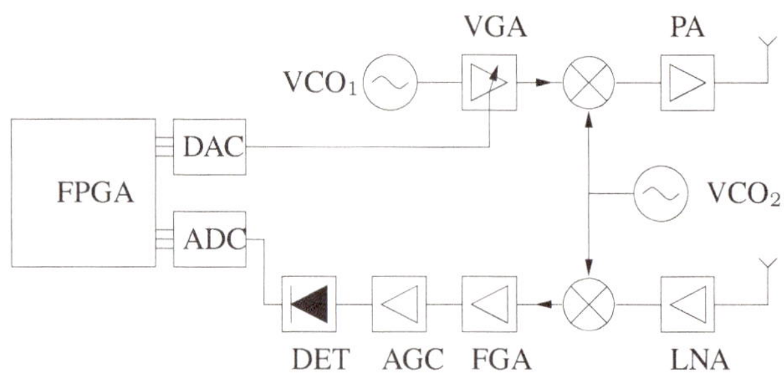

Fig. 4. Schematic capture of sensor architecture.

3 Implementation

To demonstrate the functionality of our approach we built a prototype at 2.4 GHz. This frequency is suboptimal in terms of detecting hidden persons (Fackelmeier et al., 2008), as lower frequencies have better diffraction characteristics. On the other side the ISM band at 2.4 GHz is the first band available with sufficient bandwidth.

We designed the hardware of both sensors, the car and the pedestrian system symmetrically, although the pedestrian sensor needs less processing power. Figure 4 shows the basic elements of one sensor.

We used analog amplitude modulation to get an as simple as possible transmitter and receiver design. In a later development stage it is also possible to use more complex modulation schemes as the system principle is independent on the signal modulation.

3.1 Transmitter

The analog modulation on the transmitter side is generated with a variable gain amplifier (VGA). The amplification factor is controlled by the output of the digital analog converter (DAC). This has the advandtage that we can generate a very broadband linear modulation. The output signal is then mixed by a local oscillator into the 2.4 GHz ISM Band. The frequencies of the two oscillator (VCO_1, VCO_2) have to be chosen in such a way that

$$f_{RF} = f_{VCO_1} + f_{VCO_2}. \qquad (5)$$

To avoid crosstalk between transmitter and receiver of the same system we implemented a switchable power amplifier (PA). So, after sending the data burst, the transmitter is switched off completely.

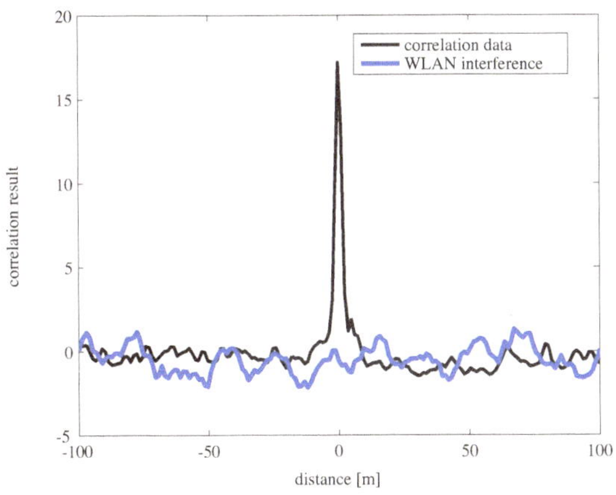

Fig. 5. correlation result of sensor data and WLAN interference.

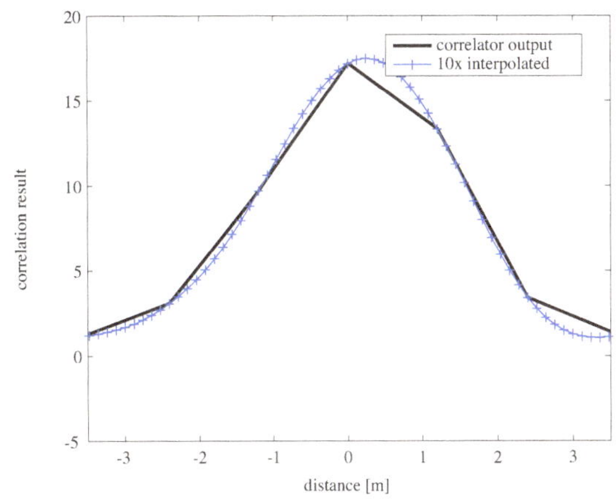

Fig. 6. comparison betweeen raw and interpolated correlation results.

3.2 Receiver

We used a full-wave envelope detector for the AM demodulation stage. This has two requirements on the IF stage, namely a tight filtering and an Automatic Gain Control (AGC). The filtering is necessary as the envelope detector is not frequency selective and therefore interference of out-of-channel communication services like digital television or GSM has to be suppressed. For the optimal performance of the detector stage the envelope demodulator has to be driven in its linear range. Thus it is essential to use adaptive preamplification in the AGC amplifier. In accordance to the desired distance range we achieved a dynamic from −80 dBm to −30 dBm at a maximum control time of 1 μs.

3.3 CPU-Unit

For the digital processing of the correlation algorithm we used a Virtex 4 FPGA. The use of a FPGA has the advantage of parallelisation. This makes it possible to carry out the correlation algorithm in real time at a clock rate of 125 MHz. It would also be possible to use a DSP or single core CPU, but this would need a more complex detection software.

For the interface to the analog front end we used an Analog to Digital Converter (ADC) and a Digital to Analog Converter (DAC) likewise with a clock rate of 125 MHz. The DAC and ADC rate has to be chosen in accordance with the data rate to fullfill the sampling theorem. We met this requirement by chosing a bit duration of 16 ns. The overall code length of one correlation burst is 256 bit. That means in combination with the AGC control time we get a minimum signal duration of 5 μs.

The pure correlation result on the car sensor at this sampling rate has a distance resolution of 1.2 m. This value is not sufficient for a automotive security application. But with the implementation of an interpolation filter we could improve it to 12 cm.

4 Measurement results

4.1 Correlation

With the correlation approach we fullfil two goals. On the one hand we suppress channel interference from other communication systems like WLAN, Bluetooth and so on. On the other hand we use it to define the time of arrival with the correlation maximum. Interference is a problem in the free accessible ISM bands as a lot of long range services use those frequencies. The ability to distinguish between different signals is therefore a big issue. Figure 5 shows the difference in the correlation result of a WLAN signal and the regular correlation data. One can easily separate the two signals by their maximum values.

But for an exact position determination this data with 1.2 m resolution is still too coarse. We improved the accuracy by implementing an additional interpolation filter. Figure 6 shows the raw correlation output and the interpolated one. With the currently used 10 times interpolation filter we get a positioning resolution of 12 cm.

4.2 Free space measurement

For the characterisation of the RTOF sensor a test scenario with a moving pedestrian and a fixed car position was used. The pedestrian moved with an approximate speed of 1 m/s straight in the direction of the car. We started at a distance of 100 m and moved directly in front of the car sensor at a distance of 2 m. As reference we used a differential GPS system (Vogel, 2007) with an accuracy down to 2 cm. Figure 7 shows an extract of the distance values of both systems.

The values from the distance sensor are clearly discretised with 12 cm discretisation step. This value is a direct result of the ADC clock rate and the interpolation filter.

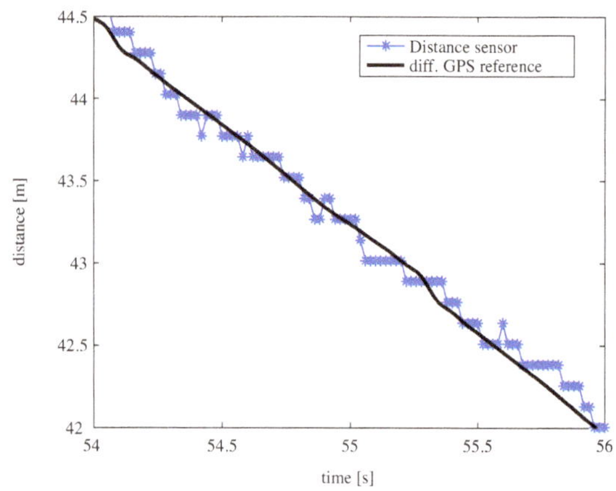

Fig. 7. Distance values in the sector from 44.5 m to 42 m.

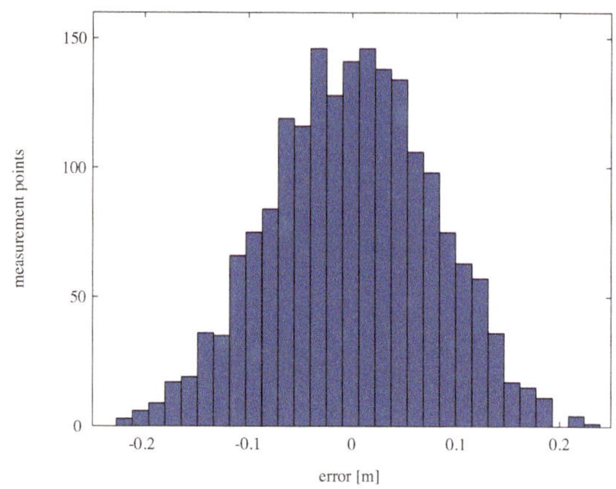

Fig. 8. Histogram of the distance error in reference to a differential GPS System.

The difference between the reference and the distance sensor is shown in the histogram Fig. 8. The error is normally distributed with a standard deviation of approximately 8 cm.

5 Conclusions

We presented an improved version of a distance measurement system for the employment in automotive applications. We used very short transmit pulses and real time correlation to avoid time consuming synchronisation methods. By employing an additional interpolation procedure in combination with an improved transmission scheme we reached a 10 times better distance resolution.

We proved our approach by means of a prototype system at 2.4 GHz. A distance resolution of 12 cm by an overall range of 450 m was achieved. In a free space measurement campaign the distance error could be determined to 8 cm. Further measurements should be done to test the system performance in applications with hidden targets.

References

Fackelmeier, A., Morhart, C., and Biebl, E. M.: Evaluation of Diffraction Effects for Identifying Hidden Targets, in: German Microwave Conference, 291–294, 2008.

Kronmüller, H.: Digitale Signalverarbeitung, Springer Verlag, 1991.

Morhart, C. and Biebl, E. M.: Ein kooperatives code-basiertes Abstandsmesssystem für eine große Anzahl simultaner Nutzer, Frequenz Journal of RF-Engineering and Telecommunications, 62, 175–179, 2008.

Rasshofer, R., Schwarz, D., Biebl, E. M., Morhart, C., Scherf, O., Zecha, S., Grünert, R., and Frühauf, H.: Pedestrian Protection Systems using Cooperative Sensor Technology, Proceedings of the 11th International Forum on Advanced Microsystems for Automotive Applications (AMAA'07), 105–115, 2007.

Vogel, K.: High-Accuracy Reference Data Acquisition for Evaluation of Active Safety Systems by means of a RTK-GNSS-Surveying System, 2007.

32

High resolution radar for automobile applications

R. Schneider and J. Wenger

DaimlerChrysler AG, Research & Technology, REM/CR, Wilhelm-Runge-Str. 11, D-89081 Ulm, Germany

Abstract. This contribution points out the potential of radar sensors with high resolution in azimuth and range in traffic environment and addresses some aspects of their practical implementation with respect to system design, hardware and software. A prototypical vehicle-based instrumentation radar is presented, which performs on-line acquisition and real-time visualisation of radar images with an update rate of up to 10 Hz. The enhanced perception capabilities of high resolution radar in traffic environment will be demonstrated with respect to new comfort and safety features, and practical realisation approaches concerning data processing and hardware including the scanning antenna will be discussed.

1 Motivation of high resolution radar for automotive application

1.1 Current status

Millimetre-wave radar has proven to be an ideal sensor in traffic environment (Wenger et al., 2000). At present some car manufacturers offer radar-based autonomous cruise control (ACC) for highway operation as an extra or will do this in the near future. Further development activities of the suppliers are primarily focused on reducing costs and size of the sensor. Although announced on their roadmaps, no additional functionality beyond ACC, like stop-and-go support, obstacle detection, and collision avoidance seems to be in sight for mass products, as this would require clearly enhanced sensor capabilities. Due to the moderate resolution of the employed sensors, insufficient information about the environment is delivered in order to enable advanced driver assistance functions like reliable obstacle warning or emergency breaking.

Conventional ACC radar sensors deliver a target list with distance, azimuth angular position, relative speed, and reflectivity of significant objects, which typically are smaller than the spatial resolution of the radar. Hence a classification of these point-targets generally is not possible. Table 1 lists some characteristics of an ACC radar sensor (Distronic by Mercedes-Benz). Especially the handling of stationary targets is difficult, as they cannot be separated by differences in their Doppler-shift.

Correspondence to: R. Schneider
(robert.r.schneider@daimlerchrysler.com)

1.2 Specific properties and potential of millimetre-wave radar imaging

The missing capability of object classification and scenario interpretation in first generation ACC sensors could be overcome using a multi or scanning-beam radar with high resolution in range and azimuth generating real-aperture radar images containing considerably more information about the ambient traffic situation. At the DaimlerChrysler Research Center in Ulm the fundamental potential of high resolution mm-wave radar with respect to automotive applications has been investigated since several years (Schneider et al., 1996, 1997; Schneider and Wenger, 1999). One of the most important results was that vehicles appear as laterally and longitudinally extended objects in high resolution radar images range versus azimuth. Caused by reflections from the road surface, the complete bottom sides can be seen, allowing for the recognition of vehicles and distinction of different classes (cars, vans, trucks) (Schneider et al., 2000). Even obscured vehicles become visible, which provides very important information for the operation in stop-and-go traffic. Another benefit is that road environment like roadsides and other ground structures (crash barriers, reflection posts, traffic signs, etc.) can clearly be identified. As the essential information about the environment is contained in a radar image with the spatial co-ordinates azimuth and range, no resolution in elevation and no large vertical antenna aperture is required. The corresponding line antenna would be well compatible with the vehicle front design.

A crucial advantage of radar sensing is its direct acquisition of velocity via Doppler processing, which could be taken

Table 1. System parameters of a typical ACC radar sensor

frequency	76–77 GHz
radar principle	pulse Doppler
number of separate beams	3
beam width (symmetrical)	3.4 deg
range resolution	5 m
Doppler resolution	1 km/h
system sensitivity	0 dBsm @ 150 m
cycle time	70 msec

Table 2. Acquisition parameters for high resolution radar data

centre frequency	76.5 GHz
field-of-view in azimuth	±10 deg
azimuth beam width	1.0 deg
elevation beam width	5.0 deg
range resolution	1 m
velocity resolution	1 km/h
system sensitivity	−20 dBsm @ 150 m

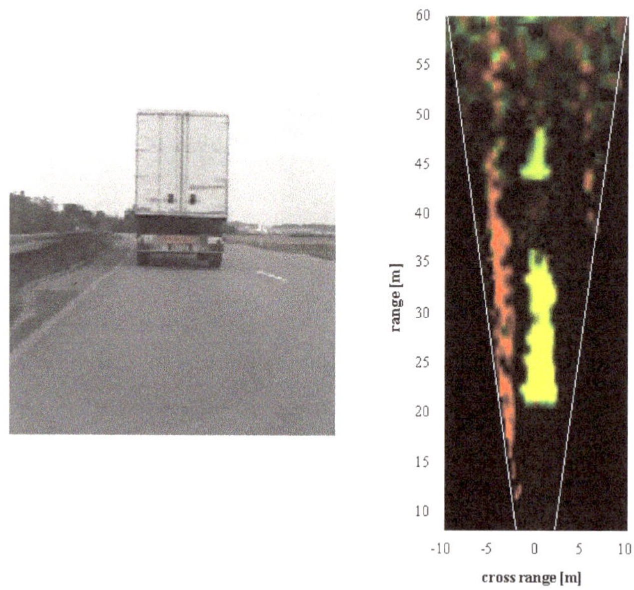

Fig. 1. Traffic scene and corresponding high resolution radar image in PPI representation.

as a third dimension in the radar images. Effective data compression is achieved by reducing the velocity information of an azimuth-range resolution cell just to a single value representing the dominant reflection object in this cell. The resulting information is described in two 2D image matrices of radar reflectivity and Doppler. For visualisation purposes both matrices can be combined into a single image using a pseudo-colour coding of Doppler and radar intensity into hue and saturation, for example.

Figure 1 shows an example of such a kind of radar image and exhibits the specific properties of high resolution radar sensing at millimetre wave radar in traffic environment. The spatial radar data acquired in polar co-ordinate grid has been transformed into a Cartesian PPI (planar position indicator) representation, which delivers a geometrically correct bird's eye view of the scene. The camera for video documentation was located close to the antenna position 60 cm above ground. The colour coding of relative Doppler shift is: yellow for the same speed, red and green for closing and opening Doppler, respectively. Radar acquisition parameters as listed in Table 2 have been applied. Details about the radar set-up will be discussed in the following sections.

Clearly visible is the left roadside with guard rail, which approaches while driving and appears as a red line. Although the right roadside is merely covered with grass, it can be recognised. The preceding truck with trailer appears yellow due to its zero relative Doppler shift, and as an extended area with true dimensions as a result from the road surface reflections. Even a car in front of the truck becomes visible by the wave duct effect. Its green colour indicates an increasing distance from the truck. Here a very attractive feature becomes evident: In platoon traffic not only the next vehicle can be tracked but also at least one ahead. Perceiving its manoeuvres could improve longitudinal vehicle control considerably.

2 Prototypical realisation of an automotive imaging radar

Based on the fundamental findings about millimetre-wave imaging an instrumentation radar system for on-line radar image acquisition and real-time visualisation was specified and realised. Although the overall system design is oriented at high performance and versatility, its basic features should have the perspective of mean to long term commercialisation. Installed in a test car the system serves for the detailed evaluation and demonstration of the capabilities of radar imaging, data acquisition, design, and test of processing algorithms, and as a basis for the implementation of new comfort or safety applications. Its operation parameters can be varied in wide ranges in order to achieve different image areas, resolution, and sensitivity.

2.1 System design

2.1.1 Radar parameters

Imaging of vehicle underbodies and detection of hidden objects via reflections from the road surface require higher sensitivity with respect to a conventional ACC radar sensor. Preliminary investigations resulted in the need of about 20 dB additional sensitivity in order to enable these features up to 150 m range, as those targets are characterised by radar cross section (RCS) values of down to −20 dBsm. An increased target dynamic of 50 dB has to be assumed. Together with the dynamic due to propagation attenuation in 5...150 m range of 60 dB an overall requirement for system dynamic results, with certainly cannot be handled by a FMCW system. Hence pulsed waveform has been chosen for the imaging radar system design because of its robustness against saturation.

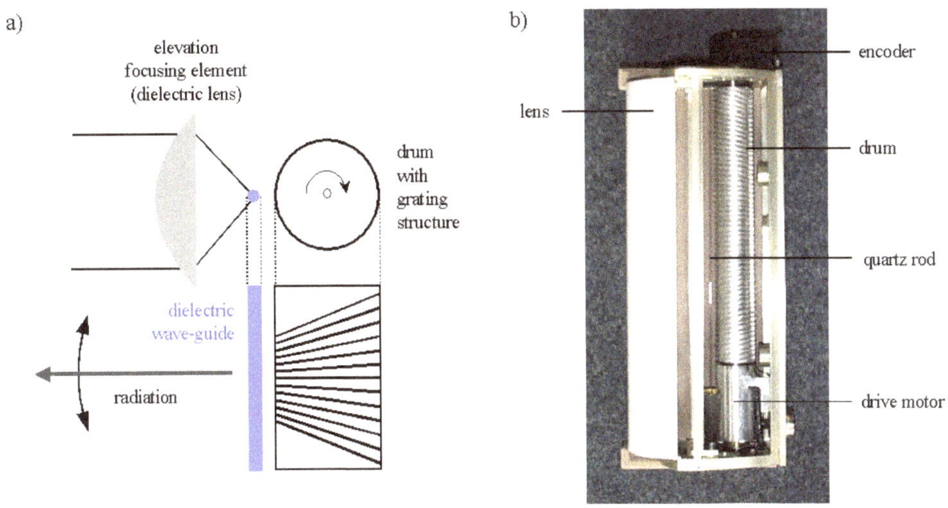

Fig. 2. Schematic diagram **(a)** and implementation **(b)** of azimuth scanning fan beam antenna.

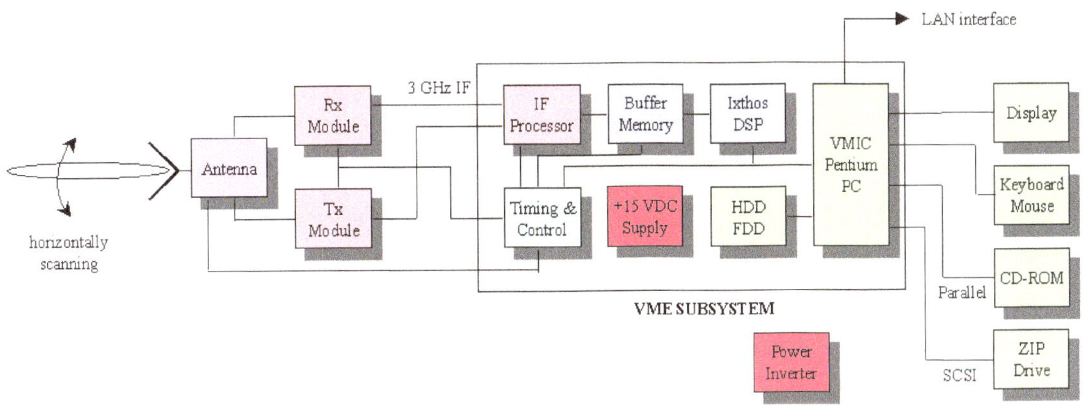

Fig. 3. Block diagram of the complete instrumentation radar system.

The radar transceiver lay-out has been based on practical imaging parameters (see Table 2). Range resolution of 1 m corresponds to 7 nsec pulse duration and to 150 MHz bandwidth of a matched IF filter. The gain of an antenna with beam widths of 1 deg×5 deg is assumed to be at least 38 dBi, which contributes considerably to the additionally required sensitivity. Speed resolution in the Doppler processing depends on the number of range bins N and the pulse repetition frequency PRF. A practical choice is N = 512 and PRF = 70 kHz, which provides sufficient unambiguous speed (500 km/h), unambiguous range (2 km), and processing gain (27 dB).

A minimum signal-to-noise ratio of 10 dB should be present in the radar images in order to achieve acceptable detection and false alarm rates. The overall system losses of radome, front-end, and signal processing are estimated not to exceed 8 dB. The transceiver design finally has to find a trade-off between transmit power and receiver noise figure F. Assuming a value of F = 8 dB achievable with state-of-the-art balanced Schottky mixers, a transmit power of 19 dBm (80 mW) would be necessary. As this hardly can be realised with today's MMIC technology, a low-noise receive amplifier (LNA) with a noise figure of 4.5 dB was provided to reduce the required transmit (peak) power to a practical value of about 35 mW. The resulting mean transmit power is below 20 μW.

The front-end design should demonstrate a long term producibility potential (cost aspect), which won't allow for parallel transceiver concepts. This means that the complete field-of-view (FOW) has to be scanned sequentially with a single antenna beam, provided that time-on-target requirements can be satisfied. The above-mentioned Doppler acquisition of 512 samples with a PRF of 70 kHz delivers full integration gain for a minimum time-on-target of about 7 msec. The required FOW is deduced from the demand of operation in urban area or narrow road curvature. Case studies show that a coverage of ±15 deg is fully sufficient, ±10 deg is acceptable with restrictions in urban (crossing) environment. Assuming an azimuth beam-width of 1 deg, the maximum achievable image update rate would be $(7\,\text{msec}\cdot 30)^{-1}$ = 4.8 Hz in the first and 7.1 Hz in the second case. Compared to a video system the update rate of an imaging radar

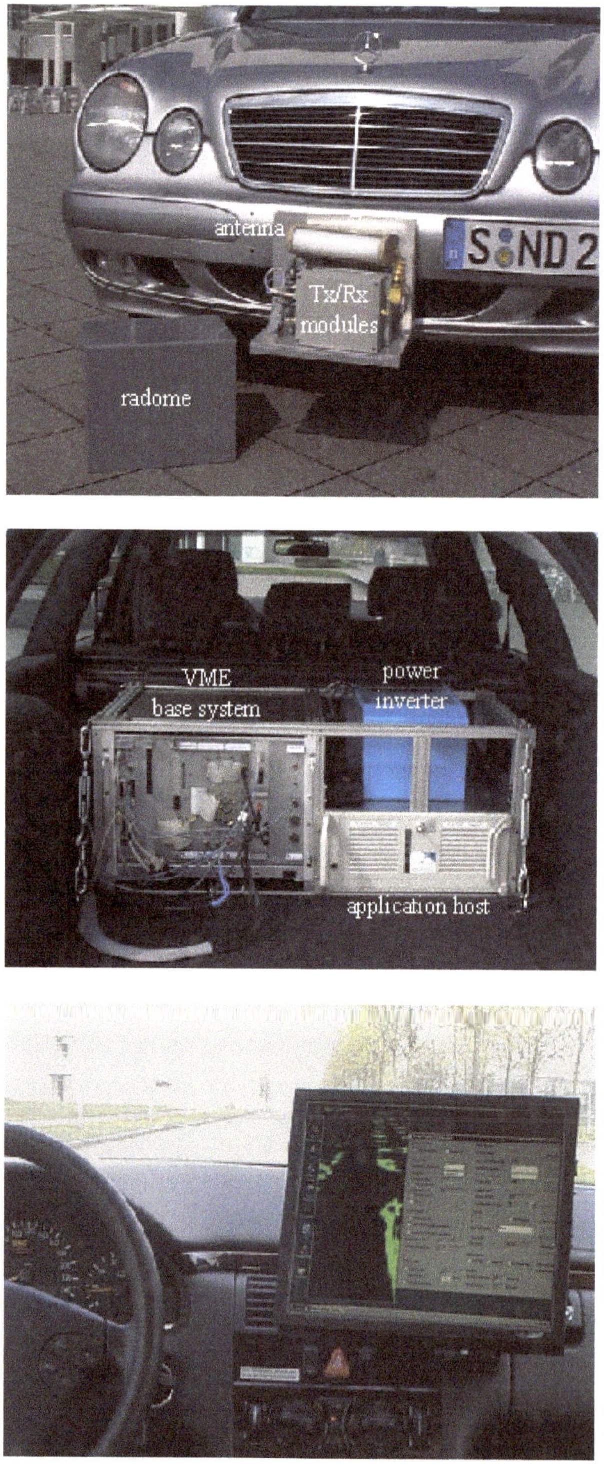

Fig. 4. Test car equipped with imaging radar: **(a)** front-end, **(b)** base system, **(c)** operation and visualisation display.

is allowed to be clearly lower, as the velocity information is directly available and intermediate images can be computed by extrapolation. Hence the mentioned update rates are assumed to be sufficient.

Table 3. Antenna characteristics

frequency	76–77 GHz
field-of-view in azimuth	± 11.5 deg
azimuth beam width	1.5 deg
elevation beam width	4.2 deg
gain	≥ 35.5 dBi
side-lobe suppression	≤ -15 dB
maximum scan rate	20 Hz
dimensions (width × height × depth)	24 cm × 6 cm × 10 cm

2.1.2 High resolution azimuth scanning antenna

The antenna forms the actual technological key component in a real aperture imaging radar. The challenge is to combine fast beam scanning of a wide FOW at good electrical properties (efficiency, side-lobes) with the potential of low-cost production, compactness, and mechanical robustness. Phased array concepts fail due to their complexity, focal plane concepts have shortcomings in achieving large FOVs, beam switching leads to high insertion loss in the required switch cascade, and pure mechanical scan isn't practicable for the large aperture.

A very attractive approach, which doesn't completely dispense with moving parts, is a tricky advancement of the image line antenna (Solbach and Wolff, 1985), which exploits the fact that inhomogeneities close to a dielectric wave-guide will cause radiation. An equidistant structure (grating) parallel to the wave-guide effects the radiation of a wave-front with a direction depending on the grating period. If now gratings with varying spacing are arranged on the circumference of a drum (see Fig. 2a), sawtooth-wise scanning of the radiated wave-front is achieved by acceleration-free rotation of the drum (Manasson et al., 1996). For beam forming in the orthogonal (elevation) plane either a cylindrical parabolic reflector or dielectric lens can be used.

Figure 2b shows the prototypic implementation of such a "spinning grating" antenna with lens (design and manufacturing by WaveBand Corp., USA), which was applied for the instrumentation radar set-up. The grating structure is realised by grooves with different pitch in a metallic cylinder resulting in a continuous beam scanning behaviour. A rigid quartz rod of about 1 mm diameter forms the dielectric wave-guide. A mode launcher allows the antenna to be connected via a standard WR-12 wave-guide flange. The drum driving is performed by a small DC gear motor (which could be integrated into the drum to increase compactness), an angular encoder serves for rotation speed control and exact reference for the beam pointing direction.

Practical experience with the antenna proves well applicability in harsh automotive environment. Some electrical and mechanical characteristics are compiled in Table 3 (verified by measurements). The azimuth beam-width is somewhat larger and accordingly the gain lower than postulated in above system design, but for availability reasons this had to be accepted.

2.2 Description of the hardware and vehicle installation

The instrumentation radar is designed as a versatile test system with real-time processing and visualisation features, which requires an open and universal hardware structure. It consists of a base unit in a standard VME chassis and the remote front-end (transmitter and receiver modules plus antenna). Figure 3 depicts the overall block diagram. The central component is an embedded Pentium III PC with Windows NT performing system controlling, radar image composition and pre-processing, visualisation, and data handling. Its main periphery is formed by timing and control electronics generating all the radar timing and a number of direct control signals, an IF processor module for converting the IF signal from the receiver (Rx) module into base-band I/Q signals and digitisation, a digital signal processor (DSP) performing the Doppler FFTs and other numerically expensive pre- processing (details see below), and a number of common devices.

System operation is done via a graphical user interface (GUI) allowing the adjustment of pertinent radar parameters like transmit pulse width, PRF, IF filter bandwidth, range sampling, number of Doppler bins, and receiver gain.

A ferrite circulator for transmit and receive duplex is mounted together with a receive LNA directly at the antenna and connects to the Tx and Rx modules. The transmit signal is generated in a phase locked dielectric resonator oscillator (DRO) at 19.125 GHz, pulse gated in a two-stage PIN-modulator, and multiplied to 76.5 GHz in an active quadrupler with an output power of about 16 dBm. The minimum pulse length is 3.5 nsec. Key component in the Rx module is a biased, balanced fundamental mixer (measured noise figure F = 7.6 dB). It converts to an IF signal of 3 GHz, which is led to the base system on a high power level in order to minimise interference in the long transmission line.

The detailed design and implementation of the VME base system and the Tx/Rx modules as well as programming the control software of the VME components were performed by Innovative Technology, USA.

Figure 4 shows the integration of the system into a test car. The front-end components are mounted centred at the front bumper in a weather-proof housing (Fig. 4a). The antenna height is positioned 60 cm above ground, which is a good compromise between exploiting multi-path effects by road surface reflections (hidden cars, vehicle underbodies) and roadside detection. Also with respect to shadowing effects of road crests (Schneider et al., 2000) the antenna position shouldn't be to low. The radome is made of 5 mm polypropylene (according to the multiple half-wavelength matching condition) and showed less than 0.2 dB two-way insertion loss.

The base system is rack-mounted together with its AC 230 V power supply in the trunk of the estate car (see Fig. 4b). An additional component placed there is a second PC provided for the intended implementation of safety or comfort applications. In the current status of the set-up it serves for synchronised video documentation and is connected to the radar control PC via Ethernet LAN.

For GUI operation of the system and the on-line radar image visualisation a display is attached at the dashboard (Fig. 4c). Note the junction on the left hand side recognisable in the instantaneous radar image.

2.3 Data processing

The imaging radar data processing is performed in the following subsequent stages: radar data pre-processing, composition of radar images, image pre-processing, and conditioning for visualisation. Most of the pre-processing is performed in the DSP card, primarily computation of the Doppler FFTs for each resolution cell and maximum search in the Doppler frequency spectra. The resulting range profiles of reflectivity and associated Doppler values have to be assembled into the actual 2D radar images. As radar acquisition is not synchronised with the antenna pointing direction, this is done based on the encoder values and interpolating onto a fix azimuth grid. The antenna scan speed is controlled in a feed-back loop including the complete processing chain, so that the azimuth increment between subsequent range profile measurements is exactly the 3 dB beam width (Nyquist condition). With the currently used antenna this results in 16 azimuth directions per scan. Range is typically sampled from 5 m to 150 m at 1 m resolution, resulting in raw radar images of 16×145 pixels. Range dependent amplitude correction is performed by R^p with an exponent $p \approx 1$ as a good compromise between point target detectability and noise limitation at larger ranges.

Some image pre-processing is performed in order to reduce artefacts caused by system deficiencies or physical effects. Cross-coupling in the transceiver or antenna leads to near range echoes at zero Doppler. They are blanked by a kind of adaptive MTI filter. Because of the relatively low side-lobe suppression of the antenna cross-range blooming of strongly reflecting objects occurs. It is reduced by appropriate lowering the level of those pixels having the same range and Doppler as this object. Another physical effect occurring in the presence of strong reflections is double bounce resulting in false echoes at twice range and twice Doppler in the images. They can be eliminated by pixel-wise checking for this condition.

In order to achieve a proper mapping of intensity values into indices suited for an image colour coding, low and high level clipping has to be performed on the logarithmic values. A heuristically optimised choice is: low level clipping at 10 dB above noise floor (this is the above specified SNR) and establishing a dynamic head room of 20 dB. The noise floor can be computed directly from the measured data by averaging the intensity of target-free resolution cells, which allows a brightness control of the radar images independent of any system parameters.

For visualisation conditioning 2D interpolation of the radar data has to be performed, which favourably is done directly in the desired co-ordinate system, either Cartesian

(PPI plot) or vanishing point projection with direct relation to optical vision. Due to the moderate FOV, trigonometric approximations can be applied in the transform formulas, which minimise the computational effort. Final step is combining intensity and Doppler matrices into a single one for pseudo-colour coding of a specific colour-map.

All of the described data processing could be implemented on the base-system PC controller without significant reduction of real-time capability. With the above mentioned radar and imaging parameters an update rate of 3 Hz results, by reduced range and Doppler settings it can be increased to 10 Hz. Additionally to visualisation the acquired data can be recorded either in MATLAB binary file or TIF image format for off-line processing.

2.4 Imaging performance

The sensitivity of the mobile instrumentation radar could be verified as predicted in the system design. Hidden vehicles proved to be detectable in all relevant situations. Roadsides can typically be recognised up to a range of 70 m, provided they consist of gravel, grass, or other vegetation (road curvature prediction based on real clothoid parameter modelling and tracking in consecutive images will work clearly farther). Radar echoes from the underbodies reliably appear in two-dimensional representation with correct dimensions and alignment in the radar images and encourage the development of corresponding classification algorithms for traffic scene interpretation.

Spurious objects are to be found regularly in the presence of guard rails, metallic fences, or kerbstones parallel to the road. Due to detour propagation vehicles appear virtually mirrored on the other side of these reflecting areas. The forward reflectivity even of low kerbstones is very high, explicable by their dihedral appearance. But here an exploitable effect becomes evident: Although nearly no backscatter can be observed from kerbstones, occurrence of multipath reflections of preceding vehicles allows localising them in the middle between direct and virtual image. This expands the capability of roadside recognition.

3 Applications

Some indications have been made in the above sections concerning new applications for an automotive high resolution radar with imaging capabilities. Currently it is presumed that the following functionalities can be achieved or at least supported by such an environmental sensor:

- improvement of ACC function (higher speed, farther range, use also off highways)
- obstacle warning (including stationary objects – for example rear end of traffic jam)
- emergency breaking
- collision avoidance

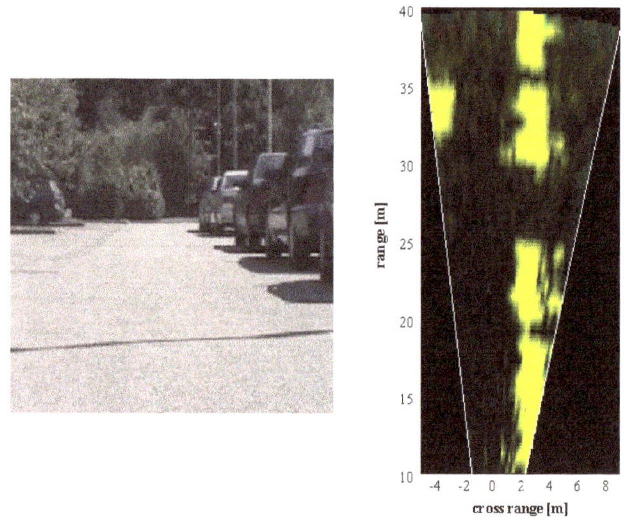

Fig. 5. Imaging capability of free parking space identification.

- stop-and-go traffic support
- detection of pedestrians
- identification of parking space
- curve warning
- visibility enhancement

Some of these are not stand-alone applications, but have to be seen in combination with others. Parking space identification would be an interesting enhancement for automatic parking systems, which currently are under development. The sensing capability of high resolution radar has been investigated by measurements in according situations and is illustrated in Fig. 5. A free parking space can be identified clearly earlier than by visual inspection and a rough size determination is possible (the free space in the analysed scene was 5 m long).

With respect to safety applications pedestrian detection also can't be an autonomous feature, as the spatial resolution of the imaging radar won't allow classification of persons. But localisation of potentially relevant objects would be a very important input for sensor fusion approaches. Radar can identify regions of interest (ROIs), in which detailed analysis based on video image processing, e.g. is performed in Heisele et al. (1996). This provides also the information redundancy, which is required in safety systems, as radar and video sensing is largely complementary.

A specific contribution of radar can be the detection of pedestrians hidden by vehicles, for example at a bus stop. Here again the wave duct effect between road surface and vehicle underbody becomes beneficial. This could help to recognise situations of persons abruptly entering the roadway (playing children). Radar measurements have been performed indicating that persons can be detected reliably in front of a bus, provided the distance is greater than the range resolution (1 m in the current system realisation).

Fig. 6. Roadside detection capability.

An interesting application based solely on radar information could be a kind of "curve assistent", which warns the driver in the danger of overshooting a curve. Exploiting the capability of continuous roadside detection even in the absence of any border structures, the road course and its curvature can be determined and related to the actual speed, which is directly available from the Doppler data. Figure 6 gives an impression of the roadside detection performance in the situation of a narrow country road. Maximum acquisition range has been 100 m. In this example a vanishing point projection has been applied to the radar image.

4 Conclusion and outlook

It could be shown that by moderate increasing of resolution, angular coverage, and sensitivity with respect to existing automotive radar sensors a new quality of environmental sensing can be established. It enables a fundamentally new approach of radar data interpretation: image processing based scenario analysis instead of point target tracking. This opens up new perspectives of comfort and safety features in future automobiles. The practical implementation of a prototypic mobile imaging radar proves technical feasibility at system design parameters, which are not contradicting to commercialisation aspects like complexity, size, and cost.

Future activities now will be hardware- and software-related as well. Mass production realisation aspects of the front-end (antenna and millimetre-wave electronics) have to be clarified. This includes also evaluation of upcoming new technologies. Autonomous implementation of the above mentioned comfort or safety features requires the development of radar image processing algorithms (based on the adaptation of established methods like edge detection, segmentation, clustering, tracking, object classification) and scene interpretation according to the particular application. An important aspect will also be fusion of radar and vision data, for which an optimised concept has to be worked out.

References

Wenger, J., Schneider, R., Stotz, M., Danner, B., Lauer, W., Nieuwenhuizen, R., and Adomat, R.: Forward Looking Radar Activities at DaimlerChrysler; Workshop on Automotive Millimetre-Wave Radar Systems (WS 3), EuMC 2000, Paris, France, pp. I,1–I,8, 3 October 2000.

Schneider, R., Wanielik, G., and Wenger, J.: Millimeterwave imaging of traffic scenarios; Intelligent Vehicles '96, Tokyo, Japan, Sept. 18–20, pp. 327–332, 1996.

Schneider, R., Wenger, J., Wanielik, G., and Neef, H.: Millimeter-wave images of traffic scenes and their automatic interpretation; Conf. Proc. Microwaves and RF, London, UK, 30 Sept.–2 Oct., pp. 239–244, 1997.

Schneider, R. and Wenger, J.: System Aspects for Future Automotive Radar; Proc. of the IEEE MTT-S Microwave Symposium, 1999.

Schneider, R., Didascalou, D., and Wiesbeck, W.: Impact of road surfaces on millimeter wave propagation; IEEE Trans. Veh. Tech. Vol. 49, No.4, pp. 1314–1320, July 2000.

Solbach, K. and Wolff, I.: Dielectric Image Line Groove Antennas for Millimeter Waves, Part II: Experimental Verification; IEEE Trans. AP, vol. AP-33, No. 7, pp. 697–706, July 1985.

Manasson, V., Sadovnik, L., and Mino, R.: MMW Scanning Antenna; IEEE AES Systems Magazine, pp. 29–33, October 1996.

Heisele, B., Neef, N., Ritter, W., Schneider, R., and Wanielik, G.: Object Detection in Traffic Scenes by a Colour Video and Radar Data Fusion Approach; First Australian Data Fusion Symposium ADF-96, Adelaide, Australia, pp. 48–52, November 1996.

Planar millimeter wave radar frontend for automotive applications

J. Grubert[1], **J. Heyen**[1], **C. Metz**[2], **L. C. Stange**[1], **and A. F. Jacob**[1]

[1]Institut für Hochfrequenztechnik, Technische Universität Braunschweig, Postfach 3329, 38023 Braunschweig, Germany
[2]Network Hardware Integration Research Department, Bell Laboratories, Murray Hill, NJ 07974, USA

Abstract. A fully integrated planar sensor for 77 GHz automotive applications is presented. The frontend consists of a transceiver multichip module and an electronically steerable microstrip patch array. The antenna feed network is based on a modified Rotman-lens and connected to the array in a multilayer approach offering higher integration. Furthermore, the frontend comprises a phase lock loop to allow proper frequency-modulated continuous wave (FMCW) radar operation. The latest experimental results verify the functionality of this advanced frontend design featuring automatic cruise control, precrash sensing and cut-in detection. These promising radar measurements give reason to a detailed theoretical investigation of system performance. Employing commercially available MMIC various circuit topologies are compared based on signal-to-noise considerations. Different scenarios for both sequential and parallel lobing hint to more advanced sensor designs and better performance. These improvements strongly depend on the availability of suitable MMIC and reliable packaging technologies. Within our present approach possible future MMIC developments are already considered and, thus, can be easily adapted by the flexible frontend design.

Es wird ein integrierter planarer Sensor für 77 GHz Radaranwendungen vorgestellt. Das Frontend besteht aus einem Sende- und Empfangs-Multi-Chip-Modul und einer elektronisch schwenkbaren Antenne. Das Speisenetzwerk der Antenne basiert auf einer modifizierten Rotman-Linse. Für eine kompakte Bauweise sind Antenne und Speisenetzwerk mehrlagig integriert. Weiterhin umfasst das Frontend eine Phasenregelschleife für eine präzise Steuerung des frequenzmodulierten Dauerstrichradars. Die aktuellen Messergebnisse bestätigen die Funktionalität dieses neuartigen Frontend-Designs, das automatische Geschwindigkeitsregelung, Kollisionswarnung sowie Nahbereichsüberwachung ermöglicht. Die Qualität der Messergebnisse hat weiterführende theoretische Untersuchungen über die potenzielle Systemleistungsfähigkeit motiviert. Unter Berücksichtigung von kommerziell erhältlichen MMICs werden verschiedene Schaltungstopologien auf der Grundlage des Signal-Rausch-Verhältnisses verglichen. Sowohl für sequenzielle als auch für parallele Ansteuerung der Antennenkeulen wird eine deutliche Leistungssteigerung ermittelt. Diese Verbesserungen hängen maßgeblich von der Verfügbarkeit geeigneter MMICs und einer zuverlässigen Aufbau- und Verbindungstechnik ab. Das vorliegende Frontend-Konzept kann auf Grund seiner Flexibilität leicht an derlei zukünftige Entwicklungen angepasst werden.

1 Introduction

Modern automobiles are equipped with an increasing number of electronic assistance systems to support the driver. In this context automotive radars gained a lot of interest in the past. This paper reports on a frequency modulated continuous wave (FMCW) sensor, which consists of a planar antenna, millimeter-wave monolithic integrated circuits (MMIC), and a phase locked loop (PLL). Printed circuit board technology and usage of MMIC offer fundamental advantages for system integration and mass production. Because of the fast progress of the MMIC performance a flexible sensor concept is useful to quickly integrate new chip sets.

2 Prototype of the frontend

2.1 Antenna

The sensor comprises a bistatic antenna system based on a planar microstrip patch array (Metz et al., 2001b). The excitation of the RX antenna is performed by a modified Rotman lens, which sets the phase and amplitude distribution to the 33 columns of the array according to the selected beam

Correspondence to: J. Grubert
(J.Grubert@TU-Braunschweig.de)

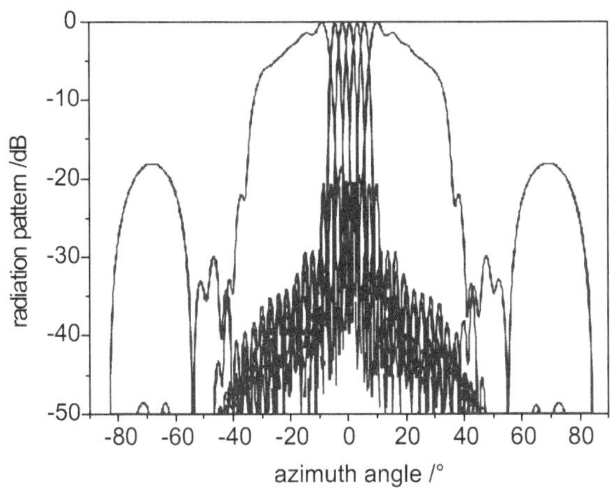

Fig. 1. Simulated RX-antenna pattern.

(H-plane). As shown in Fig. 1, the antenna system possesses five beams of 2.7° width at 26.6 dBi average gain. The distributed excitation for cut-in detection is designed to concentrate the power to the array center region. Hence, the effective array aperture is reduced, which results in a beam broadening.

The TX antenna illuminates permanently the entire field of view. With a half-power beamwidth of 22°, the resulting radiation pattern exhibits a gain of 23 dB.

Both the RX- and the TX-antenna have columns of 31 series-fed patches which determine the E-plane characteristic. Each patch column is fed at its center element by aperture coupling.

2.2 Transceiver unit

Figure 2 illustrates the architecture of the transceiver unit including PLL and antennas (Jacob et al., 2002). A voltage-controlled oscillator (VCO) generates a 12.75 GHz signal, which is amplified and tripled on-chip (CHV2242) to 38.25 GHz. The oscillator is coupled to a medium-Q resonator consisting of three coupled half-wave lines, as suggested in (Camiade et al., 2000). The oscillator exhibits a free-running phase noise of −75 dBc/Hz @ 100 kHz and a tuning range exceeding 150 MHz.

The multifunction chip (CHU2277) combines a frequency doubler and amplifiers. It has two 76.5 GHz ports with an output power of 13 dBm and 9 dBm, respectively. These feed the TX antenna and the local-oscillator (LO) port of the receiving mixer, respectively.

The integrated in-phase and quadrature mixer (CHM2177) down-converts the signal of the receiving antenna for further signal processing. Sequential lobing of the seven RX-antenna beams is provided by means of p-i-n switch MMIC. Since single-chip 77 GHz SP7T p-i-n switches are currently not available, two W6P1 p-i-n switches (Putnam et al., 1997) had to be cascaded. Hereby, additional losses in the order of 9 dB are introduced to each receiving path.

2.3 PLL

In order to suppress VCO phase noise and improve frequency chirp linearity, it is essential to employ phase-lock techniques as a part of the radar synthesizer. An especially demanding requirement for the stepped FMCW approach is a short settling time in combination with small frequency steps (Mende and Rohling, 1997).

A direct digital synthesizer (DDS) is used as a reference source for the single-loop PLL. Usually the difficulty associated with DDS is their high spurious signal content at higher output frequencies. To alleviate this, the DDS is operated at lower frequencies (4–9 MHz), where the spurious-free dynamic range is sufficiently high. This output is up-converted, filtered, and fed to the phase/frequency detector (PFD), as shown in Fig. 2.

The VCO output at 38.25 GHz is partially coupled into the feedback path and down-converted by an integrated oscillator/mixer MMIC (CHV2241). The output IF signal in the frequency range of 704–784 MHz is amplified and fed to the PFD through a prescaler/divider chain ($N = 16$) for comparison with the reference signal at 44–49 MHz.

Within 300 kHz around the carrier, the phase noise is efficiently suppressed by typically 10-15 dB, yielding a nominal value of −85 dBc/Hz @ 100 kHz. The transient behavior of the loop, determined by the loop filter bandwidth, is characterized by settling times of less than 1.5 μs.

2.4 Assembly and test

A 630-μm-thick alumina substrate carries the RF-electronics consisting of four commercially available MMICs (Camiade et al., 2000), two MMIC p-i-n switches (Putnam et al., 1997), and various passive circuit components. While all RF connections are wire bonded, the DC and IF interconnections are flip-chipped using conductive adhesives. The feed network is supported by a layer of Rohacell to avoid impairment of its electrical characteristics due to package influence.

First functionality tests were conducted in a stationary environment. The processed data is shown in Fig. 3. A corner reflector located at 60 m and a wall at about 70 m are clearly visible. A more detailed description of the measurement setup can be found in (Metz et al., 2001a).

3 Advanced RF-circuit concepts

Figure 4 depicts different standard radar frontend concepts. These were analyzed similarly to (Jacob et al., 2002), but taking into account recent MMIC developments (Camiade et al., 2002). They were compared with respect to maximum radar range. Other performance aspects as well as circuit complexity are briefly addressed in the following. The transmit and receive blocks, labeled T/R in Fig. 4, include the oscillator with its PLL as well as the receiving mixer, and are assumed to be identical in all cases.

Two antenna configurations are investigated. Monostatic systems require only a single (high gain) antenna for both

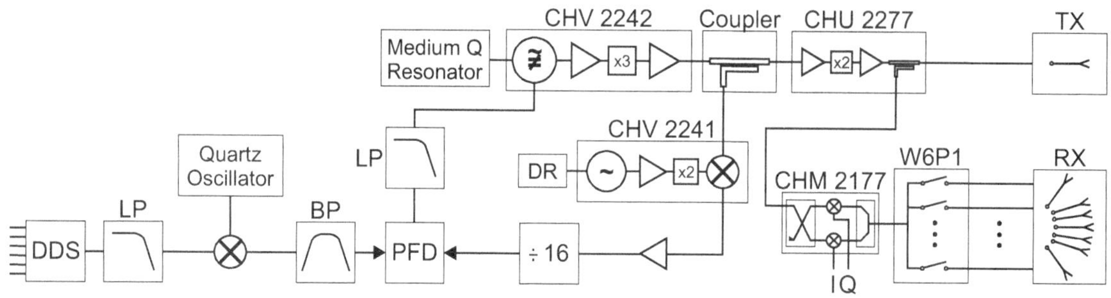

Fig. 2. Block diagram of the sensor.

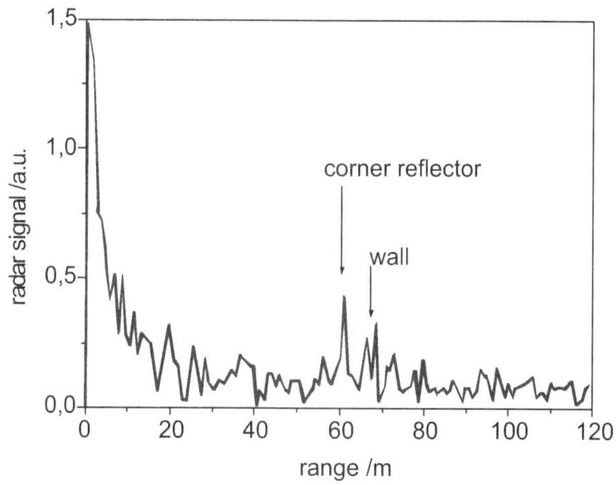

Fig. 3. Measurement result with target at 60 m and wall at 70 m.

transmit and receive operation. In Figs. 4a and c the individual beams are represented by boxes labeled T/R X. The system is small but needs a diplexer to separate the RX- and TX-paths. In monolithic integrated system this device could be a hybrid which introduces losses. Bistatic systems use two distinct antennas for transmit and receive, labeled as TX and RX, respectively (see Figs. 4b and d). The diplexer and the associated loss can thus be avoided and the RX-/TX-decoupling is improved at the expense of the required space. The TX-antenna can further be designed to simultaneously illuminate the whole field of view. This could be advantageous in terms of signal-to-noise ratio.

The surveillance of a wider angular range with a given (fine) resolution can be achieved by either sequential or parallel lobing. In the former case (Figs. 4 a and b) only one T/R-module is needed that must be sequentially connected to the different antennas by means of a multiplexer. Parallel lobing, in contrast, requires as many receivers as beams for simultaneous signal processing. Obviously, this saves time, an important feature in automotive applications. Also, the loss generally introduced by the multiplexer can be reduced. The drawback is the increased circuit complexity, i.e. the higher number of MMIC.

A comparison shows that the concept with best overall performance combines a bistatic antenna system and parallel lobing. It provides good decoupling of the TX- and RX-paths, fast signal detection because of parallel data acquisition, and large radar range because of lower losses. A possible realization based on recently presented MMIC (Camiade et al., 2000, 2002) will therefore be discussed in more detail next. Figure 5 depicts a block diagram of a possible configuration. From the sensor discussed in Sect. 2 and sketched in Fig. 2 only the integrated oscillator/mixer MMIC (CHV2241) and the multifunction chip (CHU2277) are kept. The VCO is replaced by the highly integrated version CHV2243. The output signal is amplified and partially coupled into the PLL-feedback path. The remainder is split to feed both the TX- and the LO-path. The former leads to the novel multifunction chip (CHU3077), which delivers a maximum output power of 17 dBm. The latter is connected to the multifunction chip CHU2277. Its two output ports constitute the first stage of the multiplexer. Two active power splitters (CHA2177) in the second stage provide four LO-signals. These feed dual channel mixers (CHM2378) allowing to connect eight antenna ports. To improve the signal-to-noise ratio low noise amplifiers (LNA) (CHA1077) are added in each RX-path.

4 Maximum radar range

To evaluate the system performance the frontend can be reduced to the block diagram in Fig. 6. The two-ports exhibit overall gain ($G_{TX,RX}$) and noise figure ($F_{TX,RX}$) of the entire TX- and RX-path, respectively. They also take into account loss and noise of the antennas. Their directivity is given by D_{TX} and D_{RX}, respectively, as shown in the Figure. The LO-noise is not taken into account. The power P_{RX} echoed from a target of radar cross section A at distance r follows from the well-known radar equation (Skolnik, 1980). The signal-to-noise ratio (S/N) at the mixer output is then obtained as $(S/N) = P_{RX}/(F_{RX}P_N)$. Here, $P_N = k_B T_a B$ is the noise power received by the RX-antenna (B: bandwidth; k_B: Boltzmann constant; $T_a = 290$ K: assumed noise temperature). For given minimum detectable signal to noise ratio $(S/N)_{min}$ the maximum radar range r_{max} is obtained from:

$$r_{max} = \sqrt[4]{\frac{D_{TX} D_{RX} \lambda^2 P_{TX} A}{(4\pi)^3 (S/N)_{min} F_{RX} k_B T_a B}}. \tag{1}$$

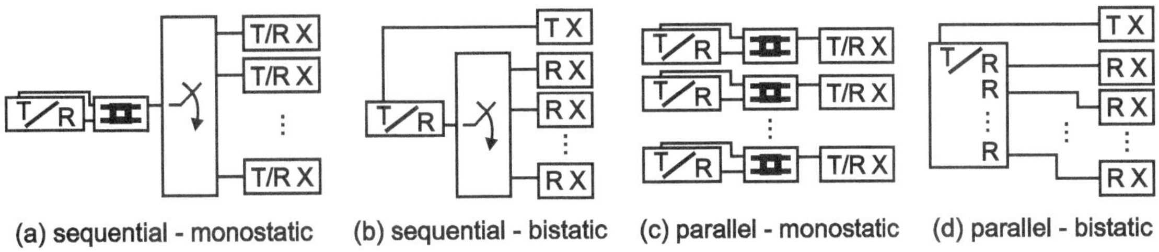

Fig. 4. Different concepts for RF-circuits. (**a**): Monostatic antenna and sequential processing. (**b**): Bistatic antenna and sequential processing. (**c**): Monostatic antenna and parallel processing. (**d**): Bistatic antenna and parallel processing.

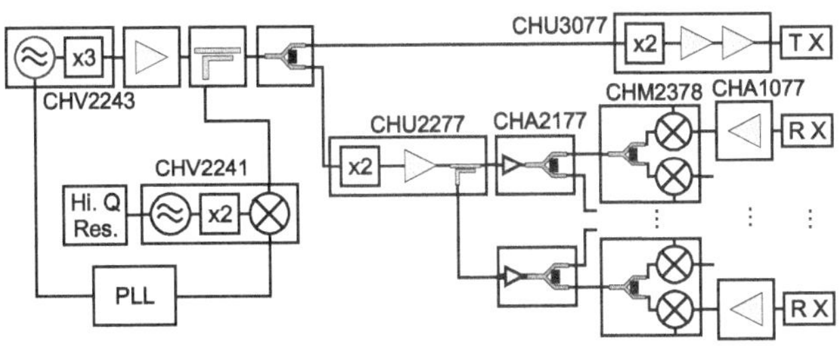

Fig. 5. Bistatic antenna and parallel processing system with latest chip generation.

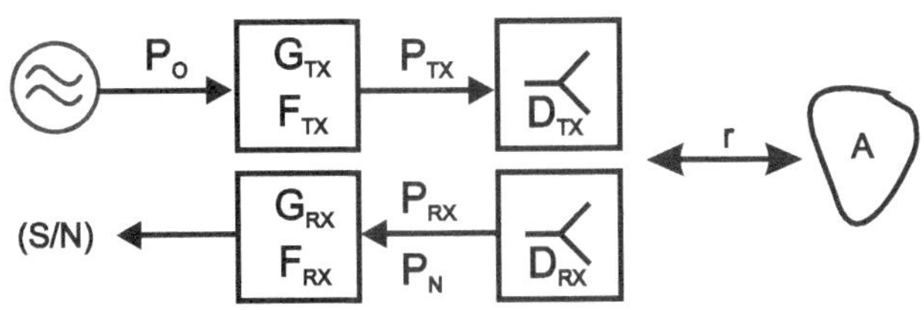

Fig. 6. Block diagram for the determination of the maximum radar range.

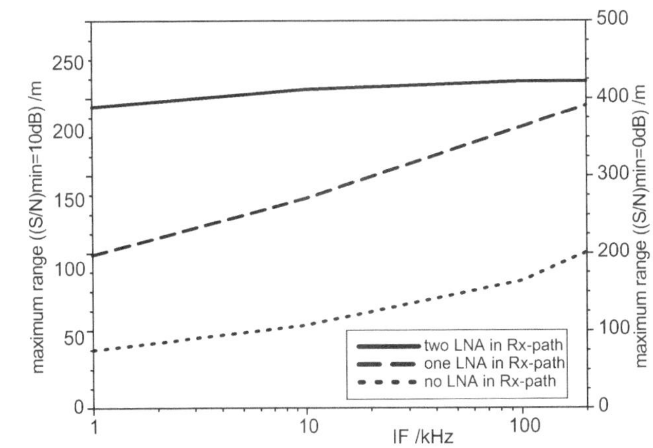

Fig. 7. Maximum radar range of the system in Fig. 5 with $(S/N)_{min}$ = 10 dB and $(S/N)_{min}$ = 0 dB.

Here, λ is the wavelength in free space. To estimate the radar range of the system in Fig. 5 the following assumptions were made. The attenuation of the bond wires and the transmit antenna reduce the output power P_O of the CHU3077 chip to P_{TX} = 15 dBm. The directivity of the TX- and RX-antenna are 27 dB and 36 dB, respectively. The losses of the RX-antenna are estimated to 10 dB. At the center frequency of 76.5 GHz the required signal bandwidth is B = 150 kHz. A worst case target with a radar cross section of $A = 1\ \text{m}^2$ is considered; this corresponds approximately to a pedestrian. The characteristics of the RX-antenna (F = 10 dB, $G = -10$ dB), the LNA (F = 4 dB, G = 17 dB (Camiade et al., 2002)), and the mixer allow to calculate the cascaded noise figure. The dependence of the mixer noise figure on output frequency (IF) is taken as specified in (UMS, 2001). Figure 7 shows achievable radar ranges for configurations with varying numbers of LNA in the RX-paths and for two minimum signal-to-noise ratios $(S/N)_{min}$. The left ordinate applies for $(S/N)_{min}$ = 10 dB, the right one for $(S/N)_{min}$ = 0 dB. The first configuration without LNA shows a maximum range which varies between 40 and 110 m (resp. 75 and 200 m) as a function of IF. Adding one LNA in each RX-path yields significant improvement with a maximum range between 110 and 220 m (resp. 200 and 390 m). A second LNA also increases the range and further reduces the IF dependence. The achievable range now varies between 220 and 240 m (resp. 390 and 420 m).

In summary, an IF of 200 kHz and a LNA in each RX-path seems to be a good compromise in terms of system performance. Indeed, a small radar cross section such as from a pedestrian is detectable at 200 m even for $(S/N)_{min}$ = 10 dB.

5 Conclusions

In this paper, a fully integrated FMCW radar sensor is presented, built from available MMIC components. Laboratory and stationary tests demonstrate the performance of this prototype. The flexible planar concept of the sensor is best suited to accommodate future MMIC developments. The latest chip generation promises significant improvement of sensor performance. Additionally, operation at higher IF appears as an effective means to substantially increase the maximum radar range.

References

Camiade, M., Domnesque, D., Quarch, Z., and Sion, A.: Fully MMIC-based front end for FMCW automotive radar at 77 GHz, Proc. 30th EuMC, Paris, France, pp. 9–12, 2000.

Camiade, M., Baglieri, D., Cortese, P., Domnesque, D., Sion, A., Beilenhoff, K., and Daembkes, H.: Monolithic microwave and MM-wave ICs for automotive applications – measures for cost reduction, IEEE MTT-S Int. Microwave Symp. Workshop Notes Automotive Radars & Prospective Circuit / Antenna Technologies from "Car Collision Avoidance" to "Autonomous Driving", Seattle, WA, 2002.

Jacob, A. F., Metz, C., Grubert, J., Heyen, J., and Stange, L. C.: Advanced Radar Concepts for Automotive Applications, IEEE MTT-S Int. Microwave Symp. Workshop Notes Automotive Radars & Prospective Circuit / Antenna Technologies from "Car Collision Avoidance" to "Autonomous Driving", Seattle, WA, 2002.

Mende, R. and Rohling, H.: A high performance AICC radar sensorconcept and results with an experimental vehicle, Radar'97, Järfälla, Sweden, pp. 21–25, 1997.

Metz, C., Grubert, J., Heyen, J., Jacob, A. F., Janot, S., Lissel, E., Oberschmidt, G., and Stange, L. C.: Fully integrated automotive radar sensor with versatile resolution, IEEE Trans. Microwave Theory Tech., 49, 12, pp. 2560–2566, Dec. 2001a.

Metz, C., Lissel, E., and Jacob, A. F.: Planar Multiresolutional Antenna for Automotive Radar, Proc. 31st EuMC, London, UK, 1, pp. 335–338, 2001b.

Putnam, J., Barter, M., Wood, K., and LeBlanc, J.: A monolithic GaAs PIN switch network for a 77 GHz automotive collision warning radar, IEEE MTT-S Int. Microwave Symp. Dig., 2, pp. 753–756, 1997.

Skolnik, M. I.: Introduction to radar systems, McGraw-Hill Inc., 1980.

United Monolithic Semiconductors S.A.S.: Data sheet CHM2378, Ref. DSCHM23781074-15-Mar.-01, United Monolithic Semiconductors S.A.S., France, 2001.

Review of singular potential integrals for method of moments solutions of surface integral equations

A. Tzoulis and T. F. Eibert

FGAN-Research Institute for High-Frequency Physics and Radar Techniques (FHR), Wachtberg-Werthhoven, Germany

Abstract. Accurate evaluation of singular potential integrals is essential for successful method of moments (MoM) solutions of surface integral equations. In mixed potential formulations for metallic and dielectric scatterers, kernels with $1/R$ and $\nabla 1/R$ singularities must be considered. Several techniques for the treatment of these singularities will be reviewed. The most common approach solves the MoM source integrals analytically for specific observation points, thus regularizing the integral. However, in the case of $\nabla 1/R$ a logarithmic singularity remains for which numerical evaluation of the testing integral is still difficult. A recently by Ylä-Oijala and Taskinen proposed remedy to this issue is discussed and evaluated within a hybrid finite element – boundary integral technique. Convergence results for the MoM coupling integrals are presented where also higher-order singularity extraction is considered.

1 Introduction

Exact solutions of electromagnetic antenna and scattering problems often rely on integral equations being solved by the method of moments (MoM) with Galerkin's method, using triangular Rao-Wilton-Glisson (RWG) vector basis functions (Rao et al., 1982). In mixed potential formulations for metallic and dielectric scatterers, the kernels of surface integrals include the Green's function of free space, as well as the gradient of Green's function. Thus, singularities of order $1/R$ and $\nabla 1/R$ must be considered, where $R=|\mathbf{r}-\mathbf{r}'|$ is the distance between observation and source points. Because of these terms, the surface integrals become singular if a testing point $\mathbf{r}$ is near the source element. In order to calculate the singular surface integrals, special methods must be used, because numerical integration routines lead to inaccurate solutions. There are many methods that can be used to evaluate singular potential integrals, such as the Duffy's transformation (Duffy, 1982; Chew et al., 2001) and the singularity extraction method (Wilton et al., 1984; Graglia, 1993; Eibert and Hansen, 1995).

In Duffy's transformation, the source triangle is divided into three subtriangles, with common vertex at the singular point. Then, the integrals over each subtriangle are transformed into integration over a square. This procedure cancels the singularity. Duffy's transformation has three main disadvantages. First, it is accurate only for sufficiently regular triangles. Second, new integration points on the source integral have to be generated for each integration point on the testing integral. This increases the computation time. Third, singularities of order $\nabla 1/R$ cannot be easily evaluated, because Duffy's transformation is derived for functions having a point singularity of order $1/R$.

In the singularity extraction method, surface integrals are regularized by extracting the singular term from the Green's function. The inner source integral of the extracted term is calculated analytically in primed coordinates for specific observation points. The remaining function is regular and the outer testing integral can be calculated numerically in unprimed coordinates. However, after the extraction of the singular term, the remaining function is not necessarily continuously differentiable, which means that a straightforward application of a numerical integration routine may lead to an inaccurate solution. Further, difficulties in integration may occur in the case of $\nabla 1/R$, when the source and test triangles have common points and are not in the same plane. In particular, after extracting the singularity and calculating the inner source integral analytically, a logarithmic singularity remains on the outer testing integral. Therefore, if higher accuracy is required, the testing integral cannot be calculated by a standard numerical calculation technique.

Ylä-Oijala and Taskinen recently proposed a remedy to this issue (Ylä-Oijala and Taskinen, 2003). The surface integrals are additionally regularized, by extracting more terms from the Green's function and its gradient. The additional extracted terms are integrated analytically over the source triangle in primed coordinates. The remaining function is at least

Correspondence to: A. Tzoulis
(tzoulis@fgan.de)

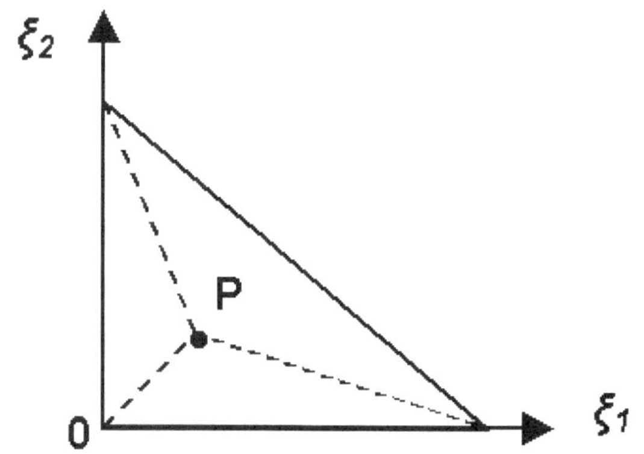

Fig. 1. Division into subtriangles.

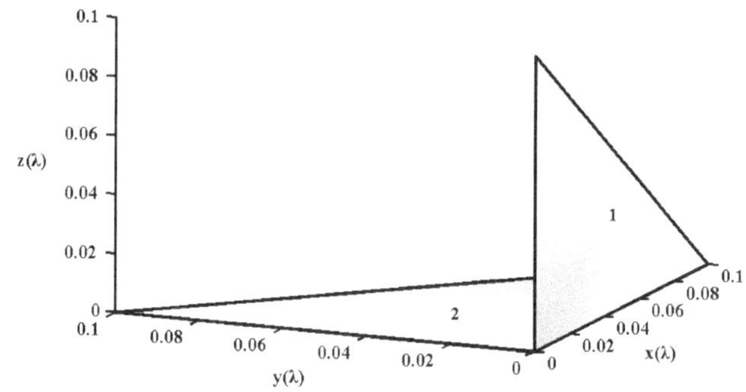

Fig. 2. Triangles for numerical examples.

once continuously differentiable and is integrated over the testing triangle numerically in unprimed coordinates. Further, in evaluating the integrals including the gradient of the Green's function, the source integral of the surface component of the singular term $\nabla 1/R$ is transformed into a line integral and the order of integration is being changed. Now, the inner testing integral is evaluated analytically in unprimed coordinates and the outer source line integral is evaluated by a standard numerical calculation technique. By these modifications logarithmic singularity can be avoided. All singularities are being extracted and calculated in closed form and numerical integration is applied only for regular and continuously differentiable functions. In this work, the above method has been evaluated within a hybrid Finite Element – Boundary Integral (FEBI) technique, using Combined Field Integral Equation (CFIE). Convergence results of the MoM coupling integrals are presented for perpendicular source and testing triangles with common edge. Using this new method, convergence of singular integrals is achieved with significantly less integration points and, because of this, accuracy, robustness and computation time of the FEBI technique is improved.

2 CFIE formulation

Consider the problem of electromagnetic scattering or radiation including dielectric and arbitrarily shaped three-dimensional bodies. According to Huygen's principle, the electric and magnetic field can be expressed as a function of the equivalent current densities on the surface S of the scatterers and the Green's function of the electromagnetic problem. Then, applying the boundary condition of the surface current densities for observation points on the surface S, integral equations can be derived with unknown quantities the tangential components of the electric and magnetic field (or equivalently the densities of the Huygen's surface currents).

For the equivalent magnetic current density $\mathbf{M}=\mathbf{E}\times\mathbf{n}$ on S, the Electric Field Integral Equation (EFIE) in mixed potential formulation is derived:

$$\begin{aligned}\frac{1}{2}(\mathbf{n}(\mathbf{r}))\times\mathbf{M}_S(\mathbf{r})=&-j\frac{\omega\mu}{4\pi}\int\!\!\int_{S'}\mathbf{J}_S(\mathbf{r}')G(\mathbf{r},\mathbf{r}')dS'\\&-j\frac{1}{4\pi\omega\varepsilon}\nabla\int\!\!\int_{S'}\nabla'_S\cdot\mathbf{J}_S(\mathbf{r}')G(\mathbf{r},\mathbf{r}')dS'\\&-\frac{1}{4\pi}\int\!\!\int_{S'}\nabla G(\mathbf{r},\mathbf{r}')\times\mathbf{M}_S(\mathbf{r}')dS'+\mathbf{E}^{\mathrm{inc}}(\mathbf{r}),\end{aligned} \tag{1}$$

with $\mathbf{r}\in S'$. Likewise, for the equivalent electric current density $\mathbf{J}=\mathbf{n}\times\mathbf{H}$ on S the Magnetic Field Integral Equation (MFIE) in mixed potential formulation is derived:

$$\begin{aligned}\frac{1}{2}\mathbf{J}_S(\mathbf{r}')=&-j\frac{\omega\mu}{4\pi}\mathbf{n}(\mathbf{r})\times\int\!\!\int_{S'}\mathbf{M}_S(\mathbf{r}')G(\mathbf{r},\mathbf{r}')dS'\\&-j\frac{1}{4\pi\omega\mu}\mathbf{n}(\mathbf{r})\times\nabla\int\!\!\int_{S'}\nabla'_S\cdot\mathbf{M}_S(\mathbf{r}')G(\mathbf{r},\mathbf{r}')dS'\\&-\frac{1}{4\pi}\mathbf{n}(\mathbf{r})\times\int\!\!\int_{S'}\nabla G(\mathbf{r},\mathbf{r}')\times\mathbf{J}_S(\mathbf{r}')dS'+\mathbf{n}(\mathbf{r})\times\mathbf{H}^{\mathrm{inc}}(\mathbf{r}),\end{aligned} \tag{2}$$

with $\mathbf{r}\in S'$. $\mathbf{E}^{\mathrm{inc}}$, $\mathbf{H}^{\mathrm{inc}}$ is the incident field, $G(\mathbf{r},\mathbf{r}')=e^{-jkR}/R$ is the Green's function of free space outside the scatterer, $\mathbf{n(r)}$ is the normal unit vector of surface S and $\nabla'_S\cdot$ denotes the surface divergence with respect to the prime coordinates.

The CFIE is the following linear combination of the EFIE and the MFIE:

$$Z_0(1-\alpha)\,\mathrm{MFIE}+\alpha\mathrm{EFIE}=0, \tag{3}$$

where α is the combination parameter, with $0\leq\alpha\leq1$, and Z_0 is the characteristic impedance of free space. Using CFIE, the internal resonance problem, which produces incorrect components in the field solution, can be avoided.

The integral equations are solved by the method of moments. Within the hybrid FEBI technique, the volume of the scatterer is modelled with tetrahedrons, which leads to a surface model with triangular elements. The surface currents are described by RWG functions (Rao et al., 1982) and the

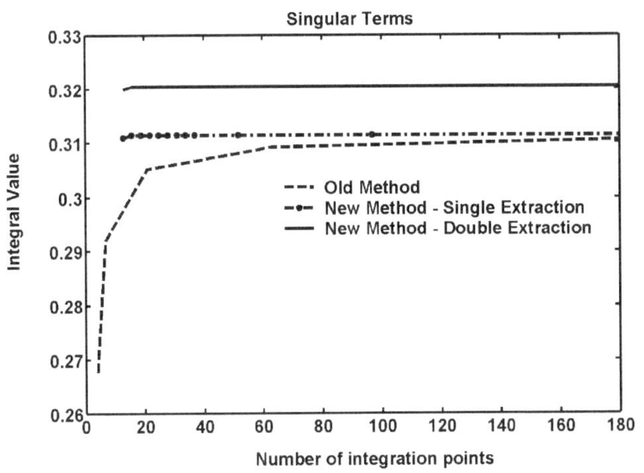

Fig. 3. Convergence of singular terms of EFIE coupling integrals with singularity $\nabla 1/R$ using the new method with single and double singularity extraction, compared with the corresponding values using the old method.

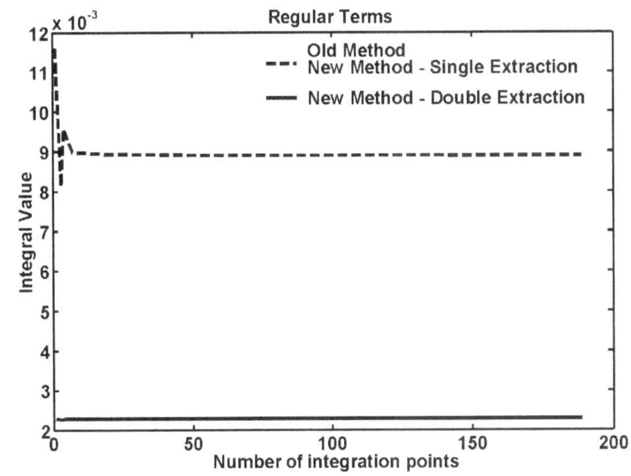

Fig. 4. Convergence of regular terms of EFIE coupling integrals with singularity $\nabla 1/R$ using the new method for single and double extraction, compared with the corresponding values using the old method.

Galerkin's method is used for the testing procedure. This means that the unknown surface currents have the form

$$\mathbf{J}_S=\sum_{n=1}^{N} i_n\boldsymbol{\beta}_n,\ \ \mathbf{M}_S=-\sum_{n=1}^{N} u_n\boldsymbol{\beta}_n. \quad (4)$$

The testing procedure is applied with RWG functions, which means that $\boldsymbol{\beta}_m=\boldsymbol{\beta}_n$. After applying this solution to the integral equation, various surface integrals must be solved. In the next section these integrals are presented and several techniques for the treatment of these integrals in singular cases are reviewed.

3 Singular integral treatments

An application of the MoM with the Galerkin's method in order to solve the CFIE using RWG basis functions requires calculation of the following integrals for the EFIE part:

$$I_1^{\mathrm{EFIE}}=\int_S\int \boldsymbol{\beta}_m(\mathbf{r})\cdot[\boldsymbol{\beta}_n(\mathbf{r}')\times\mathbf{n}(\mathbf{r})]dS, \quad (5)$$

$$I_2^{\mathrm{EFIE}}=\int_S\int \boldsymbol{\beta}_m(\mathbf{r})\cdot\left[\int_{S'}\int \boldsymbol{\beta}_n(\mathbf{r}')G(\mathbf{r},\mathbf{r}')dS'\right]dS, \quad (6)$$

$$I_3^{\mathrm{EFIE}}=\int_S\int \boldsymbol{\beta}_m(\mathbf{r})\cdot\nabla\left[\int_{S'}\int \nabla'_S\cdot\boldsymbol{\beta}_n(\mathbf{r}')G(\mathbf{r},\mathbf{r}')dS'\right]dS, \quad (7)$$

$$I_4^{\mathrm{EFIE}}=\int_S\int \boldsymbol{\beta}_m(\mathbf{r})\cdot\left[\int_{S'}\int \boldsymbol{\beta}_n(\mathbf{r}')\times\nabla G(\mathbf{r},\mathbf{r}')dS'\right]dS. \quad (8)$$

Similarly, for the MFIE part the following integrals have to be evaluated:

$$I_1^{\mathrm{MFIE}}=\int_S\int \boldsymbol{\beta}_m(\mathbf{r})\cdot\boldsymbol{\beta}_n(\mathbf{r}')ds, \quad (9)$$

$$I_2^{\mathrm{MFIE}}=\int_S\int \mathbf{n}(\mathbf{r})\times\boldsymbol{\beta}_m(\mathbf{r})\cdot\left[\int_{S'}\int \boldsymbol{\beta}_n(\mathbf{r}')G(\mathbf{r},\mathbf{r}')ds'\right]ds, \quad (10)$$

$$I_3^{\mathrm{MFIE}}=\int_S\int \mathbf{n}(\mathbf{r})\times\boldsymbol{\beta}_m(\mathbf{r})\cdot\nabla\left[\int_{S'}\int \nabla'_S\cdot\boldsymbol{\beta}_n(\mathbf{r}')G(\mathbf{r},\mathbf{r}')ds'\right]ds, \quad (11)$$

$$I_4^{\mathrm{MFIE}}=\int_S\int \mathbf{n}(\mathbf{r})\times\boldsymbol{\beta}_m(\mathbf{r})\cdot\left[\int_{S'}\int \boldsymbol{\beta}_n(\mathbf{r}')\times\nabla G(\mathbf{r},\mathbf{r}')ds'\right]ds. \quad (12)$$

The integrals I_1 for the EFIE and MFIE part are regular and can be evaluated numerically, using standard numerical integration routines without any special treatment. The integrals I_2 and I_3 for the EFIE and MFIE parts contain in the integrand the Green's function and singularities of order $1/R$ must be considered, when the observation point is near the source point. Finally, the integrals I_4 for the EFIE and MFIE parts contain in the integrand the gradient of the Green's function and singularities of order $\nabla 1/R$ must be considered. There are many methods that can be used to evaluate these integrals in the singular case, such as the Duffy's transformation and singularity extraction method.

3.1 Duffy's transformation

In this method the source triangle is first divided into three subtriangles with common vertex at singular point, as shown in Fig. 1. The inner source integral is then written in terms of these subtriangles as follows:

$$I=\int_S\int \frac{F(R,\mathbf{r},\mathbf{r}')}{R}ds=\sum_{e=1}^{3}\int_0^1\int_0^{1-\xi_1^e}\frac{F(\xi_1^e,\xi_2^e)}{R}J(\xi_1^e,\xi_2^e)d\xi_2^e d\xi_1^e, \quad (13)$$

where ξ_1^e and ξ_2^e are the parametric coordinates of the eth subtriangle and $J(\xi_1^e,\xi_2^e)$ is the corresponding Jacobian. The

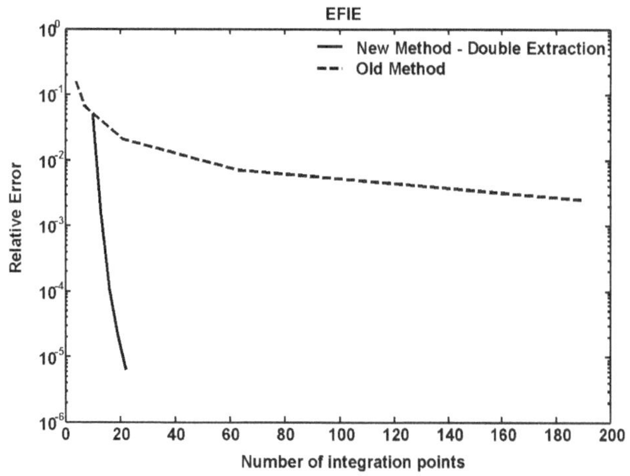

Fig. 5. Relative error of singular terms of EFIE coupling integrals with singularity $\nabla 1/R$ for the new method with double extraction and for the old method, normalized by the corresponding constant value after convergence.

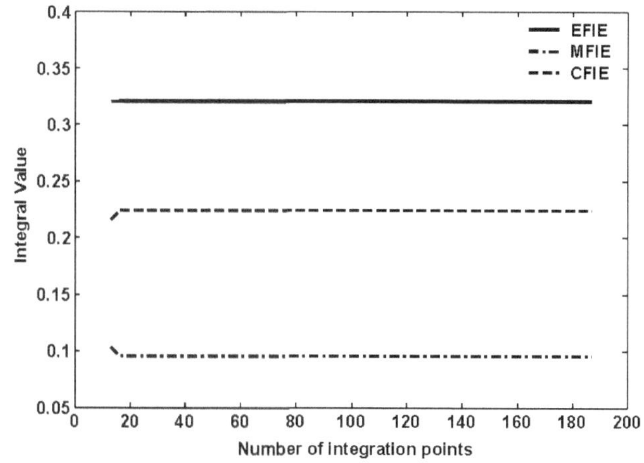

Fig. 6. Convergence of singular coupling integrals with singularity $\nabla 1/R$ for EFIE, MFIE and CFIE, using the new method with double singularity extraction.

integral over each subtriangle can be transformed into an integration over a square by introducing the transformation $\xi_2^e=(1-\xi_1^e)u$. By doing this, we have

$$I=\sum_{e=1}^{3}\int_0^1\int_0^1(1-\xi_1^e)\frac{F(\xi_1^e,(1-\xi_1^e)u)}{R}J(\xi_1^e,(1-\xi_1^e)u)d\xi_1^e du, \quad (14)$$

with

$$R=(1-\xi_1^e)\sqrt{q(\xi_1^e,u)}. \quad (15)$$

The function q is defined in terms of the coordinate ξ_1^e, the transformation variable u and the position vectors of the patch. The numerator $(1-\xi_1^e)$ cancels the singular nature of R.

3.2 Singularity extraction method

In this method, surface integrals are regularized by extracting a singular term from the Green's function as follows:

$$G(\mathbf{r},\mathbf{r}')=\left(G(\mathbf{r},\mathbf{r}')-\frac{1}{R}\right)+\frac{1}{R}. \quad (16)$$

Due to this, the gradient of the Green's function becomes

$$\nabla G(\mathbf{r},\mathbf{r}')=\nabla\left(G(\mathbf{r},\mathbf{r}')-\frac{1}{R}\right)+\nabla\frac{1}{R}. \quad (17)$$

The terms in parentheses of Eqs. (16) and (17) are regular and the extracted term includes the singularity. Applying this extraction to the integrals I_2–I_4 for the EFIE and MFIE parts, the inner source integral of the singular extracted term can be calculated in any case in closed form in primed coordinates using the formulas in (Wilton et al., 1984; Graglia, 1993; Eibert and Hansen, 1995). The remaining function is regular and the outer testing integral is being calculated numerically in unprimed coordinates. Further, the whole surface integrals I_2^{EFIE} and I_3^{EFIE} of the EFIE part can be calculated in closed form, if the test and source integrals coincide (Eibert and Hansen, 1995).

Difficulties in this approach may arise due to the following two reasons: First, after the extraction of the singular term from the Green's function, the remaining function in the parentheses of Eq. (16) has a discontinuous derivative at R=0. This discontinuity means that a straightforward application of a numerical integration routine for the testing integral may lead to an inaccurate solution. Second, difficulties may occur in the case of $\nabla 1/R$ (integrals I_4 for EFIE and MFIE part), when the source and test triangles have common points and are not in the same plane. In particular, although it is possible to calculate the inner source integral of the extracted singular term $\nabla 1/R$ analytically using the formulas in (Graglia, 1993), a logarithmic singularity still remains on the outer testing integral. Therefore, if higher accuracy is required, the testing integral cannot be calculated by a standard numerical calculation technique.

The above difficulties can be avoided by applying the singularity treatment recently proposed by Ylä-Oijala and Taskinen (Ylä-Oijala and Taskinen, 2003). In order to avoid the discontinuity of the remaining function in the parentheses of Eq. (16), an additional term can be extracted from the Green's function as follows:

$$G(\mathbf{r},\mathbf{r}')=\left(G(\mathbf{r},\mathbf{r}')-\frac{1}{R}+\frac{k^2}{2}R\right)+\frac{1}{R}-\frac{k^2}{2}R. \quad (18)$$

Due to this, the gradient of the Green's function becomes

$$\nabla G(\mathbf{r},\mathbf{r}')=\nabla\left(G(\mathbf{r},\mathbf{r}')-\frac{1}{r}+\frac{k^2}{2}R\right)+\nabla\frac{1}{R}-\frac{k^2}{2}\nabla R. \quad (19)$$

Now the terms in parentheses of Eqs. (18) and (19) have a continuous derivative at R=0 and calculation of source and testing integrals of this terms can be done easily with standard numerical procedure. The two extracted terms, i.e. the

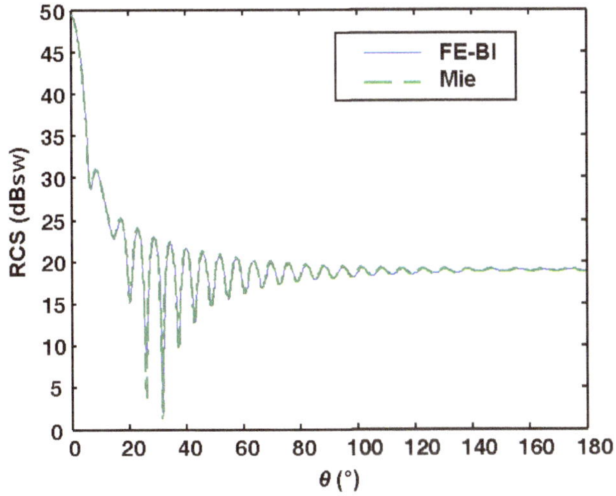

Fig. 7. Bistatic-RCS of thin-coated metallic sphere.

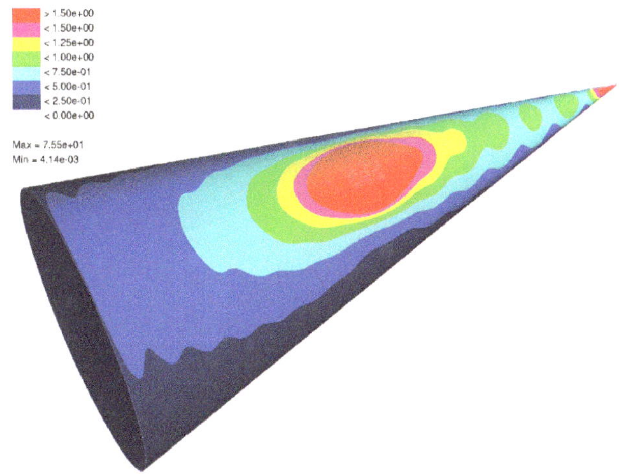

Fig. 8. Magnitude of the electric current density of a cavity-backed patch on a metallic cone.

second and third term of RHS of Eqs. (18) and (19), respectively can be integrated analytically over the source triangle in primed coordinates for all integrals $I_2 - I_4$ using the formulas presented in (Graglia, 1993) and (Ylä-Oijala and Taskinen, 2003). The formulas for analytical integration presented in (Ylä-Oijala and Taskinen, 2003) are iterative and general, allowing extraction of any number of terms from the singular kernel and integration of these terms over the source triangle in closed form.

Thereafter, the last term of Eqs. (18) and (19) can be integrated numerically over the testing triangle in unprimed coordinates for all integrals $I_2 - I_4$, because the outer integrand for this term is regular. The testing integral of the second term of Eq. (18) can be as well calculated numerically in unprimed coordinates for integrals I_2 and I_3 of the EFIE and MFIE part, because the outer integrand for this term is regular, too. The problem is to calculate the testing integral of the second term of Eq. (19), because in the outer integrand of this term a logarithmic singularity remains when the source and testing triangles have common points and are not at the same plane. This singularity exists in integrals I_4 of the EFIE and MFIE part. In particular, for analytical calculation of the source integral of the term $\nabla 1/R$ in primed coordinates the gradient is being divided into normal and surface components and the integral of the surface component over the source triangle is transformed with Gauss theorem into a line integral, which means that

$$\int\!\!\int_{S'} \nabla \frac{1}{R} dS' = \int\!\!\int_{S'} \nabla_n \frac{1}{R} dS' + \int_{\partial S'} \mathbf{u}(\mathbf{r}') \frac{1}{R} dl'. \tag{20}$$

Now, according to Eqs. (30)-(32) in Graglia (1993) the analytical expression of the line integral in the above equation has a logarithmic term, which causes the logarithmic singularity in the outer testing integral. Note that the same logarithmic singularity exists as well in the integrals I_4 when applying the single extraction of Eq. (16). In order to avoid this singularity, according to (Ylä-Oijala and Taskinen, 2003), after the source integral of the surface component of the singular term $\nabla 1/R$ is transformed into a line integral, the order of integration is being changed.

For the EFIE part, the remaining term of integral I_4^{EFIE}, which produces the logarithmic singularity after applying the singularity extraction of either Eq. (17) or Eq. (19), has the form

$$I_4^{\mathrm{EFIE}} = \int\!\!\int_{S} (\mathbf{r}-\mathbf{p}) \cdot \left[\int\!\!\int_{S'} (\mathbf{r}'-\mathbf{q}) \times \nabla' \frac{1}{R} dS' \right] dS. \tag{21}$$

where $\mathbf{p}$ is the position vector of the free vertex of the testing triangle and $\mathbf{q}$ is the position vector of the free vertex of the source triangle. Note that in the above equation the identity $\nabla(1/R) = -\nabla'(1/R)$ has been used. Replacing $(\mathbf{r}'-\mathbf{q})$ by $(\mathbf{r}'-\mathbf{r})+(\mathbf{r}-\mathbf{p})+(\mathbf{p}-\mathbf{q})$ and separating the normal and surface derivatives, Eq. (21) can be written as

$$I_4^{\mathrm{EFIE}} = \int\!\!\int_{S} (\mathbf{r}-\mathbf{p}) \cdot \left[(\mathbf{p}-\mathbf{q}) \times \int\!\!\int_{S'} \nabla'_n \frac{1}{R} dS' \right] dS$$
$$+ \int\!\!\int_{S} (\mathbf{r}-\mathbf{p}) \cdot \left[(\mathbf{p}-\mathbf{q}) \times \int\!\!\int_{S'} \nabla'_S \frac{1}{R} dS' \right] dS. \tag{22}$$

The source integral of the normal derivative is calculated analytically in primed coordinates, using Eq. (26) in Graglia (1993). In the singular case, i.e. when the testing and source triangles are not in the same plane and they have common points, the surface gradient term is dominant and numerical integration of the normal component over the testing triangle can be done with reasonable accuracy. As said before, the analytical formula for the source integral of the surface gradient term in Graglia (1993) includes a logarithmic expression, which causes the logarithmic singularity when trying to evaluate the testing integral of the surface gradient term numerically. To avoid this, after using the Gauss theorem to translate the integral of the surface gradient over the

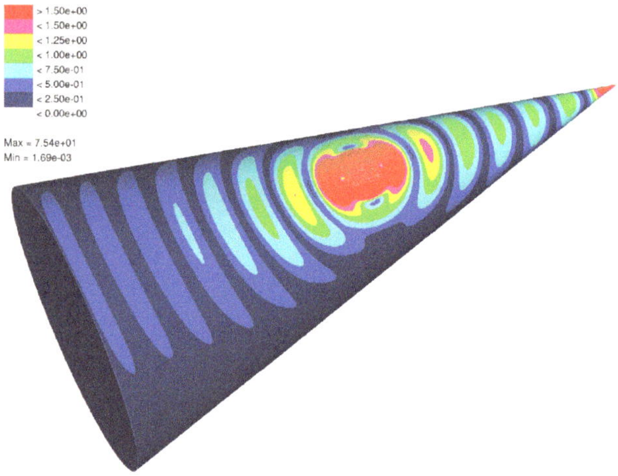

Fig. 9. Imaginary part of the electric current density of a cavity-backed patch on a metallic cone.

source triangle into a line integral, the order of integration is changed and the surface component of Eq. (22) can be written as

$$I_{4\mathrm{surf}}^{\mathrm{EFIE}} = \int_S\int (\mathbf{r}-\mathbf{p}) \cdot \left[(\mathbf{p}-\mathbf{q}) \times \int_{\partial S'} \mathbf{u}(\mathbf{r}') \frac{1}{R} dl' \right] dS$$

$$= \int_{\partial S'} \left[(\mathbf{p}-\mathbf{q}) \times \mathbf{u}(\mathbf{r}') \right] \cdot \left\{ \int_S\int \frac{(\mathbf{r}-\mathbf{p})}{R} dS \right\} dl'; \qquad (23)$$

where $\partial S'$ is the boundary of the source integral and $\mathbf{u}(\mathbf{r}')$ is the outer unit vector normal to $\partial S'$. Now, the inner testing integral is calculated analytically in unprimed coordinates using the formulas in (Ylä-Oijala and Taskinen, 2003). In those formulas the logarithmic term is cancelled. As a result, the outer source line integral has a regular integrand and allows numerical integration in primed coordinates. Hence, by considering the normal and surface gradients separately and changing the order of integration, the logarithmic singularity on the outer testing integral can be avoided.

The same procedure can be applied to the integral I_4^{MFIE} of the MFIE part. Here, the remaining term of integral I_4^{MFIE}, which produces the logarithmic singularity, after applying the singularity extraction of either Eq. (17) or Eq. (19), has the form

$$I_4^{\mathrm{MFIE}} = \int_S\int [\mathbf{n}(\mathbf{r}) \times (\mathbf{r}-\mathbf{p})] \cdot \left[\int_{S'}\int (\mathbf{r}'-\mathbf{q}) \times \nabla' \frac{1}{R} ds' \right] dS. \quad (24)$$

Applying the above procedure leads to following surface component of Eq. (24):

$$I_{4\mathrm{surf}}^{\mathrm{MFIE}} = \int_{\partial S'} \left[(\mathbf{p}-\mathbf{q}) \times \mathbf{u}(\mathbf{r}') \right] \cdot \left\{ \mathbf{n}(\mathbf{r}) \times \int_{\mathbf{S}}\int \frac{(\mathbf{r}-\mathbf{p})}{\mathbf{R}} \mathbf{dS} \right\} dl'$$

$$- \int_{\partial S'} \mathbf{u}(\mathbf{r}') \cdot \mathbf{n}(\mathbf{r}) \left\{ \int_{\mathbf{S}}\int \frac{|\mathbf{r}-\mathbf{p}|^2}{\mathbf{R}} \mathbf{dS} \right\} \mathbf{dl}'. \qquad (25)$$

Note that due to the cross product with the normal unit vector of surface S the last term of Eq. (25) must also be calculated. The inner testing integral of this term is calculated analytically in primed coordinates in (Ylä-Oijala and Taskinen, 2003) and the outer source line integral has a regular integrand and can be calculated numerically in primed coordinates.

In the next section, convergence results of the coupling integrals with singularity $\nabla 1/R$ are presented, using the above method within a hybrid FEBI technique with the CFIE. Further, numerical results of the bistatic-RCS of a thin-coated metallic sphere and of the surface currents of a cavity-backed patch on a metallic cone are presented.

4 Numerical results

The numerical treatment proposed by Ylä-Oijala and Taskinen has been applied in a hybrid FEBI technique, using the CFIE. The convergence of coupling integrals with singularity $\nabla 1/R$ has been calculated with this new method for various integral equations (EFIE, MFIE, CFIE) and for both the single extraction of Eq. (16) and the double extraction of Eq. (18). The computations have been made for the triangle combination $(1, 2)$ of Fig. 2. These triangles lie perpendicular within a tetrahedron and have a common edge.

First, the value of the EFIE coupling integrals with singularity $\nabla 1/R$ given in Eq. (8) has been calculated with the new method for single and double singularity extraction and for various integration points. The convergence of this integration is compared with the corresponding values using the old method, where only one term is extracted and integrated analytically over the source triangle and Gaussian quadrature is applied for the remaining outer testing integral. The results of this comparison are shown in Fig. 3 for the singular terms and in Fig. 4 for the regular terms of the coupling integrals. Note that the value of the regular terms of the coupling integrals with the old method is the same with the corresponding value with the new method and single extraction, because in both cases singularity was extracted by the same way and surface integrals were calculated numerically with the same routine.

It can be noticed that the coupling integrals converge faster with the new method for both singular and regular terms. Especially for the singular terms, after only a few integration points the coupling integrals using the new method have reached a constant value for single extraction, as well as a constant value for double extraction. These constant values are different because of the additional extracted term. Note that the integral value with the old method converges to the constant value of the new method with single extraction, because in both cases the same singularity extraction has been used.

The two constant values (for single and double extraction) after convergence can be used as a reference for the relative error of the value of the coupling integrals with the old method and with the new method and double extraction respectively, as shown in Fig. 5. It can be seen that the relative error for both methods becomes less by increasing the integration points, but with the old method this error still has a significant value after a lot of integration points. With the new method the relative error vanishes after only a few integrations points.

Next, convergence of the singular terms of the coupling integrals with singularity $\nabla 1/R$ has been compared for EFIE, MFIE and CFIE (with α=0.5), using the new method with double singularity extraction. The results of this comparison are shown in Fig. 6. It can be seen that for all integral equations using the new method the coupling integrals converge very fast and with about the same number of integration points.

Finally, the calculated bistatic-RCS of a thin-coated metallic sphere using the new singularity treatment within a hybrid FEBI technique with CFIE is shown in Fig. 7. The metallic kernel has a diameter of $D=10\lambda$ and the lossy thin layer a dielectric constant of $\varepsilon_r=2.5-0.05\,\mathrm{j}$. The magnitude of the electric current density of a cavity-backed patch on a metallic cone for $f=1.105\,\mathrm{GHz}$ is shown in Fig. 8. The appropriate imaginary part of the current density is shown in Fig. 9.

5 Conclusions

In this work, several techniques for the treatment of singular integrals for MoM solutions of surface integral equations have been reviewed. Especially for the integrals with singularity $\nabla 1/R$ a recently proposed method by Ylä-Oijala and Taskinen has been considered in detail, because of its capability to avoid effectively logarithmic singularity for those integrals. This method has been evaluated within a hybrid FEBI technique using CFIE where also higher-order singularity extraction was applied. The advantages of this method have been shown in convergence results for the singular case, where two triangles are not at the same plane and have common points. It was found that the coupling integrals with singularity $\nabla 1/R$ converge faster using the new method. Further, it can be noticed that the higher-order extracted terms from the singular kernel additionally regularize the coupling integrals. Because of this, accuracy, robustness and computation time of the FEBI technique is improved.

References

Chew, W. C., Jin, J.-M., Michielssen, E., and Song, J.: Fast and efficient Algorithms in Computational Eelektromagnetics, Artech House, Boston, 2001.

Duffy, M. G.: Quadrature over a pyramide or cube of integrands with a singularity at a vertex, SIAM J. Numer. Anal., 19, 1260–1262, 1982.

Eibert, T. F. and Hansen, V.: On the Calculation of Potential Integrals for Linear Source Distribution on Triangular Domains, IEEE Trans. AP, 43, 12, 1499–1502, 1995.

Graglia, R. D.: On the numerical Integration of the Linear Shape Functions Times the 3-D Greens Function or its Gradient on a Plane Triangle, IEEE Trans. AP, 41, 10, 1448–1455, 1993.

Rao, S. M., Wilton, D. R., and Glisson, A. W.: Electromagnetic scattering by surfaces of srbitrary shape, IEEE Trans. AP, 30, 409–418, 1982.

Wilton, D. R., Rao, S. M., and Glisson, A. W.: Potential Integrals for Uniform and Linear Source Distributions on Polygonal and Polyhedral Domains, IEEE Trans. AP, 32, 23, 276–281, 1984.

Ylä-Oijala, P. and Taskinsen, M.: Calculation of CFIE Impendance Matrix Elements with RWG and nxRWG Functions, IEEE Trans. AP, 51, 8, 1837–1846, 2003.

35

Fast computation of electromagnetic near-fields with the multilevel fast multipole method combining near-field and far-field translations

A. Tzoulis[1] **and T. F. Eibert**[2]

[1]FGAN-Research Institute for High-Frequency Physics and Radar Techniques (FHR), Wachtberg-Werthhoven, Germany
[2]Institute of Radio Frequency Technology, University of Stuttgart, Germany

Abstract. In Electromagnetic Compatibility (EMC) problems, computation of electromagnetic near-fields in the vicinity of complex radiation and scattering systems is often required. Numerical solution of such problems is achieved using Boundary Integral (BI) based approaches, where the involved Integral Equations (IE's) are solved with the Method of Moments (MoM). The MoM solution process is speeded up by fast IE solvers such as the Multilevel Fast Multipole Method (MLFMM). In the end the desired amplitudes of the expansion of the equivalent current densities on the discrete elements all over the Huygens' surfaces are known. Computation of the electromagnetic fields produced by the equivalent currents at observation points being in the near-field regions requires integration of the current densities over the Huygens' surfaces. Numerical evaluation of the near-field integrals using conventional integration rules can become extremely time consuming for large objects and large number of observation points. In this contribution, acceleration of the near-field integration of the equivalent current densities is provided using a postprocessing MLFMM, where near-field and far-field translations are combined in order to achieve optimum performance. The proposed approach was applied in the postprocessing stage of a powerful Finite Element Boundary Element (FEBI) method, resulting in significant decrease of the postprocessing computation time. The formulation of the proposed acceleration is presented and numerical results are shown.

1 Introduction

Complex radiation and scattering problems are solved these days routinely using Boundary Integral (BI) based methods, where the Integral Equation's (IE's) are solved by the Method of Moments (MoM) (Rao et al., 1982) using fast integral equation solvers like the Multilevel Fast Multipole Method (MLFMM) (Chew et al., 2001; Eibert, 2005). The MoM solution provides the desired amplitudes of the expansion of the equivalent current densities on the discrete elements all over the Huygens' surfaces.

In order to compute the electromagnetic fields produced by the equivalent currents on observation points in the near-field region of the involved objects, integration of the current densities over the entire Huygens' surfaces must be performed in the postprocessing stage. Such computations are needed in Electromagnetic Compatibility (EMC) problems like investigations within cars or for safety assessment in the vicinity of mobile communications base station antennas. The complexity of evaluating near-field integrals in the postprocessing stage using conventional numerical integration rules is $O(NM)$, where N is the number of equivalent current elements and M the number of observation points. It is obvious that postprocessing computations become extremely time-consuming in case of large objects and large number of observation points.

In this contribution, near-field computations in the postprocessing stage are accelerated by a Postprocessing Multilevel Fast Multipole Method (P-MLFMM). According to this, the BI-MLFMM domain of the equivalent currents is extended to also include nearby observation points and in addition to non-empty groups containing source currents also non-empty groups containing observation points are collected on each level. Near-field MLFMM translations are performed from source groups including currents to receiving groups including observation points. For far-away observation points, the field contributions are computed using far-field MLFMM translations, which are performed for each observation point on the coarsest level on which far-field condition is still satisfied. The optimum level for far-field translations is found for each observation point in the initialization step in a worst-case sense, using its shortest distance to the BI-MLFMM domain. The proposed approach was ap-

Correspondence to: A. Tzoulis (tzoulis@fgan.de)

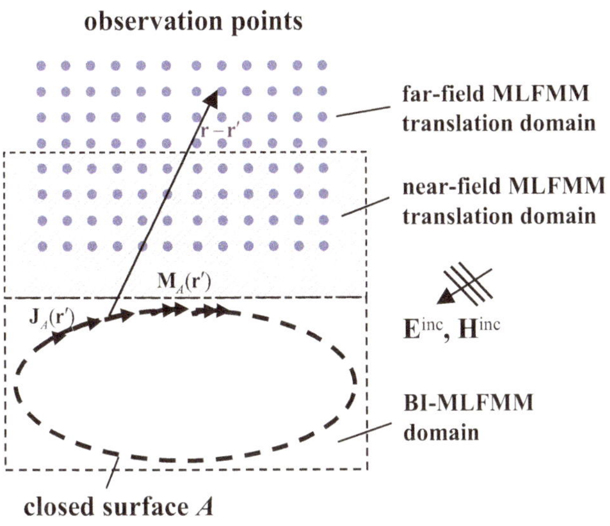

Fig. 1. Configuration for postprocessing computations of BI based methods.

plied in the postprocessing stage of a powerful Finite Element Boundary Element (FEBI) method (Jin, 1993; Volakis et al., 1998), resulting in significant decrease of the postprocessing computation time.

In the following, the formulation of the MLFMM approach for the postprocessing stage is presented and numerical results are shown.

2 Formulation

Consider the configuration of Fig. 1, where electric $\boldsymbol{J}_A(\boldsymbol{r}')$ and magnetic $\boldsymbol{M}_A(\boldsymbol{r}')$ current densities are given all over the closed Huygens' surface A and are known from the solution of any BI based method.

The electric and magnetic fields at an observation point $\boldsymbol{r}$ produced by the surface currents are given by the integral expressions

$$\boldsymbol{E}(\boldsymbol{r}) = \iint_A \Big[\bar{\mathbf{G}}_J^E(\boldsymbol{r},\boldsymbol{r}') \cdot \boldsymbol{J}_A(\boldsymbol{r}') + \bar{\mathbf{G}}_M^E(\boldsymbol{r},\boldsymbol{r}') \cdot \boldsymbol{M}_A(\boldsymbol{r}') \Big] da' + \boldsymbol{E}^{inc}(\boldsymbol{r}) \tag{1}$$

and

$$\boldsymbol{H}(\boldsymbol{r}) = \iint_A \Big[\bar{\mathbf{G}}_J^H(\boldsymbol{r},\boldsymbol{r}') \cdot \boldsymbol{J}_A(\boldsymbol{r}') + \bar{\mathbf{G}}_M^H(\boldsymbol{r},\boldsymbol{r}') \cdot \boldsymbol{M}_A(\boldsymbol{r}') \Big] da' + \boldsymbol{H}^{inc}(\boldsymbol{r}), \tag{2}$$

respectively, where $\bar{\mathbf{G}}_J^{E/H}(\boldsymbol{r},\boldsymbol{r}')$ and $\bar{\mathbf{G}}_M^{E/H}(\boldsymbol{r},\boldsymbol{r}')$ are the Green's functions of the electric or magnetic field due to electric and magnetic surface currents, respectively. The equivalent currents on A are assumed to be available in a known expansion using RWG basis functions on triangular surface elements according to

$$\boldsymbol{J}_A(\boldsymbol{r}') = \sum_n J_n \beta_n(\boldsymbol{r_n}),$$
$$\boldsymbol{M}_A(\boldsymbol{r}') = -\sum_n M_n \beta_n(\boldsymbol{r_n}). \tag{3}$$

The evaluation of the field contributions in Eqs. (1) and (2) is accelerated using MLFMM. According to this, the BI-MLFMM domain is extended towards the observation points and in addition to non-empty source groups containing currents also non-empty receiving groups containing observation points are collected on each level. The field contributions from source groups to non well-separated receiving groups are evaluated conventionally by numerical integration rules. All other field contributions at an observation point $r = r_m$ are computed using the fast multipole representations

$$\boldsymbol{E}(\boldsymbol{r_m}) = -j\frac{\omega\mu}{4\pi} \oiint e^{-j\boldsymbol{k}\cdot\boldsymbol{r}_{mm'}} T_L(\hat{k}\cdot\hat{r}_{m'n'}) \left(\bar{\mathbf{I}} - \hat{k}\hat{k}\right) \cdot \sum_n J_n \tilde{\beta}_n(\hat{k})\, d\hat{k}^2 - j\frac{k}{4\pi} \oiint \hat{k} \times e^{-j\boldsymbol{k}\cdot\boldsymbol{r}_{mm'}} T_L(\hat{k}\cdot\hat{r}_{m'n'}) \sum_n M_n \tilde{\beta}_n(\hat{k})\, d\hat{k}^2 + \boldsymbol{E}^{inc}(\boldsymbol{r_m}) \tag{4}$$

and

$$\boldsymbol{H}(\boldsymbol{r_m}) = -j\frac{k}{4\pi} \oiint \hat{k} \times e^{-j\boldsymbol{k}\cdot\boldsymbol{r}_{mm'}} T_L(\hat{k}\cdot\hat{r}_{m'n'}) \sum_n J_n \tilde{\beta}_n(\hat{k})\, d\hat{k}^2 + j\frac{\omega\varepsilon}{4\pi} \oiint e^{-j\boldsymbol{k}\cdot\boldsymbol{r}_{mm'}} T_L(\hat{k}\cdot\hat{r}_{m'n'}) \left(\bar{\mathbf{I}} - \hat{k}\hat{k}\right) \cdot \sum_n M_n \tilde{\beta}_n(\hat{k})\, d\hat{k}^2 + \boldsymbol{H}^{inc}(\boldsymbol{r_m}), \tag{5}$$

where

$$\tilde{\beta}_n(\hat{k}) = \iint_A \beta_n(\boldsymbol{r}_n) e^{j\boldsymbol{k}\cdot\boldsymbol{r}_{nn'}} da_n \tag{6}$$

is the $\hat{k}$-space representation of the source basis functions with $\boldsymbol{k} = k\hat{k}$. $T_L(\hat{k}\cdot\hat{r}_{m'n'})$ is the conventional FMM diagonal translation operator (Chew et al., 2001). This procedure is here referred to as near-field MLFMM translation, by means of plane wave representation of the radiated fields of source groups containing currents with integration over the Ewald sphere and translation of the plane waves to received field contributions at groups containing observation points. The radiation patterns of the source groups are computed in the center points using the spherical harmonics expansion of the current basis functions according to Eibert (2005).

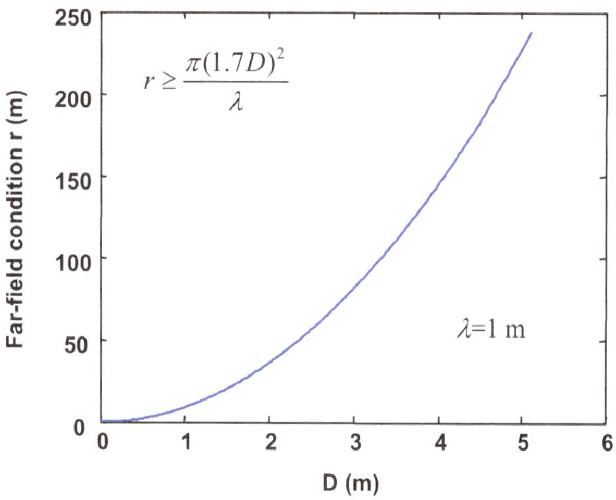

Fig. 2. Far-field condition with increasing group dimensions.

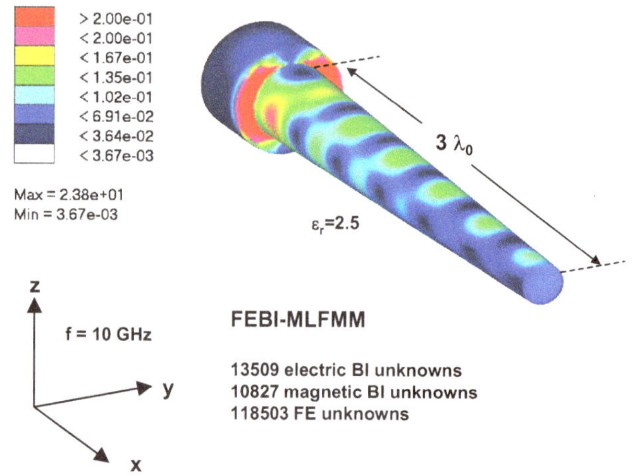

Fig. 3. Real part of surface current density on dielectric rod antenna.

Extending the near-field translation domain to all observation points can be extremely memory-consuming. Therefore, near-field MLFMM translations are applied only to nearby observation points and far-away observation points are preferably treated by far-field MLFMM translations, where the translation operator is approximated by a far-field representation only for the ray direction from the source group to the observation point. Thereby, it is assumed that the observation point lies in the far-field of the individual source group. The field contributions for an observation point $r = r_m$ within the far-field MLFMM translation domain are given by

$$\begin{aligned}\boldsymbol{E}(\boldsymbol{r_m}) = &-j\frac{\omega\mu}{4\pi}T_L^{FF}(kr_{mn'})\left(\bar{\mathbf{I}}-\hat{k}_0\hat{k}_0\right)\cdot\sum_n J_n\tilde{\beta}_n(\hat{k}_0)\\ &-j\frac{k}{4\pi}\hat{k}_0\times T_L^{FF}(kr_{mn'})\sum_n M_n\tilde{\beta}_n(\hat{k}_0)\\ &+\boldsymbol{E}^{inc}(\boldsymbol{r_m})\end{aligned} \qquad (7)$$

and

$$\begin{aligned}\boldsymbol{H}(\boldsymbol{r_m}) = &-j\frac{k}{4\pi}\hat{k}_0\times T_L^{FF}(kr_{mn'})\sum_n J_n\tilde{\beta}_n(\hat{k}_0)\\ &+j\frac{\omega\varepsilon}{4\pi}T_L^{FF}(kr_{mn'})\left(\bar{\mathbf{I}}-\hat{k}_0\hat{k}_0\right)\cdot\sum_n M_n\tilde{\beta}_n(\hat{k}_0)\\ &+\boldsymbol{H}^{inc}(\boldsymbol{r_m}),\end{aligned} \qquad (8)$$

where $T_L^{FF}(kr_{mn'})$ is the far-field MLFMM translation operator in the ray direction $\hat{k}_0 = \hat{r}_{mn'}$ (Chew et al., 2002). Improved far-field representation is obtained using the gravity center of the currents within the source groups as reference point for the radiated fields of the groups. The necessary ray direction $\hat{k}_0$, which connects the gravity center of the currents within the source group with the observation point, is computed from the neighboring sampling points of the numerical integration in the $\hat{k}$-space by interpolation.

The condition used to perform far-field translations is

$$r_{far} \geq \frac{\pi(1.7D_L)^2}{\lambda} \qquad (9)$$

and is derived from the criterion for far-field translations according to the Fast Far-Field Approximation (FAFFA) in (Chew et al., 2002). D_L is the group dimension on level L and λ is the wavelength. The condition increases with the square of group dimensions, as shown in Fig. 2, which means that enforcing satisfaction of this condition for the whole scatterer or antenna would be very inefficient. It is rather assumed that far-field condition is satisfied by the individual source groups on the various levels. For each observation point, far-field translations are performed on the coarsest level L, on which condition (9) is still satisfied. This level is found for each observation point in the initialization step in a worst-case sense using its shortest distance to the BI-MLFMM domain. Improved accuracy is provided by choosing a more stringent far-field condition.

The proposed acceleration of near-field computations was applied at the postprocessing stage of a powerful FEBI method. In the next section numerical results showing the savings in computation time will be presented.

3 Numerical examples

First numerical example is the dielectric rod antenna shown in Fig. 3. The operating frequency of the antenna was 10 GHz and a delta-gap voltage source was used for excitation. In the same figure, the real part of the equivalent surface current density derived from the FEBI solution can be seen, using a tetrahedral FE volume mesh with 118503 unknowns and a triangular surface mesh with 13509 electric and 10827 magnetic unknowns. In the postprocessing stage the electric field was first observed on 6901 observation points along the antenna axis, as shown in Fig. 4. The computations were performed with the postprocessing MLFMM approach and with

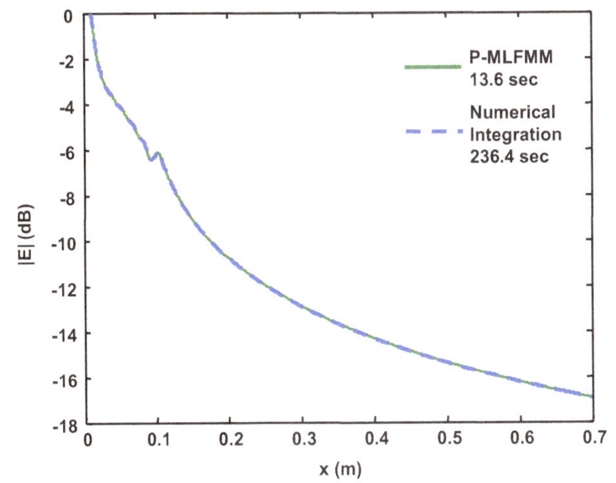

Fig. 4. Electric field along the axis of dielectric rod antenna.

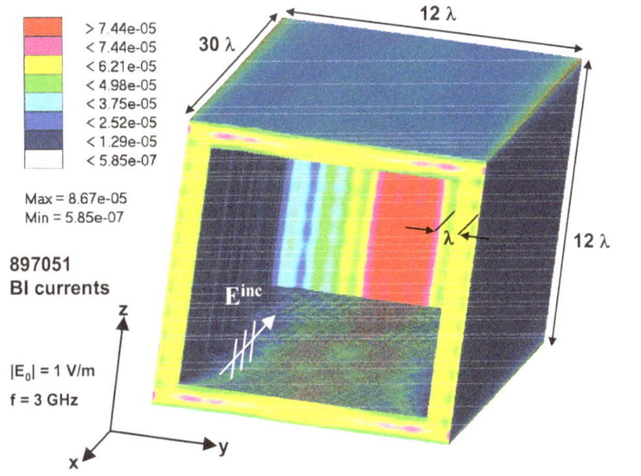

Fig. 6. Magnitude of surface current density on open metallic 30 λ cuboid.

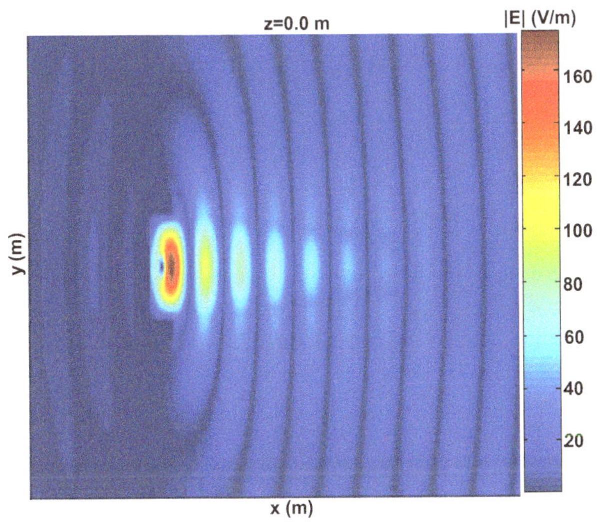

Fig. 5. Instantaneous electric field for $t = 0$ s on the xy-plane.

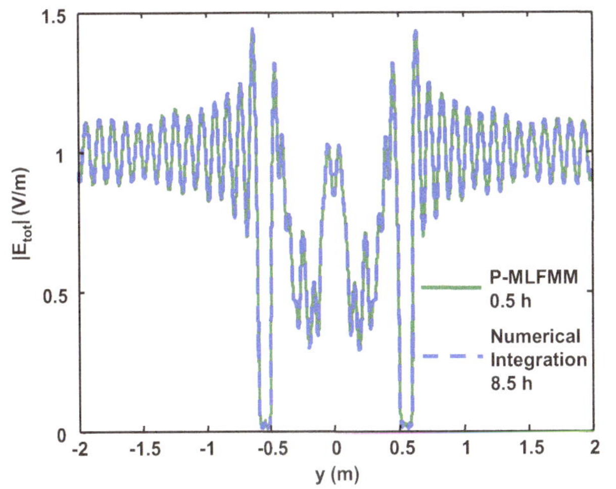

Fig. 7. Total electric field along the y-axis for open metallic cuboid.

conventional near-field integration. It can be seen that results show excellent agreement and the P-MLFMM computations need much less computation time. Further, the instantaneous electric field for $t = 0$ s can be seen in Fig. 5 on the xy-plane, cutting through the antenna in the middle along its axis. These computations were performed on 100701 observation points with the P-MLFMM approach in about 7 min on an AMD Athlon 2800+ PC combining near-field and far-field translations. The wave radiated by the antenna mainly at the transition region between metallic mounting and dielectric material with very low radiation in the backward direction can be clearly seen.

Next example is the open metallic cuboid shown in Fig. 6. The length of the structure is 30 λ and the square cross section has an inner edge length of 10 λ. The object is placed symmetrically to the xy-plane. In the same figure the magnitude of the surface current density on the metallic cuboid can be seen, which is the BI solution of 897051 surface currents for vertically polarized plane wave excitation traveling towards the opening of the cuboid. In the postprocessing stage, the electric field was observed along the y-axis on 10001 observation points by performing the computations with the fast postprocessing approach and with conventional near-field integration. The results are shown in Fig. 7, where excellent agreement and saving of large amount of computation time can be seen. Further, the electric field was observed on a $2 \times 4\,m^2$ window in the xy-plane with 1 mill. observation points. The computations were performed with the P-MLFMM approach in about 8 h combining near-field and far-field translations. The estimated computation time with conventional numerical integration would be about 850 h, which is unacceptably high. The results are shown in Fig. 8, where the expected distribution of the electric field inside, outside and within the metallic walls of the cuboid can be seen.

Last example is the full-scale model of a Mig-29 shown

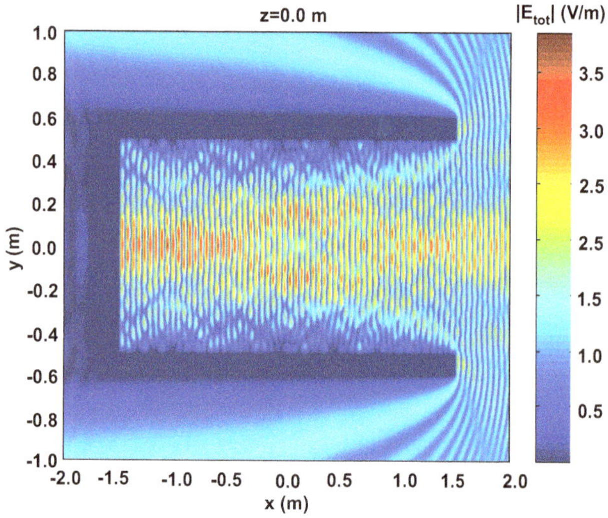

Fig. 8. Total electric field through the open metallic cuboid.

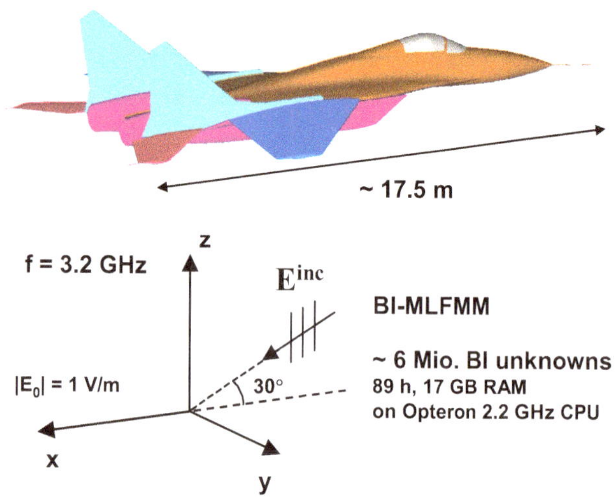

Fig. 9. Full-scale Mig-29 model for $f = 3.2\,\text{GHz}$.

in Fig. 9. The model is about 17.5 m long and was excited by a vertically polarized plane wave traveling 150° from the x-axis towards the nose of the aircraft. The BI solution of this problem was performed on an AMD Opteron 2.2 GHz CPU using about 6 mill. unknowns. In the postprocessing stage the near-field was observed through the aircraft on a plane parallel to the xy-plane on about 2.2 mill. observation points. These computations were performed with the P-MLFMM approach in about 3 h, combining near-field and far-field translations. The magnitude of the total electric near-field through the Mig-29 is shown in Fig. 10, where the expected distribution of the electric field can be seen.

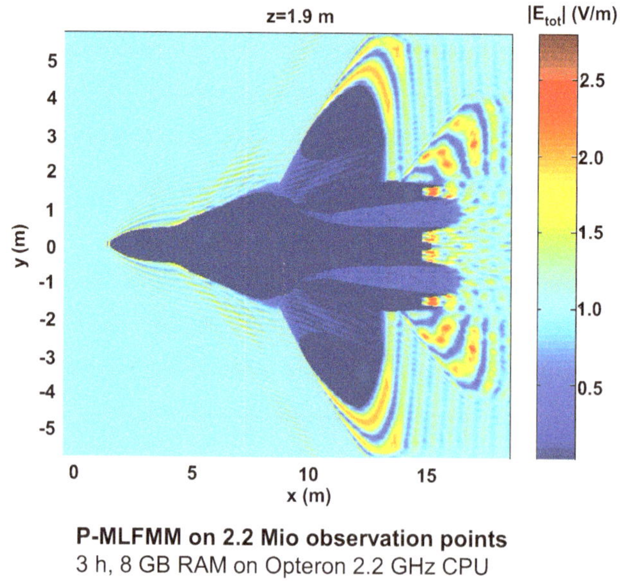

Fig. 10. Total electric field through Mig-29.

4 Conclusions

In this contribution, the multilevel fast multipole method was used to accelerate near-field computations in the postprocessing stage of boundary integral based methods for numerical solution of electromagnetic radiation and scattering problems, where optimum performance is achieved combining near-field and far-field translations. The proposed P-MLFMM was applied in the postprocessing stage of a powerful FEBI method and numerical results were shown, where saving of large amount of computation time in the postprocessing stage could be clearly seen.

References

Chew, W. C., Jin, J.-M., and Michielssen, E.: Fast and efficient algorithms in computational electromagnetics, Boston: Artech House, 2001.

Chew, W. C., Cui, T. J., and Song, J. M.: A FAFFA-MLFMA algorithm for electromagnetic scattering, IEEE Trans. Antennas Propagat., 50, 1641–1649, 2002.

Eibert, T. F.: A Diagonalized Multilevel Fast Multipole Method with Spherical Harmonics Expansion of the *k*-Space Integrals, IEEE Trans. Antennas Propagat., 53, 814–817, 2005.

Jin, J.-M.: The Finite Element Method in Electromagnetics, 2nd Ed., John Wiley and Sons, Inc., New York, 2002.

Rao, S. M., Wilton, D. R., and Glisson, A. W.: Electromagnetic scattering by surfaces of arbitrary shape, IEEE Trans. AP, 30, 409–418, 1982.

Volakis, J. L., Chatterjee, A., and Kempel, L.-C.: Finite Element Method for Electromagnetics, IEEE, Inc., New York, 1998.

Telemetric surface acoustic wave sensor for humidity

M. Dierkes and U. Hilleringmann

Faculty for Computer Science, Electrical Engineering and Mathematics, Sensor Technology Department, University of Paderborn, Warburger Str. 100, 33098 Paderborn, Germany

Abstract. Surface acoustic wave sensors consist of a piezoelectric substrate with metal interdigital transducers (IDT) on top. The acoustic waves are generated on the surface of the substrate by a radio wave, as it is well known in band pass filters. The devices can be used as wireless telemetric sensors for temperature and humidity, transmitting the sensed signal as a shift of the sensor's resonance frequency.

1 Introduction

The insulation capability of the oil in high voltage power transformers strongly depends on its humidity content. As known from literature, the maximum electrical field before spark discharge occurs descents dramatically, if the water content of the transformer oil rises to about 50 ppm. It is the aim of the power plants owners to get information about the current oil condition continuously. Due to the extreme environment inside the transformer it is not possible to use sensors with wires for power supply and data transfer.

The approach of this work is to use surface acoustic wave (SAW) sensors to measure the humidity content of the oil by telemetric data acquisition. The sensor consists of an interdigital transducer (IDT) on top of a piezoelectric quartzite chip, which converts an electrical input signal of a radio wave into an acoustical wave propagating at the surface of the sensor material. The surface wave is transformed back into an electromagnetic wave by another IDT located on the SAW sensor (delay line), or by the same IDT in case of a resonator structure (Fig. 1). In both cases the transmitted frequency can be detected by high sensitive receivers.

The velocity of the acoustic wave respective its frequency depends on the density of the surface material of the sensor. The captured humidity reduces the density, so the transmitted frequency received by the detector drops. The paper describes the principles of the SAW sensor for applications in liquids, its integration technology, and the first measurement results.

Correspondence to: M. Dierkes
(dierkes@sensorik.upb.de)

2 Acoustic waves

There are different modes of SAWs, which can be selected by the set up of the sensor structure. Most common in SAW band pass filters are Rayleigh waves (Fig. 2). These waves consist of oscillations partly perpendicular to the sensor surface, partly in the propagation direction. The first part of the wave results in an energy transfer out of the wave into the ambient, so it is suitable for gases only. In case of liquids the propagation loss is very high, so this kind of wave is not suitable for sensors operating in oil.

More useful seems to be a transversal acoustic wave, which is called a "Love"-wave. This kind of wave consists of oscillations parallel to the sensor surface, but perpendicular to the propagation direction (Fig. 3). Because there is no amplitude of the wave perpendicular to the surface plane, the propagation loss for this kind of wave is very low even in liquids like oil.

The wavelength of the stimulated acoustic wave corresponds with the periodicity of the IDT finger structure. Only those waves with a wave length of two times the pitch of the IDT will cause a surface acoustic wave with the well defined frequency of the resonance mode. This means the resonance frequency can be defined by the layout of the IDT structure.

Due to the humidity dependence of the resonance frequency, the transmitted signal from the sensor is a measure for the ambient humidity.

3 Sensor integration

As it is necessary to convert the electric signal into a surface wave, piezoelectric substrates like quartzite or Lithiumniobate are suitable for SAW sensor applications. Quartzite seems to be the adequate material due to its compatibility to standard silicon technology, which is used to integrate the

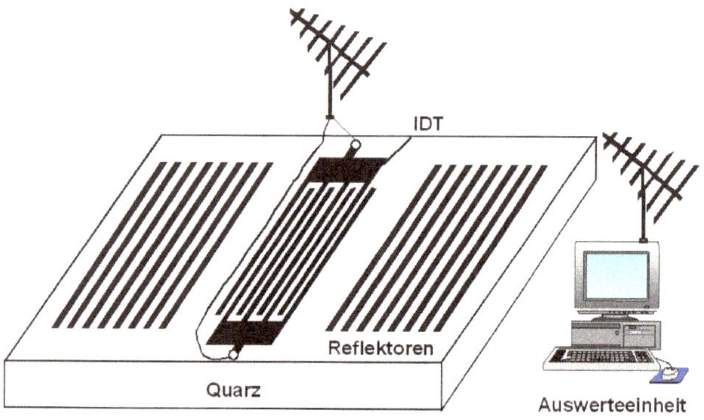

Fig. 1. SAW sensor system consisting of an SAW sensor, antennas and a receiver unit.

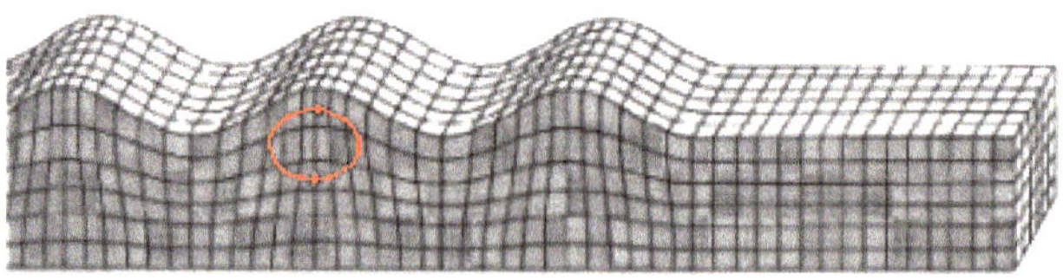

Fig. 2. Sketch of a Rayleigh wave propagating at the surface of a solid.

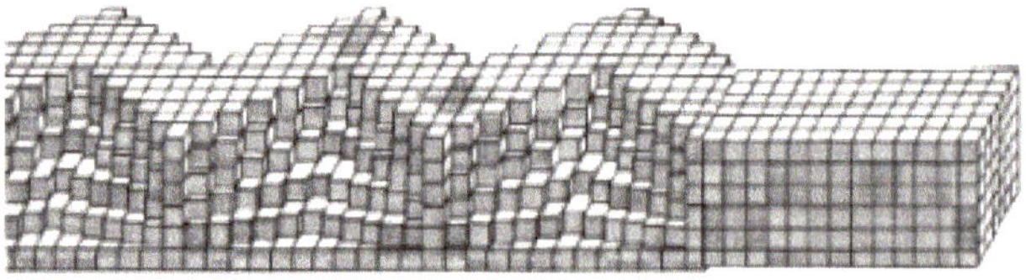

Fig. 3. Sketch of a love-type surface wave propagating at the surface of a solid.

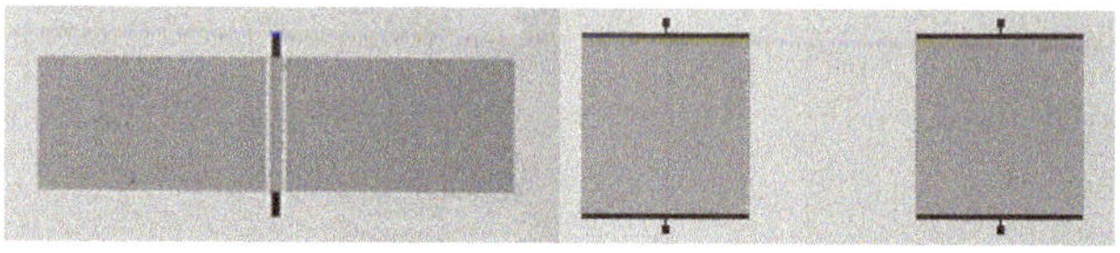

Fig. 4. Photograph of the SAW sensors, left a resonator, right a delay line device.

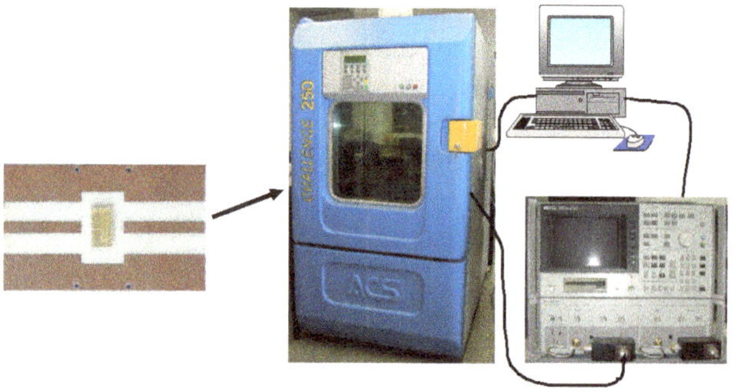

Fig. 5. Test device mounted on a circuit board for a test in the climatic chamber.

sensor device of this paper. The process starts with the deposition of aluminium on top of the cleaned quartzite wafer (100 mm in diameter) either by e-beam evaporation or by magnetron sputtering. Next a photo lithography step using vacuum contact and UV exposure defines the metal structures on the surface. Typical structure sizes are in the range of 1 μm. The aluminium layer is etched by chlorine chemistry applying reactive ion etching. Then the resist is removed in oxygen plasma, so the IDT structures are completely finished (Fig. 4).

The stimulation of a love wave in the sensor device requires a top layer on the substrate surface, where the velocity of the acoustic wave is lower than in the quartzite itself. On the other side, the humidity sensor needs a humidity sensitive layer on top. Silicon oxide can meet both requirements in case of a PECVD oxide. Due to its porous character it is able to adsorb humidity. Nevertheless, it is not known if an oxide layer is suitable in oil ambient.

4 Measurement set-up

First tests of the sensor have been performed in ambient atmosphere at well defined temperature and humidity. The test equipment consists of a network analyser and a climatic chamber, which are controlled by a computer (Fig. 5).

The network analyser sweeps the frequency from 1 kHz up to 500 MHz and transmits the electromagnetic signal to the sensor device. In case of resonance with the IDT structure, only little energy is reflected. So the scattering parameter S_{11} strongly drops in case of the resonance frequency at the input.

The device under test is mounted on a 50Ω circuit board while the electric connections consist of ultrasonic bonds. Up to now no broadcast transmitting/receiving has been realized.

5 Results

The temperature dependence of the sensor's resonance frequency is depicted in Fig. 6. Due to the raising temperature the velocity of the acoustic surface wave in the sensor drops continuously, so the resonance frequency descents with increasing temperature.

The regression line of the frequency-temperature diagram results in −47 kHz/K. This high sensitivity allows a resolution in the mK range.

The dependence of the resonance frequency on the relative humidity is shown in Fig. 7. An increase in the humidity concentration results in a reduced density of the sensing material, so the acoustic wave velocity drops again. The sensitivity is about −74 kHz/%.

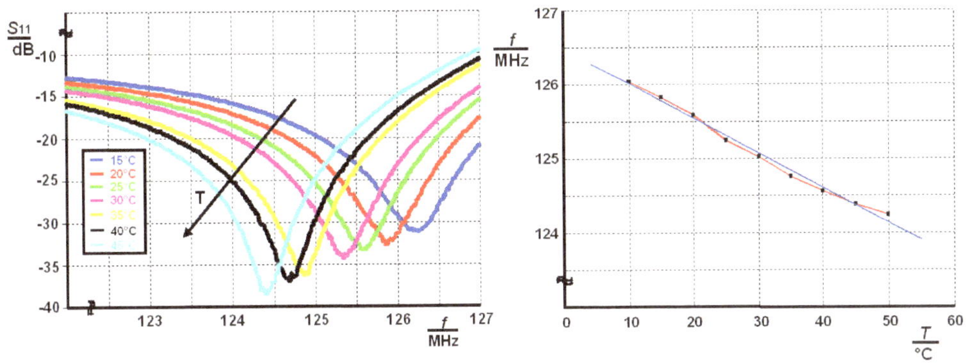

Fig. 6. Temperature dependence of the scatter parameter S_{11}, reflecting the resonance frequency of the sensor.

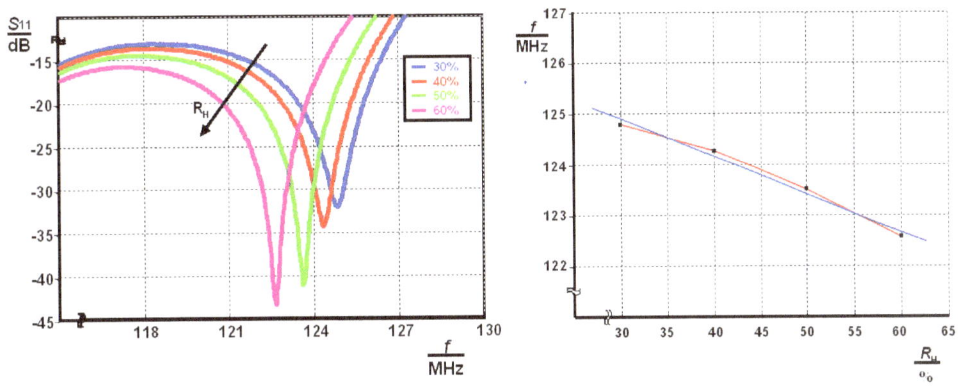

Fig. 7. The S_{11} parameter reflecting the resonance frequency of the SAW sensor versus the input frequency with the relative humidity as the parameter.

6 Conclusion

SAW sensors strongly react on humidity and temperature changes in the ambient atmosphere. Sensitivities of about −47 kHz/K for temperature and −74 kHz/% for humidity were achieved with an resonator type SAW sensor consisting of 20 IDT fingers and 500 reflectors. Although these results are satisfactory, it is not sure to reach the aim of detecting 50 ppm of humidity in transformer oil.

Acknowledgements. The authors thank Werner Büttner, Thomas Diekmann, Heinz Funke, Ralf Otterbach and Christoph Pannemann for their qualified discussions and their support during integration of the devices. They are grateful to the German DFG for financial support of this work (grant Hi 551/5-1).

References

Reindl, L., Scholl, G., and Schmidt, F.: Funksensorik und Identifikation mit OFW Sensoren, VDI Berichte Nr. 1530, p. 799–810, 2000.

Vellekoop, M. J.: Acoustic wave sensors and their technology, Ultrasonics, 36, p. 7–14, 1998.

Jakoby, B. and Vellekoop, M. J.: Properties of Love Waves: Applications in Sensors, Smart Materials Structures, p. 668–679, 1997.

Du, J., Harding, G. L., Dencher, P. R., Barnett, D., and Howe, E.: A study of Love wave acoustics sensors, Sensor and Actuators A56, p. 211–219, 1996.

Leidl, A., Filkorn, M., Rahammer, S., Endres, H.-E., and Drost, S.: Flüssigkeitsanalyse mit Oberflächensensoren, VDI Berichte Nr. 1255, p. 295–300, 1996.

Zinke, O. and Brunswig, H.: Hochfrequenztechnik, Springer Berlin, 5, Band 1A, p. 224–256, 1995.

Flottmeyer, F., Lange, G., Miksa, T., Neldner, W., and Sundermann, U.: Betrachtung zum Problem der Nutzungsdauer von Hochspannungs-Transformatoren und -Wandler, ETG-Fachbericht 55, VDE-Verlag, 1995.

Buff, W., Plath, F., Schmeckebier, O., Rusko, M., Vandahl, T., Luck, H., and Möller, F.: Remote Sensor System Using Passive SAW Sensors, Proc. IEEE Ultrasonics Symp., p. 585–588, 1994.

Andle, J. C., Vetelino, J. F.: Acoustic wave biosensors, Sensor and Actuators A44, p. 167–176, 1994.

Wang, Z., Cheeke, J. D. N, and Jen, C. H. : Sensitivity analysis for Love mode acoustic gravimetric sensors, Appl. Phys. Lett., 64 (22), p. 2940–2942, 1994.

Kovacs, G., Lubking, G. W., Vellekoop, M. J., and Venema, A.: Love waves for (bio)chemical sensing in Liquids, IEEE Ultrasonics Symposium, p. 281–285, 1992.

Meinke, H. and Gundlach, F. W.: Taschenbuch der Hochfrequenztechnik, Springer Berlin, 4, p. L65–L74, 1986.

Auld, B. A.: Acoustics Fields and Waves in Solids, Vol. II, Wiley, New York, 1973.

Characterisation of human gait using a continuous-wave radar at 24 GHz

C. Hornsteiner and J. Detlefsen

Technische Universität München, Lehrstuhl für Hochfrequenztechnik, Fachgebiet Hochfrequente Felder und Schaltungen, 80333 München, Germany

Abstract. Human locomotion consists of a complex movement of various parts of the body. The reflections generated by body parts with different relative velocities result in different Doppler shifts which can be detected as a superposition with a Continuous-Wave (CW) Radar. A time-frequency transform like the short-time Fourier transform (STFT) of the radar signal allows a representation of the signal in both time- and frequency domain (spectrogram). It can be shown that even during one gait cycle the velocity of the torso, which constitutes the major part of the reflection, is not constant. Further a smaller portion of the signal is reflected from the legs. The velocity of the legs varies in a wide range from zero (foot is on the ground) to a velocity which is higher than that of the torso. The two dominant parameters which characterise the human gait are the step rate and the mean velocity. Both parameters can be deduced from suitable portions of the spectrogram. The statistical evaluation of the two parameters has the potential to be included for discrimination purposes either between different persons or between humans and other moving objects.

1 Human gait analysis

The human gait was investigated for a long time beginning in the nineteenths century under scientific considerations. With the upcoming improvements in photography it was possible to analyse the movement of different body parts with a series of pictures. In our days the photo camera was replaced by a CCD camera and with several viewpoints from different angles it is possible to extract 3-dimensional data of the scene. A model which used such a motion capture system has been developed by Zonghua and Troje (2004). To get periodic data the test persons were placed on a treadmill. For identification of different body parts for the later processing each persons wore 15 reflective markers (s. Fig. 1). The 3-dimensional position was then expressed in a Fourier series

Correspondence to: C. Hornsteiner
(christoph.hornsteiner@tum.de)

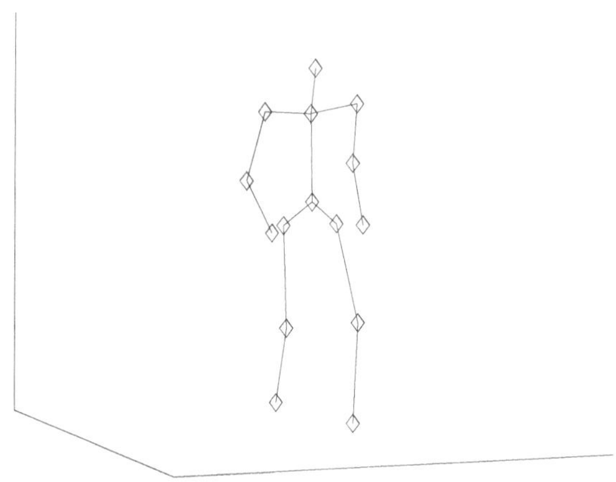

Fig. 1. Gait model of Zonghua and Troje (2004) with 15 points.

$$p(t) = p_0 + p_1 \sin(\omega t) + q_1 \cos(\omega t) + \\ + p_2 \sin(2\omega t) + q_2 \cos(2\omega t) \quad (1)$$

for each coordinate axis x, y and z. With this representation it is easily possible to look at the process from every viewpoint. To get a velocity representation of the movement Eq. (1) has to be derived. This can be done either analytically by taking the derivative from each summand or numerically by taking the derivative of the discretised time data of $p(t)$. The largest velocity components arise in the direction of locomotion. Therefore, this direction is of main interest for the measurement with a radar sensor. The obtained velocity over time for all 15 marker points is displayed in Fig. 2.

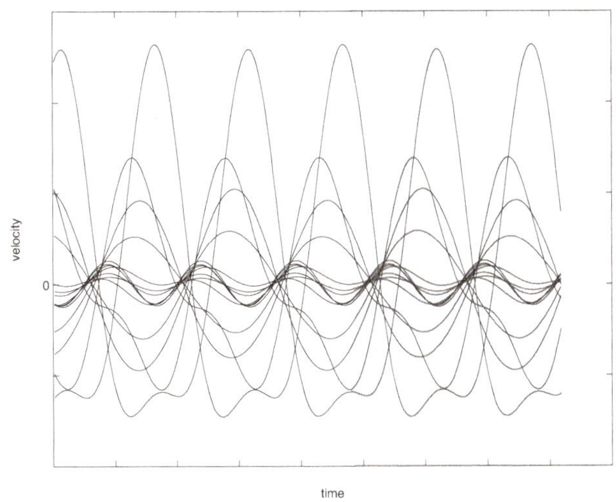

Fig. 2. Velocity in the direction of locomotion of all 15 markers over time.

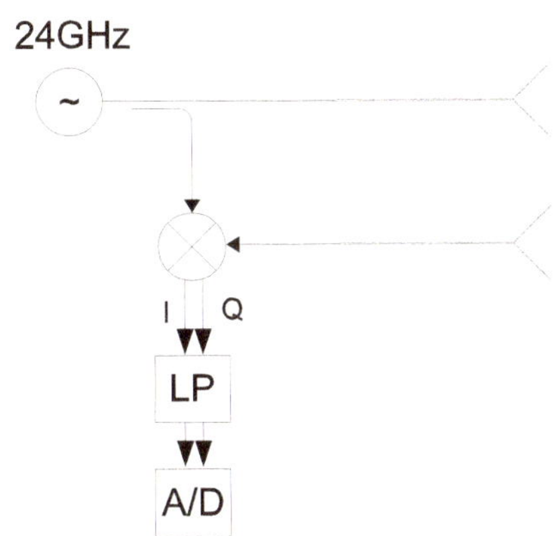

Fig. 3. Block diagram of a CW-Radar with I/Q-Demodulation.

Although the treadmill is running with constant velocity, the velocity of the torso is not constant. It is changing nearly sinusoidal with the step frequency. The highest velocities are expectedly observed from the legs. While one leg attains its maximum velocity the other is in its minimum. The velocity contribution of the knees and arms lies in between torso and legs.

2 Doppler radar

For the measurements a CW-Radar at a carrier frequency of 24 GHz was used. With the Doppler formula

$$f_D = 2 f_c \frac{v}{c} \tag{2}$$

it can be seen that velocity v is proportional to the measured frequency f_D. Because of increasing Doppler frequencies f_D with increasing carrier frequency f_c at constant velocity, the choice of a high carrier frequency leads to a better velocity resolution for a constant time interval. The basic block diagram of the radar sensor can be seen in Fig. 3. With the use of a I/Q-demodulation it is possible to discriminate between positive and negative frequencies and therefore between approach and receding motion. A spectral transform is used to derive the frequency and so the velocity. By doing this with the complete time signal the best frequency resolution is obtained giving no time resolution. In the short-time Fourier transform (STFT) only parts of the signal are transformed to frequency domain subsequently applying a windowing function $w(t)$ which is moved over the time signal $x(t)$.

$$\mathrm{STFT}(\tau, f) = \int_{-\infty}^{+\infty} x(t) \cdot w(t-\tau) \cdot \exp(-j2\pi f t)\mathrm{d}t$$

The so called spectrogram are the absolute values of the STFT displayed on a logarithmic scale.

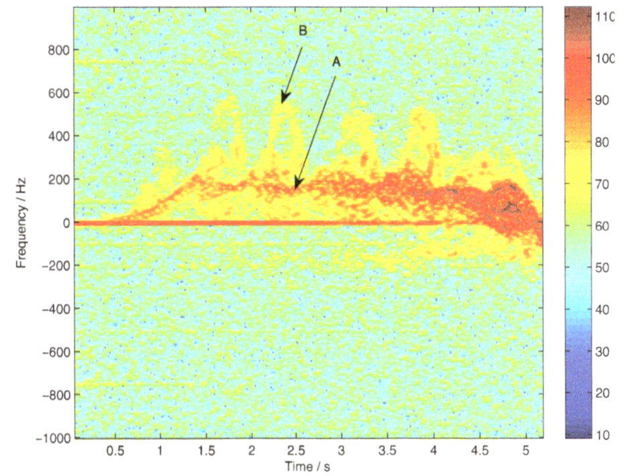

Fig. 4. Spectrogram of a walking person (window length 128 ms).

3 Feature extraction

The spectrogram of a measured motion can be seen in Fig. 4. The spectral amplitude corresponds to the radar cross section of the moving parts. The main contribution (A) comes from the torso, but since it is non-rigid there is no distinct line but a spectral spread. The contribution of the legs (B) is much smaller but it can be seen that it is still above noise level. The visible velocity response of the legs is in accordance to the model. Lower frequency components occur while one foot is on the ground and the other reaches its maximum. The velocity of the torso movement (A) may be detected by taking the frequency at which the maximum amplitude occurs. Due to the spectral spread this curve will include errors if the

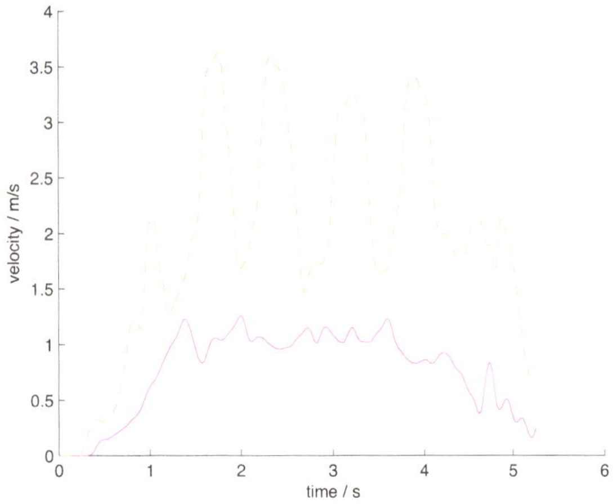

Fig. 5. Extracted velocity pattern of torso (-) and legs (- -).

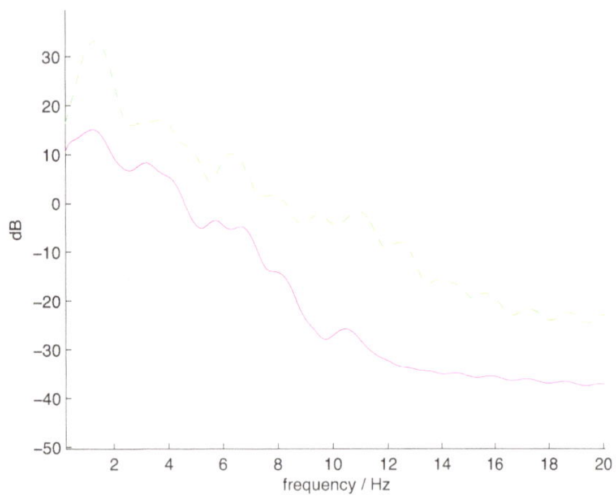

Fig. 6. Spectrum of velocity pattern from torso (-) and legs (- -).

detected velocity is not identical to the mean velocity of the torso.

The spectral components of the moving legs (B) above the torso are surrounded mostly by noise. Detection of signal above noise level is the standard process in signal theory. The resulting velocity patterns for torso and legs are displayed in Fig. 5. In both graphs the step frequency is visible. The legs with a high velocity deviation and the torso with lower velocity deviation and also visible error. For the robust detection of periodic oscillations it will be changed from time domain to frequency domain. To avoid low frequency components from acceleration process a high pass filter is applied to the signal. The resulting spectrum is shown in Fig. 6. As expected the magnitude at the step frequency of 1.3 Hz is for the legs more than 10 dB higher than those of the torso. Consequently the leg movement will be used for further evaluation.

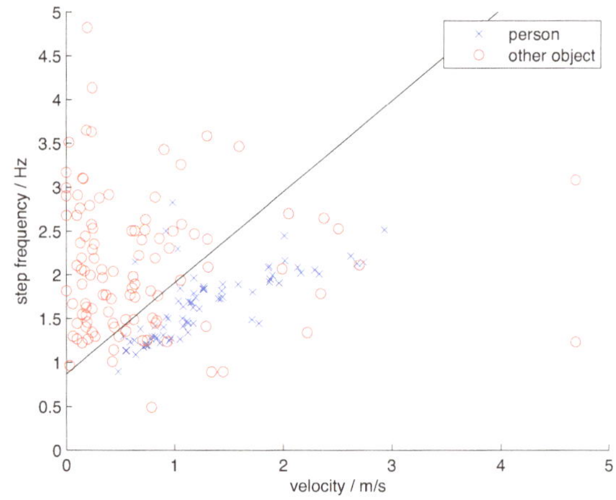

Fig. 7. Scatterplot with linear decision boundary.

Table 1. Confusion matrix with actual classes (I: Person, II: Other object) as columns and predicted classes as rows for linear discrimination.

	I	II
I	0.385	0.143
II	0.022	0.450

4 Statistical evaluation

Due to the periodic nature the step frequency is a feature candidate for a statistical evaluation. Because of known dependencies between step frequency and velocity for human gait, it is expected to be useful to combine these both features. Other objects which also produce a Doppler spectrum will not show these dependencies and should not be localized in the feature space. For discrimination two classes are created with class one including all walking persons and class two including all other objects. Class one consists of 74 data points and class two of 108 data points. The scatter plot shows a localisation of class one which makes it possible to separate both classes. The easiest way to do this is to construct a linear separation boundary using methods of linear discrimination analysis. The scatterplot with the decision boundary can be seen in Fig. 7 and the resulting confusion matrix in Table 1. The resubstitution misclassification rate with this trained classifier is 16.5%. The linear classifier is not able to discriminate between points of class two lying below the localized class one. This can be improved by the construction of a nonlinear classifier. A widely used method to do this is the use of support vector machines (SVM). The nonlinearity is obtained with the use of nonlinear kernel functions. The kernel functions used here were radial basis functions. The resulting decision boundary with this classifier can be seen in

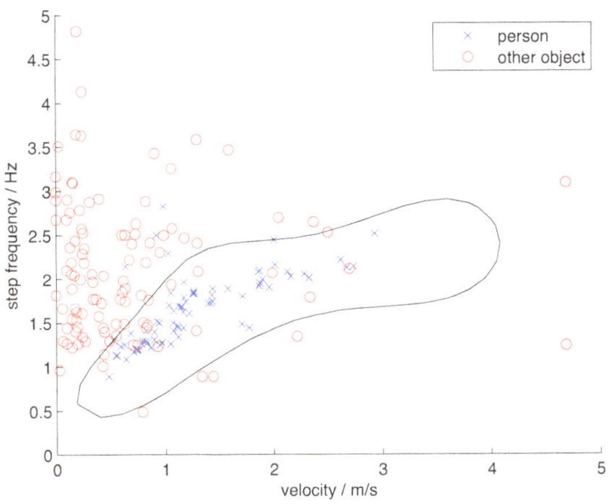

Fig. 8. Scatterplot with nonlinear decision boundary.

Table 2. Confusion matrix with actual classes (I: Person, II: Other object) as columns and predicted classes as rows for discrimination with SVM.

	I	II
I	0.385	0.088
II	0.022	0.505

Fig. 8 and in Table 2 the associated confusion matrix. With this classifier the resubstitution misclassification rate sinks to 11%.

5 Summary

The movement of different body parts can be made visible with a short-time Fourier transform of a CW radar signal. From this display it is possible to extract different velocity curves. The main parameters describing the human gait are step rate and mean velocity which could be obtained from this curves. The two dimensional scatter plot of these two features shows the possibility to discriminate between human gait and other objects. This discrimination was first done by a linear classifier and then improved with a nonlinear classifier using support vector machines.

References

Fölster, F., Rohling, H., and Ritter, H.: Observation of a walking pedestrian with a 24GHz automotive radar sensor, GeMiC 2006, Session 10b-4, 2006.

Gandhi, T., Gandhi, T., and Trivedi, M. M.: Pedestrian Protection Systems: Issues, Survey, and Challenges, Intelligent Transportation Systems, IEEE Transactions, 8, 413–430, 1524–9050, 2007.

Geisheimer, J. L., Marshall, W. S., and Greneker, E.: A continuous-wave (CW) radar for gait analysis, in: Signals, Systems and Computers, 2001, Conference Record of the Thirty-Fifth Asilomar Conference, 1, 834–838, 2001.

Otero: Application of a continous wave radar for human gait recognition, The MITRE Corporation, 2005.

Zonghua, Z. and Troje, N. F.: 3D Periodic Human Motion Reconstruction from 2D Motion Sequences, in: Computer Vision and Pattern Recognition Workshop, 2004 Conference, pp. 186–186, 2004.

38

Analysis and design of a composite left-/right-handed leaky wave antenna based on the H_{10} rectangular waveguide mode

Y. Weitsch and T. F. Eibert

Institute of Radio Frequency Technology, Universität Stuttgart, 70550 Stuttgart, Germany

Abstract. The presented "open" composite left-handed/right-handed (CRHLH) substrate integrated waveguide performs well as a low-profile leaky wave antenna. This design is distinguished due to the fact that it is derived from the approved equivalent circuit model of the H_{10} rectangular hollow waveguide mode. The wave propagation behaviour is visualised by the dispersion diagram calculated by two different methods, infinite periodic full-wave simulation and Matrix-Pencil analysis of driven field solutions. The periodic configuration is also analysed in terms of the Bloch impedance. Although FR-4 serves as substrate the antenna features an efficiency of about 50% to 60%. The radiation performance demonstrates nearly backfire to almost endfire scanning capability of the antenna by mere frequency variation. Broadside radiation is possible due to the balanced state at 4 GHz.

1 Introduction

The presented leaky-wave antenna is not only attractive as conformal antenna. Having a low profile, it achieves a relatively high gain due to its large aperture. Leaky-wave antennas belong to the traveling wave type. The wave while propagating along the antenna structure radiates a certain part of the transported energy more or less in a continuous manner. The radiation direction is controllable by mere variation of frequency. Hence, neither additional phase shifters nor other complex feeding techniques are necessary. In terms of conventional right-handed (RH) configurations, only endfire radiation is feasible. This restriction has been relieved with the discovery of metamaterials. In particular, combined right-handed/left-handed (CRHLH) (Lai and Itoh, 2004; Caloz and Itoh, 2005) structures offer the possibility of radiation into the entire upper hemisphere. Additionally, the naturally occurring parasitic effects (of right-handed nature) are also taken into account as they are not suppressible. A peculiarity is broadside radiation resulting from the antenna being in the so-called balanced state. Figure 1 depicts the equivalent circuit of a combined RH/LH configuration. While the equivalent circuit (EQC) of a RH transmission line is well known, the one for the LH case is obtained by duality. A LH system means that the phase and the group velocity are anti-parallel. Both EQCs are valid for infinitesimal lengths Δz or rather if the line can be treated as uniformly homogeneous. For the same reason, a metamaterial can only be defined as such if its unit sections are much smaller than the wavelength of operation. Then, it deserves the name metamaterial with effective medium parameters disclosing novel characteristics on the macroscopic level. In the microscopic view, one unit section is normally repeated periodically facilitating the production process. The same holds for the dispersion analysis of the configuration which can be restricted to the examination of the irreducible Brillouin zone under the constraint of periodic boundary conditions. The total composition can be evaluated according to Floquet's theorem (Collin, 1991).

One of the common approaches to realise the desired properties in a metamaterial is to employ lumped or distributed circuit elements (Eleftheriades et al., 2002) to set up the EQC. Short-circuited or open transmission lines may also serve as the required inductors and capacitors. Conversely, this work begins with the established EQC given in Fig. 2 of the H_{10} rectangular hollow waveguide mode (Fig. 3) (Marcuvitz, 1951). Important is that this EQC is very close to the aimed EQC of the CRHLH configuration since only the series capacitors are missing. Hence, only a small modification is sufficient to achieve the desired structure. One implementation is a rectangular waveguide with corrugated side walls like analysed in Eshrah and Kishk (2006). In order to simplify the production of the waveguide and being cost

Correspondence to: Y. Weitsch
(mail@ihf.uni-stuttgart.de)

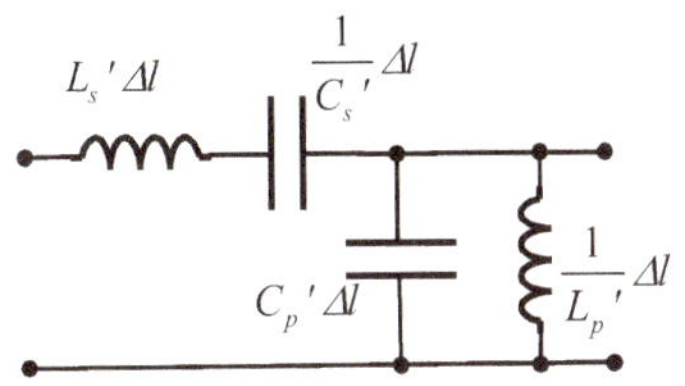

Fig. 1. Equivalent circuit model of a short section Δl of a composite LH/RH transmission line.

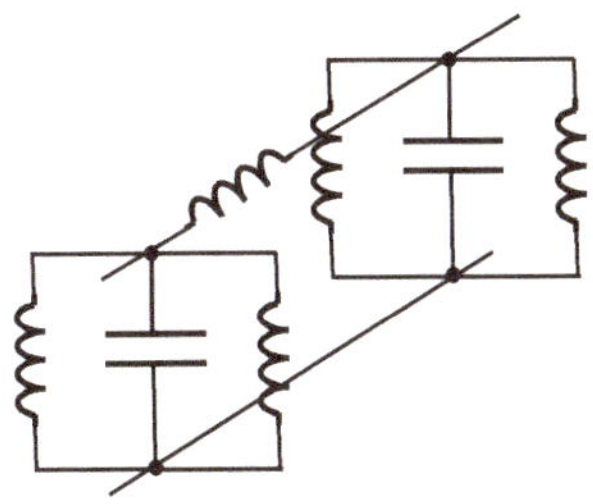

Fig. 2. Equivalent circuit model of the H_{10} rectangular hollow waveguide mode.

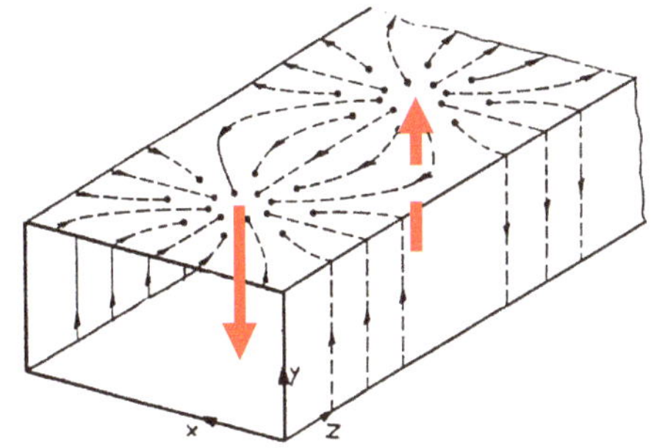

Fig. 3. Illustration of surface wall and displacement currents of the H_{10} rectangular hollow waveguide mode.

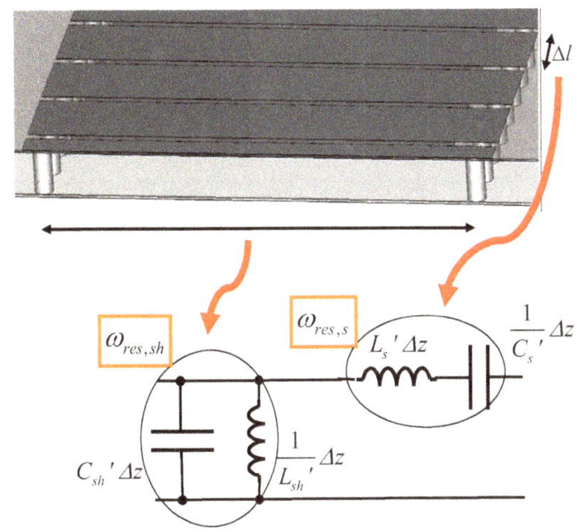

Fig. 4. Correspondence of SIW geometries and circuit model equivalences.

effective, we utilise a substrate-integrated waveguide (SIW) (Deslandes and Wu, 2005) in which the vertical side walls are replaced by rows of vias. Yet we aim at an "open" configuration with spurious radiation enabling the design of a composite RH/LH leaky-wave antenna.

In the following, we elucidate the design steps towards an ultra-wideband leaky-wave antenna with ample beam scanning capability including broadside direction. The description of the antenna involves an explanation of the balancing process precondition for broadside radiation. Likewise we attain matching. Since the antenna is constructed periodically, an analysis of the Bloch impedance characteristic for periodic configurations follows. The wave nature is investigated and illustrated by the dispersion diagram validated by two separate techniques, full-wave simulation of the infinite periodic configuration and the Matrix-Pencil estimation technique. The latter is applied to the field distribution of the gained driven solution. The measurements of a prototype reinforce the simulation results obtained by CST Microwave Studio.

2 Antenna description

Starting point is the EQC illustrated in Fig. 2 of the H_{10} rectangular waveguide mode as depicted in Fig. 3. The lacking series capacitor may be implemented by slots in the top metallic plate perpendicular to the direction of the propagating wave. In contrast to conventional slotted waveguide designs, the slots are distanced much smaller than the wavelength not to influence higher order modes but to interact with the first mode able to propagate. One cut is very narrow to achieve a considerable capacitance value. A second level of metallic strips just below the upper metallization bridges the gaps and enhances the capacitance and we arrive at especially short sections (Figs. 4 and 5). Kapton foil acts as separation layer. The upper metallic bars are short-circuited with the metallic ground plane by via holes which now define the transverse dimensions of the waveguide. The horizontal geometry of the SIW not only prescribes the cut-off frequency of the original waveguide below which effective propagation was impossible. It also fixes the value of the shunt resonance of the EQC. The proportions in longitudinal direction determine the series resonance. For this reason, finding the balanced condition is very convenient since the important design parameters are decoupled from each other. So, balancing is typically achieved by adapting the length Δl of one unit cell and the overlapping area covered by the metallic bars on the top and the metallic tongue slightly below (Figs. 4 and 5). Having reached the balanced state,

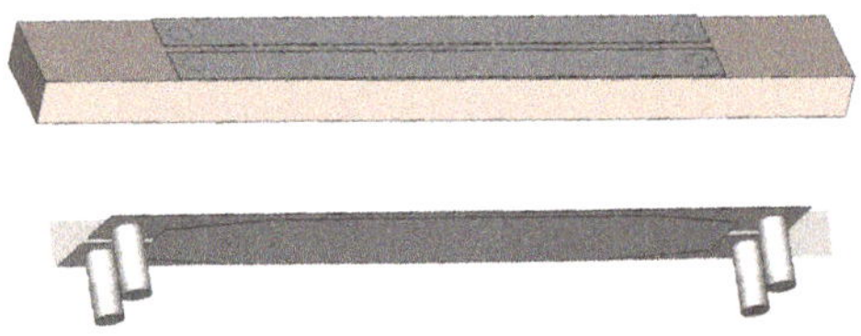

Fig. 5. Elementary cell of the periodic architecture.

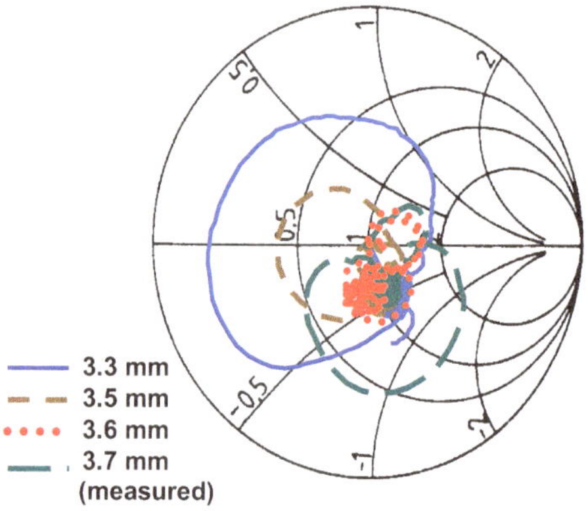

Fig. 6. S_{11} of structures with different unit cell size lengths.

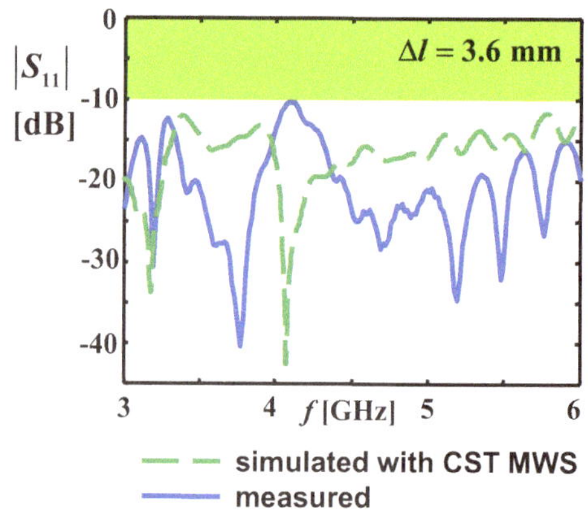

Fig. 7. S_{11} of the LH / RH transmission line.

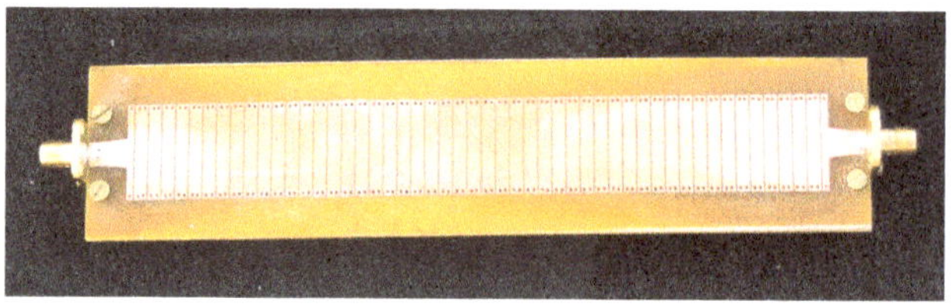

Fig. 8. Fabricated leaky-wave antenna.

the series resonant frequency equates the resonant frequency of the shunt circuit or, if referring to a transmission line the characteristic impedance as issued in Caloz and Itoh (2005), ideally becomes independent from frequency. The effect of varying the length Δl is visible in the Smith Chart shown in Fig. 6. Broadband matching is observed with a length of 3.6 mm as the curve of the reflection coefficient concentrates well around the matching point (red dotted line). This is moreover demonstrated with the magnitude of the input reflection coefficient shown in Fig. 7. The tapered transition section connecting the periodic cell configuration to a 50 Ω microstrip line is thereby included. The final prototype is pictured in Fig. 8 being composed of 50 periods.

3 Bloch impedance

Since the traits in a periodic structure are repeated periodically with p, e.g along z-direction, the solution of the corresponding eigenvalue problem $\psi(z)$ can be represented as the product of a periodic function $\psi_p(z)$ and a propagation factor $e^{-\gamma p}$ according to Bloch theorem:

$$\psi(z) = \psi_p(z) e^{-\gamma p} \quad \text{with} \quad \psi_p(z+p) = \psi_p(z)\,. \tag{1}$$

From an electromagnetic point of view, ψ may describe the magnetic or the electric field in the strucuture. In order to gain an equivalent circuit model, it is required to find equivalent voltages and currents representing the waves in the structure. A possible definition of voltage V and current I is given by

$$V = \int_{C_v} \boldsymbol{E} \cdot \boldsymbol{n}_{\text{ground}}\, ds\,, \tag{2}$$

$$I = \int_{C_i} (\boldsymbol{n}_g \times \boldsymbol{H}) \cdot \boldsymbol{n}_{\text{port}}\, ds\,, \tag{3}$$

according to a cross section as depicted in Fig. 9. Since these definitions are kind of arbitrary for non-TEM (transverse-electromagnetic)-waves (Marcuvitz, 1951), it is useful to also consider the power P transported by the wave along the waveguide, which can be computed via

$$P = \frac{1}{2} \iint_S (\boldsymbol{E} \times \boldsymbol{H}^*) \cdot \boldsymbol{n}_{\text{port}}\, dS\,. \tag{4}$$

The required field data is extractable from the driven simulation. Based on these definitions, an characteristic impedance, the so-called Bloch impedance Z_B (Collin, 1991), can be calculated on three different ways:

$$Z_{B,PI} = \frac{2P}{|I|^2}\,, \quad Z_{B,PV} = \frac{|V|^2}{2P}\,, \quad Z_{B,VI} = \frac{V}{I}\,. \tag{5}$$

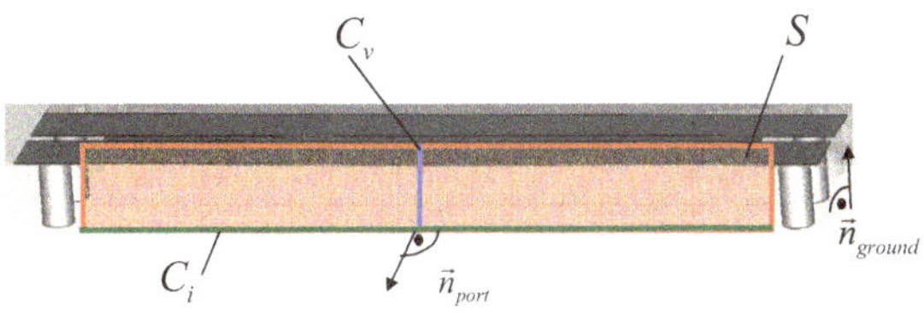

Fig. 9. Cross section within elementary cell.

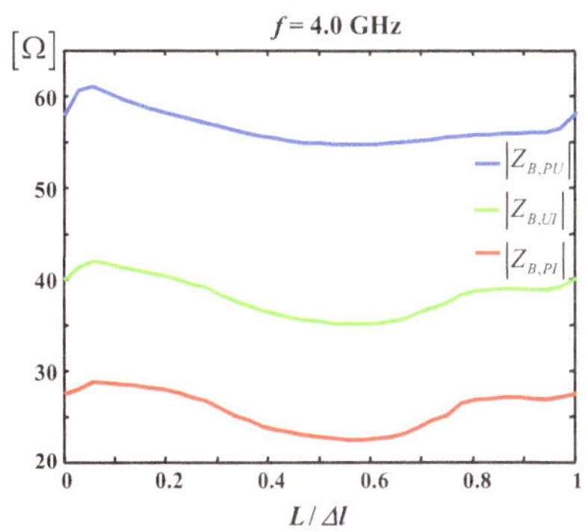

Fig. 10. Bloch impedance calculated along unit cell length according to the three definitions.

For TEM-waveguides, where the transverse fields are gradient fields, it is known that these three definitions give the same result for the characteristic impedance. However, for our case of a non-TEM-waveguide different results are obtained, where $Z_{B,VI}$ will not even lead to an equivalent circuit model in which the same power is transported as in the original waveguide (see Fig. 10). Similarly, different equivalent circuit models will be obtained, all describing the propagation behaviour of the considered wave, if the definitions remain consistent within the equivalent circuit. Since real power flow together with the longitudinal surface wall current components appear as physically most self-evident quantities, an characteristic impedance according to $Z_{B,PI}$ is usually preferred.

The Bloch impedance is identical at each periodic terminal plane. Since the location of a terminal can be set arbitrarily within one period, also considered as unit cell, it varies therein and this is the difference between the characteristic impedance of a homogeneous waveguide and the Bloch impedance in a waveguide composed of periodically repeated inhomogeneous unit cells. In Fig. 11, Z_B is calculated over frequency at the input interface. It is noted that the imaginary part changes from positive to negative values at about 4 GHz around the transition frequency.

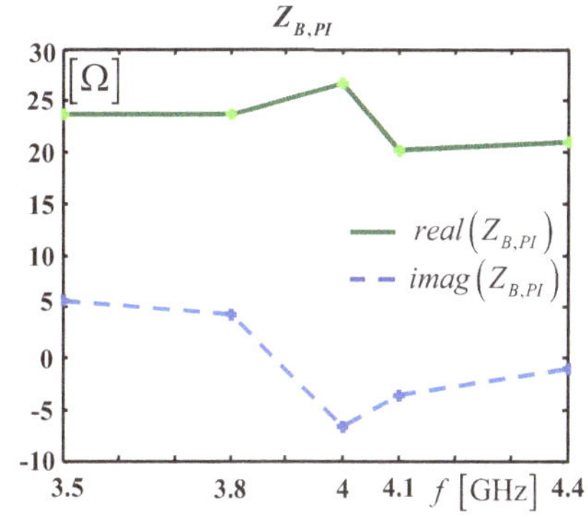

Fig. 11. Complex Bloch impedance.

4 Dispersion diagram and radiation performance

Together with the Bloch impedance, the propagation constant $\gamma = \alpha + j\beta$ sufficiently describes the eigenmode. Its imaginary part β, as solution of the eigenvalue equation or of the characteristic polynomial, is obtained by two techniques, the full-wave simulation of the infinite periodic configuration (by Ansoft HFSS) and from the field distribution of the driven problem by the Matrix-Pencil estimation method (Yingbo and Sarkar, 1990). The latter assumes a parametric signal model and approximates the field solution along a straight curve in propagation direction by a series of complex exponentials representing the various modes in the waveguide:

$$y_k = x_k + n_k = \sum_{t=1}^{M} A_t e^{\gamma_t \kappa k} + n_k \tag{6}$$

The field samples y_k are retrieved along the line from the field solution with the sample distance κ, while they may be overlayed by noise n_k. Subsequently, the so-called matrix pencils (special matrices) are formed helping to formulate an eigenvalue problem. Via singular value decomposition (SVD), the eigenvalues corresponding to the propagation constants are found and the modes having the strongest impact are filtered out. An estimation of the amplitudes follows according to the least mean square principle.

The obtained dispersion results are summarized in Figs. 12 and 13.

Both dispersion diagrams visualise the eigenfrequencies $f(\beta)$ belonging to the eigenvalues and verify the same dispersion behaviour. It is to emphasize that the antenna operates in the fundamental mode and not in higher space harmonics. If the eigenwave propagates in z-direction and attenuation along the waveguide is neglected, its spatial dependence in the far-field can be formulated by

$$\psi(x, z) \sim e^{-j\beta_z z} e^{-k_x x} \tag{7}$$

in which k_x is the wavenumber component pointing in the air above and denotes radiation or attenuation, dependent on the

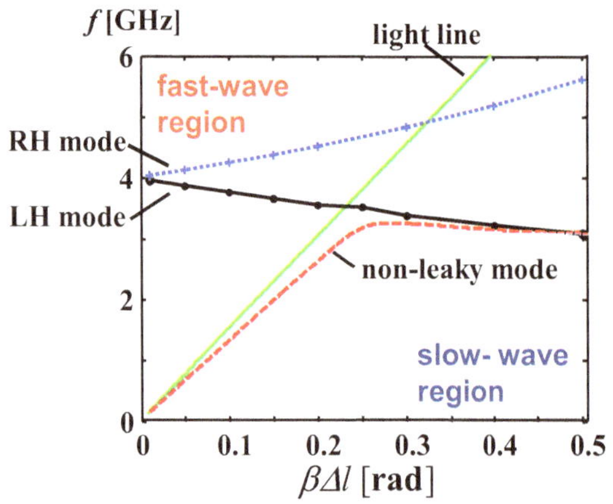

Fig. 12. Dispersion diagram computed by Ansoft HFSS.

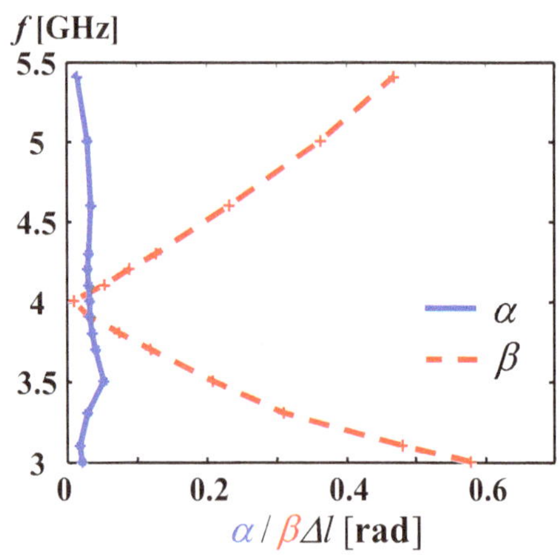

Fig. 13. Dispersion diagram obtained by the Matrix-Pencil method.

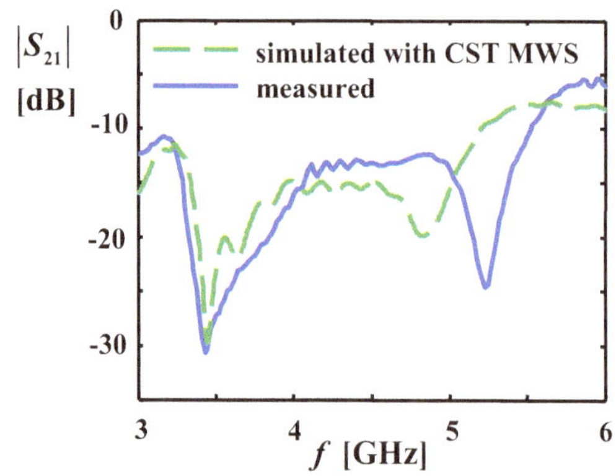

Fig. 14. Transmission factor for the LH/RH transmission line.

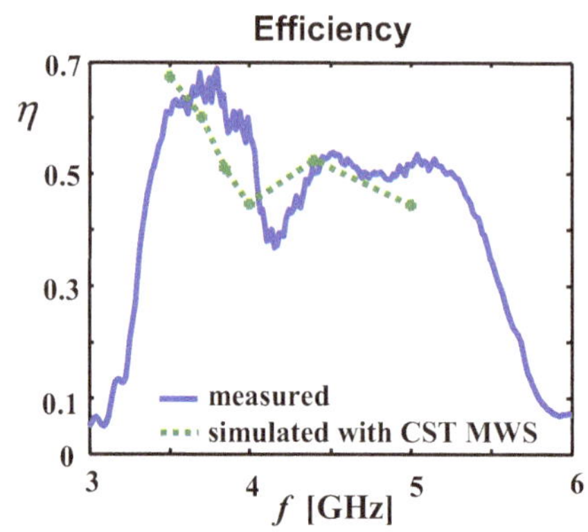

Fig. 15. Efficiency achieved with the dielectric FR-4.

kind of solution to

$$k_x = \sqrt{k_0^2 - \beta_z^2} \tag{8}$$

being real or imaginary. In the dispersion diagram in Fig. 12, the lowest mode found just represents a plane wave (non-leaky) traveling above the antenna structure. For increasing frequencies, the wave starts interaction with the waveguide to perceive in the bend down of the curve. During its course, its β_z values are larger than the phase constant of free space and thus no radiation in space occurs. The fundamental waveguide mode can be separated into two parts, a LH and a RH one. As anti-parallel group and phase velocity is exhibited only in the LH range, the slope of the curve belonging to the LH domain is negative. The upper twig of the curve is RH. Only in a balanced CRHLH structure like here, the RH mode continuously converts into the LH mode at the transition frequency, of about 4 GHz in our case, with vanishing β_z corresponding to infinite wavelength. If β_z increases with rising frequency in the RH case and reducing frequency in the LH case, radiation in the upper hemisphere takes place as long as β_z still remains less than the phase constant of free space and a real k_x exists. Due to the anti-parallel feature in the LH domain, the phase constant points to the reverse direction resulting in a backfire beam. The endfire position is reached while the antenna is operated in the RH mode. Broadside radiation is only effective with β_z=0 and k_x is identical with k_0. While β_z further increases, it crosses the light line and becomes too large again in order to permit a real k_x. Consequently, the wave is solely guided again. In contrast to the dispersion diagram obtained by full-wave simulation, the estimation technique also reveals the attenuation constant α. The attenuation of the wave along the waveguide not only represents the contribution of losses but also alludes to the amount of power leakage due to radiation. To shed light on the radiation performance, the transmission factor S_{21} is given in Fig. 14. Its course proves broadband radiation as it is less than -10 dB from about 3 GHz up to 5.3 GHz. So, the accepted energy at the input is sufficiently consumed by leakage. To remark is the good accordance of the simulations and the measurements. Utilizing the relatively lossy substrate

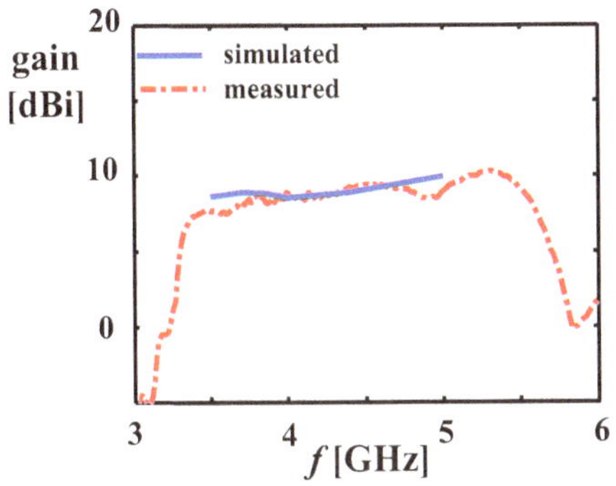

Fig. 16. Maximum gain in plane ϕ=0° for each frequency.

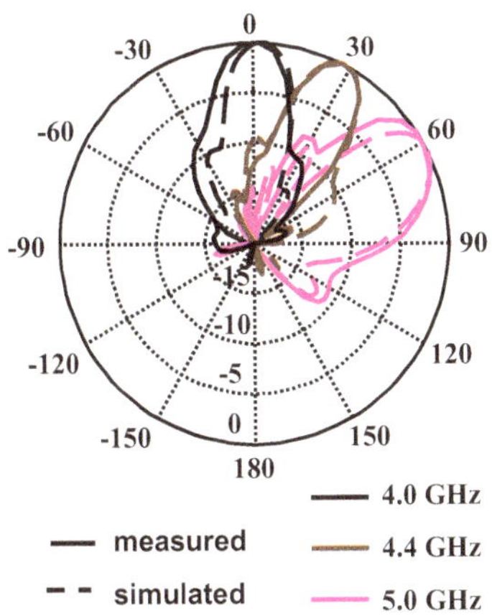

Fig. 17. Simulated and measured radiation patterns [dB] overlayed.

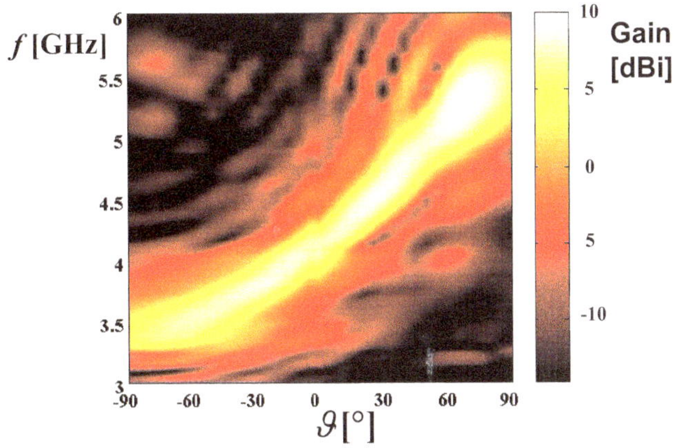

Fig. 18. Frequency-dependent scan performance from −90° to +90°.

FR-4, an efficiency of 50% to 60% (Fig. 15) is attained. An augmentation in efficiency is conceivable by means of a low-loss substrate. Maximum gain is evaluated according to IEEE definition and illustrated in Fig. 16. For broadside radiation at 4 GHz, a gain of 8.64 dBi is reached. The maximum gain of 10.15 dBi is measured at 5.3 GHz, which corresponds to a scan angle of over 70°. In Fig. 17, several radiation pattern cuts stemming from measurements and from simulations are superimposed and reveal good agreement. The side lobes are roughly around -10 dB with respect to the main lobe. Last but not least, in Fig. 18 the frequency-dependent scan behaviour is depicted verifying the contiguous radiation performance.

5 Conclusions

A broadband conformal leaky-wave antenna was presented which is continuously scannable from nearly endfire to backfire including broadside radiation. Its efficiency reaches about 50% to 60% despite of the low-cost and lossy dielectric substrate FR-4. In contrast to other configurations, the presented design is based on the H_{10} rectangular hollow waveguide mode, which was upgraded to a composite left-/righ-handed mode by introducing transverse slots with narrow spacing. It was optimised for broadside radiation at around 4 GHz. Simulation and measurement results are in good accordance and demonstrate good radiation behaviour and beam scanning capability of the leaky-wave antenna.

References

Caloz, C. and Itoh, T.: Electromagnetic Metamaterials: Transmission Line Theory and Microwave Applications, Wiley, New York, 2005.

Collin, R. E.: Field Theory of Guided Waves, 2nd ed., IEEE Press, New York, 1991.

Deslandes, D. and Wu, K.: Analysis and Design of Current Probe Transition From Grounded Coplanar to Substrate Integrated Rectangular Waveguides, IEEE Trans. Microwave Theory Techniques, 53(8), 2487–494, 2005.

Eleftheriades, G. V., Iyer, A. K., and Kremer, P. C.: Planar negative refractive index media using periodically L-C loaded transmission lines, IEEE Trans. Microwave Theory and Techniques, 50(12), 2702–2712, 2002.

Eshrah, I. A. and Kishk, A. A.: Analysis of Left-handed Rectangular Waveguides with Dielectric-filled Corrugations Using the Asymptotic Corrugation Boundary Conditions, IEE Proc.-Microw. Antennas Propag., 153(3), 221–225, 2006.

Lai, A. and Itoh, T.: CompositeRight/Left-Handed Transmission Line Metamaterials, IEEE Microwave Mag., pp. 34–50, 2004.

Marcuvitz, N.: Waveguide Handbook, McGrall-Hill, New York, 1951.

Yingbo, H. and Sarkar, T. K.: Matrix Pencil Method for Estimating Parameters of Exponentially Damped/Undamped Sinusoids in Noise, IEEE Transactions on Acoustics, Speech and Signal Processing, 38(5), 1990.

Design of Sievenpiper HIS for use in planar broadband antennas by means of effective medium theory

S.K. Hampel, O. Schmitz, O. Klemp, and H. Eul

Leibniz Universität Hannover, Institut für Hochfrequenztechnik und Funksysteme, Appelstr. 9A, 30167 Hannover, Germany

Abstract. The claim for multistandard operating handsets of small physical size as well as the ever increasing demand for higher data rates require new broadband operating antennas. Because of the widespread use of especially planar broadband antennas a lot of factors influencing the characteristic antenna parameters have to be regarded. Furthermore, aspects regarding the electromagnetic compatibility inside the handheld as well as the protection of biological systems, e.g. the user of a mobilephone, have to be payed attention to. An electromagnetic structure which allows for protection by means of shielding as well as enhances the antennas efficiency by providing unique electromagnetic properties are the so called Sievenpiper High Impedance Surfaces (HIS) invented by Sievenpiper (1999). This paper will present the theory and the well known design equations for those structures. An investigation by means of simulation tools and measurement setups will be done to approve the accuracy of the theoretical results. Here measurement results of the impedance and radiation properties of a planar log.-per. four-arm antenna equiped in conjunction with a fabricated prototype Sievenpiper HIS will be presented.

1 Introduction

The trend of digital convergence with its multiradio applications keeps on raising the bar for research and application engineers. On the one hand more spectral effiency and interoperability is demanded by the customers by simultaneously needs for increasing battery life and decreasing physical size of the handheld. On the other hand the complexity of the transceiver circuitry and with it the production costs raise due to operating in different communication standards with multiple antennas. Therefor it is necessary to use broadband antennas with an easy impedance matching network and constant antenna parameters regarding frequency, e.g. input impedance or radiation pattern, to keep circuitry complexity down. An antenna which provides such properties is for example the so called planar log.-per. trapezoidal antenna (TA) as proposed in Klemp et al. (2005) which is a representative for the group of frequency independent antennas. These planar broadband antennas exhibit a bidirectional radiation characteristic which can cause perturbation due to parasitic effects such as surface wave propagation or electromagnetic wave interaction with different sections of a handset. Because of that volumetric effects there is a need for so called spatial filters to ease the perturbation occuring in the communication terminal while conserving the radiation efficiency of the antenna as well as shielding the environment against unwanted radiation.

The Sievenpiper HIS offers attractive boundary conditions to overcome the mentioned obligations. This type of spatial filter is a metallo-dielectric material with 2D periodicity where all physical sizes of a so called unit cell must be much smaller than the corresponding wavelength in the operational bandwidth. In general the structure consists of periodic cylindrical vias placed in a grounded dielectric slab with a metal frequency selective surface layer (FSS) on top.

Because of combining the reflection properties of artificial magnetic conductors (AMC) and the ability of suppressing surface wave propagation like electromagnetic band-gap structures (EBG) (see Sievenpiper, 2003) this structures seem to be adequate for use in mobile communication handhelds. The resulting reflection phase of $\phi_\Gamma = 0°$ with an absolute reflection coeficient of $|\Gamma| = 1$ at the surface allows using the HIS as near field reflector while incident and reflected wave will not cancel out each other. Due to the in-phase image current the close proximity of the Sievenpiper HIS reflector to radiating elements results in low profile antennas with an enhanced efficiency compared to fully metallic reflectors or absorber materials. Furthermore, the suppression of surface

Correspondence to: S. K. Hampel
(hampel@hft.uni-hannover.de)

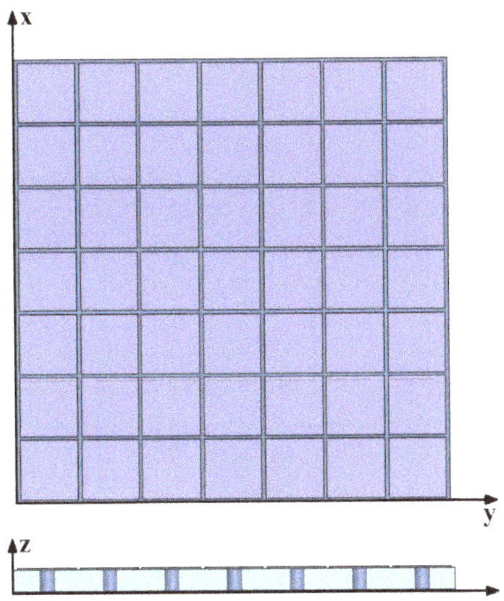

Fig. 1. Sievenpiper HIS placed in the xy-plane displayed in top- and side-view.

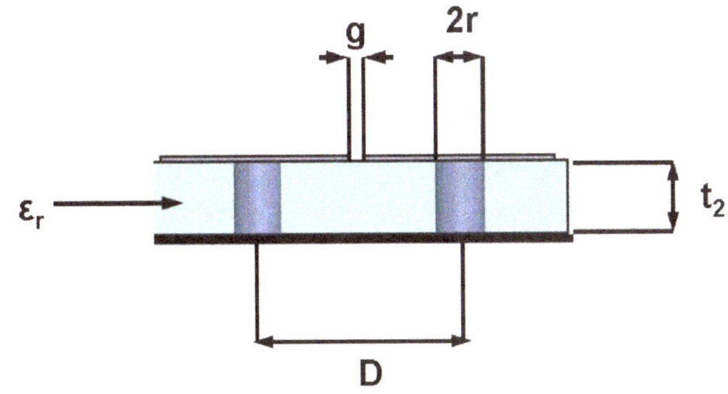

Fig. 2. Side-view of a zoomed region of the Sievenpiper HIS from Fig. 1 containing two unit cells with each a size of D.

waves eliminates the transport of interfering power to circuitry elements in the transceiver which can cause malfunctions. Moreover a smooth and uneffected radiation pattern can be expected because of avoiding parasitic radition from finite groundplane edges.

The organization of the paper is as follows. In Sect. 2 the underlying theory of the Sievenpiper HIS and the equations for designing the geometry parameters of the structure for a given frequency of operation are presented. Section 3 deals with simulation models and simulation results for verifying the derived equations. Here the simulated reflection phase properties for orthogonal incidence and the influence of the HIS on matching and radiation properties of a reference dipole are presented. In the last section measurement results of a planar broadband log.-per. four arm trapezoidal antenna in conjunction with a fabricated prototype Sievenpiper HIS are highlighted. The properties under consideration are the input impedance as well as the radiation pattern.

2 Theoretical description of reflection properties

The underlying theory of Sievenpiper HIS is in general based on the effective medium theory (see Merill, 1999) and moreover on the transmission line theory. These theories can be used when describing the (homogenized) electromagnetic behaviour of small inclusions in a host material or the quasi-static properties of devices much smaller than the wavelength respectively. On the one hand the structures can be seen as composite materials consisting of a periodic FSS on a conductor backed substrate, acting as a boundary with such a high surface impedance, leading to a vanishing magnetic field in theory, therefore called artificial magnetic conductor (AMC).

On the other hand the cylindrical vias embedded periodically in the substrate build the so called Brown's rodded medium or the "Fakir's bed of Nails" which solution can be found in Rotman (1962) and King et al. (1983). They are acting like a plasma below its plasma frequency, suppressing the propagation of surface waves comparable to the double negative materials (DNG) or metamaterials (MM), first mentioned by Veselago (1968). Because of exhibiting frequencies where no wave propagation is possible a so called band-gap occurs. Due to that textures which offer this kind of filter mechanism they are called electromagnetic band-gap structures (EBG).

Due to the very complex mechanisms yielding to the surface wave suppression and the complications in finding adequate equations describing the interaction of electromagnetic waves and the periodic structure, here only the reflection properties are highlighted. Furthermore, the propagation of surface waves in single antenna stand-alone applications is less critical than in antenna array arrangements and will be neglegted in a first approximation. Further information about and insight into the surface wave properties of these structures can be found in Clavijo (2003) and Sievenpiper (1999).

For determining the reflection properties of the exemplary structure in Fig. 1 the transmission line theory can be used, no matter what kind of patch geometry will be offered by the FSS (triangle, square, hexagon,...). The only restriction is the physical size of the geometry with regard to the corresponding wavelength at the operational frequencies. All describing parameters of a so called unit cell shown in Fig. 2 as substrate-heigth t_2, via-radius r, periodicity D or gap-width g have to be much smaller than the wavelentgh, i.e. $t_2, r, D, g << \lambda$. For an orthogonal incident wave travelling in negative z-direction adjacent edges of the FSS build a shunt capacitance. The resulting admittance of two adjacent metal stripes seen by the wave, can be obtained by means of calculus of variations and is well known in literature (see Marcuvitz, 1996). Calculating the edge capacitance from this

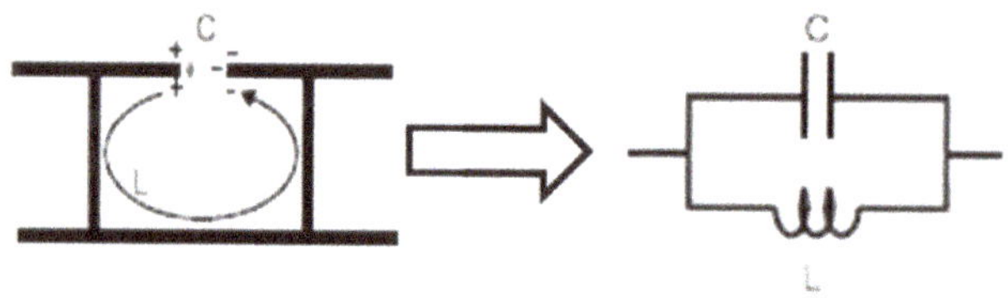

Fig. 3. Equivalent LC circuit for a unitcell.

admittance leads to

$$C_{G_{TM}} = \epsilon_{eff}\epsilon_0 \frac{2D}{\pi} \ln\left(\frac{2D}{\pi g}\right) \quad (1)$$

for TM polarization and

$$C_{G_{TE}} = \epsilon_{eff}\epsilon_0 \frac{2D}{\pi} \ln\left(\frac{2D}{\pi g}\right) \cos^2\theta_i \quad (2)$$

for TE polarization (see Simovski, 2005). Here $\epsilon_{eff}=\sqrt{(\epsilon_1+\epsilon_2)/2}$ indicates the effective permittivity of the two half spaces under consideration ($z<0$, $z>0$) and θ_i is the angle of incidence.

The conductor backed host-medium itself offers an inductive surface impedance, which is typical for substrate layers as can be found in Collin (1991) for example. The via array stabilizes the surface impedance of the substrate by mimicing a TEM waveguide for TM polarized waves. The expression for the surface impedance for both TE and TM polarization is therefore equivalent and as a first approximation independant of the incidence angle (see Clavijo, 2003). This yields to:

$$Z_{D_{TE}} = Z_{D_{TM}} = j\omega\mu_0 t_2 \quad (3)$$

Both surface impedances are in parallel for an incident wave and can be combined by means of the well known equation for parallel resistors in lumped networks expressed in Eq. (4).

$$Z_S(\omega) = \frac{Z_G(\omega) Z_D(\omega)}{Z_G(\omega) + Z_D(\omega)} \quad (4)$$

For orthogonal incidence as well as for small angles of incidence the capacitive impedance for the FSS layer is given by $Z_{G_{TM}}=Z_{G_{TE}}=Z_G=1/j\omega C_G$. While neglecting losses the two impedances build an ideal parallel resonant circuit as shown in Fig. 3. At resonant frequency the denominator vanishes and the impedance goes for infinity. The equation for the reflection coefficient known from the transmission-line theory (see Zinke and Brunswig, 2000) by use of the derived expressions for the resulting surface impedance yields

$$r = \frac{j\omega t_2 \mu_r \mu_0 - Z_0\left(1-\left(\frac{\omega}{\omega_0}\right)^2\right)}{j\omega t_2 \mu_r \mu_0 + Z_0\left(1-\left(\frac{\omega}{\omega_0}\right)^2\right)} \quad (5)$$

with $\omega_0=1/\sqrt{LC}$ representingthe resonant frequency of the circuit.

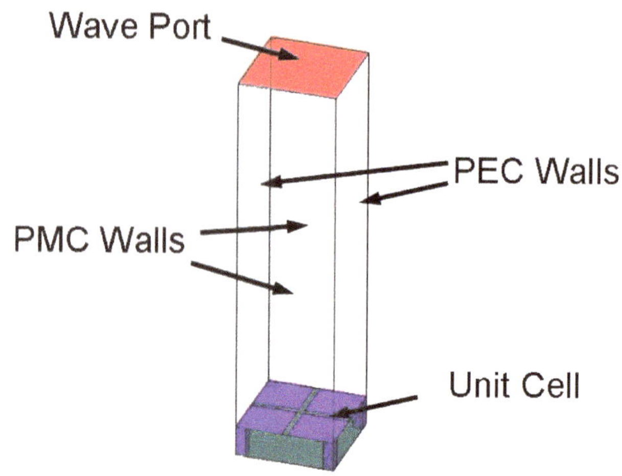

Fig. 4. HFSS simulation model for determining the reflection properties of a HIS: TEM waveguide with periodic boundary conditions.

Due to the metallic surface of the structure the absolute value of the reflection coefficient remains 1, whereas the occuring reflection phase decides on whether more constructive or more destructive interference takes place. Because of the frequency dependence of the reflection coefficient an operational bandwidth can be given which is located in between the frequencies where the reflection phase ϕ_Γ equals $\pm\pi/2$. Under this constraints and with the definition of a relative bandwidth $B_{rel}=f_u/f_l$, the ratio of the upper frequency ($\Phi_\Gamma=-\pi/2$) to the lower frequency ($\Phi_\Gamma=+\pi/2$) of the operational bandwidth, Eq. (5) can be written as follows:

$$\frac{B_{rel}-1}{\sqrt{B_{rel}}} = 2\pi\mu_r \frac{t_2}{\lambda_{res}} \quad (6)$$

Here λ_{res} stands for the wavelength at resonant frequency. As can be seen, this transcendent expression and the bandwidth respectively depend mainly on the substrate height.

3 Simulation results

This section deals with simulation results for evaluating the derived design equations by means of the simulation tool *HFSS* of Ansoft Corporation. First of all the reflection properties of a Sievenpiper HIS will be investigated. The used simulation model is shown in Fig. 4 and consists of an ideal TEM waveguide. This is realized by the use of so called periodic boundary conditions where adjacent walls offer perfect electric and perfect magnetic boundary conditions alternating circular around the longitudinal axis. The TEM wave is excited by a wave port traveling down the longitudinal axis of the waveguide. At the end a unit cell of the HIS under test is placed. The parameters of the designed HIS are D=8.5 mm, g=0.5 mm, t_2=3 mm and r=1 mm. The host substrate is a FR4 with a relative permittivity of ϵ_r=4.4 resulting in a theoretical resonant frequency of f_{res}=4.65 GHz

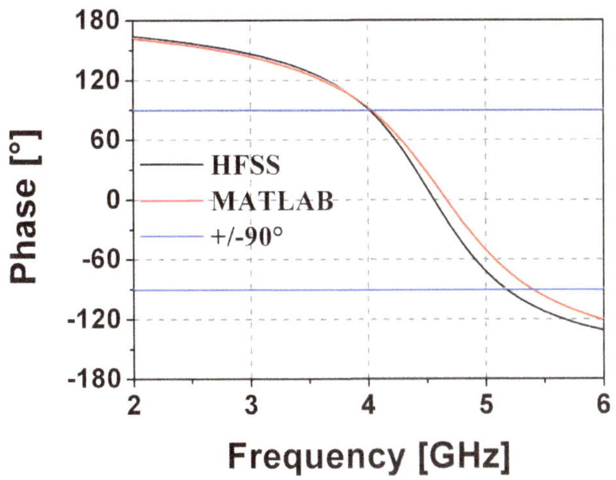

Fig. 5. Reflection phase of the considered HIS over frequency.

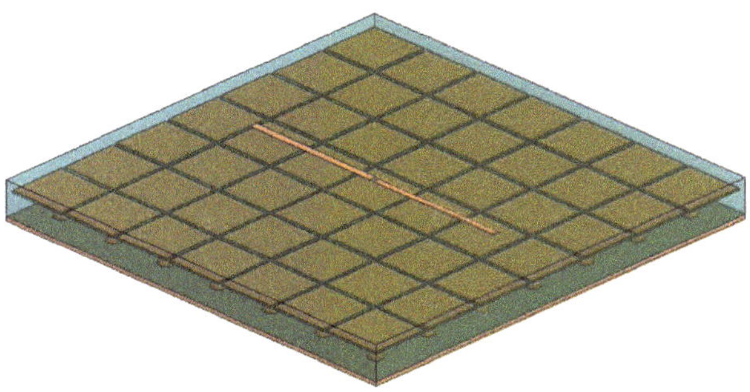

Fig. 6. Flat dipole at FR4 substrate placed 1.5 mm above a Sievenpiper HIS.

with bandedges at f_l=4 GHz and f_u=5.2 GHz. The reflection phase of this structure can be easily calculated when using this simulation model by deembeding of the reflection coefficient of the wave port onto the surface of the HIS. The curves over frequency derived by the simulation and the presented equations from Sect. 2 implemented in MATLAB are shown in Fig. 5. A good agreement of the two curves can be observed. Only the resonant frequency and the frequency where $\Phi_\Gamma=-\pi/2$ reveal small differences of 100 MHz (2.2%) and 200 MHz (3.9%) respectively. But under consideration of the simplifications done by deriving the equations these results can be seen as a proof for the accuracy of the design formulas for determining the reflection properties of a Sievenpiper HIS. The only drawback of this simulation model is the invariable orthogonal incidence of the propagating wave. Moreover due to the wide spectrum of different modes existing in the reactive near field of a radiating antenna and their corresponding incidence angles, small deviations from the desired functionality and operational frequency must be taken into account.

The next step while analyzing the fuctionality of the this

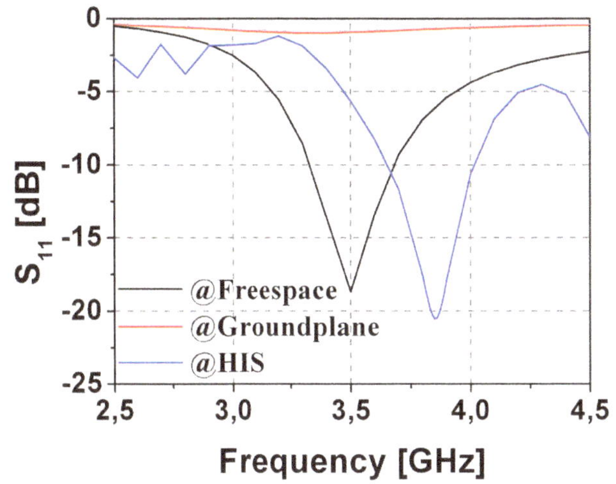

Fig. 7. Return loss for a flat dipole placed above a perfect conducting plate, above a HIS and in freespace.

geometry is placing an antenna above the structure and investigating the influence of the HIS on its characteristic properties and vice versa. For the purpose of being able to evaluate the effect of the Sievenpiper HIS on the antenna correctly it is important to use a simple and well known antenna geometry. Therefore the use of a dipole as reference antenna geometry is convenient and often used in literature while evaluating miscellaneous applications. Figure 6 shows the configuration of the dipole in conjunction with the filter structure.

Considered parameters for evaluating the efficiency of this arrangement and especially the HIS are the return loss of the antenna as well as its gain. The dipole length is designed for resonating at 3.5 GHz and it is placed on a FR4 substrate with a thickness of 1.5 mm ($\approx\lambda_{eff}/30$@3.5 GHz).

Applying the same parameter values to the HIS structure as in the case of investigating the reflection properties the resonant frequency of the surface impedance is shifted down to 3.65 GHz based on the derived equations and 3.55 GHz resulting from the simulation. This shift is founded by placing the antenna substrate on top of the HIS which affects the capacity due to replacing the effective medium consisting of air and FR4 by a homogeneous medium consisting exclusively of FR4. This increases the shunt capacitance of adjacent edges and hence reduces the resonant frequency. Figure 7 displays the curves for the return loss. To be able to evaluate the composite structure of antenna and HIS the curves for the dipole placed in freespace as well as for the case of a conductor backed antenna substrate with 1.5 mm thickness are also stated. While regarding the run of the curve for the dipole in conjunction with the metallic groundplane it is obvious that this configuration does not accept or radiate any power respectively. The dipole at freespace reveals the expected characteristic with a minimum of S_{11}=−18 dB at 3.5 GHz, the designed resonant frequency. The dipole placed above the Sievenpiper HIS offers almost the same characteristic as the arrangement in freespace does. The main difference is a shift of the whole curve up to higher frequencies with a better return loss of $S_{11}=-21$ dB at 3.8 GHz. The frequency shift can be explained as follows. The dipole is excited by a so called lump port which mimics an ideal source arranged between the two arms of the dipole. The arms form a parasitic capacitance with the metallic surface of the HIS which is in parallel to the antenna arms. So the dipole seems to be electrically shorter and resonates at a higher frequency. Investigations with two coaxial cables as feeder for the dipole have shown the opposite effect of building a series resonance, due to the fact that the outer conductor was connected to the groundplane of the HIS. Moreover a variation of the antenna substrat-height shows a decreasing frequency of the point with minimal return loss, while increasing the substrate thickness. This can be regarded as reducing the parasitic parallel capacitance which is assumed to be proportional to the inverse of the substrate height (comparable to a parallel-plate capacitor).

Furthermore the phase shift of a wave travelling through the antenna substrate and back has to be regarded. So the optimal constructive interference of the radiated and reflected wave has to be expected for frequencies where $\phi_\Gamma < 0°$, which is above the resonance of the Sievenpiper HIS. The discrepancies of the two curves above 4.25 GHz is related to an occuring edge resonance of the HIS groundplane at about 4.65 GHz. The corresponding characteristic of the absolut gain in the main direction, i.e. in z-direction, is presented in Fig. 8.

Due to the results for the return loss the curve of the dipole placed 1.5 mm over a metal plate will not be analyzed here because as mentioned before this arrangement does not accept any power and hence can not radiate effectively.

The dipole in freespace reveals an almost constant gain over frequency. At the point of minimum return loss the absolut gain is G_i=2.5 dB which is in good agreement with the theoretical result for an ideally thin wire dipole which offers G_i=2.15 dB.

As can be seen in the frequency range of good matching, i.e. $S_{11} < -10$ dB, from 3.6 GHz to 4 GHz the absolute gain of the dipole in conjunction with Sievenpiper HIS is much higher than for the dipole in freespace. This observed effect can be seen as a proof for the predicted reflection properties of the HIS since the in-phase reflection of an incident wave leads to constructive interference. This superimposition of a directly radiated wave and the reflected wave ideally doubles the electric field strength in the primary half space of radiation and hence leads to an increase in absolute gain of 3 dB. This effect can be observed in Fig. 8 at the frequency point with best matching at 3.8 GHz. Here the value for the absolute gain is $G_{i_{3.8\,\mathrm{GHz}}}$=5.8 dB. The boost of the gain by a value higher than 3 dB, exactly 3.3 dB, is founded in numerical uncertainty.

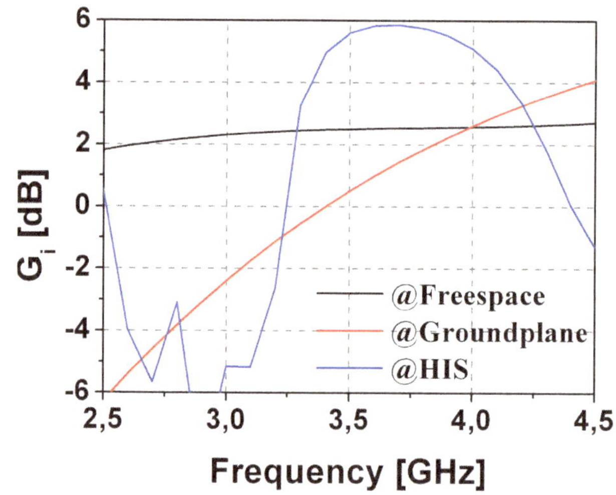

Fig. 8. Absolut antenna gain in the main direction for a flat dipole placed above a perfect conducting plate, a HIS and in freespace.

Finally the radiation patterns of the two arrangements in freespace and over the Sievenpiper HIS are evaluated. Figure 9 shows both the horizontal and the vertical diagramm of the dipole for the two dispositions. Here again the improvement of the gain is obvious. Furthermore it can be seen that the front to back ratio is increased due to the shielding effect of the HIS structure with respect to the lower half space. By the way the smooth pattern is an indication for no or at least negligible parasitic radiation from the HIS structure or its groundplane which normally generates sidelobes in the pattern.

After having highlighted the correct function of the HIS and the accuracy of the design equations, now an evaluation by means of measurement results will be done.

4 Measurement results

The former section which dealed with simulation results revealed the accuracy of the derived equations of Sect. 2 for designing Sievenpiper HIS with respect to its reflection properties. The next step treated in this section is to fabricate a prototype and measure its influence on important antenna parameters of a planar broadband antenna, while using the HIS as near field reflector, as done before in the simulation.

The used braodband antenna is the so called planar log.-per. four-arm trapezoidal antenna (TA) with $\widehat{w}$=45° and M=5 periods as presented in Klemp et al. (2005).

The mentioned trapezoidal antenna is depicted in Fig. 10.

Due to its four arms this antenna geometry offers the possibility of dual linear or circular polarized radiation. For dual linear polarization two opposite arms are excited in antiphase. Hence differential excitation of adjacent arm pairs with no phase difference between the upper and right (or left)

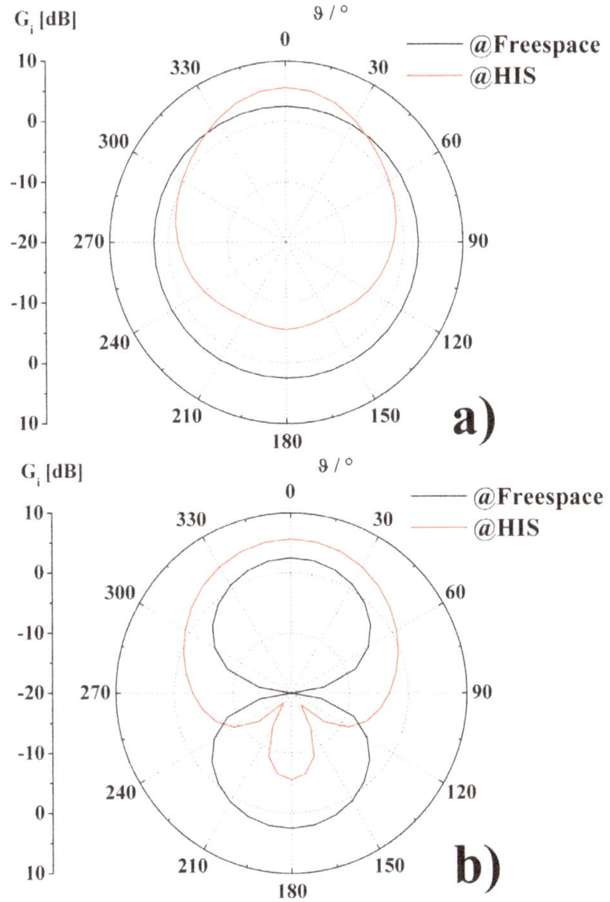

Fig. 9. Horizontal and vertical dipole characteristic for arrangement in freespace and in conjunction with HIS for cuts at **a)** ϕ=0° and **b)** ϕ=90°.

arm and the lower and left (or right) arm results in a radiating configuration which is similar to a cross dipole. Attention has to be paid to the fact that horizontal excitation yields vertical polarized radiation and vice versa due to the shape of the arms.

Another main feature of this antenna is its quasi-frequency independent input inpedance as well as the almost constant radiation properties like gain and effective antenna area respectively due to its selfcomplemetarity and selfsimilarity. The theoretical values are known from literature and former investigations and publications (see Klemp et al., 2005) and must be constant over frequency in the range of approximately 2 GHz up to 6 GHz for the regarded standard TA.

The antenna was fabricated on 1.5 mm FR4 which results in an approximately purely real input impedance of $Re\{Z_{in}\}=77\Omega$ and a vanishing imaginary part.

The prototype of the Sievenpiper HIS used as near field reflector in conjunction with the introduced planar broadband antenna offers the same parameters as stated in the simulation model which yields a theoretical resonant frequency of

Fig. 10. Standard planar broadband four arm log.-per. trapezoidal antenna.

Fig. 11. Fabricated prototype of Sievenpiper HIS with theoretical resonance frequency at f_{res}=3.65 GHz and a relative bandwidth of B_{rel}=1.26 ranging from f_l=3.25 GHz to f_u=4.1 GHz.

f_{res}=3.65 GHz in homogeneous FR4 substrate with an permittivity of approximately ϵ_r=4.4. A picture of the prototype is shown in Fig. 11 with drill holes in the center for the four semi ridgid feed lines of the antenna.

The characteristic of the real part as well as the imaginary part of the input impedance of opposite arms excited in a differential way is plotted in Fig. 12. As in the simulation of the reference dipole, here the curves for the TA on a grounded dielectric FR4 slab, on a microwave absorber as well as on the fabricated HIS are shown over frequency.

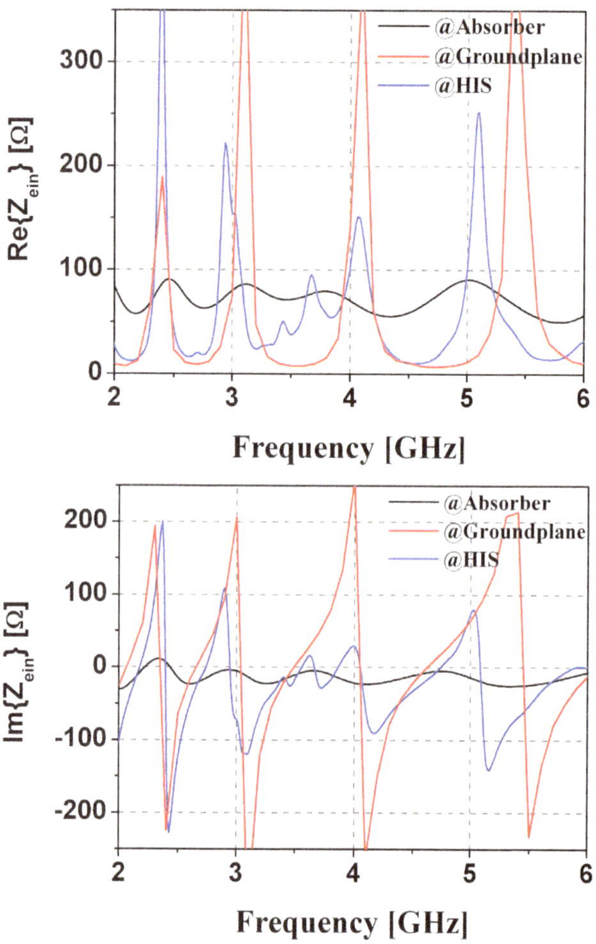

Fig. 12. Real and imaginary part of the input impedance over frequency of one arm pair for the TA in conjunction with a metal plate, a microwave absorber as well as the fabricated HIS.

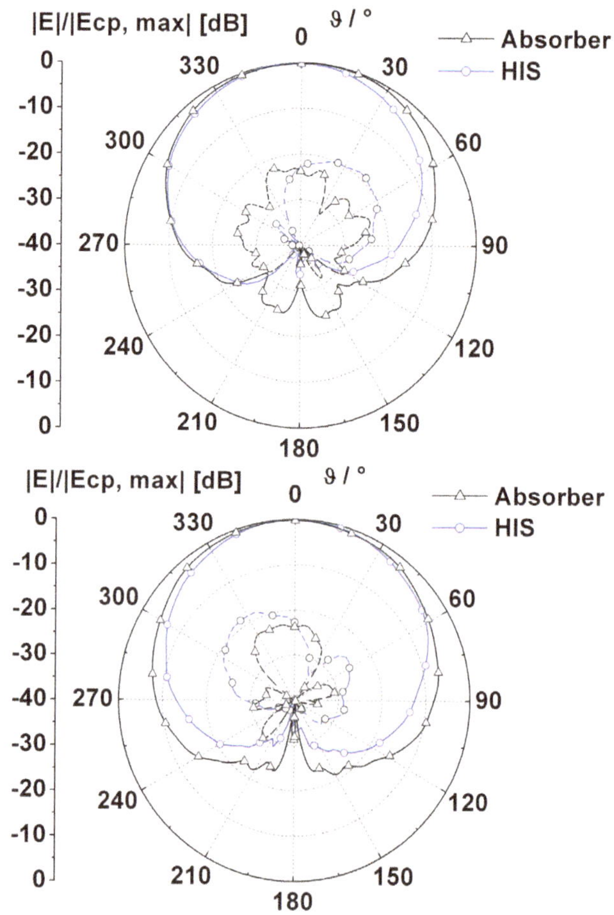

Fig. 13. Horizontal and vertical diagramm of the relative radiation pattern for the TA in conjunction with a microwave absorber and a Sievenpiper HIS at 3.6 GHz.

The curve for the TA on the conductor backed substrate reveals strong resonance behaviour with no remarkable matching point in both real and imaginary part. When considering the characteristics of the impedance, while using the filter structure, a strong effect of the HIS becomes obvious between 3 GHz and 4 GHz. This effect is expressed in variations from the resonant behaviour occuring outside this frequency band. Further investigations between 3.5 GHz and 4 GHz show similarities between the reference curve when using a microwave absorber and the one with the near field reflector. The imaginary part vanishes, whereas the real part alternates with an amplitude of only 20Ω around a theoretical (virtual) mean value of 77Ω offered by the curve while using the microwave absorber. An optimal match is observed at 3.6 GHz which coincides with the theoretical predicted behaviour of the HIS structure.

Now the influence of the reflector on the radiation characteristic at a specific frequency point will be highlighted. As mentioned before a quasi optimal match occurs at 3.6 GHz. Therefore it is straightforward to evaluate the radiation pattern at this frequency. The resulting pattern for the TA positioned at a microwave absorber as well as over the HIS prototype are presented in Fig. 13.

The measured characteristics reveal the benefit when using the HIS structure. It is obvious that the Sievenpiper HIS increases the front-to-back ratio because of the shielding effect, while simultaneously narrows the beamwidth due to the constructive interference. Moreover the unwanted cross polarized component is kept very low as in the case when using the microwave absorber. Only at wider angles a deviation between the two curves with a 5 dB better performance for the microwave absorber has to be mentioned concerning the suppression of the cross polarized component. But due to a cross polarization discrimination of better than 15 dB for the combination of antenna and HIS in the whole primary half space, this effect is of less importance. Furthermore, the pattern measured with the absorber reveals ripple for the cross polarized component in the upper horizontal diagramm which is founded in induced currents radiating from the finite groundplane edges whereas the use of the HIS shows a smooth shape due to suppressing this currents and surface waves, respectively.

5 Conclusions

In this paper we presented the theory for describing a new kind of spatial filter known as Sievenpiper HIS. The unique in-phase reflection characteristics offered by these composite metallo-dielectric structures were explained and the well known design formulas based on the transmission-line theory were presented. The accuracy of the describing equations as well as the predicted filter properties have been verified by means of results derived from simulation models in the 3D fullwave solver $HFSS$ of Ansoft Corporation. Furthermore, the influence of the introduced HIS in conjuction with a reference antenna has also been stated by use of the simulation tool $HFSS$. Especially the input matching and the absolute antenna gain has been evaluated. Here also some parasitic effects occuring due to interaction in the reactive antenna near filed have been explained. Finally a prototype HIS was fabricated and measurements have been done where a planar broadband log.-per. four-arm trapezoidal antenna was placed over the prototype. The derived results in terms of arm input impedance over frequency as well as relative radiation patterns at a specific frequency were stated and evaluated. All the derived results showed good agreement with the expected behaviour based on the theoretical description of the HIS with only small deviations.

References

Clavijo, S.: Design Methodology for Sievenpiper High-Impedance Surfaces: An Artificial Magnetic Conductor for Positive Gain Electrically Small Antennas, IEEE Transactions on Antennas and Propagation, 51, 2678–2690, 2003.

Collin, R. E.: Field Theory Of Guided Waves, IEEE Press, New York, 1991.

King, R., Thiel, D., and Park, K.: The synthesis of surface reactances using an artificial dielectric, IEEE Transactions on Antennas Propagation, 31, 471–476, 1983.

Klemp, O., Schultz, M., and Eul, H.: Novel Logarithmically Periodic Antennas for Broadband Polarization Diversity Reception, AEÜ International Journal on Electronics and Communications, 1, 317–326, 2005.

Marcuvitz, N.: Waveguide Handbook, Peter Peregrinus Ltd, London, 1986.

Merill, W. M.: Effective Medium Theories for Artificial Materials Composed of Multiple Sizes of Spherical Inclusions in a Host Continuum, IEEE Transactions on Antennas and Propagation, 47, 142–148, 1999.

Rotman, W.: Plasma Simulation by Artificial DIelectrics and Parallel-Plate Media, IRE Transactions on Antennas and Propagation, 82–95, 1962.

Sievenpiper, D. F.: High-Impedance Electromagnetic Surfaces, Dissertation, University of California, Los Angeles, 1999.

Sievenpiper, D. F.: High-Impedance Electromagnetic Surfaces with a Forbidden Frequency Band, IEEE Transactions on Microwave Theory and Techniques, 51, 2059–2074, 2003.

Simovski, C. R.: High-Impedance Surfaces Having Stable Resonance With Respect to Polarisation and Incidence Angle, IEEE Transactions on Antennas and Propagation, 53, 908–914, 2005.

Veselago, V.: The electrodynamics of substances with simultaneously negative values ϵ and μ, Sov. Phys. Uspheki, 10, 509–514, 1968.

Zinke, O. and Brunswig, H.: Hochfrequenztechnik 1, Springer Verlag, Berlin, 2000.

On the applicability of conventional transmission line theory within cavities

F. Gronwald

Institute for Fundamental Electrical Engineering and EMC, Otto-von-Guericke-University Magdeburg, Universitätsplatz 2, 39106 Magdeburg, Germany

Abstract. We investigate whether or not conventional transmission line theory needs to be modified if transmission lines are considered that are located in a cavity rather than in free space. Our analysis is based on coupled Pocklington's equations that can be reduced to integral equations for the antenna mode and the transmission line mode. Under the usual assumptions of conventional transmission line theory these modes do approximately decouple within a cavity. As a result, cavity properties will primarily influence the antenna mode but not the transmission line mode.

1 Introduction

Interior problems of Electromagnetic Compatibility analysis involve electric and electronic components that are located within cavities (Tesche et al., 1997; Lee, 1995). Usually, transmission lines will constitute a part of these components. To model the electromagnetic propagation along transmission lines we have to resort to the Maxwell theory. It is desirable to simplify Maxwell's equations to Telegrapher equations since solutions of Telegrapher equations are fairly easy to obtain. But in case of interior problems we have to examine if these simplification can be made inside a cavity and this is the subject of this paper. Our strategy will be to exhibit the steps that are necessary to derive conventional transmission line theory from integral equations of antenna theory. There already is a number of such derivations (see, for example, King, 1955; Tkachenko et al., 1995; Tesche et al., 1997; Haase et al., 2004). These approaches use electric field integral equations as physical basis but differ in the assumptions and approximations that are made in order to arrive at the conventional transmission line theory. Also they assume, implicitly or explicitly, that the transmission lines are located in free space. The distinctive feature of our approach is that we take advantage of a separation of antenna mode currents and transmission line mode currents right from the beginning.

Conventional transmission lines are metallic structures that transmit electromagnetic signals and energy. In this respect they are very similar to systems of transmitting and receiving antennas. However, the physical mechanisms that govern the electromagnetic transmission along transmission lines is quite different if compared to the electromagnetic transmission between pairs of antennas, compare Fig. 1.

In between a pair of antennas the electromagnetic transmission results from a propagating electromagnetic field which, for practical purposes, can often be approximated by a radiation field. This does not mean that in such a situation no Coulomb fields are present. Coulomb fields will be related to the electric charges that move along the antennas and constitute their near-fields. But in many cases the transmitting and receiving antennas are sufficiently far apart such that the main coupling is mediated by the radiation field which resembles a freely propagating electromagnetic field. Electric charges are not involved in the actual electromagnetic transmission that happens in between the antennas. They only are required at the beginning and at the end of the transmission in order to, respectively, generate and receive the transmitting electromagnetic field.

The electromagnetic transmission along a transmission line does involve electric charges. These charges are located on the transmission line which normally consists of a highly conducting material. They are accompanied by Coulomb fields which dominate their mutual electromagnetic interaction at short distances. While the electric charges get accelerated they will produce radiation fields. In particular, this will happen at high frequencies or if the transmission line is strongly curved or bent. Normally, such a creation of radiation fields by the electric charges on the transmission line is an unwanted effect which influences the properties of the transmission line. For many situations this influence is small

Correspondence to: F. Gronwald
(Frank.Gronwald@et.uni-magdeburg.de)

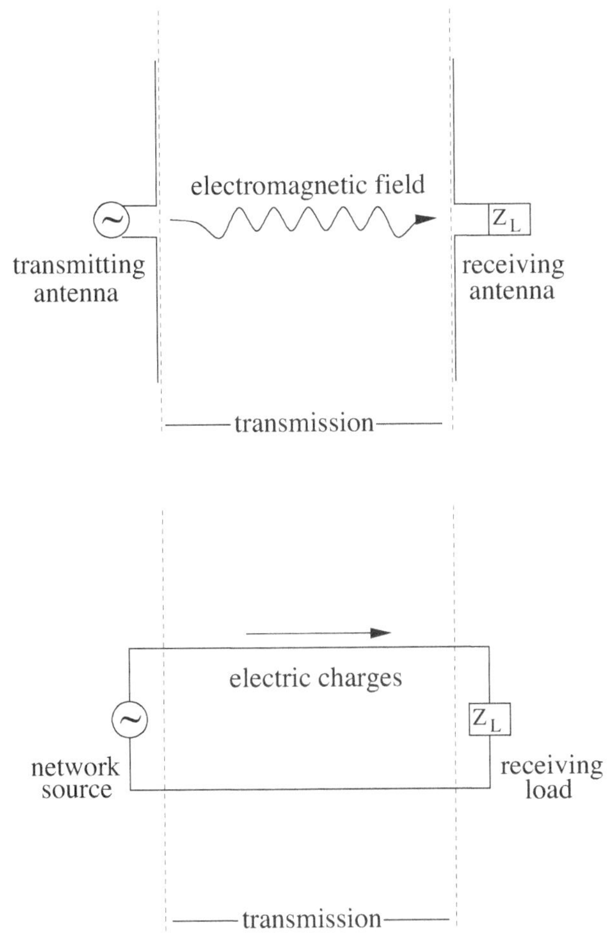

Fig. 1. Electromagnetic transmission by means of a pair of antennas (upper part) and a transmission line (lower part). In between the antennas an electromagnetic field mediates the actual transmission while the transmission line provides electric charges that mediate the transmission between the source and the load.

and negligible. Therefore, in the conventional transmission line theory focus is put on the electric charges and their accompanying Coulomb fields.

The conventional transmission line theory can be derived from the Maxwell theory as a limiting case and contains the electric current, representing electric charges, and the electric voltage, representing the associated Coulomb fields, as main physical quantities. Clearly, these two quantities are not independent of each other. They are related by the Telegrapher equations, which constitute a set of coupled first order differential equations, and are much easier to solve than the Maxwell equations.

In the derivation of the Telegrapher equations from the Maxwell equations it is customary to consider some, a priori, arbitrary transmission line and to assume a number of restrictions (Tesche et al., 1997):

1. The conductors are geometrically uniform, i.e. the transmission line is not curved or bent.

2. The distance between the conductors of the transmission line is small compared to the wavelength of the exciting electromagnetic field.

3. The thickness of the conductors of the transmission is small if compared to the wavelength of the exciting electromagnetic field.

4. The conductors are perfectly conducting.

The second and third of these restrictions are not clearly cut since "smallness" with respect to a wavelength is not a precise notion. The reason for these approximate criteria is that, in fact, one would like to remove the influence of radiation fields on the transmission line. However, Coulomb fields and radiation fields are inseparably intertwined. Therefore, in the conventional transmission line theory, one only takes into account electromagnetic interactions between electric charges at short distances where Coulomb fields dominate and radiation fields can be neglected. Also the first and fourth restriction are put forth to avoid an influence of radiation fields on the transmission line. In contrast to the second and third restriction they can be formulated in a mathematically exact way with no approximations involved. Since in the derivation of the Telegrapher equations approximations are inevitable it is often acceptable to relax the first and fourth conditions to some degree and allow for transmission lines which are slightly bent, i.e. which are characterized by radii of curvature that are large compared to the wavelength of the exciting field, and which are good conducting rather than perfectly conducting, i.e. which are characterized by a conductivity σ that fulfills the requirement $|\sigma| \gg |\varepsilon\omega|$.

It has been mentioned that in the derivation of the conventional transmission line theory it usually is assumed that the transmission line is located in free space. It follows that in the derivation of the Telegrapher equations the Green's function of free space is employed. If we want to consider a transmission line within a resonating environment we may employ a cavity's Green's function rather than the Green's function of free space. Thus, it is necessary to check if this modification has an influence on the validity of the Telegrapher equations of the conventional transmission line theory.

We stress that the motivation to work within the framework of transmission line theory stems from the simplicity of the Telegrapher equations if compared to the Maxwell equations. Alternatively, we can always work on the level of electric field integral equations and directly apply approximate analytic methods or numerical methods.

2 Coupled Pocklington's equations, antenna and transmission line mode

We consider a set of coupled Pocklington's equations that models the electromagnetic coupling to a system of wires and represents a transmission line. For concreteness we consider two wires and assume a thin-wire approximation. Then the corresponding coupled Pocklington's equations are given by Nakano (1996)

$$j\omega\mu\left[\int_{\text{wire 1}} \overline{\boldsymbol{G}}^{E}(\tau_1,\tau_1')\boldsymbol{I}_1(\tau_1')\,d\tau_1' + \int_{\text{wire 2}} \overline{\boldsymbol{G}}^{E}(\tau_1,\tau_2')\boldsymbol{I}_2(\tau_2')\,d\tau_2'\right]\cdot \boldsymbol{e}_{\tau_1} = E_{\tan}^{\text{inc}}(\tau_1)\,, \quad (1)$$

$$j\omega\mu\left[\int_{\text{wire 1}} \overline{\boldsymbol{G}}^{E}(\tau_2,\tau_1')\boldsymbol{I}_1(\tau_1')\,d\tau_1' + \int_{\text{wire 2}} \overline{\boldsymbol{G}}^{E}(\tau_2,\tau_2')\boldsymbol{I}_2(\tau_2')\,d\tau_2'\right]\cdot \boldsymbol{e}_{\tau_2} = E_{\tan}^{\text{inc}}(\tau_2)\,. \quad (2)$$

Here we introduced the variables τ_1, τ_2 that parameterize the length of wire 1 and wire 2, respectively. Fixed values of these variables represent fixed wire positions. The unit vectors $\boldsymbol{e}_{\tau_1}$, $\boldsymbol{e}_{\tau_1}$ are tangent to the line-like wires at τ_1, τ_2. The currents $\boldsymbol{I}_1(\tau_1)$, $\boldsymbol{I}_2(\tau_2)$ result from the thin-wire approximation and are defined by

$$\boldsymbol{I}_i(\tau_i) := I_i\boldsymbol{e}_{\tau_i} \quad (3)$$

for $i = 1, 2$. The scalar I_i is the value of the electric current at the wire position τ_i. Finally, we denote by $\overline{\boldsymbol{G}}^{E}$ the dyadic Green's function for the electric field (Tai, 1994).

If the wires form a transmission line we expect that they can be parameterized by a common coordinate ξ with $\xi = \xi_0$ at the beginning and $\xi = \xi_L$ at the end of the line, compare Fig. 2. We take this coordinate as a common integration variable and write Eqs. (1) and (2) as

$$j\omega\mu\left[\int_{\xi_0}^{\xi_L}\left(\overline{\boldsymbol{G}}^{E}(\tau_1,\tau_1')\boldsymbol{I}_1(\tau_1')\frac{\partial\tau_1'}{\partial\xi'} + \overline{\boldsymbol{G}}^{E}(\tau_1,\tau_2')\boldsymbol{I}_2(\tau_2')\frac{\partial\tau_1'}{\partial\xi'}\right)d\xi'\right]\cdot\boldsymbol{e}_{\tau_1} = E_{\tan}^{\text{inc}}(\tau_1)\,, \quad (4)$$

$$j\omega\mu\left[\int_{\xi_0}^{\xi_L}\left(\overline{\boldsymbol{G}}^{E}(\tau_2,\tau_1')\boldsymbol{I}_1(\tau_1')\frac{\partial\tau_1'}{\partial\xi'} + \overline{\boldsymbol{G}}^{E}(\tau_2,\tau_2')\boldsymbol{I}_2(\tau_2')\frac{\partial\tau_2'}{\partial\xi'}\right)d\xi'\right]\cdot\boldsymbol{e}_{\tau_2} = E_{\tan}^{\text{inc}}(\tau_2)\,. \quad (5)$$

The variables τ_1, τ_2 are now understood as functions of the parameter ξ.

Next we introduce two currents $\boldsymbol{I}_{\text{A}}$ and $\boldsymbol{I}_{\text{TL}}$ as linear combinations of $\boldsymbol{I}_1$ and $\boldsymbol{I}_2$,

$$\boldsymbol{I}_{\text{A}} := \tfrac{1}{2}(\boldsymbol{I}_1 + \boldsymbol{I}_2)\,, \quad (6)$$

$$\boldsymbol{I}_{\text{TL}} := \tfrac{1}{2}(\boldsymbol{I}_1 - \boldsymbol{I}_2)\,. \quad (7)$$

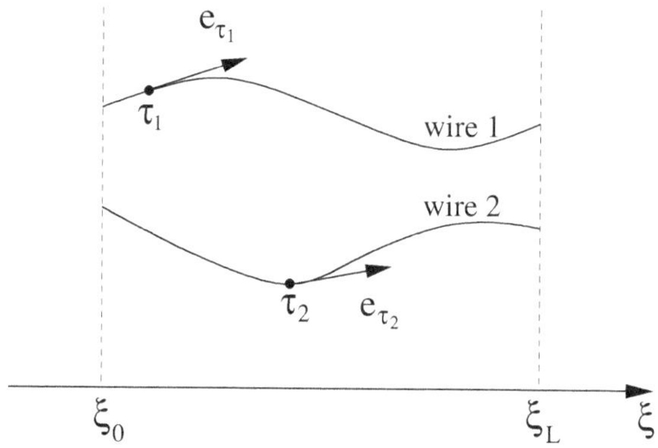

Fig. 2. Introduction of a common variable ξ which parameterizes the wires of a transmission line.

The inverse equations are

$$\boldsymbol{I}_1 = \boldsymbol{I}_{\text{A}} + \boldsymbol{I}_{\text{TL}}\,, \quad (8)$$

$$\boldsymbol{I}_2 = \boldsymbol{I}_{\text{A}} - \boldsymbol{I}_{\text{TL}}\,. \quad (9)$$

These identifications are well-known from the conventional transmission line theory where $\boldsymbol{I}_{\text{A}}$ represents the so-called "antenna mode" or "common mode" and $\boldsymbol{I}_{\text{TL}}$ represents the so-called "tranmission line mode" or "differential mode". In our present context these identifications are still formal. We note that neither $\boldsymbol{I}_{\text{A}}$ nor $\boldsymbol{I}_{\text{TL}}$ need to be tangent to one of the wires. But it is clear that we still may split $\boldsymbol{I}_{\text{A}}$ and $\boldsymbol{I}_{\text{TL}}$ into a component and a unit vector,

$$\boldsymbol{I}_{\text{A}} = \frac{1}{2}\left(I_1\boldsymbol{e}_{\tau_1} + I_2\boldsymbol{e}_{\tau_2}\right) =: I_{\text{A}}\boldsymbol{e}_{I_A}\,, \quad (10)$$

$$\boldsymbol{I}_{\text{TL}} = \frac{1}{2}\left(I_1\boldsymbol{e}_{\tau_1} - I_2\boldsymbol{e}_{\tau_2}\right) =: I_{\text{TL}}\boldsymbol{e}_{I_{TL}}\,. \quad (11)$$

If the relations (8) and (9) are inserted into Eqs. (4) and (5) it is simple to find

$$j\omega\mu\left[\int_{\xi_0}^{\xi_L}\left(\left[\overline{\boldsymbol{G}}^{E}(\tau_1,\tau_1')\frac{\partial\tau_1'}{\partial\xi'} + \overline{\boldsymbol{G}}^{E}(\tau_1,\tau_2')\frac{\partial\tau_2'}{\partial\xi'}\right]\boldsymbol{I}_{\text{A}}(\xi') + \left[\overline{\boldsymbol{G}}^{E}(\tau_1,\tau_1')\frac{\partial\tau_1'}{\partial\xi'} - \overline{\boldsymbol{G}}^{E}(\tau_1,\tau_2')\frac{\partial\tau_2'}{\partial\xi'}\right]\boldsymbol{I}_{\text{TL}}(\xi')\right)d\xi'\right]\cdot\boldsymbol{e}_{\tau_1} = E_{\tan}^{\text{inc}}(\tau_1)\,, \quad (12)$$

$$j\omega\mu\left[\int_{\xi_0}^{\xi_L}\left(\left[\overline{\boldsymbol{G}}^{E}(\tau_2,\tau_1')\frac{\partial\tau_1'}{\partial\xi'} + \overline{\boldsymbol{G}}^{E}(\tau_2,\tau_2')\frac{\partial\tau_2'}{\partial\xi'}\right]\boldsymbol{I}_{\text{A}}(\xi') + \left[\overline{\boldsymbol{G}}^{E}(\tau_2,\tau_1')\frac{\partial\tau_1'}{\partial\xi'} - \overline{\boldsymbol{G}}^{E}(\tau_2,\tau_2')\frac{\partial\tau_2'}{\partial\xi'}\right]\boldsymbol{I}_{\text{TL}}(\xi')\right)d\xi'\right]\cdot\boldsymbol{e}_{\tau_2} = E_{\tan}^{\text{inc}}(\tau_2)\,. \quad (13)$$

We both add and subtract these equations and obtain

$$j\omega\mu \int_{\xi_0}^{\xi_L} \Big(G^E_{+\mathrm{A}}(\tau_1,\tau_2,\tau_1',\tau_2')I_\mathrm{A}(\xi') + G^E_{+\mathrm{TL}}(\tau_1,\tau_2,\tau_1',\tau_2')I_\mathrm{TL}(\xi')\Big)\, d\xi' = E^{\mathrm{inc}}_{\tan}(\tau_1) + E^{\mathrm{inc}}_{\tan}(\tau_2)\,, \tag{14}$$

$$j\omega\mu \int_{\xi_0}^{\xi_L} \Big(G^E_{-\mathrm{A}}(\tau_1,\tau_2,\tau_1',\tau_2')I_\mathrm{A}(\xi') + G^E_{-\mathrm{TL}}(\tau_1,\tau_2,\tau_1',\tau_2')I_\mathrm{TL}(\xi')\Big)\, d\xi' = E^{\mathrm{inc}}_{\tan}(\tau_1) - E^{\mathrm{inc}}_{\tan}(\tau_2)\,, \tag{15}$$

where we introduced the abbreviations

$$G^E_{+\mathrm{A}}(\tau_1,\tau_2,\tau_1',\tau_2') := \boldsymbol{e}_{\tau_1}\cdot\left(\overline{\boldsymbol{G}}^E(\tau_1,\tau_1')\frac{\partial\tau_1'}{\partial\xi'} + \overline{\boldsymbol{G}}^E(\tau_1,\tau_2')\frac{\partial\tau_2'}{\partial\xi'}\right)\cdot\boldsymbol{e}_{I_A} + \boldsymbol{e}_{\tau_2}\cdot\left(\overline{\boldsymbol{G}}^E(\tau_2,\tau_1')\frac{\partial\tau_1'}{\partial\xi'} + \overline{\boldsymbol{G}}^E(\tau_2,\tau_2')\frac{\partial\tau_2'}{\partial\xi'}\right)\cdot\boldsymbol{e}_{I_A}\,, \tag{16}$$

$$G^E_{+\mathrm{TL}}(\tau_1,\tau_2,\tau_1',\tau_2') := \boldsymbol{e}_{\tau_1}\cdot\left(\overline{\boldsymbol{G}}^E(\tau_1,\tau_1')\frac{\partial\tau_1'}{\partial\xi'} - \overline{\boldsymbol{G}}^E(\tau_1,\tau_2')\frac{\partial\tau_2'}{\partial\xi'}\right)\cdot\boldsymbol{e}_{I_TL} + \boldsymbol{e}_{\tau_2}\cdot\left(\overline{\boldsymbol{G}}^E(\tau_2,\tau_1')\frac{\partial\tau_1'}{\partial\xi'} - \overline{\boldsymbol{G}}^E(\tau_2,\tau_2')\frac{\partial\tau_2'}{\partial\xi'}\right)\cdot\boldsymbol{e}_{I_TL}\,, \tag{17}$$

$$G^E_{-\mathrm{A}}(\tau_1,\tau_2,\tau_1',\tau_2') := \boldsymbol{e}_{\tau_1}\cdot\left(\overline{\boldsymbol{G}}^E(\tau_1,\tau_1')\frac{\partial\tau_1'}{\partial\xi'} + \overline{\boldsymbol{G}}^E(\tau_1,\tau_2')\frac{\partial\tau_2'}{\partial\xi'}\right)\cdot\boldsymbol{e}_{I_A} - \boldsymbol{e}_{\tau_2}\cdot\left(\overline{\boldsymbol{G}}^E(\tau_2,\tau_1')\frac{\partial\tau_1'}{\partial\xi'} + \overline{\boldsymbol{G}}^E(\tau_2,\tau_2')\frac{\partial\tau_2'}{\partial\xi'}\right)\cdot\boldsymbol{e}_{I_A}\,, \tag{18}$$

$$G^E_{-\mathrm{TL}}(\tau_1,\tau_2,\tau_1',\tau_2') := \boldsymbol{e}_{\tau_1}\cdot\left(\overline{\boldsymbol{G}}^E(\tau_1,\tau_1')\frac{\partial\tau_1'}{\partial\xi'} - \overline{\boldsymbol{G}}^E(\tau_1,\tau_2')\frac{\partial\tau_2'}{\partial\xi'}\right)\cdot\boldsymbol{e}_{I_TL} - \boldsymbol{e}_{\tau_2}\cdot\left(\overline{\boldsymbol{G}}^E(\tau_2,\tau_1')\frac{\partial\tau_1'}{\partial\xi'} - \overline{\boldsymbol{G}}^E(\tau_2,\tau_2')\frac{\partial\tau_2'}{\partial\xi'}\right)\cdot\boldsymbol{e}_{I_TL}\,. \tag{19}$$

3 Decoupling of antenna and transmission line mode in free space

The expressions we obtained so far look more complicated than the original Eqs. (1) and (2) that we started from. To nevertheless appreciate this form of the coupled Pocklington's equations we specialize to the case of straight and parallel wires. Then we may align a Cartesian coordinate system

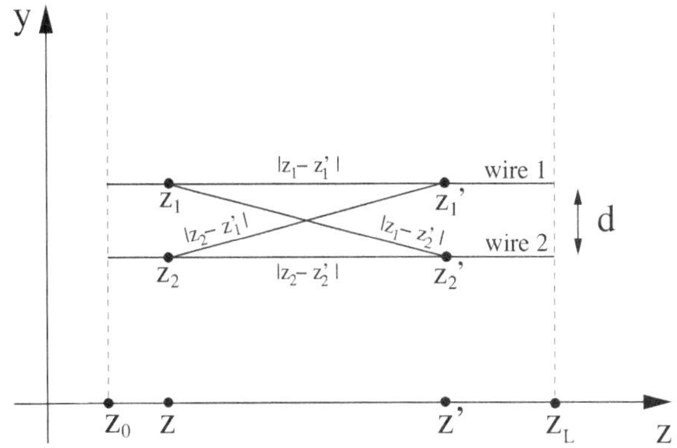

Fig. 3. Geometry of a straight two-wire transmission line.

such that the z-axis is parallel to the wires and may choose $\xi = z$. This leads to the simplifications

$$\boldsymbol{e}_{\tau_1} = \boldsymbol{e}_{\tau_2} = \boldsymbol{e}_{I_A} = \boldsymbol{e}_{I_{TL}}\,, \tag{20}$$

$$\boldsymbol{e}_{\tau_{1,2}}\cdot\overline{\boldsymbol{G}}^E\cdot\boldsymbol{e}_{I_{\mathrm{A,TL}}} = G^E_{zz}\,, \tag{21}$$

$$\frac{\partial\tau_1}{\partial\xi} = \frac{\partial\tau_2}{\partial\xi} = 1\,. \tag{22}$$

Accordingly, Eqs. (16) – (19) reduce to

$$G^E_{+\mathrm{A}}(z_1,z_2,z_1',z_2') = G^E_{zz}(z_1,z_1') + G^E_{zz}(z_1,z_2') + G^E_{zz}(z_2,z_1') + G^E_{zz}(z_2,z_2')\,, \tag{23}$$

$$G^E_{+\mathrm{TL}}(z_1,z_2,z_1',z_2') = G^E_{zz}(z_1,z_1') - G^E_{zz}(z_1,z_2') + G^E_{zz}(z_2,z_1') - G^E_{zz}(z_2,z_2')\,, \tag{24}$$

$$G^E_{-\mathrm{A}}(z_1,z_2,z_1',z_2') = G^E_{zz}(z_1,z_1') + G^E_{zz}(z_1,z_2') - G^E_{zz}(z_2,z_1') - G^E_{zz}(z_2,z_2')\,, \tag{25}$$

$$G^E_{-\mathrm{TL}}(z_1,z_2,z_1',z_2') = G^E_{zz}(z_1,z_1') - G^E_{zz}(z_1,z_2') - G^E_{zz}(z_2,z_1') + G^E_{zz}(z_2,z_2')\,, \tag{26}$$

Let us now suppose that the transmission line is located in free space such that the Green's function is that of free space, $G^E_{zz} = G^E_{\mathrm{free}\ zz}$. It is, in particular, translation invariant,

$$G^E_{\mathrm{free}\ zz}(z,z') = G^E_{\mathrm{free}\ zz}(|z-z'|)\,. \tag{27}$$

For straight, parallel wires we have, compare Fig. 3,

$$|z_1 - z_1'| = |z_2 - z_2'| \implies G^E_{\mathrm{free}\ zz}(z_1,z_1') = G^E_{\mathrm{free}\ zz}(z_2,z_2')\,, \tag{28}$$

$$|z_1 - z_2'| = |z_2 - z_1'| \implies G^E_{\mathrm{free}\ zz}(z_1,z_2') = G^E_{\mathrm{free}\ zz}(z_2,z_1')\,. \tag{29}$$

It follows that the relations (23) – (26) reduce to

$$G^E_{+\mathrm{A}}(z_1, z_2, z_1', z_2') = 2\left(G^E_{\mathrm{free}\ zz}(z_1, z_1') + G^E_{\mathrm{free}\ zz}(z_1, z_2')\right), \tag{30}$$

$$G^E_{+\mathrm{TL}}(z_1, z_2, z_1', z_2') = 0\,, \tag{31}$$

$$G^E_{-\mathrm{A}}\ (z_1, z_2, z_1', z_2') = 0\,, \tag{32}$$

$$G^E_{-\mathrm{TL}}(z_1, z_2, z_1', z_2') = 2\left(G^E_{\mathrm{free}\ zz}(z_1, z_1') - G^E_{\mathrm{free}\ zz}(z_1, z_2')\right), \tag{33}$$

and in view of the integral equation system (14), (15) it is recognized that the antenna mode I_A and the transmission line mode I_TL completely decouple,

$$j\omega\mu \int_{z_0}^{z_L} G^E_{+\mathrm{A}}(z_1, z_2, z_1', z_2') I_\mathrm{A}(z')\, dz' = +E^{\mathrm{inc}}_{\mathrm{tan}}(z_1) + E^{\mathrm{inc}}_{\mathrm{tan}}(z_2)\,, \tag{34}$$

$$j\omega\mu \int_{z_0}^{z_L} G^E_{-\mathrm{TL}}(z_1, z_2, z_1', z_2') I_\mathrm{TL}(z')\, dz' = +E^{\mathrm{inc}}_{\mathrm{tan}}(z_1) - E^{\mathrm{inc}}_{\mathrm{tan}}(z_2)\,. \tag{35}$$

4 Reduction to conventional transmission line theory in free space

The antenna mode current I_A in Eq. (34) vanishes at the beginning and at the end of the transmission line. This is analogous to the boundary conditions of the antenna current on a single wire antenna. Also the Green's function $G^E_{+\mathrm{A}}(z_1, z_2, z_1', z_2')$ is similar to the kernel of the Pocklington's equation for a single wire antenna since in Eq. (30) the terms $G^E_{\mathrm{free}\ zz}(z_1, z_1')$ and $G^E_{\mathrm{free}\ zz}(z_1, z_2')$ add up and only considerably differ if the distance $|z - z'|$ is smaller or of the order of the distance d between wire 1 and wire 2. It follows that Eq. (34) can be solved with methods of antenna theory in cavities (Gronwald, 2005).

The conventional transmission line theory is contained in Eq. (35). To explicitly see this we first note that a Pocklington's equation

$$j\omega\mu \int_{z_L}^{z_o} G^E_{zz}(z, z') I(z')\, dz' = E^{\mathrm{inc}}_{\mathrm{tan}}(z) \tag{36}$$

is equivalent to a mixed potential integral equation (Nakano, 1996)

$$\frac{1}{j\omega\varepsilon} \int_{z_0}^{z_L} \left[\frac{\partial G^\phi(z, z')}{\partial z} \frac{\partial I(z')}{\partial z'} + k^2 G^A_{zz}(z, z') I(z') \right] dz' = -E^{\mathrm{inc}}_z(z) \tag{37}$$

with $G^\phi(z, z')$ and $G^A_{zz}(z, z')$ indicating the Green's functions for the scalar potential ϕ and the magnetic vector potential $\boldsymbol{A}$ in the Lorenz gauge, respectively. We introduce a per-unit-length charge q' by the continuity equation

$$\frac{\partial I}{\partial z} + j\omega q' = 0 \tag{38}$$

and define a potential $V^{q'}$ by

$$V^{q'} := \frac{1}{\varepsilon} \int_{z_0}^{z_L} G^\phi(z, z') q'(z')\, dz'\,. \tag{39}$$

Furthermore, in view of Eq. (33), we introduce the combinations

$$G^\phi_{-\mathrm{TL}}(z, z') = 2\big(G^\phi_{\mathrm{free}}(z_1, z_1') - G^\phi_{\mathrm{free}}(z_1, z_2')\big)\,, \tag{40}$$

$$G^A_{-\mathrm{TL}}(z, z') = 2\big(G^A_{\mathrm{free}\ zz}(z_1, z_1') - G^A_{\mathrm{free}\ zz}(z_1, z_2')\big)\,, \tag{41}$$

and note that $G^\phi_{-\mathrm{TL}}$ and $G^A_{-\mathrm{TL}}$ are localized functions that are characterized by a sharp peak in the domain where the distance $|z - z'|$ is small. This feature is often used in the derivation of the conventional transmission line theory (see Tkachenko et al., 1995, for example). It leads to the simplifications

$$\int_{z_0}^{z_L} G^\phi_{-\mathrm{TL}}(z, z') q'(z')\, dz' \approx q'(z) \int_{z_0}^{z_L} G^\phi_{-\mathrm{TL}}(z, z')\, dz' \tag{42}$$

$$\int_{z_0}^{z_L} G^A_{-\mathrm{TL}}(z, z') I(z')\, dz' \approx I(z) \int_{z_0}^{z_L} G^\phi_{-\mathrm{TL}}(z, z')\, dz' \tag{43}$$

The mixed potential integral Eq. (37) and the continuity Eq. (38) can now be written as conventional Telegrapher equations

$$\frac{\partial V^{q'}}{\partial z}(z) + j\omega L' I(z) = -\big(E^{\mathrm{inc}}_{\mathrm{tan}}(z_1) - E^{\mathrm{inc}}_{\mathrm{tan}}(z_2)\big)\,, \tag{44}$$

$$\frac{\partial I}{\partial z}(z) + j\omega C' V^{q'}(z) = 0\,, \tag{45}$$

with

$$C' := \frac{\varepsilon}{\int_{z_0}^{z_L} G^\phi_{-\mathrm{TL}}(z, z')\, dz'} \approx \frac{\pi\varepsilon}{\ln(d/\rho)}\,, \tag{46}$$

$$L' := \mu \int_{z_0}^{z_L} G^A_{-\mathrm{TL}}(z, z')\, dz' \approx \frac{\mu}{\pi} \ln(d/\rho)\,. \tag{47}$$

Here the distance between the wires is, as before, denoted by d and the wire radius is denoted by ρ.

5 Transmission lines in cavities

In the derivation of the Telegrapher Eqs. (44) and (45) from the Pocklington's Eqs. (1) and (2) we employed three times the properties of free space Green's functions:

1. To decouple for straight, parallel wires the antenna mode from the transmission line mode we used translation invariance of the free space Green's function $G^E_{\text{free } zz}$ in Eq. (27).

2. To pull in Eqs. (42) and (43) the charge density q' and the current I out of the integrals we used the strong Coulomb singularity that is contained in the free space Green's functions G^ϕ_{free} and $G^A_{\text{free } zz}$.

3. To calculate the per-unit-length parameter L' and C' we used the explicit mathematical expression for the free space Green's functions G^ϕ_{free} and $G^A_{\text{free } zz}$.

Within a cavity the Green's functions can always be written as a sum of a free space part and a boundary part which takes into account the effect of the cavity walls,

$$G_{\text{cav}} = G_{\text{free}} + \tilde{G}\,. \tag{48}$$

We now state that the steps 1.–3. can approximately be performed within a cavity:
1. The decoupling of antenna and transmission line mode requires the kernels $G^E_{+\text{TL}}(z_1, z_2, z'_1, z'_2)$ and $G^E_{-\text{A}}(z_1, z_2, z'_1, z'_2)$ to vanish. From the relations (24) and (25), and Fig. 3 with $d = y_1 - y_2$ it follows that it is meaningful to consider the Taylor expansions

$$\tilde{G}^E_{zz}(z_1, z'_2) \approx \tilde{G}^E_{zz}(z_1, z'_1) - \frac{\partial \tilde{G}^E_{zz}}{\partial y}(z_1, z'_1)\, d \tag{49}$$

$$\tilde{G}^E_{zz}(z_1, z'_2) \approx \tilde{G}^E_{zz}(z_2, z'_2) + \frac{\partial \tilde{G}^E_{zz}}{\partial y}(z_2, z'_2)\, d \tag{50}$$

$$\tilde{G}^E_{zz}(z_2, z'_1) \approx \tilde{G}^E_{zz}(z_2, z'_2) + \frac{\partial \tilde{G}^E_{zz}}{\partial y}(z_2, z'_2)\, d \tag{51}$$

$$\tilde{G}^E_{zz}(z_2, z'_1) \approx \tilde{G}^E_{zz}(z_1, z'_1) - \frac{\partial \tilde{G}^E_{zz}}{\partial y}(z_1, z'_1)\, d \tag{52}$$

since then

$$\tilde{G}^E_{+\text{TL}}(z_1, z_2, z'_1, z'_2) = \tilde{G}^E_{-\text{A}}(z_1, z_2, z'_1, z'_2) \approx \left(\frac{\partial \tilde{G}^E_{zz}}{\partial y}(z_1, z'_1) + \frac{\partial \tilde{G}^E_{zz}}{\partial y}(z_2, z'_2) \right) d\,. \tag{53}$$

If the terms involving the derivative of the Green's function are small the kernels $G^E_{+\text{TL}}(z_1, z_2, z'_1, z'_2)$ and $G^E_{-\text{A}}(z_1, z_2, z'_1, z'_2)$ are small as well and, as a result, antenna and transmission line mode approximately decouple. For a general discussion of this point one might represent the cavities Green's function G_{cav} by means of an expansion in rotational and irrotational eigenvectors (Tai, 1994).

The rotational eigenvectors are solutions of sourceless Helmholtz equations and their spatial variation is of the order of the wavelength considered. Then the contributions of these rotational eigenvectors to the first order terms in Eqs. (49)–(52) will be of the order of kd. Thus, according to the usual assumption $kd \ll 1$ of conventional transmission line theory, these contributions will be small.

The irrotational eigenvectors are solutions of electrostatic Poisson equations and contain the Coulomb singularity. Their spatial variation diverges close to a Coulomb singularity and, otherwise, decays quickly. Since $\tilde{G} = G_{\text{cav}} - G_{\text{free}}$ contains no Coulomb singularity we expect that, in general, the irrotational eigenvectors will lead to no significant spatial variation of $\tilde{G}$. There is only one situation where this argument does not apply and this is when the distance of one of the wires to a cavity wall is of the order or smaller than the wire distance d. In this case there can be a dominant Coulomb interaction with the cavity wall (that is, with the mirrored wires) which is embedded in $\tilde{G}$ since G_{free} does not take into account scattering contributions from the cavity walls. Then the spatial variation of $\tilde{G}$ might become large and one would need to actually calculate the derivatives of Eq. (53) for the specific configuration in order to see if they still lead to only small corrections.
2. We need to reconsider the approximations (42) and (43) with

$$G^\phi_{-\text{TL}}(z, z') = 2\big(G^\phi_{\text{cav}}(z_1, z'_1) - G^\phi_{\text{cav}}(z_1, z'_2)\big)\,, \tag{54}$$

$$G^A_{-\text{TL}}(z, z') = 2\big(G^A_{\text{cav } zz}(z_1, z'_1) - G^A_{\text{cav } zz}(z_1, z'_2)\big)\,. \tag{55}$$

The approximations (42) and (43) are valid since the differences of the form $G(z_1, z'_1) - G(z_1, z'_2)$ approximately cancel if $|z - z'|$ is much larger than the separation d. The sharp peak of $G^\phi_{-\text{TL}}(z, z')$ and $G^A_{-\text{TL}}(z, z')$ at $z = z'$ is due to the fact that for $z \to z'$ we have, with $\rho \ll d$,

$$G^\phi_{-\text{TL}}(z, z') = G^A_{-\text{TL}}(z, z') \approx \frac{1}{2\pi}\left(\frac{1}{\rho} - \frac{1}{d}\right) \gg 1\,. \tag{56}$$

This feature is unaffected by the presence of the cavity. Rotational contributions of $\tilde{G}$ to G_{cav} will approximately cancel within the differences (54) and (55), and Coulomb interactions with the cavity walls will not significantly change the property (56).
3. The electrostatic calculation that led to the values of the per-unit-length parameters (47) and (46) consists of the evaluation of the integrals $\int_{-L/2}^{L/2} G^{\phi,A}_{-\text{TL}}(z, z')\, dz'$. Within a cavity these parameters will significantly change only if one or both of the wires is close to a cavity wall such that the Coulomb field in the vicinity of the transmission lines is significantly perturbed by the presence of the cavity. In this case the parameters L', C' have to be calculated from an electrostatic calculation which takes into account the cavity wall.

6 Conclusions

Conventional transmission line is applicable not only in free space but also within a cavity. The dynamical (rotational) electromagnetic fields within the cavity will mainly couple to the antenna mode current which approximately is decoupled from the transmission line mode. In particular, resonances of the cavity will strongly interact with the antenna mode but not with the transmission line mode. Some care must be taken if the presence of the cavity has a significant influence on the (irrotational) Coulomb fields in the vicinity of the transmission line. In this case, quantitative changes of the per-unit-length parameters are expected and need to be calculated from the actual transmission line configuration.

References

Gronwald, F.: The influence of electromagnetic singularities on an active dipole antenna within a cavity, Adv. in Radio Science, 1, 57–61, 2003.

Gronwald, F.: Calculation of mutual antenna coupling within rectangular enclosures, IEEE Trans. Electromagn. Compat., 47, 1021–1025, 2005.

Haase, H., Steinmetz, T., and Nitsch, J.: New propagation models for electromagnetic waves along uniform and nonuniform cables, IEEE Trans. Electromagn. Compat., 46, 345–352, 2004.

King, R. W. P.: Transmission Line Theory, McGraw-Hill, New York 1955.

Lee, K. S. H.: EMP Interaction: Principles, Techniques, and Reference Data, revised printing, Taylor and Francis, Washington D.C. 1995.

Nakano, H.: Antenna Analysis using Integral Equations, in Analysis Methods for Electromagnetic Wave Problems, 2, edited by Yamashita, E., Artech House, Boston, 1996.

Tai, C.-T.: Dyadic Green Functions in Electromagnetic Theory, IEEE Press, New York, 1994.

Tesche, F. M., Ianoz, M. V., and Karlsson, T.: EMC Analysis Methods and Computational Methods, John Wiley and Sons, New York, 1997.

Tkachenko, S., Rachidi, F., and Ianoz, M.: Electromagnetic Field Coupling to a Line of Finite Length: Theory and Fast Iterative Solutions in Frequency and Time Domains, IEEE Trans. Electromagn. Compat., 37, 509–518, 1995.

Optimized tracking for cooperative sensor systems in multipath environments

D. Schwarz[1], **R. H. Rasshofer**[1], **and E. M. Biebl**[2]

[1]BMW Group Research and Technology, Munich, Germany
[2]Fachgebiet Höchstfrequenztechnik, Technische Universität München, Germany

Abstract. In a cooperative sensor system for pedestrian protection, a pedestrian and other road users exchange data by means of radio frequency communication. In the proposed system, the pedestrian carries a transponder which is interrogated by a vehicle and sends an anonymous identification (ID) sequence. By decoding the ID, the interrogation unit in the vehicle detects the presence of the transponder. Evaluating the incident wave of the transponder's answer, a localisation is possible.

In the proposed localization system, the measurement results can be distorted by multipath propagation. Multipath errors result if signals of the same transponder arrive simultaneously at the receiver unit from different directions. In this case, erroneous distances and angles are measured. Because the signals arriving from different directions contain the same transponder ID, they can be assigned to their origin. One of the challenges in post-processing for signal improvement is enhancing the selection of the correct position information by making assumptions about the pedestrian's movement and by knowing the vehicle's current driving parameters. Additionally, information contained in multipath signals is used to form a better estimate for the true position of the transponder. To overcome the problems related to multipath propagation effects, a method is proposed that estimates the origin of a multipath signal and maps the distorted position information back to the true position. A fusion of tracked direct positions and mapped multipath signals leads to an improvement in positioning accuracy.

1 Introduction

Pedestrian detection approaches based on lidar (Fürstenberg, 2004), radar, or vision-based systems (Gavrila et al., 2004) offer highly accurate positioning information of objects or persons in the vehicle's surroundings. The reliability of these systems has been improved rapidly using new detection, classification and fusion algorithms (Paraschoudis et al., 2006). Current systems are already capable to cover a large number of typical accident scenarios. In scenarios where pedestrians are fully or partially hidden to these sensors either their detection or their correct classification might fail. The results of accident studies show that 40% of all fatal pedestrian accidents in Germany are caused by a late recognition of the situation by the driver (German In-Depth-Accident Study, 2005). In these cases, a cooperative sensor system could support active safety systems with valuable information. A pedestrian that is occluded to the view of the driver and sensors built in the vehicle can be detected earlier which provides more time for triggering collision avoidance and collision mitigation systems.

1.1 Related work

In Shi et al. (2005) a proposed cooperative system determines the DoA of an incident transponder signal. The distance is not measured directly but through triangulation using the detected DoAs at two subsequent locations while the vehicle is moving.

Meinken et al. (2007) describe a cooperative system for traffic safety based on communication and sensor technologies. The concept relies on interactions between an in-vehicle module and a user's device. Positioning is done by two communication-based distance measurements between the rear-view mirrors and the transponder (Sikora, 2006). Through triangulation the position can be calculated.

Other previous work that mention cooperative systems use infrastructure devices as repeaters (Kubota et al., 2006) but do not use a localisation system integrated into a vehicle.

1.2 Content

First, the integrated localisation system in the test vehicle is introduced and measurement principles are described in more detail. The influences of multipath propagation on measurement results are pointed out in Sect. 3. To overcome the drawbacks caused by multipath effects, requirements on

Correspondence to: D. Schwarz
(Daniel.Schwarz@bmw.de)

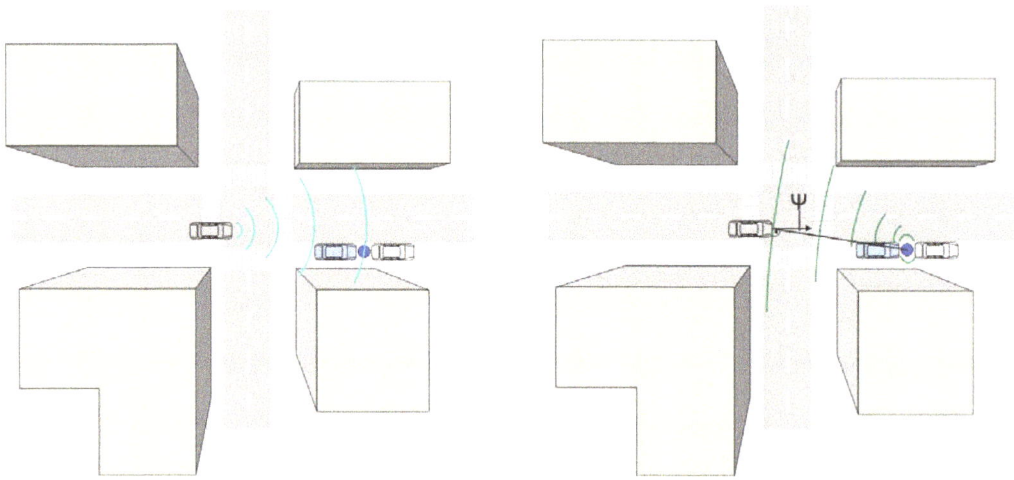

Fig. 1. Interrogation pulse of vehicle in an urban scenario (left), answer pulse of a transponder hidden to the driver between parked cars.

Fig. 2. 3-D model of antenna array integrated behind the front window of test vehicle.

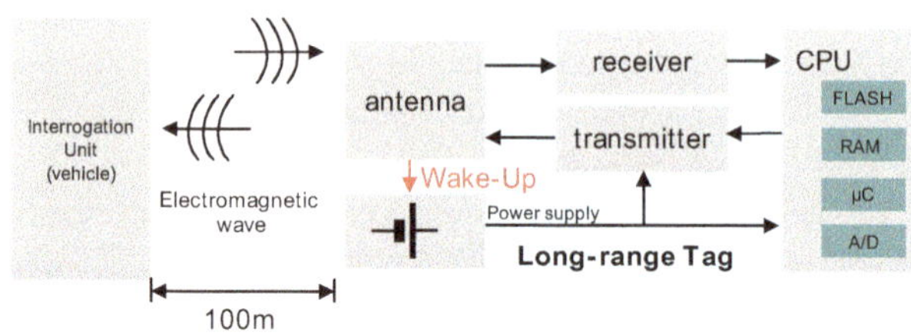

Fig. 3. Interaction between interrogation unit and transponder.

a robust tracking algorithm are deduced, and in Sect. 4 an example for an algorithm which is able to deal with multipath effects is introduced. An outlook on further work and a conclusion are presented in the final sections.

2 System overview

2.1 On board unit

A central part of the system is an antenna array which is integrated behind the wind shield of the vehicle. The antenna array is used to send an interrogation pulse and to receive the answer from the transponder to be localized. Data samples of the single antenna elements are used for the detection of an object's presence and determination of the object position. The driver's field of view is not affected by the integrated antennas and no modifications need to be made to the car body or design of the test vehicle (see Fig. 2).

The initial design of the system will utilize ISM bands to ease international licensing. As each ISM band offers different channel properties, three different ISM bands have been experimentally investigated (434 MHz, 868 MHz and 2.4 GHz). Communication with a transponder being optically hidden behind a car represents one of the most challenging scenarios. With all three frequencies the received signal strength from a transponder showed to be strong enough for a reliable detection, even in worst case scenarios. To minimize the antenna array's size, 2.4 GHz has been chosen as operating frequency. Switching to the IEEE 802.11p standard at 5.9 GHz, intended for Car-to-Car (C2C) communication, even would lead to a further reduction of the antenna size.

2.1.1 Direction-of-Arrival (DoA)

Using six antennas in the antenna array, an estimation of the DoA can be obtained. If a valid signal sent by a transponder is identified, the amplitudes and phases of the antenna voltages are sampled and used to derive the DoA. Different well-known high resolution DoA estimation algorithms like ESPRIT (Estimation of Signal Parameters via Rotational

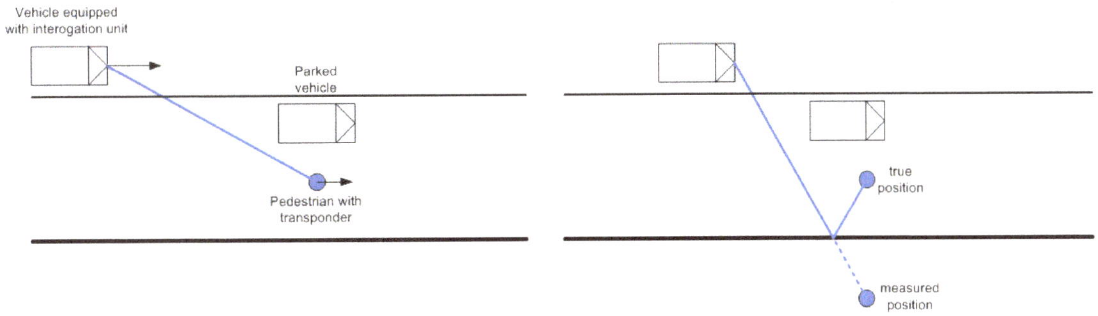

Fig. 4. Correct position of pedestrian is measured (left), multipath signals caused by reflection at a wall lead to different ToF and DoA measurements (right).

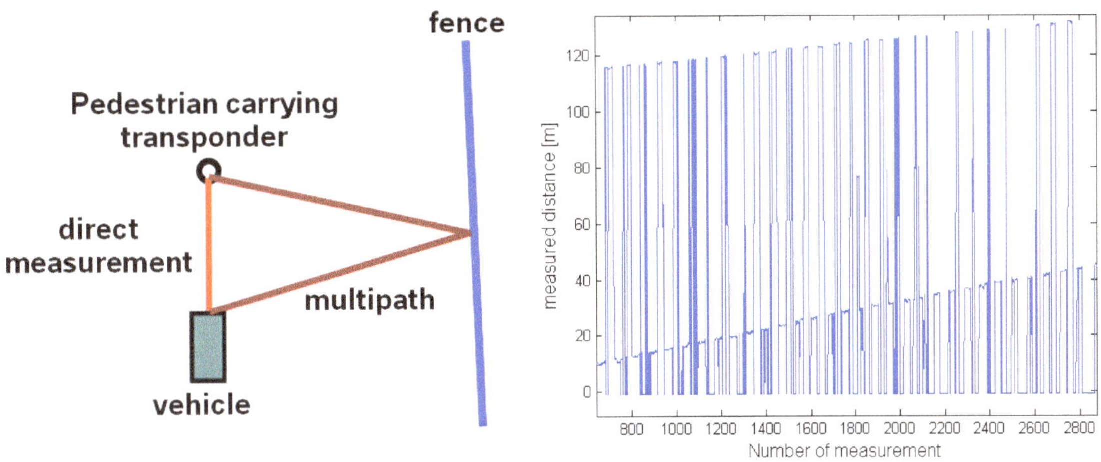

Fig. 5. Measurement setup for cooperative sensor system (left), distance measurement results: multipath propagation causes ambiguity, distance ranges 10–40 m or 110–130 m (right).

Invariant Techniques) or MUSIC (Multiple Signal Classification) (Schmidt, 1987) are investigated and applied to the raw data to obtain reliable azimuth angle data. In the first measurements, the transponder's azimuth angle could be estimated with an accuracy of about 1° in different distances in the case of a Line-of-Sight (LoS) connection. As expected, the accuracy of the angular measurement decreases without a LoS connection. Tests show the possibility of improving the localisation accuracy, depending on SNR and multipath scenario, by applying suitable tracking algorithms in a post-processing step.

2.1.2 Time-of-Flight (ToF)

The tag's distance can be calculated by a ToF measurement, determining the round trip delay of the interrogation pulse. Furthermore, the change of the measured azimuth angle can be investigated knowing the vehicle's velocity to achieve more accurate positioning (Shi et al., 2005). The ToF measurement is done by means of a subcarrier technique. The received interrogation sequence is decoded and, after a certain time-delay, a response is sent containing information about the time delay and the ID-code (Rasshofer et al., 2007). Knowing the delay time and the time between interrogation pulse and received answer, the distance can be calculated. Since each tag uses a different time-delay, TDMA techniques are applied to allow for separation of the responses from several tags. Without superposition of different responses the precision of both ToF and DoA measurements is enhanced.

2.2 Transponder

If the transponder identifies a valid interrogation pulse, it changes into an active state and starts transmitting. This wake up mechanism is suitable to reduce the power consumption and to extend battery lifetime. Another promising approach features the integration of acceleration sensors to shut down the transponder if it is not moved for a longer time and thus is unlikely to be carried by a pedestrian.

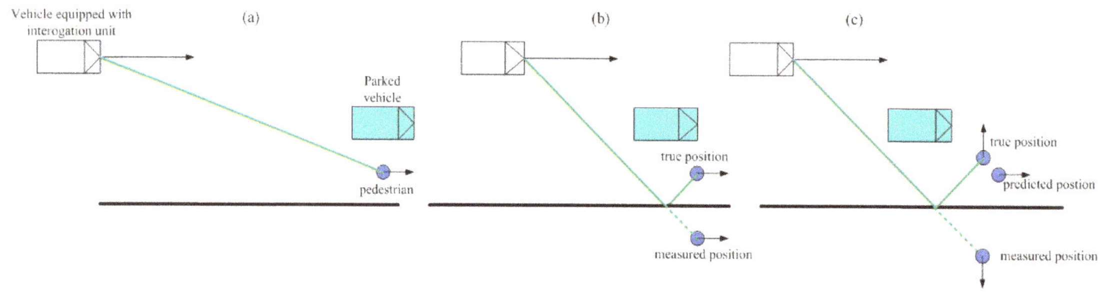

Fig. 6. Example for simulation setup.

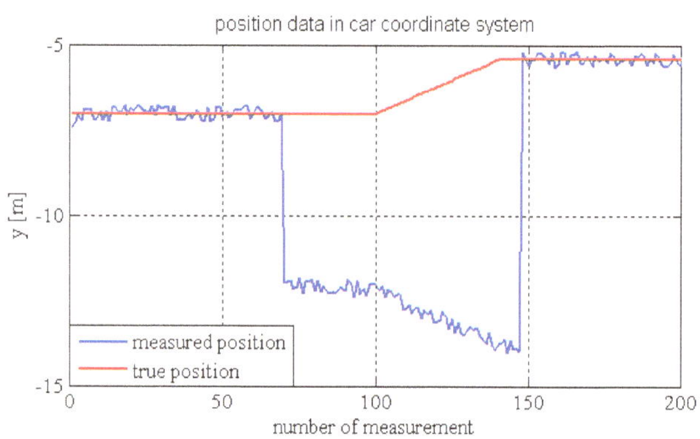

Fig. 7. Simulation data of scenario shown in Fig. 6.

3 Multipath effects in the cooperative sensor system

A tracking system for cooperative sensor data needs to deal with multipath propagation of the transponder signals. As sudden strong interferences in the measured position caused by these effects can not be modelled as Gaussian noise, a standard Kalman-Filter (Kalman, 1960) is not the optimal solution for tracking purposes.

In a first approach, the position data corrupted by multipath phenomena are tried to be removed by a tracking procedure. However, measurements showed that there is useful information contained in multipath signals, including information about the moving direction of the transponder, which, if also utilized, can lead to a better positioning accuracy.

4 Adapted tracking algorithms

4.1 Kalman-Filter with plausibility check

Looking at Fig. 5, a first approach to compensate for multipath effects could be to sort out all measurements influenced by multipath propagation. This is done by a plausibility check of the data when looking at several subsequent measurements. The remaining measurement points are – in this simple case – sufficient for a stable tracking with a Kalman-Filter using a system model with constant velocity (CV) for the pedestrian.

However, it can be seen easily that the erroneous measurement values in a range between 110 and 130 m are increasing similar to the correct measurement points. They indicate a movement of the pedestrian away from the car. The existence of the multipath data in this case can almost completely be explained by a reflection of the transponder signal at a fence near by the measurement setup. As influences on the DoA measurement showed similar results and some interferences could be explained by a simple reflection at planar obstacles, a new approach has been chosen. For further studies and the development of algorithms simulations have been conducted to generate a large number of scenarios.

4.2 Mapping of multipath signals

As an example for the various simulations conducted, a setup as shown in Fig. 6 should be considered. A moving vehicle determines the position of a pedestrian moving with a different velocity using the cooperative sensor system. After some simulation time the pedestrian is occluded to the vehicle and due to multipath signals the wrong position is observed in subsequent measurements.

During the occlusion – if multipath signals are neglected – only a prediction of the pedestrian's movement can be done by the last estimated velocity vector. Changes in the moving direction can not be recognized in this case. Figure 7 shows position data received in this scenario.

In this approach, a rule-based algorithm detects the sudden change of position data. The position of the pedestrian is predicted using a Kalman-Filter with constant velocity model and for the multipath data, a new track is generated. With the simple assumption that one reflection plane is responsible for the multipath signal, the algorithm tries to calculate the parameters of this plane. As measurements showed, this assumption holds true in many scenarios. The movement of the pedestrian can be predicted quite accurately in the first few hundred milliseconds due to physical limitations like e.g. maximum possible acceleration. If a relation between the predicted position and the tracked multipath position can be found and the reflection plane is observed to be stable for a

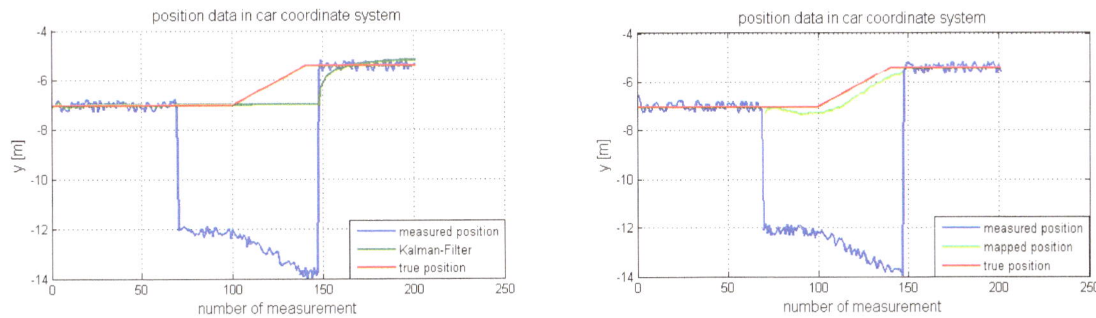

Fig. 8. Comparison between prediction using a Kalman-Filter (left) and the proposed algorithm for multipath data (right).

few cycles, one can start to map back the data to their origin. In the following cycles, the position of the plane in the car coordinate system needs to be updated taking into account the driving parameters of the vehicle, like e.g. velocity or yaw rate. Results are shown in Fig. 8.

This way information, like changes in moving direction, contained in multipath data is interpreted and used for a more robust and stable estimation of the pedestrian's position.

5 Further work

Future work should start by adopting algorithmic parameters using measurement data from test scenarios. Increased detection accuracy of a sudden change between a direct signal and a multipath signal should be achieved by taking the received signal strength as additional input.

In the proposed system one position information per cycle was used as input. In the next step, several hypotheses for the position should be used as the measurement principles are able to provide more than one value per cycle. This data can be handled with particle filters based on Sequential Monte Carlo Methods (SMCM) (Doucet et al., 2001) combined with the proposed mapping algorithm.

6 Conclusions

A system for pedestrian detection based on cooperative sensor technology has been introduced. The influence of multipath propagation on the system was investigated using ToF and DoA measurement results. A tracking algorithm takes these effects into account and uses a model for the origin of multipath signals. Simulations showed the feasibility of the multipath compensation by mapping erroneous position information back to their origin for an improvement of positioning accuracy.

Acknowledgements. The cooperative sensor system presented in this publication is part of the main efforts in the AMULETT-project which is co-funded by the Bavarian Ministry of Economic Affairs, Infrastructure, Transport and Technology (BayStMWIVT). The project aims to avoid accidents involving vulnerable road users such as pedestrians and bicyclists by means of cooperative sensor technology.

References

Doucet, A., de Freitas, N., and Gordon, N. (Eds.): Sequential Monte Carlo Methods in Practice, Springer, 2001.

Fürstenberg, K. Ch.: Pedestrian Protection using Laserscanners, IEEE Conference on Intelligent Transportation Systems, 436–437, 2005.

Gavrila, D. M., Giebel, J., and Munder, S.: Vision-Based Pedestrian Detection: The PROTECTOR System, IEEE Intelligent Vehicles Symposium, 13–18, 2004.

German In-Depth-Accident Study: http://www.gidas.org, 2005.

Kalman, R. E.: A New Approach to Linear Filtering and Prediction Problems, J. Basic Eng.-T. ASME, 82 (Series D), 35–45, 1960.

Kubota, S., Okamoto, Y., and Oda, H.: Safety Driving Support System Using RFID for Prevention of Pedestrian-involved Accidents, 6th International conference on ITS Telecommunications Proceedings, 226–229, 2006.

Meinken, K., Andreone, L., Guarise, A., and Sikora, A.: WATCH-OVER – The Concept Of A Cooperative System For Vehicle To Vulnerable Road Users Communication, 20th International Technical Conference on the Enhanced Safety of Vehicles, 07-0126, 2007.

Paraschoudis, V., Tatschke, T., Walchshaeusl, L., and Lindl, R.: Autonomous braking for collision mitigation purposes by means of multi sensor perception, FISITA World Automotive Congress, F2006V181, 2006.

Rasshofer, R., Schwarz, D., Biebl, E., Morhart, C., Scherf, O., Zecha, S., Grünert, R., and Frühauf, H.: Pedestrian Protection Systems using Cooperative Sensor Technology, Proceedings of the 11th International Forum on Advanced Microsystems for Automotive Applications (AMAA'07), Berlin, 135–145, 2007.

Schmidt, R. O.: Multiple Emitter Location and Signal Parameter Estimation, IEEE T. Antenn. Propag., 34, 276–280, 1987.

Sikora, A.: Vehicle-to-Vulnerable road user cooperative communication and sensing technologies to improve transport safety, D4.1 – Communication technologies specifications, 2006.

Shi, J., Müller, H. C., and Marx, M.: Pedestrian Detection and Localization Using Antenna Array and Sequential Triangulation, IEEE Conference on Intelligent Transportation Systems, Vienna, 126–130, 2005.

MIMO performance of a planar logarithmically periodic antenna with respect to measured channel matrices

H. Rabe[1], D. Kornek[1], M. Stege[2], and I. Rolfes[1]

[1]Institute for Radiofrequency and Microwave Engineering, Hanover, Germany
[2]Signalion GmbH, Dresden, Germany

Abstract. The increasing interest in wireless transmission of highest data rates for multimedia applications (e.g. HDTV) demands the use of communication systems as e.g. described in the IEEE 802.11n draft specification for WLAN including spatial multiplexing or transmit diversity to achieve a constant high data rate and a small outage probability. In a wireless communications system the transmission of parallel data stream leads to multiple input/multiple output (MIMO) systems, whose key parameters heavily depend on the properties of the mobile channel. Assuming an uncorrelated channel matrix the correlation between the multiplexed data streams is caused by the coupling of the antennas, so that the radiation element becomes an even more important part of the system. Previous work in this research area (Klemp and Eul, 2006) has shown that planar log.-per four arm antennas are promising candidates for MIMO applications providing two nearly decorrelated radiators, which cover a wide frequency range including both WLAN bands at 2.4 GHz and 5.4 GHz. Up to now the MIMO performance of this antenna is mainly analyzed by simulations. In this contribution measured channel matrices in a real office environment are studied in terms of the antenna's MIMO performance such as outage probability. The obtained results recorded by using a commercial platform are compared to the simulated ones.

1 Introduction

The possibility of achieving a remarkable performance gain of data rate and link reliability by using spatial multiplexing in wireless communication systems lead to the first WLAN standard 802.11n that uses MIMO techniques. However, there are no concrete rules for antenna designers to generate antennas that offer a good MIMO performance preferably in as many channel scenarios as possible. The used antenna structures, the placing as well as the arrangement can have significant influence on the overall system performance. This contribution shows, how polarization diversity can be used to increase orthogonality of the single subchannels from a theoretical point of view by means of a stochastic channel model together with a narrow band assumption (see Sect. 2) for a 2×2 MIMO system. To prove this statement several measurements of channel matrices have been made in an office environment with planar logarithmically periodic antennas (LPs) as well as monopole antennas at typical WLAN frequencies at 2.4 GHz and 5.4 GHz. The measurements have been taken with the HaLo220 testbed that allows the transmission and reception of two RF signals simultaneously. A LP antenna enables the radiation of two linearly polarized waves with an axial ratio of better than 30 dB. The performance differences between a setup with LP antennas and with standard monopole antennas are discussed in Sect. 3.2. The paper ends with a conclusion.

Correspondence to: H. Rabe
(rabe@hft.uni-hannover.de)

2 MIMO channel

To make reliable MIMO measurements of channel matrices some general boundaries should be taken into account. Sections 2.1 and 2.2 describe the necessity of considering the coherence time and the coherence bandwidth in order to fulfil the narrow band assumption and to preserve stationarity of the channel for short time periods. By considering the aforementioned conditions the channel can be expressed as a channel matrix **H** consisting of complex values shown in Sect. 2.3. This section further discusses the influence of the antennas on the channel matrix separately for LOS (Line Of Sight) and NLOS (Non Line Of Sight) scenarios.

2.1 Coherence time

In general, the impulse response of a radio channel is assumed to be linear but not stationary. The movement of persons and objects in the channel causes more or less rapidly changing wave propagation conditions. In the case of a WLAN system, which is intended to operate in an indoor environment, the velocity of the moving scatterers rarely exceeds v_{sc}=5 km/h (Jiménez, 2002) in most cases. Depending on the center frequency the phase of the transmission path is being changed due to a moving scatterer. Regarding a short time period only, the transmission phase can be assumed to be stationary as well as the whole channel impulse response. The time interval, in which the correlation of the channel response does not rise above 0.5 is defined as coherence time T_C in this paper. During the coherence time interval T_C the channel impulse response can be described as one complex delta peak that scales the obtained signal in the equivalent baseband domain. As a rule of thumb formula Eq. (1) can be used to determine T_C.

$$T_C = \frac{v_{sc}}{\lambda} \tag{1}$$

The coherence time decreases with shrinking wavelength so the worst case is a transmission at the highest used frequency. For the high WLAN frequency band around 5.4 GHz the coherence time can be estimated to T_C=40 ms. However, this formula is not adequate in many cases. Another approach (Rappaport, 1996), that is said to be more appropriate determines a smaller value of T_C=18 ms which was assumed to be valid for the examined WLAN channels.

2.2 Coherence bandwidth

It is furthermore assumed, that every frequency component in the baseband signal is distorted the same way so the frequency response of the channel is ideally flat. It is important for accurate measurements that this requirement is fulfilled by the radiated signals. Analogous to the coherence time one can determine a coherence bandwidth B_C which is directly connected to the delay spread σ_τ of the channel. The longer the dominant propagation pathes are the more sensitive the phase difference is between adjacent signal frequencies. The coherence bandwidth can be estimated by expression (2) according to Sklar (2001) and is related to the delay spread σ_τ that includes the path lengths of the channel implicitly.

$$B_C = \frac{1}{50\sigma_\tau} \tag{2}$$

For indoor channels the delay spread strongly depends on the scenario. When transmitter and receiver are both situated in a small room, where the path lengths are supposed to be short, there will be no large spread in the arrival of the signal echoes. Even if the path length are long the delay spread can be short if a strong line of sight component or another dominant propagation path exists. A worst case scenario would be a channel with two similarly weighted pathes of different length, for example a LOS and a reflected component coming from a wall. In this paper the delay spread is assumed to stay below σ_τ=100 ns, which is more than sufficient for the office environment where the measurement took place and about the double of the value measured by McDonnel et al. (1998). Therefore, the coherence bandwidth can be estimated from expression (2) to 200 kHz. This restriction will be considered in Sect. 3.

2.3 MIMO channel matrix

With the preceding simplifications of a narrow band channel the impulse response $h(t)$ degenerates to one complex value between transmit and receive antenna within a coherence time interval. With multiple transmit and receive antennas each single impulse response value can be composed in the MIMO matrix $\mathbf{H}$ as shown in Fig. 1. The receiver signal expands to a receive vector $\boldsymbol{y}$, which M elements are a linear combination of the N elements in the transmit vector s. The relation between transmit and receive vectors is expressed in Eq. (3).

$$\begin{pmatrix} y_1 \\ y_2 \\ \vdots \\ y_m \end{pmatrix} = \begin{pmatrix} h_{11} & h_{12} & \dots & h_{1n} \\ h_{21} & h_{22} & \dots & h_{2n} \\ \vdots & \vdots & \ddots & \vdots \\ h_{m1} & h_{m2} & h_{m3} & h_{mn} \end{pmatrix} \cdot \begin{pmatrix} s_1 \\ s_2 \\ \vdots \\ s_n \end{pmatrix} \tag{3}$$

The channel capacity C_{MIMO} in the case of equally divided power P_T over the transmit antennas can be expressed as follows according to Foschini and Gans (1998).

$$C_{\text{MIMO}} = \log_2 det\left(\overline{\mathbf{E}} - \frac{P_T}{M\sigma_n^2} H H^H\right)$$
$$\Leftrightarrow C_{\text{MIMO}} = \sum_i \log_2\left(1 + \frac{P_T}{M\sigma_n^2}\lambda_i\right) \tag{4}$$

$\overline{\mathbf{E}}$ is the identity matrix, λ_i is the ith eigenvalue of the matrix product between H and its hermitian H^H and σ_n^2 is the noise power which is assumed to be white gaussian noise. From Eq. (4) one can see that the spectral efficiency is increased by using higher transmit power or by trying to achieve a high rank of the matrix product HH^H or the matrix $\mathbf{H}$, respectively. For WLAN indoor channels rayleigh fading is assumed so the matrix $\mathbf{H}$ is filled with independent and identically-distributed (IID) complex random variables for each coherence time. Unfortunately it is very hard to achieve the data rate which is offered by the pure rayleigh fading channel model. In fact the signals are transmitted and received at the feeding points of the antennas and will be somehow correlated (see Schumacher and Kermoal, 2002) which avoids reaching the theoretical limit. The model which is used in 802.11n for modeling some of these effects is a stochastic channel model with one or more azimuthal probability density functions describing the incoming and outgoing waves from an antenna array for a few coherence times.

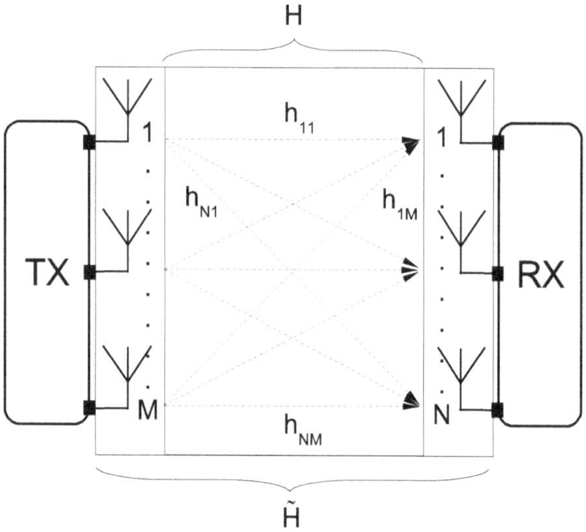

Fig. 1. The MIMO channel matrix $\mathbf{H}$ itself and the matrix $\tilde{\mathbf{H}}$ including correlation due to the involved antennas.

Considering the resulting correlation of the signals on the feeding point extends the channel matrix $\mathbf{H}$ to the matrix $\tilde{\mathbf{H}}$ as presented in Fig. 1. In addition to the stochastic modeling of the WLAN channel the official model considers a LOS component which is added to the pure stochastic channel. Therefore the channel matrix $\tilde{\mathbf{H}}$ can be thought of being composed from a sum of two matrices $\tilde{\mathbf{H}}_{\mathrm{LOS}}$ and $\tilde{\mathbf{H}}_{\mathrm{NLOS}}$ with different weights as expressed in Eq. (5).

$$\tilde{\mathbf{H}} = \sqrt{P} \cdot \left(\sqrt{\frac{K}{K+1}} \tilde{\mathbf{H}}_{\mathrm{LOS}} + \sqrt{\frac{1}{K+1}} \tilde{\mathbf{H}}_{\mathrm{NLOS}} \right) \qquad (5)$$

The rice factor K determines these weights and is fixed according to a certain channel scenario while the NLOS component is following the fluctuations of the channel modeled by IID values. The power P describes the power of the transmitted signals and scales the whole channel matrix. For an antenna designer it is desirable to benefit from spatial multiplexing offered by the NLOS and LOS component of the channel as well. The following two Sects. 2.3.1 and 2.3.2 describe which antenna parameters are important to look at in order to increase the channel capacity.

2.3.1 Antennas in a LOS scenario

The LOS component describes a fixed propagation path between transmit and receive antenna. This fact prevents the use of pattern diversity (see Sect. 2.3.2) in this case, which aims at exciting the eigenpathes of the channel using different propagation pathes. A more promising approach is the use of polarization diversity. Radiating and receiving two orthogonally polarized waves with a low crosstalk allows the transmission of two independent data channels. Figure 2 reveals this relation. A transmit antenna radiates an ideally

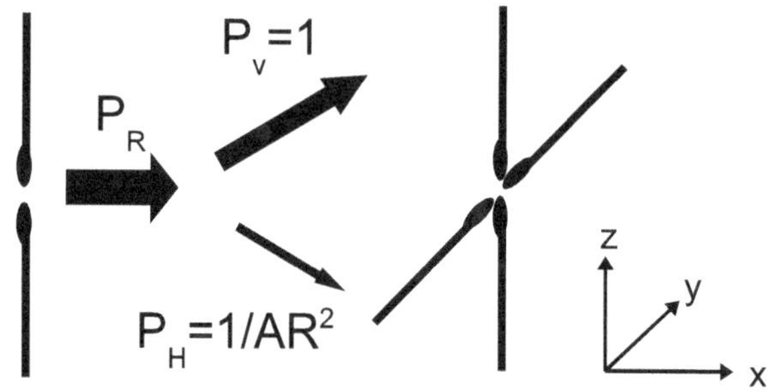

Fig. 2. Reception of an ideally linearly polarized wave by a cross dipole with finite axial ratio.

vertical polarized wave with the power P_R. The receiving dipole antennas are oriented along the z- and the y-axis, respectively. In general, the antennas are coupled, so the horizontally placed dipole in the x-y plane will receive a little amount of the vertical polarized wave anyway. This property is described by the axial ratio AR, which is expressed as the ratio between the maximum and the minimum absolute value of the electric field strengths in the polarization ellipse. Assigning a certain AR to the receiving antennas in Fig. 2 and normalizing the vertically received power P_H to 1 the horizontally received can be determined to $P_H=1/AR^2$. In this symmetrical case, the same relation can be shown in the case of receiving a horizontal wave with the vertically oriented dipole, where the received power is also determined by means of the AR. The higher the AR is, the higher the suppression of the unwanted orthogonal polarization direction is. According to the channel scenario in Fig. 2 the channel matrix $\tilde{\mathbf{H}}$ for the 2×2 case of infinite AR on transmitter side and a finite AR on receiver side, the channel matrix is proportional to expression (6).

$$\tilde{\mathbf{H}}_{\mathrm{LOS}} \approx \begin{pmatrix} 1 & 1/AR \\ 1/AR & 1 \end{pmatrix} \qquad (6)$$

The equation shows that the channel matrix approaches more and more the identity matrix as the AR increases. The eigenvalues of $\tilde{H}\tilde{H}^H$ and the spectral efficiency, respectively, are increasing similarly. If an antenna array offers a high AR over a wide angular spread, the channel matrix rank will stay high for many LOS cases.

2.3.2 Antennas in a NLOS scenario

For the NLOS components of the channel it is necessary to unite the stochastic channel model including angle of arrival (AoA), angle of departure (AoD) and the according angular spreads with the complex far field patterns of the single antenna elements in the array. The established model in (Schumacher and Kermoal, 2002) describes a way to include the antenna influences by stamping a correlation on the elements of the pure channel matrix $\mathbf{H}$ resulting in the matrix $\tilde{\mathbf{H}}$. The

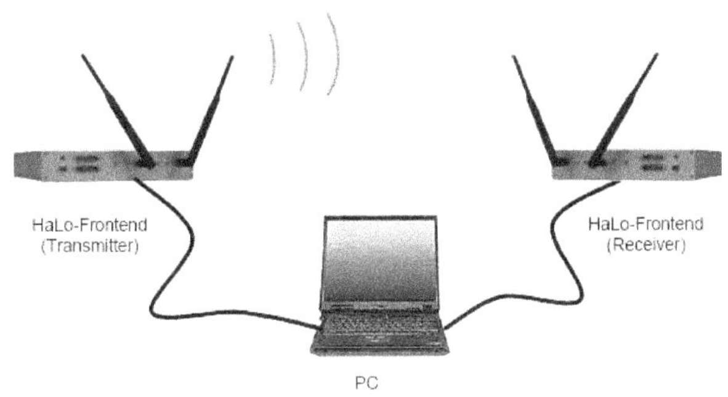

Fig. 3. The HaLo220 MIMO prototyping system.

envelope correlation coefficient ρ_e can be approximated by the square of the absolute value of the correlation coefficient $|\rho_{ij}|^2$ between the antennas i and j in an antenna array. It determines the degree of linear independency of the columns in the channel matrix and is expressed in formula 7 according to Fujimoto and James.

$$\rho_e = \mid \rho_{ij} \mid^2 = \frac{\mid R_{ij} \mid^2}{\sigma_i^2 \cdot \sigma_j^2} \tag{7}$$

The variance σ^2 is the part of the radiated or received power which contributes to the transmission in a certain channel scenario. The covariance R_{ij} expresses the amount of power which is commonly radiated or received by the antennas i and j and which can not be distinguished anymore in the receiver signals. The goal is to minimize the covariance R_{ij} which can only be done on antenna side. For the stochastic channel model, the covariance follows Eq. (8).

$$R_{ij} \approx \int_0^{2\pi} \int_0^{\pi} (C_{\vartheta_i}(\vartheta,\varphi) \cdot C^*_{\vartheta_j}(\vartheta,\varphi) + XPR \cdot C_{\varphi_i}(\vartheta,\varphi)$$
$$\ldots\ C^*_{\varphi_j}(\vartheta,\varphi)) \cdot p_{\vartheta,\varphi}(\vartheta,\varphi) \cdot e^{j\mathbf{k}r_{ij}} \cdot sin(\vartheta) d\vartheta d\varphi \tag{8}$$

The expression contents the far field patterns C for each antenna i in an array in ϑ and φ polarization in dependence of the azimuthal and elevational directions. The patterns of the antenna elements i and j are multiplied for each polarization direction and can be interpreted as the commonly radiated part of the power in one spatial direction. The cross-polarization ratio XPR is a property of the propagation channel and is a measure for the conversion of a linearly polarized wave into its orthogonal polarization. If the ratio is XPR=1, the power of a linearly polarized wave is equally distributed into its co- and cross polarization direction. This is generally assumed in indoor WLAN channels and can be explained by the numerous reflections in the propagation pathes. The product of the patterns is further on multiplied by the propability density function emphasizing the spatial direction of the dominant propagation pathes. Another factor is the exponential term expressing the phase delay between the antenna signals in a specific room direction. As mentioned before it is important to minimize the the covariance between each antenna element in an antenna array for MIMO applications. Three antenna parameters out of Eq. (8) can be tuned to achieve this goal. One parameter is the pattern diversity. If the antennas "look" in different directions the product of the patterns is minimized. Another approach is separating the antennas in order to benefit from the phase difference between the signals. The final possibility is to use antennas with different polarizations. In this case, the pattern product can be minimized. If this condition can be kept over a large angular spread, the correlation is small in many channel scenarios.

It was shown, how the LOS and the NLOS components of the channel matrix $\tilde{\mathbf{H}}$ can be minimized. In both cases it can be pointed out that polarization diversity helps to exploit the spatial diversity of the channel. This statement has been proven by measured channel matrices for different antenna setups as described in the following section.

3 Measurement

The measurement of MIMO channel matrices in WLAN channels requires a possibility for simultaneously transmitting and receiving RF signals in both frequency bands around 2.4 GHz and 5.4 GHz. The equipment used and the applied measurement method is explained in Sect. 3.1. The Sect. 3.1.1 shortly introduces the logarithmic periodic antenna used for the measurement. The measurement results taken in the office environment are described in Sect. 3.1.2 are shown and discussed in Sect. 3.2.

3.1 Measurement hardware

The HaLo220 system consists of two router like devices equipped with antennas and connected to a PC via USB as shown in Fig. 3. Each device contents two dual band transceiver units that can be configured as receiver or transmitter. A fast memory allows the playback or the acquisition of baseband data with different bandwidths up to 40 MHz. The synchronous transmission and reception of data makes it possible to measure the complex impulse response of the channel. However it is necessary to estimate the CIR (channel impulse response) for each path of the 2×2 system by considering the restrictions of the coherence time and bandwidth from Sect. 2. The method applied to separate the single pathes is an evaluation of the transmission of two CW signals with a slight frequency shift Δf of 200 kHz. The receiver starts recording with a sample rate of f_s=5 MSamples/s, when the signal level crosses an adjustable trigger level as the CW signals impinge. After acquisition, n=50 000 samples of the received complex baseband data are transferred to the PC where an FFT is applied. The duration of the transmission is t_{tr}=10 ms=n/f_s, which lies within the coherence time interval of 18 ms according to Sect. 2.1. The FFT for each of the two receiver signals shows two peaks with different magnitudes and angles according to the transmitted CW signals. By determining these peaks and their values the channel matrix can be composed for every coherence time and the channel capacity can be calculated according to Eq. (4). Because of the flat fading channel the complex values in the frequency domain result in a delta peak in the time domain with the same value. To get trustable predictions of the channel behavior, 1000 measurements for each channel scenario have been taken.

3.1.1 Antenna hardware

To demonstrate the influence of polarization diversity on the examined channels two different types of antennas were used. The first type is the well known $\lambda/4$ monopole that can be mounted on the HaLo system directly. Two of these elements were used in a distance of about λ for f=2.4 GHz and were both aligned vertically. This setup offers weak polarization diversity. In contrast, the logarithmic periodical antenna shown in Fig. 4 provides the radiation of two linearly polarized waves with high AR. This planar antenna type is a self-complementary structure applied on $FR4$ substrate consisting of four orthogonally placed arms as shown in Fig. 4 (Klemp and Eul, 2006). It has originally been designed for broadband applications. Exciting an opposite pair of arms with a differential signal generates a linear polarized wave radiating normally to the substrate plane in both directions. The feeding points for each element are in the center of the antenna and are connected via a semirigid coaxial waveguide from the backside. To reduce interactions with the coaxial cables, the feeding network and the generation of mantle modes a $\lambda/4$ reflector is placed on the backside of the antenna. Hence, the gain in main beam direction is increased to approximately G=5 dBi for a narrow band around the center frequency in a full wave simulation at 2.4 GHz. To switch between the two measurement frequencies the distance of the reflector can be adjusted. In this configuration, the simulated AR of the orthogonal arm pairs is better than 30 dB for both frequencies.

3.1.2 Office scenario

A floor plan of the examined office environment is illustrated in Fig. 5. The transmit and receive antennas have been placed in the same height of about 1.2 m above the ground and in a distance of 5 m. To separate between a channel which is dominated by the LOS component and the NLOS component, respectively, a metal plate is placed in front of the transmit antennas. The plate causes a reflection of the transmitted signal to the back of the room. To avoid the direct backscattering from the plate back into the LP antenna it was twisted a little bit (see the left image in Fig. 5). In the monopole setup the LP antennas were just replaced by two vertically oriented monopoles at the same place that are directly connected to the HaLo housing. Three different an-

Fig. 4. Logarithmically periodic antenna.

tenna setups were examined. In two cases similar antenna types were used on transmitter and receiver side, monopole and LP antennas respectively. In the third case two different types of antennas were used, an LP antenna on receiver side and two monopoles on transmitter side.

3.2 Measurement results

The measurement results are presented as a cumulative distribution function (cdf) of the measured channel capacities for 1000 coherence time intervals. The cdf can be interpreted as the probability of the channel capacity to fall below the value on the abscissa. It is also called the outage rate. Figure 6 shows the results for the LOS scenario (without metal plate) for the two subchannels for each of the three antenna configurations. In the case of using two LP antennas both subchannels are strong. The curves show a high slope because of the dominating static LOS component of the channel. There is no significant change in the channel capacity and due to the good polarization decoupling the channel matrix has high rank which results in a good quality of both subchannels. In comparison to the other antenna setups the sum capacity of the $LP \times LP$ case is the best. Replacing the transmit antenna with two vertically oriented monopoles causes a decrease of the stronger subchannel due to the missing gain into the direction of the receiving antenna. The influence on the weaker subchannel is more significant. It decreases from about 10 bit/s/Hz to 2.5 bit/s/Hz at an outage rate of -1. Both signals are transmitted with vertical polarization and will be received by the vertical polarized arm pair of the LP antenna with more signal power than by the horizontally polarized arm pair. By means of the channel matrix the elements in the second column are much lower than the ones in the first column. The subchannels can hardly be seperated in this case. Most of the signal power received in the horizontal direction is caused by reflections on the walls which avoids the weaker subchannel to vanish completely. This relation results from the smaller slope of the weak compared to the

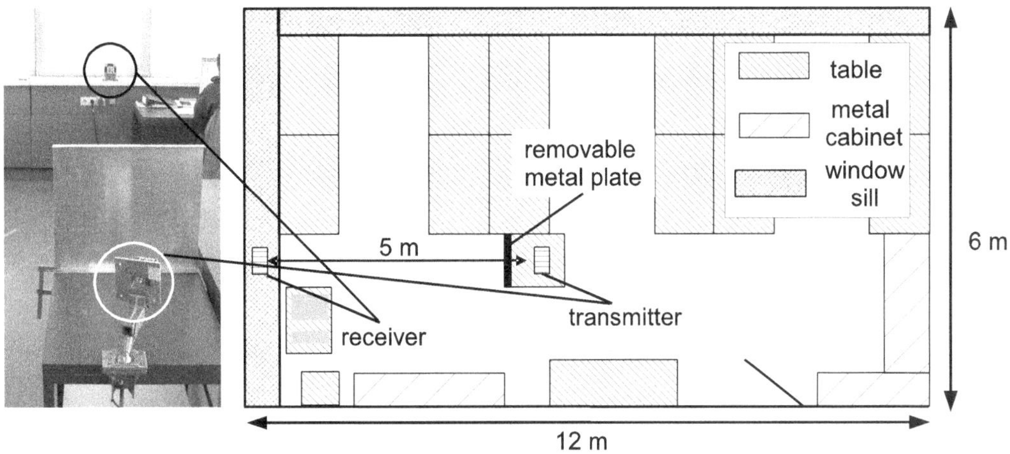

Fig. 5. Floor plan of the office room.

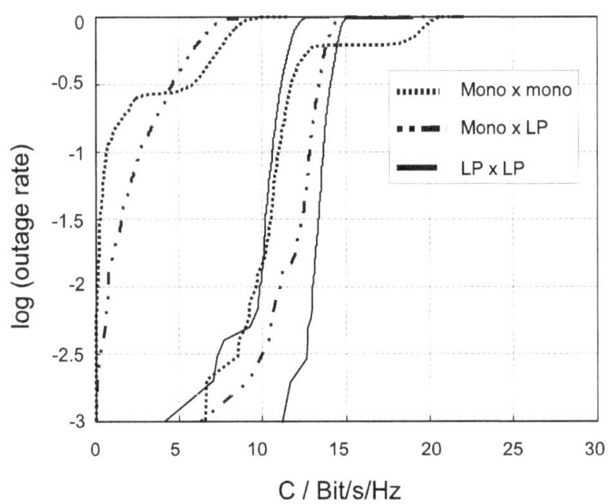

Fig. 6. Measured cdf of the subchannels at 2.4 GHz for three antenna setups in the LOS scenario.

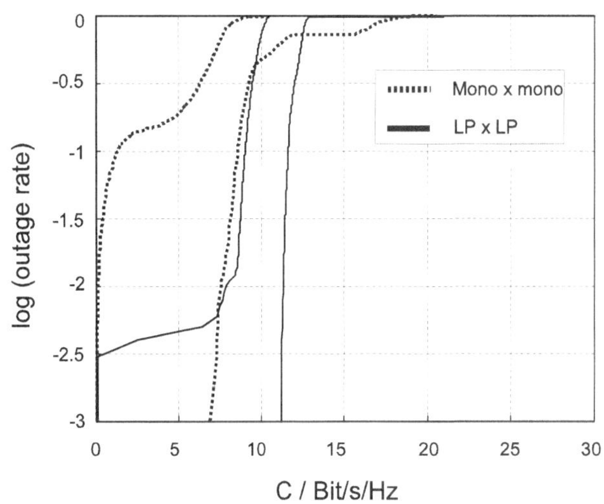

Fig. 7. Measured cdf of the subchannels at 5.4 GHz for two antenna setups.

strong subchannel which implies a higher weight of rayleigh fading in the channel according to a decreasing rice factor K in Eq. (5). In the last antenna setup with monopole antennas on transmit and receiver side this effect gets even stronger. The very constant far field pattern of the single antenna elements in azimuthal direction leads to multiple transmission pathes that are frequently changing. The weaker LOS path results in a SNR loss which can be seen by the vertically shifted curve of the stronger subchannel with and without LP antenna. The higher influence of the stochastic properties of the channel results in a higher spread of the measured channel capacities.

Similar results can be obtained by the measurement at 5.4 GHz in Fig. 7. All mentionable effects can be shown regarding the setups of two LPs and two monopoles. In comparison to the measurement at 2.4 GHz the curves are shifted horizontally because of the higher path loss. The shape of the curves, however, is kept as the propagation properties seem to be similar. The subchannels of the LOS dominated channel in the $LP \times LP$ setup is expressed by the high slope of the curves. Both subchannels are strong due to the polarization decoupling which drops in the case of two monopole antennas on each side. Here again the stochastic influences of the channel decrease the slope of the curves. Up to now it can be pointed out that polarization diversity leads to a better decoupling of the subchannels for both frequencies in the LOS case. The good SNR of the channel is due to the gain in main beam direction resulting in a high capacity of both subchannels.

The next step is to analyze the results for the NLOS case. According to Sect. 2.3.2 polarization diversity will lead to

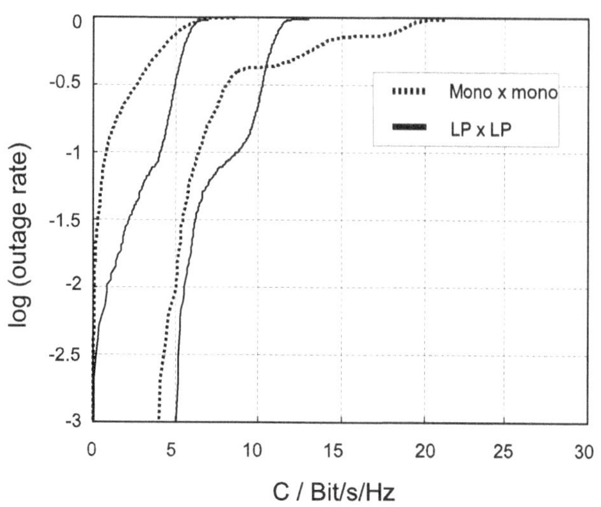

Fig. 8. Measured cdf of the subchannels at 5.4 GHz for two antenna setups in the NLOS scenario.

small correlation of the antenna elements and therefore to a decoupling of the subchannels. The comparison of the subchannel capacities for the setup of two LP antennas and four monopole antennas for 5.4 GHz are shown in Fig. 8. For both antenna setups the capacities are smaller compared to the ones obtained in the LOS scenarios because of the missing LOS component which has been faded out by placing the metal plate in front of the transmit antennas. Starting with the LP antenna setup it can be determined that the smaller subchannel is not that strong anymore in comparison to one in the LOS case. This is due to the depolarization effect of the channel considered in expression (8) with the variable XPR. The slope of the curves is decreased because of the changing channel conditions as already seen in the changing weight of the Rice factor between the monopole and the LP setup in the LOS case. In the NLOS case the stronger subchannels are similar except for the slope. It seems that in the monopole setup the propagation pathes are more fluctuating than in the LP channel. Although this property would lead to a decorrelation in conjunction with the distance of about λ for the monopoles it can be pointed out that the weaker subchannel of the LP setup is still better than in the monpole setup. This emphasizes the role of polarization decoupling in the NLOS case.

4 Conclusions

The contribution highlighted the antenna properties leading to a good MIMO performance for LOS as well as for NLOS channels assuming a stochastic channel model and a narrow band propagation channel. It has been determined that polarization diversity can improve orthogonality of the subchannels. This issue has been investigated using different antenna setups with high and low polarization decoupling. The LP antenna offers a high depolarization and leads to a good performance compared to a monopole setup that is used in the most present WLAN devices. In future WLAN applications even more antennas will have to be placed in order to exploit the spatial diversity properties of the channel. Due to the unaltered geometrical restrictions antenna designers will be forced to consider every possibility to decrease the correlation between the single antenna elements. Applying planar antennas with different polarization and pattern properties can be a solution of this problem as presented in this paper.

References

Foschini, G. J. and Gans, M. J.: On Limits of Wireless Communications in a Fading Environment when Using Multiple Antennas, Wireless Personal Communications, 6, 311–335, 1998.

Fujimoto, K. and James, J. R.: Mobile antenna systems handbook, Artech House, 2001.

Jiménez, V. P. G.: Channel Estimation for Bit-loading in OFDM-based WLAN, University Carlos III of Madrid, Spain, 2002.

Klemp, O. and Eul, H.: On the Application of Multimode Diversity Reception using Miniaturized Wideband Log.-Per, Planar Antennas Proceedings of the 9th European Conference on Wireless Technology, 2006.

McDonnell, J. T. E., Spiller, T. P., and Wilkinson, T. A.: RMS Delay Spread in Indoor LOS Environments at 5.2 GHz, HP Laboratories Bristol, 1998.

Rappaport, T. S.: Wireless Communications: Principles and Practice, Prentice Hall PTR, 1996.

Schumacher, L. and Kermoal, J. P.: A Stochastic MIMO Radio Channel Model With Experimental Validation, IEEE J. selected areas in communications, 20(6), 1211–1226, 2002.

Sklar, B.: Rayleigh Fading Channels in Mobile Digital Communication System Part 1: Characterization, IEEE Communications Magazine, 1997.

43

On UWB beamforming

T. Kaiser

Duisburg-Essen University, Faculty of Engineering, Department of Communication Systems, Bismarckstraße 81, 47048 Duisburg, Germany

Abstract. Ultra-Wideband (UWB) communication systems and Multi-Input-Multi-Output (MIMO) techniques rank among the few emerging key technologies in wireless communications. For that reason the marriage of these two complementary approaches should deserve attention. Apparently, the extremely large ultra-wide bandwidth creates rich multipath diversity which calls, at a first glance, additional antenna elements into question. However, another point of view is as follows. The attenuation by solid materials usually increases with increasing frequency; e.g. frequencies above, say, 10 GHz are considered to be blocked by walls etc. Since UWB can occupy more than 7 GHz of bandwidth (according to FCC regularisation) the performance of a communication link can be physically extended only by adding spatial information, i.e. multiple antennas, even if such extension may play a minor role. From this point of view UWB& MIMO presents an upper physical bound for indoor communications and is therefore at least worth to be investigated. In order to see the forest for the trees, we will focus in this limited contribution on beamforming among all alternative MIMO techniques (like space time coding or spatial multiplexing).

1 Introduction

In order to avoid strong interference of conventional narrowband transmission systems by ultra-wideband signals, the power spectral density of UWB systems is fairly limited, which leads – despite the enormous bandwidth – to a rather restricted coverage. This weakness is opposite to one of the main strengths of MIMO techniques; they are able to increase the range. Such a reversal indicates the potential of a marriage of these two complementary techniques.

Numerous other benefits can be envisaged. For example, the capability of MIMO systems to spatially distinguish among wavefronts impinging from different directions not only facilitates the equalizer design of UWB systems by reduction of delay spread, but rather enables a further increase of data rate – Gbit/s over air becomes feasible. Moreover, by MIMO techniques, narrowband and broadband interferers can be spatially suppressed so that the number of concurrent users might be significantly increased. Last but not least a reduction of electromagnetic radiation can be expected from UWB & MIMO, which in turn may also save battery life.

However, a large number of different challenges do resist. For example, digital beamforming seems to be prevented due to the extremely high sampling rate. In contrast, analog beamforming requires adjustable true time delays, such delays exhibit noticeable tolerance and therefore less precision. Besides these technology barriers, numerous challenges have to be overcome in fundamental processing of UWB signals by MIMO techniques.

2 An analogy

Without reinventing the wheel, some ideas can be borrowed from speech or also from radar processing. The following table shows the similarities and differences of broadband beamforming for speech processing and UWB communication systems.

Apparently, the relative bandwidth is of the same order and the delay spread differ by a factor 10^6, which is caused by the ratio between speed of propagation of light and sonic waves. Furthermore, the environment is the same as well as the propagation phenomena, e.g. multipath or broadband reflection by obstacles. Because fading is not so severe in UWB communications (Cramer et al., 1999; Lee et al., 2000) it is reasonable to assume some degree of signal correlation among the antenna elements which results in regular shaped wavefronts. The concrete type of the wavefront, i.e. spherical or planar, depends on the ratio between antenna aperture and transmitter-receiver distance. In order to guarantee a useful transmission, the signal to noise ratio at the receiver has to

Correspondence to: T. Kaiser
(thomas.kaiser@uni-duisburg.de)

Table 1. An analogy between speech processing and UWB transmission.

Property	Speech Processing	UWB transmission
bandwidth (rel.)	≈10 kHz (180%)	≈10 GHz (110%)
delay spread	≈100 ms	≈100 ns
environment	indoor, non-stationary	indoor, non-stationary
challenges	highly reverberant	multipath propagation
wavefront shape	spherical or planar	spherical or planar
typ. SNR at receiver	moderately positive (6–12 dB)	moderately positive (6–12 dB)
receiver type	waveform estimation	detection (matched filter!)
type of interferers	broadband	broadband and narrowband
type of noise	correlated (e.g. ventilator)	uncorrelated
other issues	signal duration unknown	signal duration known

be moderately positive, say a 6–12 dB. This can be achieved either in case of limited range or by spreading techniques.

All these mentioned aspects show the similarities between speech and UWB spatial processing. The main differences result from the fact that the UWB receiver knows the principal waveform whereas in speech processing not only the waveform is completely unknown but also when speech begins and when it ends. From this point of view, UWB seems to be easier to deal with because the transmit signal can be designed adequately. For example, in order to estimate the direction-of-arrival θ of an impinging wavefront (see Fig. 1), the transmit signal can be repeated periodically so that at the receiver the sampling rate can be distinctly reduced by almost arbitrary low undersampling techniques. Moreover, for UWB transmission a matched filter (MF) can be principally deployed that further boosts up the SNR. However, this approach tends to be a bit overrated, since in dense UWB multipath environments numerous reflections occur and each individual reflection will modify the transmitted impulse shape simply because of the non-flat frequency response of the reflective material. One last major difference concerns the kind of noise and interferers. While in speech processing broadband interferers dominate and the noise is most often correlated (e.g. ventilator), in UWB transmission the interferers can be broadband as well as narrowband and the noise is mainly of thermal type and hence uncorrelated.

In conclusion, because of its affinity, some concepts – like broadband beamforming – can be borrowed from speech processing.

3 Broadband beamforming

As mentioned previously we will focus on beamforming of UWB signals. The following figure illustrates a conventional filter and sum beamformer.

Basically, each filter can be substituted by a delay in order to compensate for the different signal travel time among the antenna elements and to coherently summing up all branches. This so-called array gain enhances the SNR by a factor of N (or $10\log_{10}(N)$ on a decibel scale) under perfect conditions.

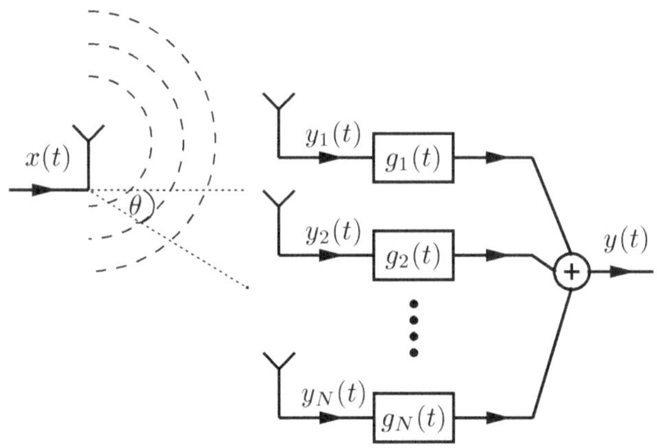

Fig. 1. A filter and sum broadband beamformer.

However, interferers cannot be suppressed only by delays. For that reason, the use of filters becomes justified. Figure 2 shows the beampattern defined as

$$B(\theta, \theta_0) = \max_t |y(t, \theta, \theta_0)|^2$$

for a family of sinusoidal signals

$$x(t) = \sin(2\pi f t), \quad f = 3.1\text{GHz}...10.6\text{GHz}$$

and $N = 8$ antennas, a steering direction of $\theta_0 = 0°$, an antenna spacing $d = \lambda_c/2$, where $\lambda_c = c_0/f_c$, c_0 is speed of light and $f_c = 6.85$ GHz is the center frequency according to FCC[1]-regularisation.

Observe that while the location of the mainlobe is frequency-independent, the locations of the sidelobes and the nulls are frequency-dependent. This means that broadband interferers cannot be completely suppressed by a simple delay and sum beamformer. Filters represent a remedy and allow an almost arbitrary high interferer cancellation just by adequately increase of the filter orders. In principle, conventional beamformer techniques (e.g. the minimum variance

[1] Federal Commission for Communications

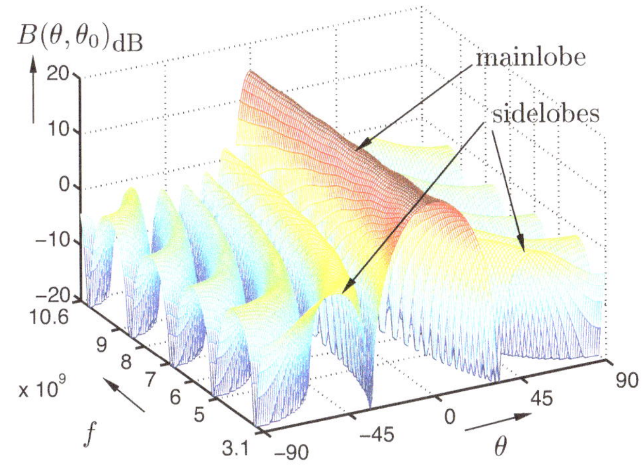

Fig. 2. Beampattern of a delay and sum broadband beamformer.

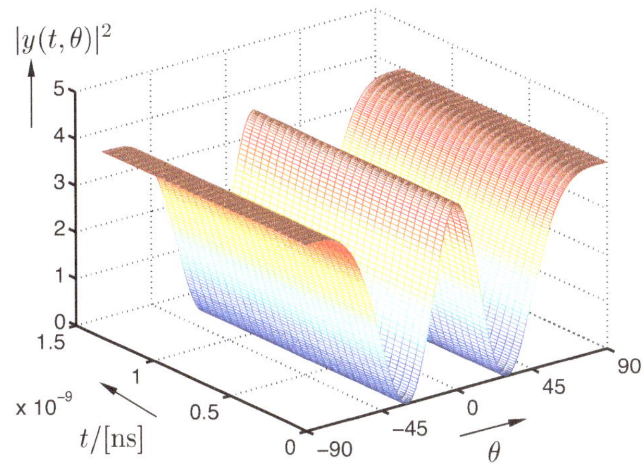

Fig. 3. Squared magnitude of narrowband beamformer output for $d = \lambda_c$, $N = 2$.

distortionless beamformer (MVDR)) can be deployed in order to design the filter and sum beamformer simply by calculating the weight vector for a set of suitable frequencies and transforming the resulting weight matrix into time domain.

UWB measurements have shown, at least under line-of-sight (LOS) conditions, the propagation of individual echoes arriving at the receiver in so-called clusters (see for example the descriptive videos at Whyless, www.whyless.org/public/wp5.htm). Due to their ultrashort duration, these echoes are resolvable in time so that they also become resolvable in angle; hence, direction-of-arrival estimation can be principally performed. Moreover, the recorded signals seem to be delayed replicas of each other so that beamforming becomes feasible. Even for a spacing of a few ten centimeters, the relevant echoes seem to be stongly correlated. In addition, a small spacing (e.g. 4 cm) does not cause serious coupling (Sibille and Bories, 2003).

(Ultra)-broadband beamforming offers an additional benefit opposite to narrowband beamforming. Narrowband beamforming suffers from so-called grating lobes. They occur for antenna spacings d larger than $\lambda_c/2$ and they lead to an ambiguity of the direction-of-arrival. For that reason, common narrowband antenna elements show a spacing of half the carrier wavelength or even smaller than this. Taking into account that the mainlobe width depends not only on the number of antennas but also on the spacing, narrowband beams can only be further narrowed by increasing the number of antennas. In turn, in (ultra)-broadband processing also the spacing can be increased further without ambiguity. In order to illustrate this fact, the following figures (Figs. 3–6) show the squared magnitude of the beamformer output and the beampattern for an antenna spacing $d = \lambda_c$ in case of two antennas, lower $f_l = 3.1\,\mathrm{GHz}$ and upper cut-off frequencies $f_u = 10.6\,\mathrm{GHz}$ ($\rightarrow f_c = 6.85\,\mathrm{GHz}$), and an input signal $x(t) = \mathrm{e}^{-\mathrm{j}2\pi f_c t}$ (narrowband) or $x(t) = \mathrm{e}^{-2\pi((f_u-f_l)t)^2}\mathrm{e}^{-\mathrm{j}2\pi f_c t}$ (broadband).

It can be seen that the narrowband beamformer will suffer from the mentioned ambiguity since it cannot distinguish

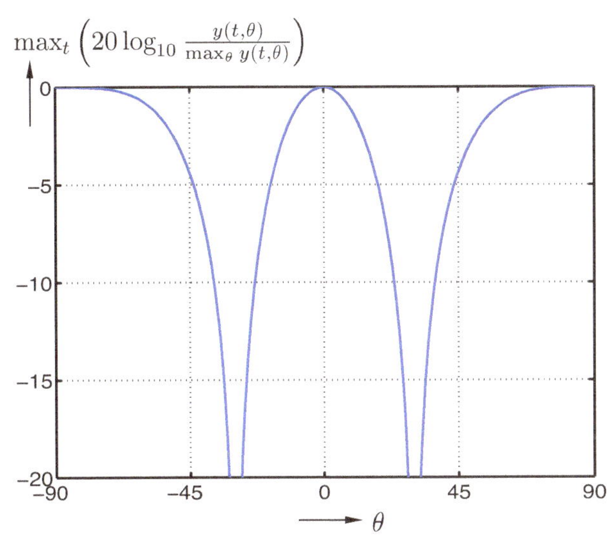

Fig. 4. Narrowband normalized beampattern for $d = \lambda_c$, $N = 2$ in logarithmic scale.

among the three possible direction-of-arrivals -90^0, 0^0 and $+90^0$. In contrast, the broadband beamformer can clearly resolve the 0^0-DoA, because of no ringing of the received impulse-shaped wavefront.

It is also of interest to study the effects caused by increasing the spacing and the number of antennas, see Figs. 7, 8.

Observe that extending the spacing leads to a narrowing of the mainlobe, whereas more antennas not only further decrease the mainlobe width but also increase the main-to-sidelobe ratio. For example, in case of $d = 2\lambda_c$, $N = 4$, the mainlobe width is approximately 15^0 so that with such a linear array up to $180^0/15^0 = 12$ spatially seperated users can be ideally served.

Note also that for $d = 2\lambda_c$, $N = 2$ the array size is 8.76 cm, which might be implementable on a terminal. In turn, for $d = 2\lambda_c$, $N = 4$ the array size becomes 26.28 cm, even this might be implementable – now on a laptop.

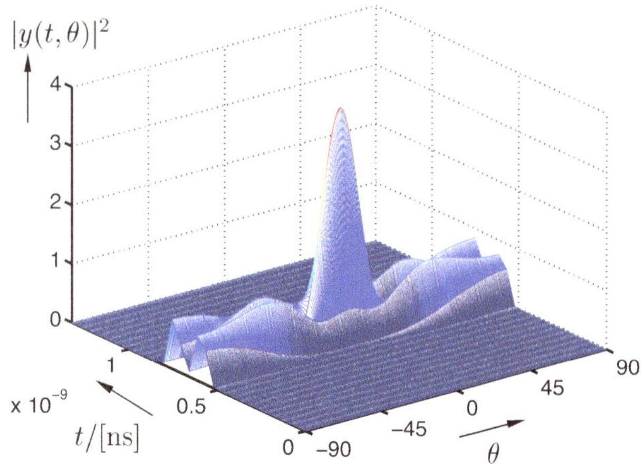

Fig. 5. Squared magnitude of broadband beamformer output for $d = \lambda_c$, $N = 2$.

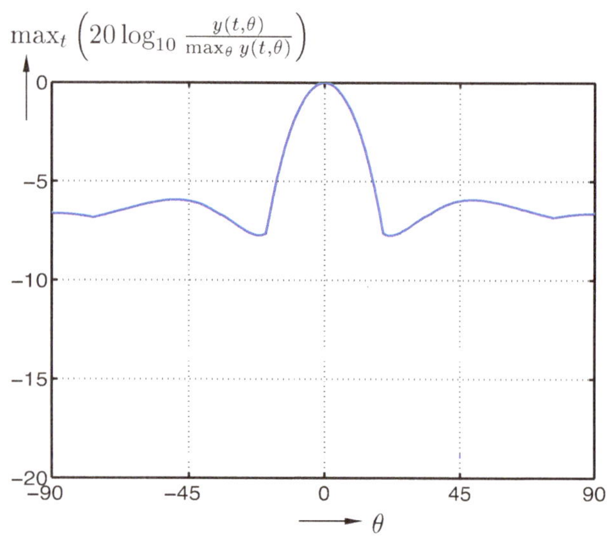

Fig. 6. Broadband normalized beampattern for $d = \lambda_c$, $N = 2$ in logarithmic scale.

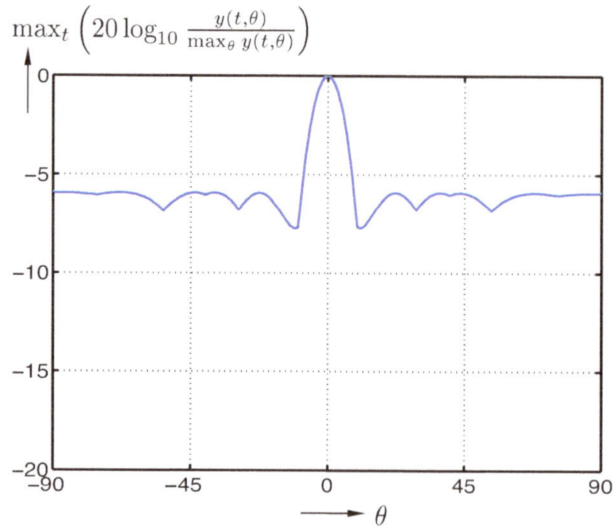

Fig. 7. Broadband normalized beampattern for $d = 2\lambda_c$, $N = 2$ in logarithmic scale.

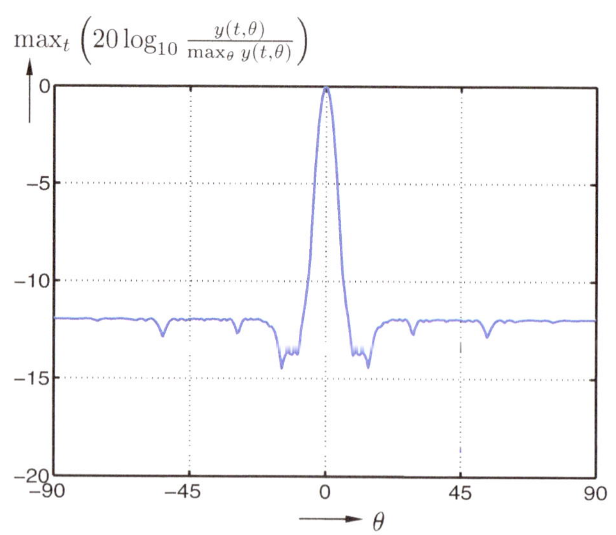

Fig. 8. Broadband normalized beampattern for $d = 2\lambda_c$, $N = 4$ in logarithmic scale.

4 Conclusion and future work

In this contribution, we motivate the marriage of UWB and multi-antenna technique by focussing on beamforming. Beside the complementary classes of multi-antenna approaches, i.e. spatial multiplexing and space-time coding, there are numerous open issues in this new and highly multi-disciplinary field:

- adjustable precise analog true time delays (or even filters)
- direction of arrival, time delay estimation
- beamforming for UWB channels (model & data)
- exploiting multipath propagation than combating
- pulse shaping, modulation for UWB multi-antenna systems
- localisation, synchronisation with UWB multi-antenna systems
- UWB antenna arrays, where the frequency responses among the elements may differ, but should overlap
- broadband beamforming with low number of antennas
- transmit beamforming.

Of further interest is how the channel capacity is affected by UWB transmission. First work towards this relevant topic can be found in Feng et al. (2004a, 2004b). It seems that the number of antennas has the same impact than the bandwidth, in fact, the equation known from narrowband MIMO systems for the ergodic channel capacity

$$C_e = BN\log_2(1 + \mathrm{SNR})$$

also holds approximately true for UWB channels. Such result further motivate the merge of these two key technologies in wireless communications.

References

Cramer, J. M., Scholtz, R. A., and Win, M. Z.: On the analysis of UWB communcation channels, IEEE Military Communication Conference Proceedings (MILCOM), 2, 1191–1195, 1999.

Lee, H., Haan, H., Shin, Y., and Im, S.: Multipath Characteristics of Impulse Radio Channels, Proc. IEEE VTC, 15–18 May, Tokyo, Japan, 2487–2491, 2000.

Sibille, A. and Bories, S.: Space diversity for UWB communications, EPMCC, 22–25 April, Glasgow, UK, 2003.

www.whyless.org/public/wp5.htm.

Feng, Z., Kaiser, T., and Czylwik, A.: On the Evaluation of Channel Capacity of Multi-Antenna UWB Indoor Wireless Systems – Part I: Frequency Flat Case, IEEE Transactions on Communications, submitted, 2004a.

Feng, Z., Kaiser, T., and Czylwik, A.: On the Evaluation of Channel Capacity of Multi-Antenna UWB Indoor Wireless Systems – Part II: Frequency Selective Case, IEEE Transactions on Communications, submitted, 2004b.

Permissions

All chapters in this book were first published in ARS, by Copernicus Publications; hereby published with permission under the Creative Commons Attribution License or equivalent. Every chapter published in this book has been scrutinized by our experts. Their significance has been extensively debated. The topics covered herein carry significant findings which will fuel the growth of the discipline. They may even be implemented as practical applications or may be referred to as a beginning point for another development.

The contributors of this book come from diverse backgrounds, making this book a truly international effort. This book will bring forth new frontiers with its revolutionizing research information and detailed analysis of the nascent developments around the world.

We would like to thank all the contributing authors for lending their expertise to make the book truly unique. They have played a crucial role in the development of this book. Without their invaluable contributions this book wouldn't have been possible. They have made vital efforts to compile up to date information on the varied aspects of this subject to make this book a valuable addition to the collection of many professionals and students.

This book was conceptualized with the vision of imparting up-to-date information and advanced data in this field. To ensure the same, a matchless editorial board was set up. Every individual on the board went through rigorous rounds of assessment to prove their worth. After which they invested a large part of their time researching and compiling the most relevant data for our readers.

The editorial board has been involved in producing this book since its inception. They have spent rigorous hours researching and exploring the diverse topics which have resulted in the successful publishing of this book. They have passed on their knowledge of decades through this book. To expedite this challenging task, the publisher supported the team at every step. A small team of assistant editors was also appointed to further simplify the editing procedure and attain best results for the readers.

Apart from the editorial board, the designing team has also invested a significant amount of their time in understanding the subject and creating the most relevant covers. They scrutinized every image to scout for the most suitable representation of the subject and create an appropriate cover for the book.

The publishing team has been an ardent support to the editorial, designing and production team. Their endless efforts to recruit the best for this project, has resulted in the accomplishment of this book. They are a veteran in the field of academics and their pool of knowledge is as vast as their experience in printing. Their expertise and guidance has proved useful at every step. Their uncompromising quality standards have made this book an exceptional effort. Their encouragement from time to time has been an inspiration for everyone.

The publisher and the editorial board hope that this book will prove to be a valuable piece of knowledge for researchers, students, practitioners and scholars across the globe.

List of Contributors

W. Pascher
Allg. und Theoretische Elektrotechnik, FernUniversität, D-58084 Hagen, Germany

J. H. den Besten
Opto-Electronic Devices group, Eindhoven University of Technology, Eindhoven, The Netherlands

D. Caprioli
Opto-Electronic Devices group, Eindhoven University of Technology, Eindhoven, The Netherlands

R. van Dijk
TNO Physics and Electronics Laboratory, The Hague, The Netherlands

X. J. M. Leijtens
Opto-Electronic Devices group, Eindhoven University of Technology, Eindhoven, The Netherlands

M. K. Smit
Opto-Electronic Devices group, Eindhoven University of Technology, Eindhoven, The Netherlands

R. Schuhmann
Technische Universität Darmstadt, Dept. of Electrical Engineering and Information Technology, Computational Electromagnetics Laboratory (TEMF), Schloßgartenstr. 8, 64289 Darmstadt, Germany

T. Weiland
Technische Universität Darmstadt, Dept. of Electrical Engineering and Information Technology, Computational Electromagnetics Laboratory (TEMF), Schloßgartenstr. 8, 64289 Darmstadt, Germany

G. Hebermehl
Greifswalder Str. 147, 10409 Berlin, Germany

J. Schefter
Weierstrass Institute for Applied Analysis and Stochastics (WIAS), Mohrenstr. 39, 10117 Berlin, Germany

R. Schlundt
Weierstrass Institute for Applied Analysis and Stochastics (WIAS), Mohrenstr. 39, 10117 Berlin, Germany

Th. Tischler
Ferdinand-Braun-Institut für Höchstfrequenztechnik (FBH), Albert-Einstein-Str. 11, 12489 Berlin, Germany

H. Zscheile
Ferdinand-Braun-Institut für Höchstfrequenztechnik (FBH), Albert-Einstein-Str. 11, 12489 Berlin, Germany

W. Heinrich
Ferdinand-Braun-Institut für Höchstfrequenztechnik (FBH), Albert-Einstein-Str. 11, 12489 Berlin, Germany

P. Leuchtmann
Institut für Feldtheorie und Höchstfrequenztechnik, ETH Zürich, CH-8092 Zürich, Switzerland

C. Fumeaux
Institut für Feldtheorie und Höchstfrequenztechnik, ETH Zürich, CH-8092 Zürich, Switzerland

D. Baumann
Institut für Feldtheorie und Höchstfrequenztechnik, ETH Zürich, CH-8092 Zürich, Switzerland

A. Becker
Chair of Electromagnetic Theory, University of Wuppertal, Germany

V. Hansen
Chair of Electromagnetic Theory, University of Wuppertal, Germany

G. Armbrecht
Leibniz Universität Hannover, Institut für Hochfrequenztechnik und Funksysteme, Appelstr. 9A, 30167 Hannover, Germany

O. Klemp
Leibniz Universität Hannover, Institut für Hochfrequenztechnik und Funksysteme, Appelstr. 9A, 30167 Hannover, Germany

I. Rolfes
Leibniz Universität Hannover, Institut für Hochfrequenztechnik und Funksysteme, Appelstr. 9A, 30167 Hannover, Germany

G. Fischer
Lucent Technologies, Bell Labs Advanced Technologies, Thurn-und-Taxis-Straße 10, 90411 Nuremberg, Germany

F. Pivit
Institut für Höchstfrequenztechnik und Elektronik, Universität Karlsruhe, Kaiserstraße 12, 76128 Karlsruhe, Germany

W. Wiebeck
Institut für Höchstfrequenztechnik und Elektronik, Universität Karlsruhe, Kaiserstraße 12, 76128 Karlsruhe, Germany

G. Del Galdo
Communications Research Laboratory, Ilmenau University of Technology, P.O. Box 100565, 98684 Ilmenau, Germany

M. Haardt
Communications Research Laboratory, Ilmenau University of Technology, P.O. Box 100565, 98684 Ilmenau, Germany

C. Schneider
Communications Research Laboratory, Ilmenau University of Technology, P.O. Box 100565, 98684 Ilmenau, Germany

F. Gerbl
Fachgebiet Höchstfrequenztechnik, Technische Universität München, Munich, Germany

E. M. Biebl
Fachgebiet Höchstfrequenztechnik, Technische Universität München, Munich, Germany

A. Tzoulis
FGAN-FHR, Neuenahrer Str. 20, 53343 Wachtberg, Germany

T. F. Eibert
Institute of Radio Frequency Technology, Universität Stuttgart, 70550 Stuttgart, Germany

T. Vaupel
FGAN-FHR, Neuenahrer Str. 20, 53343 Wachtberg, Germany

V. Hansen
Department of Electromagnetic Theory, University of Wuppertal, Germany

H. Buddendick
Institute of Radio Frequency Technology, Universität Stuttgart, 70550 Stuttgart, Germany

T. F. Eibert
Lehrstuhl für Hochfrequenztechnik, Technische Universität München, 80290 München, Germany

C. H. Schmidt
Institute of Radio Frequency Technology, Universität Stuttgart, Germany

T. F. Eibert
Lehrstuhl für Hochfrequenztechnik, Technische Universität München, Germany

R. Schuhmann
Univ. Paderborn, EIM-E, FG Theoretische Elektrotechnik, Warburger Straße 100, 33098 Paderborn, Germany

B. Bandlow
Univ. Paderborn, EIM-E, FG Theoretische Elektrotechnik, Warburger Straße 100, 33098 Paderborn, Germany

G. Lubkowski
TU Darmstadt, Inst. für Theorie Elektromagnetischer Felder, Schloßgartenstraße 8, 64289 Darmstadt, Germany

T. Weiland
TU Darmstadt, Inst. für Theorie Elektromagnetischer Felder, Schloßgartenstraße 8, 64289 Darmstadt, Germany

B. Bandlow
University Paderborn, EIM-E, FG Theoretische Elektrotechnik, Warburger Straße 100, 33098 Paderborn, Germany

R. Schuhmann
University Paderborn, EIM-E, FG Theoretische Elektrotechnik, Warburger Straße 100, 33098 Paderborn, Germany

M. Clemens
Technische Universität Darmstadt, Dept. of Electrical Engineering and Information Technology, Computational Electromagnetics Laboratory (TEMF), Schloßgartenstr. 8, 64289 Darmstadt, Germany

M. Wilke
Technische Universität Darmstadt, Dept. of Electrical Engineering and Information Technology, Computational Electromagnetics Laboratory (TEMF), Schloßgartenstr. 8, 64289 Darmstadt, Germany

T. Weiland
Technische Universität Darmstadt, Dept. of Electrical Engineering and Information Technology, Computational Electromagnetics Laboratory (TEMF), Schloßgartenstr. 8, 64289 Darmstadt, Germany

A. Becker
Lehrstuhl für Theoretische Elektrotechnik der Universität Wuppertal, Rainer-Gruenter-Str. 21, 42119 Wuppertal, Germany

V. Hansen
Lehrstuhl für Theoretische Elektrotechnik der Universität Wuppertal, Rainer-Gruenter-Str. 21, 42119 Wuppertal, Germany

G. Kotyrba
Fachbereich Elektrotechnik und Informationstechnik, Bergische Universität Wuppertal, Rainer-Gruenter-Str. 21, 42119 Wuppertal, Germany

H. Chaloupka
Fachbereich Elektrotechnik und Informationstechnik, Bergische Universität Wuppertal, Rainer-Gruenter-Str. 21, 42119 Wuppertal, Germany

C.Waldschmidt
Institut für Höchstfrequenztechnik und Elektronik (IHE), Universität Karlsruhe, Germany

C. Kuhnert
Institut für Höchstfrequenztechnik und Elektronik (IHE), Universität Karlsruhe, Germany

T. Fügen
Institut für Höchstfrequenztechnik und Elektronik (IHE), Universität Karlsruhe, Germany

W. Wiesbeck
Institut für Höchstfrequenztechnik und Elektronik (IHE), Universität Karlsruhe, Germany

D. Bechler
Institut für Nachrichtentechnik, Universität Karlsruhe, Kaiserstr. 12, 76128 Karlsruhe, Germany

M. Grimm
Institut für Nachrichtentechnik, Universität Karlsruhe, Kaiserstr. 12, 76128 Karlsruhe, Germany

K. Kroschel
Institut für Nachrichtentechnik, Universität Karlsruhe, Kaiserstr. 12, 76128 Karlsruhe, Germany

S. Bertl
Technische Universität München, Lehrstuhl für Hochfrequenztechnik, Fachgebiet Hochfrequente Felder und Schaltungen, 80333 München, Germany

A. Dallinger
Technische Universität München, Lehrstuhl für Hochfrequenztechnik, Fachgebiet Hochfrequente Felder und Schaltungen, 80333 München, Germany

J. Detlefsen
Technische Universität München, Lehrstuhl für Hochfrequenztechnik, Fachgebiet Hochfrequente Felder und Schaltungen, 80333 München, Germany

R. Herschmann
Leibniz Universität Hannover, Institute of Radiofrequency and Microwave Engineering, Appelstraße 9a, 30167 Hannover, Germany
Smart Devices GmbH & Co. KG, Schönebecker Allee 2, 30823 Garbsen, Germany

M. Camp
Leibniz Universität Hannover, Institute of Radiofrequency and Microwave Engineering, Appelstraße 9a, 30167 Hannover, Germany
Smart Devices GmbH & Co. KG, Schönebecker Allee 2, 30823 Garbsen, Germany

H. Eul
Leibniz Universität Hannover, Institute of Radiofrequency and Microwave Engineering, Appelstraße 9a, 30167 Hannover, Germany

A. Dekorsy
Lucent Technologies Networks System GmbH, Thurn-und-Taxis-Strasse 10, 90411 Nuremberg, Germany

M. Schacht
Lucent Technologies Networks System GmbH, Thurn-und-Taxis-Strasse 10, 90411 Nuremberg, Germany

S. Brueck
Lucent Technologies Networks System GmbH, Thurn-und-Taxis-Strasse 10, 90411 Nuremberg, Germany

G. Fischer
Lucent Technologies Networks System GmbH, Thurn-und-Taxis-Strasse 10, 90411 Nuremberg, Germany

L. Häring
Department of Communication Systems, University of Duisburg-Essen, Bismarckstr. 81, 47057 Duisburg, Germany

A. Czylwik
Department of Communication Systems, University of Duisburg-Essen, Bismarckstr. 81, 47057 Duisburg, Germany

S. K. Hampel
Leibniz Universität Hannover, Institut für Hochfrequenztechnik und Funksysteme, Appelstr. 9A, 30167 Hannover, Germany

O. Schmitz
Leibniz Universität Hannover, Institut für Hochfrequenztechnik und Funksysteme, Appelstr. 9A, 30167 Hannover, Germany

I. Rolfes
Leibniz Universität Hannover, Institut für Hochfrequenztechnik und Funksysteme, Appelstr. 9A, 30167 Hannover, Germany

T. Ringbeck
PMDTechnologies GmbH, Am Eichenhang 50, 57076 Siegen, Germany

T. Möller
PMDTechnologies GmbH, Am Eichenhang 50, 57076 Siegen, Germany

B. Hagebeuker
PMDTechnologies GmbH, Am Eichenhang 50, 57076 Siegen, Germany

V. Jungnickel
Fraunhofer Institute for Telecommunications, Heinrich-Hertz Institut, Einsteinufer 37, D-10587 Berlin, Germany

T. Haustein
Fraunhofer Institute for Telecommunications, Heinrich-Hertz Institut, Einsteinufer 37, D-10587 Berlin, Germany

A. Forck
Fraunhofer Institute for Telecommunications, Heinrich-Hertz Institut, Einsteinufer 37, D-10587 Berlin, Germany

U. Kruege
Fraunhofer Institute for Telecommunications, Heinrich-Hertz Institut, Einsteinufer 37, D-10587 Berlin, Germany

V. Pohl
Fraunhofer Institute for Telecommunications, Heinrich-Hertz Institut, Einsteinufer 37, D-10587 Berlin, Germany

C. von Helmolt
Fraunhofer Institute for Telecommunications, Heinrich-Hertz Institut, Einsteinufer 37, D-10587 Berlin, Germany

H.-L. Bloecher
Daimler AG, Group Research and Advanced Engineering, 89081 Ulm, Germany

A. Sailer
University of Ulm, Institute of Microwave Techniques, 89081 Ulm, Germany

G. Rollmann
GR-Consulting/SARA, 71732 Tamm, Germany

J. Dickmann
Daimler AG, Group Research and Advanced Engineering, 89081 Ulm, Germany

A. Fackelmeier
Fachgebiet Höchstfrequenztechnik, Technische Universität München, Munich, Germany

C. Morhart
Fachgebiet Höchstfrequenztechnik, Technische Universität München, Munich, Germany

E. M. Biebl
Fachgebiet Höchstfrequenztechnik, Technische Universität München, Munich, Germany

J. Adam
Institute for Electrical and Information Engineering, Christian-Albrechts-Universität zu Kiel, Germany

L. Klinkenbusch
Institute for Electrical and Information Engineering, Christian-Albrechts-Universität zu Kiel, Germany

C. Morhart
Fachgebiet Höchstfrequenztechnik, Technische Universität München, Munich, Germany

E. M. Biebl
Fachgebiet Höchstfrequenztechnik, Technische Universität München, Munich, Germany

R. Schneider
DaimlerChrysler AG, Research & Technology, REM/CR, Wilhelm-Runge-Str. 11, D-89081 Ulm, Germany

J. Wenger
DaimlerChrysler AG, Research & Technology, REM/CR, Wilhelm-Runge-Str. 11, D-89081 Ulm, Germany

J. Grubert
Institut für Hochfrequenztechnik, Technische Universität Braunschweig, Postfach 3329, 38023 Braunschweig, Germany

J. Heyen
Institut für Hochfrequenztechnik, Technische Universität Braunschweig, Postfach 3329, 38023 Braunschweig, Germany

C. Metz
Network Hardware Integration Research Department, Bell Laboratories, Murray Hill, NJ 07974, USA

L. C. Stange
Institut für Hochfrequenztechnik, Technische Universität Braunschweig, Postfach 3329, 38023 Braunschweig, Germany

A. F. Jacob
Institut für Hochfrequenztechnik, Technische Universität Braunschweig, Postfach 3329, 38023 Braunschweig, Germany

A. Tzoulis
FGAN-Research Institute for High-Frequency Physics and Radar Techniques (FHR), Wachtberg-Werthhoven, Germany

T. F. Eibert
FGAN-Research Institute for High-Frequency Physics and Radar Techniques (FHR), Wachtberg-Werthhoven, Germany

A. Tzoulis
FGAN-Research Institute for High-Frequency Physics and Radar Techniques (FHR), Wachtberg-Werthhoven, Germany

T. F. Eibert
Institute of Radio Frequency Technology, University of Stuttgart, Germany

M. Dierkes
Faculty for Computer Science, Electrical Engineering and Mathematics, Sensor Technology Department, University of Paderborn, Warburger Str. 100, 33098 Paderborn, Germany

U. Hilleringmann
Faculty for Computer Science, Electrical Engineering and Mathematics, Sensor Technology Department, University of Paderborn, Warburger Str. 100, 33098 Paderborn, Germany

C. Hornsteiner
Technische Universität München, Lehrstuhl für Hochfrequenztechnik, Fachgebiet Hochfrequente Felder und Schaltungen, 80333 München, Germany

J. Detlefsen
Technische Universität München, Lehrstuhl für Hochfrequenztechnik, Fachgebiet Hochfrequente Felder und Schaltungen, 80333 München, Germany

Y. Weitsch
Institute of Radio Frequency Technology, Universität Stuttgart, 70550 Stuttgart, Germany

T. F. Eibert
Institute of Radio Frequency Technology, Universität Stuttgart, 70550 Stuttgart, Germany

S. K. Hampel
Leibniz Universität Hannover, Institut für Hochfrequenztechnik und Funksysteme, Appelstr. 9A, 30167 Hannover, Germany

O. Schmitz
Leibniz Universität Hannover, Institut für Hochfrequenztechnik und Funksysteme, Appelstr. 9A, 30167 Hannover, Germany

O. Klemp
Leibniz Universität Hannover, Institut für Hochfrequenztechnik und Funksysteme, Appelstr. 9A, 30167 Hannover, Germany

H. Eul
Leibniz Universität Hannover, Institut für Hochfrequenztechnik und Funksysteme, Appelstr. 9A, 30167 Hannover, Germany

F. Gronwald
Institute for Fundamental Electrical Engineering and EMC, Otto-von-Guericke-University Magdeburg, Universitätsplatz 2, 39106 Magdeburg, Germany

D. Schwarz
BMW Group Research and Technology, Munich, Germany

R. H. Rasshofer
BMW Group Research and Technology, Munich, Germany

E. M. Biebl
Fachgebiet Höchstfrequenztechnik, Technische Universität München, Germany

H. Rabe
Institute for Radiofrequency and Microwave Engineering, Hanover, Germany

D. Kornek
Institute for Radiofrequency and Microwave Engineering, Hanover, Germany

M. Stege
Signalion GmbH, Dresden, Germany

I. Rolfes
Institute for Radiofrequency and Microwave Engineering, Hanover, Germany

T. Kaiser
Duisburg-Essen University, Faculty of Engineering, Department of Communication Systems, Bismarckstraße 81, 47048 Duisburg, Germany